DIFFERENTIAL GEOMETRY

Symposium in Pure Mathematics, ...

PROCEEDINGS OF SYMPOSIA
IN PURE MATHEMATICS

VOLUME XXVII, PART 2

DIFFERENTIAL GEOMETRY

AMERICAN MATHEMATICAL SOCIETY
PROVIDENCE, RHODE ISLAND
1975

PROCEEDINGS OF THE SYMPOSIUM IN PURE MATHEMATICS
OF THE AMERICAN MATHEMATICAL SOCIETY

HELD AT STANFORD UNIVERSITY
STANFORD, CALIFORNIA
JULY 30-AUGUST 17, 1973

EDITED BY

S. S. CHERN and R. OSSERMAN

Prepared by the American Mathematical Society
with the partial support of National Science Foundation Grant GP-37243

Library of Congress Cataloging in Publication Data CIP

Symposium in Pure Mathematics, Stanford University, 1973.
Differential geometry.

(Proceedings of symposia in pure mathematics; v. 27, pt. 1–2)
"Final versions of talks given at the AMS Summer Research Institute on Differential Geometry."
Includes bibliographies and indexes.
1. Geometry, Differential–Congresses.
I. Chern, Shiing-Shen, 1911– II. Osserman, Robert. III. American Mathematical Society.
IV. Series.
QA641.S88 1973 516′.36 75-6593
ISBN 0-8218-0248-8 (v. 2)

Printed in the United States of America

CONTENTS

*General lecture given at the Institute.

*General lecture given at the Institue.

Indexes

*General lecture given at the Institute.

Preface

The papers in these PROCEEDINGS represent the final versions of talks given at the AMS Summer Research Institute on Differential Geometry, which took place at Stanford University, Stanford, California, from July 30 to August 17, 1973. This Institute was made possible by a grant from the National Science Foundation. The organizing committee consisted of Raoul H. Bott, Eugenio Calabi, S. S. Chern, Leon W. Green, Shoshichi Kobayashi, Tilla K. Milnor, Barrett O'Neill, Robert Osserman, James Simons, I. M. Singer, with the coeditors serving as cochairmen.

The activities were divided between general lectures and seminar talks. In these PROCEEDINGS the general lectures have been distributed among the various seminars, according to their subject matter. Each part of the PROCEEDINGS consists of a group of seminars, whose titles and chairmen are as follows:

Part 1:
- Riemannian geometry (J. Cheeger)
- Submanifolds (K. Nomizu)
- Foliations (B. L. Reinhart)
- Algebraic and piecewise-linear topology (T. F. Banchoff and H. R. Gluck)
- Miscellaneous (B. O'Neill and J. Simons)

Part 2:
- Complex differential geometry (S. Kobayashi)
- Partial differential equations (J. L. Kazdan and F. W. Warner)
- Homogeneous spaces (J. Wolf)
- Relativity (T. Frankel)

Generally papers are included in the seminars in which they were presented, although in certain cases the contents would make them more appropriate in another section. In cases where a complete version of the talk appears elsewhere, only an abstract is included here, together with a reference to the full paper.

A list of open problems submitted by participants was compiled by Leon Green. These are included at the end of Part 1 in the Miscellaneous Section.

We should like to think the seminar chairmen, and also the secretarial staff: Dorothy Smith and Muriel Toupin of AMS as well as Catherine Lowe and Elizabeth Plowman of the Stanford Mathematics Department, all of whose tireless efforts were a large factor in the success of the Institute.

S. S. Chern
Robert Osserman

January 1975

COMPLEX DIFFERENTIAL GEOMETRY

Proceedings of Symposia in Pure Mathematics
Volume 27, 1975

RIEMANN-ROCH THEOREM FOR SINGULAR VARIETIES

PAUL BAUM*

This expository note reports on joint work with W. Fulton and R. MacPherson. Our Riemann-Roch (R.-R.) theorem will be stated. There will also be a few comments on further developments.

1. Hirzebruch's formula. Let M be a compact complex-analytic manifold. On M let E be a holomorphic vector-bundle. Denote by $\dot{E}$ the sheaf of germs of holomorphic sections of E. The cohomology groups $H^i(M, \dot{E})$ are finite-dimensional and vanish for $i > n = \dim_C M$. Define $\chi(M, \dot{E})$ by

$$(1.1) \qquad \chi(M, \dot{E}) = \sum_{i=0}^{n} (-1)^i \dim_C H^i(M, \dot{E}).$$

The Hirzebruch R.-R. theorem **[14]** computes $\chi(M, \dot{E})$. To recall Hirzebruch's formula, let $x_1, x_2, \cdots, x_n$ be the formal Chern roots of the holomorphic tangent bundle T of M. Let $y_1, y_2, \cdots, y_r$ be the formal Chern roots of E. In $H^*(M; \boldsymbol{Q})$ set

$$(1.2) \qquad \operatorname{td}(M) = \prod_{i=1}^{n}\left[\frac{x_i}{1 - e^{-x_i}}\right],$$

$$(1.3) \qquad \operatorname{ch}(E) = e^{y_1} + e^{y_2} + \cdots + e^{y_r}.$$

$\operatorname{td}(M)$ is the Todd class of M. $\operatorname{ch}(E)$ is the Chern character of E. Hirzebruch's theorem is

---

AMS (MOS) subject classifications (1970). Primary 32C35, 58G10, 14F05, 55F50; Secondary 53C05.

*Research partially supported by the National Science Foundation under grant GP 22580.

(1.4) $$\chi(M, \dot{E}) = (\mathrm{ch}(E) \cup \mathrm{td}(M))\,[M].$$

This can be viewed as a special case of the Atiyah-Singer index theorem **[6]**. On M there is the $\bar{\partial}$ complex of E:

$$0 \to C^\infty(E) \to C^\infty(\bar{T}^* \otimes E) \to \cdots \to C^\infty(\Lambda^n \bar{T}^* \otimes E) \to 0.$$

By the Dolbeault isomorphism **[11]** the ith homology of this complex is $H^i(M, \dot{E})$. The complex is elliptic. Applying Atiyah-Singer gives (1.4).

To extend (1.4) to singular varieties, recall that the cap product gives a pairing

(1.5) $$H^*(M; \boldsymbol{Q}) \otimes H_*(M; \boldsymbol{Q}) \to H_*(M; \boldsymbol{Q}).$$

In $H_*(M; \boldsymbol{Q})$ let $\tau(M)$ be the Poincaré dual of $\mathrm{td}(M)$:

(1.6) $$\tau(M) = \mathrm{td}(M) \cap [M].$$

Then (1.4) can be rewritten

(1.7) $$\chi(M, \dot{E}) = [\mathrm{ch}(E) \cap \tau(M)]_0$$

where $[\mathrm{ch}(E) \cap \tau(M)]_0$ is the zero-dimensional component of $\mathrm{ch}(E) \cap \tau(M)$. The aim of this note is to define for a compact complex projective variety X (which may be singular) a "Todd class" $\tau(X) \in H_*(X; \boldsymbol{Q})$. This Todd class $\tau(X)$ will have the following property.

For any holomorphic vector-bundle E on X define $\chi(X, \dot{E})$ by

(1.8) $$\chi(X, \dot{E}) = \sum_{i=0}^{n} (-1)^i \dim_{\boldsymbol{C}} H^i(X, \dot{E}), \qquad n = \dim_{\boldsymbol{C}} X.$$

Then

(1.9) $$\chi(X, \dot{E}) = [\mathrm{ch}(E) \cap \tau(X)]_0.$$

2. Homology and exact sequences of vector-bundles. To define $\tau(X)$ for singular X, a direct geometric approach to the Todd class is needed. The first step in doing this is to relate homology and exact sequences of vector-bundles. This will be done here using differential forms and the Chern-Weil curvature theory of characteristic classes.

To fix the notation, let E be a C^∞ complex vector-bundle on the oriented C^∞ manifold W. A connection for E is a differential operator D

(2.1) $$D\colon C^\infty(E) \to C^\infty(T^*_{\boldsymbol{C}} W \otimes E)$$

such that

(2.2) $$D(fs) = df \otimes s + fDs.$$

In (2.1) and (2.2) $C^\infty(\ )$ denotes the vector-space of all C^∞ sections of ( ). $T^*_{\boldsymbol{C}} W$ is the complexified cotangent bundle of W. $s \in C^\infty(E)$. $f\colon W \to \boldsymbol{C}$ is any C^∞ function.

If K is the curvature of D and $r =$ fibre $\dim_{\boldsymbol{C}} E$, let $\mathrm{ch}(K)$ be the differential form on W:

$$\text{(2.3)}\qquad \begin{aligned} \operatorname{ch}(K) = r &+ \frac{\sqrt{-1}}{2\pi}\operatorname{Trace}(K) + \frac{1}{2!}\left[\frac{\sqrt{-1}}{2\pi}\right]^2 \operatorname{Trace}(K \wedge K) \\ &+ \frac{1}{3!}\left[\frac{\sqrt{-1}}{2\pi}\right]^3 \operatorname{Trace}(K \wedge K \wedge K) + \cdots. \end{aligned}$$

According to the Chern-Weil curvature theory of characteristic classes [10], $\operatorname{ch}(K)$ is closed and the cohomology class determined by $\operatorname{ch}(K)$ is $\operatorname{ch}(E)$.

Suppose now that, on W,

$$0 \to E_q \to E_{q-1} \to \cdots \to E_0 \to 0$$

is an exact sequence of C^∞ complex vector-bundles. Denote the ith map of the sequence by $\eta_i : E_i \to E_{i-1}$. Let $D_q, D_{q-1}, \cdots, D_0$ be connections for $E_q, E_{q-1}, \cdots, E_0$. Assume that $D_q, D_{q-1}, \cdots, D_0$ are compatible with the exact sequence in the sense that for each $i = q, q-1, \cdots, 1$ there is commutativity in the following diagram:

$$\begin{array}{ccc} C^\infty(E_i) & \xrightarrow{\ \eta_i\ } & C^\infty(E_{i-1}) \\ \Big\downarrow D_i & & \Big\downarrow D_{i-1} \\ C^\infty(T^*_{\mathbf{C}}W \otimes E_i) & \xrightarrow{\ 1 \otimes \eta_i\ } & C^\infty(T^*_{\mathbf{C}}W \otimes E_{i-1}) \end{array}$$

Let K_i be the curvature of D_i. Under these conditions we then have

$$\text{(2.4)}\qquad \sum_{i=0}^{q} (-1)^i \operatorname{ch}(K_i) = 0.$$

The proof of (2.4) is not at all difficult. Note that (2.4) is an equality of *differential forms*.

Let Z be a compact subset of W. Assume that Z has the property:

(2.5) If V is any open subset of W with $V \supset Z$, then there is an open subset U of W such that $V \supset U \supset Z$ and such that Z is a deformation retract of U.

For example, if W admits a triangulation with Z a PL subcomplex of the triangulation, then Z satisfies (2.5).

Let there be given on W a complex L of C^∞ vector-bundles,

$$L = \{0 \to F_l \to F_{l-1} \to \cdots \to F_0 \to 0\},$$

with

$$\text{(2.6)}\qquad 0 \to F_l \to F_{l-1} \to \cdots \to F_0 \to 0 \text{ is exact on } W - Z.$$

Choose an open subset U of W and a compact set Δ of W such that

$$\text{(2.7)}\qquad U \supset \Delta \supset Z,$$

$$\text{(2.8)}\qquad Z \text{ is a deformation retract of } U,$$

$$\text{(2.9)}\qquad Z \text{ is contained in the interior of } \Delta.$$

Next, on W choose connections $D_l, D_{l-1}, \cdots, D_0$ for $F_l, F_{l-1}, \cdots, F_0$ such that

$$\text{(2.10)} \quad \begin{array}{l} \text{On } W - \Delta,\ D_l, D_{l-1}, \cdots, D_0 \text{ are compatible with the exact sequence} \\ 0 \to F_l|W - \Delta \to F_{l-1}|W - \Delta \to \cdots \to F_0|W - \Delta \to 0. \end{array}$$

On W consider the differential form $\sum_{i=0}^{l}(-1)^i \operatorname{ch}(K_i)$. Here K_i is the curvature of D_i. Then according to (2.4) and (2.10),

$$\text{(2.11)} \qquad \sum_{i=0}^{l}(-1)^i \operatorname{ch}(K_i) \text{ vanishes on } W - \Delta.$$

Thus $\sum_{i=0}^{l}(-1)^i \operatorname{ch}(K_i)$ determines an element in the cohomology of U with complex coefficients and compact supports. Denote this cohomology by $H_c^*(U; \boldsymbol{C})$. There are isomorphisms

$$\text{(2.12)} \qquad H_c^*(U; \boldsymbol{C}) \to H_*(U; \boldsymbol{C}) \leftarrow H_*(Z; \boldsymbol{C}).$$

The first isomorphism is Poincaré duality. The second is induced by the inclusion of Z in U. Hence $\sum_{i=0}^{l}(-1)^i \operatorname{ch}(K_i)$ determines an element of $H_*(Z; \boldsymbol{C})$. Denote this element of $H_*(Z; \boldsymbol{C})$ by ch.(L). It can be readily verified that ch. (L) depends *only* on W, Z, and L. ch.(L) is independent of how $U, \Delta, D_l, D_{l-1}, \cdots, D_0$ were chosen. Moreover, ch.(L) is a rational homology class:

$$\text{(2.13)} \qquad \text{ch.}\,(L) \in H_*(Z; \boldsymbol{Q}).$$

3. Todd class of a vector-bundle. The operation ch. of § 2 can be used to obtain the Todd class of a vector-bundle.

Let M be a compact oriented C^∞ manifold. On M let E be a C^∞ complex vector-bundle. The Todd class of E is defined by

$$\text{(3.1)} \qquad \operatorname{td}(E) = \prod_{i=1}^{r}\left[\frac{x_i}{1 - e^{-x_i}}\right]$$

where $x_1, x_2, \cdots, x_r$ are the formal Chern roots of E:

$$\text{(3.2)} \qquad \operatorname{td}(E) \in H^*(M; \boldsymbol{Q}).$$

Denote by $\pi\colon E \to M$ the projection of E onto M. For $i = 0, 1, \cdots, r = $ fibre $\dim_C E$, let $\Lambda^i E^*$ be the ith exterior power of the dual vector-bundle E^*. Let $\pi^*\Lambda^i E^*$ be the pull-back via π of $\Lambda^i E^*$. Thus $\pi^*\Lambda^i E^*$ is a vector-bundle on E. Contraction gives a vector-bundle map

$$\text{(3.3)} \qquad \pi^*\Lambda^i E^* \to \pi^*\Lambda^{i-1}E^*.$$

In detail, let $p \in M$, $u \in E_p$, and $\phi_1, \phi_2, \cdots, \phi_i \in E_p^*$. Then the map of (3.3) is:

$$\text{(3.4)} \qquad (u, \phi_1 \wedge \phi_2 \wedge \cdots \wedge \phi_i) \to \left[u, \sum_{j=1}^{i}(-1)^{j+1}\phi_j(u)\phi_1 \wedge \cdots \wedge \hat{\phi}_j \wedge \cdots \wedge \phi_i\right].$$

Denote by $\hat{E}$ the complex of vector-bundles

$$\text{(3.5)} \qquad 0 \to \pi^*\Lambda^r E^* \to \pi^*\Lambda^{r-1}E^* \to \cdots \to \pi^*\Lambda^0 E^* \to 0.$$

The zero section of E gives an inclusion $M \subset E$.

(3.6) $\hat{E}$ is exact on $E - M$.

The operation ch. of § 2 can be applied to $\hat{E}$. In the ring $H^*(M; \boldsymbol{Q})$, td(E) is invertible. Let 1/td(E) $\cap$ [M] be the Poincaré dual of 1/td(E). Then

(3.7) $$\text{ch.}(\hat{E}) = 1/\text{td}(E) \cap [M].$$

(3.7) is well known to the topologists. A proof of (3.7) using differential forms is given in § 9 below.

4. Todd class of a complex manifold. Let M be a compact complex-analytic manifold. Denote the holomorphic tangent bundle of M by T. Choose a C^∞ embedding of M in S^{2N} where S^{2N} is the $2N$-sphere and N is large. Let ν be the normal bundle of M in S^{2N}. Then for N large ν will have the structure of a C^∞ complex vector-bundle on M with

(4.1) $T \oplus \nu$ is isomorphic to the trivial complex vector-bundle $M \times \boldsymbol{C}^N$.

Quite generally, for two vector-bundles E, F,

(4.2) $$\text{td}(E \oplus F) = \text{td}(E) \cup \text{td}(F).$$

So (4.1) implies

(4.3) $$\text{td}(T) = 1/\text{td}(\nu).$$

Applying ch. to $\hat{\nu}$ and using (3.7) gives

(4.4) $$\text{ch.}(\hat{\nu}) = \text{td}(M) \cap [M].$$

(4.4) gives the desired direct geometric approach to td(M).

5. Todd class of a projective variety. Let X be a compact complex projective variety. Given a coherent complex-analytic sheaf $\mathscr{F}$ on X, make the following three choices:

(5.1) Choose a holomorphic embedding j: $X \to M$, where M is a nonsingular compact complex projective variety. For example, M could be $\boldsymbol{CP}^n$.

(5.2) On M, choose a holomorphic vector-bundle resolution $0 \to \dot{E}_q \to \dot{E}_{q-1} \to \cdots \to \dot{E}_0 \to j_*\mathscr{F} \to 0$ of the direct image sheaf $j_*\mathscr{F}$.

(5.3) Choose a C^∞ embedding of M in S^{2N}, N large.

Let ν be the normal bundle of M in S^{2N}. The embedding j:$X \to M$ and the zero section of ν give inclusions

(5.4) $$X \subset M \subset \nu.$$

Denote by $\pi: \nu \to M$ the projection of ν onto M. On M, let L be the complex of vector-bundles obtained by deleting $j_* \mathscr{F}$ from the resolution (5.2):

$$(5.5) \qquad L = \{0 \to E_q \to E_{q-1} \to \cdots \to E_0 \to 0\}.$$

Then

$$(5.6) \qquad L \text{ is exact on } M - X.$$

On ν, let $\pi^* L$ be the pull-back via π of L:

$$(5.7) \qquad \pi^* L = \{0 \to \pi^* E_q \to \pi^* E_{q-1} \to \cdots \to \pi^* E_0 \to 0\}.$$

$$(5.8) \qquad \pi^* L \text{ is exact on } \pi^{-1}(M - X).$$

On ν there is the vector-bundle complex $\hat{\nu}$.

$$(5.9) \qquad \hat{\nu} \text{ is exact on } \nu - M.$$

Form the tensor-product complex $\pi^* L \otimes \hat{\nu}$. This is exact where either $\pi^* L$ or $\hat{\nu}$ is exact. So by (5.8) and (5.9), $\pi^* L \otimes \hat{\nu}$ is exact on $\pi^{-1}(M - X) \cup (\nu - M) = \nu - X$:

$$(5.10) \qquad \pi^* L \otimes \hat{\nu} \text{ is exact on } \nu - X.$$

Define $\gamma_X(\mathscr{F})$ by

$$(5.11) \qquad \gamma_X(\mathscr{F}) = \mathrm{ch.}(\pi^* L \otimes \hat{\nu}).$$

Then

$$(5.12) \qquad \gamma_X(\mathscr{F}) \in H_*(X; \boldsymbol{Q}).$$

$$(5.13) \qquad \gamma_X(\mathscr{F}) \text{ depends only on } X \text{ and } \mathscr{F}.$$

$\gamma_X(\mathscr{F})$ is independent of how the choices (5.1), (5.2), (5.3) were made. This assertion (5.13) is the main technical point of the present note. The proof will be given elsewhere.

To define the "Todd class" $\tau(X)$, let $\mathscr{O}_X$ be the structure sheaf of X and set

$$(5.14) \qquad \tau(X) = \gamma_X(\mathscr{O}_X).$$

(4.4) implies

$$(5.15) \qquad \text{If } X \text{ is nonsingular } \tau(X) = \mathrm{td}(X) \cap [X].$$

If X is singular and $\rho: \tilde{X} \to X$ is a resolution of singularities of X in the sense of Hironaka [**13**], then $\rho_* \tau(\tilde{X})$ is usually *not* equal to $\tau(X)$. This will be clear from the functorial properties of $\gamma_X(\mathscr{F})$.

6. Functorial properties of $\gamma_X(\mathscr{F})$. On X let $0 \to \mathscr{F}_1 \to \mathscr{F}_2 \to \mathscr{F}_3 \to 0$ be a short exact sequence of coherent complex-analytic sheaves. Then in $H_*(X; \boldsymbol{Q})$,

$$(6.1) \qquad \gamma_X(\mathscr{F}_2) = \gamma_X(\mathscr{F}_1) + \gamma_X(\mathscr{F}_3).$$

So if $K_0^\omega(X)$ denotes the Grothendieck group of coherent complex-analytic sheaves on X, then γ_X gives an additive homomorphism of $K_0^\omega(X)$ to $H_*(X; \boldsymbol{Q})$:

(6.2) $$\gamma_X\colon K_0^{\omega}(X) \to H_*(X; \mathbf{Q}).$$

This homomorphism is natural. If Y is another compact complex projective variety and $f\colon X \to Y$ is a holomorphic map then there is commutativity in the diagram:

(6.3) $$\begin{array}{ccc} K_0^{\omega}(X) & \xrightarrow{\ f_!\ } & K_0^{\omega}(Y) \\ \downarrow{\scriptstyle \gamma_X} & & \downarrow{\scriptstyle \gamma_Y} \\ H_*(X; \mathbf{Q}) & \xrightarrow[\ f_*\ ]{} & H_*(Y; \mathbf{Q}) \end{array}$$

Here $f_!$ is as defined by Grothendieck [8]

(6.4) $$f_!(\mathscr{F}) = \sum_{i=0}^{n} (-1)^i R^i f_* \mathscr{F}, \qquad n = \dim_{\mathbf{C}} X.$$

Suppose Y is a point. Let $g\colon X \to \cdot$ be the map of X to a point. There is the evident isomorphism

(6.5) $$K_0^{\omega}(\cdot) \cong \mathbf{Z}.$$

Then for any coherent complex-analytic sheaf $\mathscr{F}$ on X,

(6.6) $$g_!(\mathscr{F}) = \sum_{i=0}^{n} (-1)^i \dim_{\mathbf{C}} H^i(X, \mathscr{F}), \qquad n = \dim_{\mathbf{C}} X.$$

Let $K_0^{\omega}(X)$ be the Grothendieck group of holomorphic vector-bundles on X. $(E, \mathscr{F}) \to \dot{E} \otimes \mathscr{F}$ gives a pairing

(6.7) $$K_{\omega}^0(X) \otimes K_0^{\omega}(X) \to K_0^{\omega}(X).$$

For homology and cohomology there is the usual cap product pairing

(6.8) $$H^*(X; \mathbf{Q}) \otimes H_*(X; \mathbf{Q}) \to H_*(X; \mathbf{Q}).$$

There is commutativity in the following diagram:

(6.9) $$\begin{array}{ccc} K_{\omega}^0(X) \otimes K_0^{\omega}(X) & \longrightarrow & K_0^{\omega}(X) \\ \downarrow{\scriptstyle \mathrm{ch} \otimes \gamma_X} & & \downarrow{\scriptstyle \gamma_X} \\ H^*(X; \mathbf{Q}) \otimes H_*(X; \mathbf{Q}) & \longrightarrow & H_*(X; \mathbf{Q}) \end{array}$$

If E is a holomorphic vector-bundle on X, then the commutativity of (6.9) implies

(6.10) $$\gamma_X(\dot{E}) = \mathrm{ch}(E) \cap \tau(X).$$

From (6.6) and (6.3) it then follows that

(6.11) $$\chi(X, \dot{E}) = [\mathrm{ch}(E) \cap \tau(X)]_0.$$

(6.11) extends Hirzebruch's formula (1.4) to X which may be singular.

REMARK. If X is singular, let $\rho: \tilde{X} \to X$ be a resolution of singularities of X in the sense of Hironaka **[13]**. According to (6.3) there is commutativity in the following diagram:

$$
\text{(6.12)} \qquad
\begin{array}{ccc}
K_0^{\omega}(\tilde{X}) & \xrightarrow{\ \rho_!\ } & K_0^{\omega}(X) \\
\Big\downarrow \gamma_{\tilde{X}} & & \Big\downarrow \gamma_X \\
H_*(\tilde{X}; \boldsymbol{Q}) & \xrightarrow[\ \rho_*\ ]{} & H_*(X; \boldsymbol{Q})
\end{array}
$$

Thus if it were the case that $\rho_!(\mathcal{O}_{\tilde{X}}) = \mathcal{O}_X$, then also $\rho_*\tau(\tilde{X}) = \tau(X)$. But usually $\rho_!(\mathcal{O}_{\tilde{X}})$ is not equal to $\mathcal{O}_X$. And usually $\rho_*\tau(\tilde{X})$ is not equal to $\tau(X)$.

7. *K*-theory. M. Atiyah suggested that the natural homomorphism $\gamma_X: K_0^{\omega}(X) \to H_*(X; \boldsymbol{Q})$ should factor through $K_0^t(X)$. Here K_*^t denotes the homology theory that goes with the well-known cohomology K-theory. Atiyah's suggestion does work. There is a natural homomorphism $\phi_X: K_0(X) \to K_0^t(X)$. The operator ch. of § 2 gives a natural transformation of homology theories

$$\text{(7.1)} \qquad \text{ch.}: K_*^t \to H_*(\ ; \boldsymbol{Q}).$$

The composition

$$\text{(7.2)} \qquad K_0(X) \xrightarrow{\ \phi_X\ } K_0^t(X) \xrightarrow{\ \text{ch.}\ } H_*(X; \boldsymbol{Q}) \text{ is } \gamma_X.$$

To describe ϕ_X two basic properties of $K_0^t(X)$ are needed. In order to state these, let $i: X \to S^{2N}$ be a topological embedding of X into an even-dimensional sphere S^{2N}. View i as an inclusion $X \subset S^{2N}$. Assume that S^{2N} can be triangulated with X a PL subcomplex of the triangulation. Let $K_t^0(S^{2N}, S^{2N} - X)$ denote the usual cohomology K-theory group of the pair $(S^{2N}, S^{2N} - X)$. The two properties needed are the following:

(7.3) There is a K-theory "Lefschetz duality" isomorphism $K_0^t(X) \cong K_t^0(S^{2N}, S^{2N} - X)$.

(7.4) If U is an open set of S^{2N} with $U \subset X$, then a complex $Q = \{0 \to F_l \to F_{l-1} \to \cdots \to F_0 \to 0\}$ of vector-bundles on U determines an element of $K_t^0(S^{2N}, S^{2N} - X)$ provided Q is exact on $U - X$.

For (7.3) see **[1]**, **[17]**. For (7.4) see **[4]**. Note that excision gives an isomorphism $K_0^t(U, U - X) \cong K_t^0(S^{2N}, S^{2N} - X)$. By Part II of **[4]**, Q determines an element of $K_t^0(U, U - X)$.

Let $\mathscr{F}$ be a coherent complex-analytic sheaf on X. Make choices as in (5.1), (5.2), (5.3). Let M, L, ν be as in (5.1)—(5.5). Recall that according to (5.11), $\gamma_X(\mathscr{F})$ is defined by

(7.5) $$\gamma_X(F) = \mathrm{ch.}(\pi^*L \otimes \hat{\nu}).$$

A tubular neighborhood of M in S^{2N} is homeomorphic to ν, so there are inclusions

(7.6) $$X \subset M \subset \nu \subset S^{2N}.$$

Hence by (7.4), $\pi^*L \otimes \hat{\nu}$ determines an element of $K_t^0(S^{2N}, S^{2N} - X)$. Applying the Lefschetz duality isomorphism of (7.3) this gives an element of $K_0^t(X)$. Denote this element of $K_0^t(X)$ by $[\pi^*L \otimes \hat{\nu}]$. $\phi_X(\mathscr{F})$ is defined by

(7.7) $$\phi_X(\mathscr{F}) = [\pi^*L \otimes \hat{\nu}].$$

Then

(7.8) $$\phi_X(\mathscr{F}) \text{ depends only on } X \text{ and } \mathscr{F}.$$

$\phi_X(\mathscr{F})$ is independent of how the choices (5.1), (5.2), (5.3) were made.

The homomorphism $\phi_X: K_0^\omega(X) \to K_0^t(X)$ is functorial and has a module property analogous to (6.9). This K-theory formulation of R.-R. is more precise and more natural than the statement of R.-R. using ordinary homology theory. So Atiyah's point of view gives a better understanding of R.-R. In essence R.-R. just asserts that any compact complex projective variety X is canonically oriented in the K_*^t-theory. The orientation class is $\phi_X(\mathscr{O}_X)$.

Let E be a holomorphic vector-bundle on X. What is the K-theory formula for $\chi(X, \dot{E})$? To state this, let $i: X \to S^{2N}$ be a topological embedding, as above, of X into S^{2N}. (7.3) gives an isomorphism

(7.9) $$K_0^t(X) \cong K_t^0(S^{2N}, S^{2N} - X).$$

The inclusion of pairs $(S^{2N}, S^{2N} - X) \supset (S^{2N}, \phi)$ gives a homomorphism

(7.10) $$K_t^0(S^{2N}, S^{2N} - X) \to K_t^0(S^{2N}).$$

Bott periodicity [2], [9] gives an isomorphism

(7.11) $$K_t^0(S^{2N}) \cong \mathbf{Z} \oplus \mathbf{Z}.$$

Thus, given a virtual vector-bundle ξ on S^{2N}, Bott periodicity associates to ξ two integers $\alpha(\xi)$ and $\beta(\xi)$. One of these, say $\alpha(\xi)$, is just the fibre dimension of ξ. So $\beta(\xi)$ is the main part of Bott's theorem. Starting, then, with a holomorphic vector-bundle E on X, $\phi_X(\dot{E}) \in K_0^t(X)$. The maps

(7.12) $$K^t(X) \cong K_t^0(S^{2N}, S^{2N} - X) \to K_t^0(S^{2N}) \xrightarrow{\beta} \mathbf{Z}$$

assign an integer to $\phi_X(\dot{E})$. This integer is $\chi(X, \dot{E})$.

8. Further developments. These are:

(i) explicit computation of $\tau(X)$ where X is locally a complete intersection;

(ii) holomorphic Lefschetz fixed point formula for an automorphism of finite order;

(iii) R.-R. in higher K-theory;

(iv) R.-R. in abstract algebraic geometry.

For (i), recall that (1.2) gives an explicit formula for the Todd class of a nonsingular variety. If X is singular but is locally a complete intersection, then X has a virtual tangent bundle $TX \in K_t^0(X)$ with

$$\tau(X) = \mathrm{td}(TX) \cap [X]. \tag{8.1}$$

Compare **[12]**.

For (ii) let $\alpha\colon X \to X$ be a holomorphic automorphism of X of finite order. For each $i = 0, 1, \cdots, n = \dim_{\mathbf{C}} X$, α induces a linear transformation

$$\alpha_i\colon H^i(X, \mathcal{O}_X) \to H^i(X, \mathcal{O}_X). \tag{8.2}$$

Define $L(\alpha)$ by

$$L(\alpha) = \sum_{i=0}^{n} (-1)^i \operatorname{Trace}(\alpha_i). \tag{8.3}$$

If α has only isolated fixed points, then the approach to R.-R. of this note leads to an explicit formula for $L(\alpha)$ in terms of the local behavior of α near its fixed points. The statement and proof of this formula took shape during this conference. Thanks go especially to B. Bennett and M. Rosen for several very helpful comments. If X is nonsingular, then the formula is the same as that given by Atiyah-Bott **[3]** and by Atiyah-Segal **[5]**.

For (iii) let $K_i^\omega(X)$ be the ith Quillen **[16]** higher K-theory group of the category of coherent complex-analytic sheaves on X. Let $K_i^t(X)$ be the ith Atiyah-Hirzebruch K_*^t group of X. We conjecture that an appropriate "functorialization" of the definition of ϕ_X given above will give natural maps

$$\phi_i^X\colon K_i^\omega(X) \to K_i^t(X), \qquad i = 0, 1, 2, \cdots. \tag{8.4}$$

For X nonsingular this question is not difficult. So for X nonsingular there is a definition of these ϕ_i^X. For X singular we have a proposed definition of these maps ϕ_i^X.

For (iv) let X be a complete projective algebraic variety defined over an algebraically closed field. Let $K_0^a(X)$ be the Grothendieck group of coherent algebraic sheaves on X. Let $A(X)$ be the Chow group of rational equivalence classes of cycles on X. Then there is a natural R.-R. map

$$\psi_X\colon K_0^a(X) \to A(X) \otimes_{\mathbf{Z}} \mathbf{Q}. \tag{8.5}$$

ψ_X is obtained by making the construction of γ_X given in § 5 above purely algebraic. This is done by using a Grassmannian-graph construction as in **[15]**.

9. Proof of (3.7). Let M be a C^∞ manifold. On M let E be a C^∞ complex vector-bundle. Consider $\hat{E}$.

$$\hat{E} = \{0 \to \pi^*\Lambda^r E^* \to \pi^*\Lambda^{r-1}E^* \to \cdots \to \pi^*\Lambda^0 E^* \to 0\}. \tag{9.1}$$

Choose connections $D_r, D_{r-1}, \cdots, D_0$ for $\pi^*\Lambda^r E^*, \pi^*\Lambda^{r-1}E^*, \cdots, \pi^*\Lambda^0 E^*$ so as to have the following:

(9.2) There exists a positive definite Hermitian inner product for E with $D_r, D_{r-1}, \cdots, D_0$ compatible with $\hat{E}$ on the complement of the unit ball bundle $B(E)$.

Let K_i be the curvature of D_i. The differential form $\sum_{i=0}^{r}(-1)^i \operatorname{ch}(K_i)$ can be integrated over the fibre of E to obtain a closed differential form on M. Denote this differential form on M by

$$\int_{\text{Fibre}} \sum_{i=0}^{r}(-1)^i \operatorname{ch}(K_i).$$

Denote by $t(E)$ the element of $H^*(M; \boldsymbol{C})$ determined by

$$\int_{\text{Fibre}} \sum_{i=0}^{r}(-1)^i \operatorname{ch}(K_i).$$

Then the following are proved by straightforward verifications:

(9.3) $t(E)$ is well defined. $t(E)$ depends only on E and M. $t(E)$ does not depend on the choice of connections $D_r, D_{r-1}, \cdots, D_0$.

(9.4) If E_1, E_2 are two vector-bundles on M, then $t(E_1 \oplus E_2) = t(E_1) \cup t(E_2)$.

(9.5) $t(E)$ is natural with respect to vector-bundle pull-back. If $f\colon M_1 \to M_2$ is a C^∞ map and E is a vector-bundle on M_2, then $t(f^*E) = f^*t(E)$.

To prove (3.6) one has to show

(9.6) $$t(E) = 1/\operatorname{td}(E).$$

From (9.3), (9.4), and (9.5) it follows that (9.6) need only be proved when E has fibre dimension one. So assume

(9.7) $$\text{fibre} \dim_{\boldsymbol{C}} E = 1.$$

In this case $\hat{E}$ is

(9.8) $$\hat{E} = \{0 \to \pi^*E^* \to E \times \boldsymbol{C} \to 0\}.$$

Here $E \times \boldsymbol{C}$ is the trivial line bundle on E. On E this trivial line bundle $E \times \boldsymbol{C}$ has the obvious section γ defined by

(9.9) $$\gamma(u) = (u, 1), \qquad u \in E.$$

Define a connection D_0 for $E \times \boldsymbol{C}$ by requiring

(9.10) $$D_0\gamma = 0.$$

The zero section of E gives an inclusion $M \subset E$. On $E - M$ define a connection

∇ for $\pi^*E^* | E - M$ by requiring

$$\text{(9.11)} \qquad \text{On } E - M,\ \nabla,\ D_0 \text{ are compatible with } \hat{E} | E - M.$$

On M choose a connection δ for E^*. Also choose a positive definite Hermitian inner-product $\langle\ ,\ \rangle$ for E. Denote the unit ball bundle by $B(E)$. Let $\psi\colon E \to [0, 1]$ be a C^∞ function with

$$\text{(9.12)} \qquad \psi = 1 \text{ on } E - B(E),$$

$$\text{(9.13)} \qquad \psi(u) = \langle u, u\rangle \text{ on an open set containing } M.$$

On E define a connection D_1 for π^*E^* by

$$\text{(9.14)} \qquad D_1 = \psi\nabla + (1 - \psi)\,\pi^*(\delta).$$

Let K_1 be the curvature of D_1. K_1 is a 2-form on E. We wish to compute the element in $H^*(M; \boldsymbol{C})$ given by $\int_{\text{Fibre}} - \operatorname{ch}(K_1)$. To do this let U be an open set of M on which there is a C^∞ section s of E with

$$\text{(9.15)} \qquad \langle s(x), s(x)\rangle = 1, \qquad x \in U.$$

Define $z\colon \pi^{-1}(U) \to \boldsymbol{C}$ by

$$\text{(9.16)} \qquad u = z(u)s(\pi u), \qquad u \in \pi^{-1}(U).$$

On U let θ be the connection form of δ with respect to s:

$$\text{(9.17)} \qquad \delta s = \theta \otimes s.$$

On $\pi^{-1}(U)$ let $\tilde{\theta}$ be the connection form of D_1 with respect to π^*s. Then (9.14) implies

$$\text{(9.18)} \qquad \tilde{\theta} = \psi\ dz/z + (1 - \psi)\pi^*\theta.$$

On U, $K_1 = d\tilde{\theta}$. So on U,

$$\text{(9.19)} \qquad K_1 = d\psi \wedge dz/z + (1 - d\psi)\,\pi^*\theta + (1 - \psi)\,d\pi^*\theta.$$

By letting the open set of (9.13) become larger and larger, and then passing to a limit we may assume

$$\text{(9.20)} \qquad \psi(u) = \langle u, u\rangle \quad \text{on } B(E),$$

$$\text{(9.21)} \qquad \psi = 1 \qquad \text{on } E - B(E).$$

For a detailed account of how such a limiting process is carried out see § 8, especially (8.85) to (8.93) of [7]. Granted (9.20) and (9.21), on $\pi^{-1}(U) \cap B(E)$, $\psi = z\bar{z}$. Therefore, on $\pi^{-1}(U) \cap B(E)$,

$$\text{(9.22)} \qquad K_1 = d\bar{z}\,dz + (1 - z\,d\bar{z} - \bar{z}\,dz)\,\pi^*\theta + (1 - z\bar{z})\,\pi^*d\theta.$$

$\operatorname{ch}(K_1)$ is given by

$$\text{(9.23)} \qquad \operatorname{ch}(K_1) = 1 + \frac{\sqrt{-1}}{2\pi}K_1 + \left(\frac{\sqrt{-1}}{2\pi}\right)^2 \cdot \frac{1}{2!}K_1^2 + \left(\frac{\sqrt{-1}}{2\pi}\right)^3 \cdot \frac{1}{3!}K_1^3 + \cdots.$$

For $l = 1, 2, 3, \cdots$ use (9.22) to find K_1^l. The only term which has a nonzero integral over the fibre is $l\, d\bar{z}\, dz\, (1 - z\bar{z})^{l-1}(\pi^* d\theta)^{l-1}$. Therefore

$$(9.24) \qquad \int_{\text{Fibre}} - \operatorname{ch}(K_1) = \sum_{l \geqq 1} \int_{\text{Fibre}} \frac{1}{(l-1)!}\left[\frac{\sqrt{-1}}{2\pi}\right]^l dz\, d\bar{z}(1 - z\bar{z})^{l-1}(\pi^* d\theta)^{l-1}.$$

In $\boldsymbol{C}$,

$$(9.25) \qquad \int_{\text{unit disc}} (1 - z\bar{z})^{l-1}\, dz\, d\bar{z} = 2\pi/l\sqrt{-1}, \quad l = 1,2,\cdots.$$

(9.25) is very easily proved by computing the integral in polar coordinates. Using (9.25) in (9.24) gives

$$(9.26) \qquad \int_{\text{Fibre}} - \operatorname{ch}(K_1) = \sum_{l \geqq 1} \frac{1}{l!}\left[\frac{\sqrt{-1}}{2\pi}\right]^{l-1} (d\theta)^{l-1}.$$

Now $d\theta$ is the curvature form of δ. Thus the element of $H^*(M; \boldsymbol{C})$ determined by $\sqrt{-1}/2\pi\, d\theta$ is $c_1(E^*) = -c_1(E)$. So the element of $H^*(M; \boldsymbol{C})$ determined by $\int_{\text{Fibre}} -\operatorname{ch}(K_1)$ is

$$\sum_{l \geqq 1} \frac{(-1)^{l-1} c_1(E)^{l-1}}{l!} = \frac{1 - e^{-c_1(E)}}{c_1(E)} = \frac{1}{\operatorname{td}(E)}.$$

This concludes the proof.

References

1. M. F. Atiyah, *Global theory of elliptic operators*, Proc. Internat. Conf. on Functional Analysis and Related Topics (Tokyo, 1968), Univ. of Tokyo Press, Tokyo, 1970, pp. 21–30.

2. M. F. Atiyah and R. Bott, *On the periodicity theorem for complex vector-bundles*, Acta Math. **112** (1964), 229–247.

3. ———, *A Lefschetz fixed point formula for elliptic complexes*. II. *Applications*, Ann. of Math. (2) **88** (1968), 451–491.

4. M. F. Atiyah, R. Bott and A. Shapiro, *Clifford modules*, Topology **3** (1964), suppl. 1, 3–38.

5. M. F. Atiyah and G. B. Segal, *The index of elliptic operators*. II, Ann. of Math. (2) **87** (1968), 531–545.

6. M. F. Atiyah and I. M. Singer, *The index of elliptic operators*. I, Ann. of Math. (2) **87** (1968), 484–530.

7. P. Baum and R. Bott, *Singularities of holomorphic foliations*, J. Differential Geometry **7** (1972), 279–342.

8. A. Borel and J.-P. Serre, *Le théorème de Riemann-Roch*, Bull. Soc. Math. France **86** (1958), 97–136. MR **22** #6817.

9. R. Bott, *The stable homotopy of the classical groups*, Ann. of Math. (2) **70** (1959), 313–337. MR **22** #987.

10. R. Bott and S. S. Chern, *Hermitian vector-bundles and the equidistribution of the zeroes of their holomorphic sections*, Acta Math. **114** (1965), 71–112.

11. P. Dolbeault, *Formes différentielles et cohomologie sur une variété analytique complexe*. I, Ann. of Math. (2) **64** (1956), 83–130; II, Ann. of Math. (2) **65** (1957), 282–330. MR **18**, 670; **19**, 171.

12. A. Grothendieck, P. Berthelot and L. Illusie, *Théorie des intersections et théorème de Riemann-Roch* (SGA 6), Lecture Notes in Math., vol. 225, Springer-Verlag, Berlin and New York, 1971.

13. H. Hironaka, *Resolution of singularities of an algebraic variety of characteristic zero*. I, II, Ann. of Math. (2) **79** (1964), 109–326.

14. F. Hirzebruch, *Topological methods in algebraic geometry*, Springer-Verlag, New York-Heidelberg, 1966.

15. R. MacPherson, *Chern classes for singular varieties*, Ann. of Math. **100** (1974), 423–432.

16. D. Quillen, *Algebraic K-theory* (Battelle Institute Conference, 1972), *Higher K-theories*, Lectures Notes in Math.,no. 341, Springer-Verlag, New York, 1973, pp. 85–147.

17. G. W. Whitehead, *Generalized homology theories*, Trans. Amer. Math. Soc. **102** (1962), 227–283. MR **25** #573.

BROWN UNIVERSITY

Proceedings of Symposia in Pure Mathematics
Volume 27, 1975

A COSTRUCTION OF NONHOMOGENEOUS EINSTEIN METRICS

E. CALABI*

A continuing interest in the problem of the existence, in a compact complex manifold admitting some Kähler metric, of a new Kähler metric with prescribed volume element [4], [5], has prompted the author to present this method of constructing in a simple way a class of Einstein metrics, and especially Ricci flat metrics in certain types of open manifolds.

The manifolds under consideration are complex tubular domains $M = \frac{1}{2}D \oplus i\boldsymbol{R}^n \subset \boldsymbol{C}^n$, where D denotes any connected, open subset of $\boldsymbol{R}^n$. The resulting complex manifolds are domains of holomorphy, if and only if D is convex [3, pp. 90–92]; the metrics constructed in M are Kähler metrics, invariant under the group of translations along $i\boldsymbol{R}^n$. Among the metrics constructed in M it is possible to impose additional restrictions, such as Einstein's equation $R_{ij} = Kg_{ij}$ ($K =$ constant), or, in particular, Ricci flat metrics, $R_{ij} = 0$. In the Ricci flat case it turns out that M with the metric of the type considered can never be complete, except in the trivial case where $D = \boldsymbol{R}^n$ and $M = \boldsymbol{C}^n$ with a flat, translation invariant metric.

Let $f : D \to \boldsymbol{R}$ be a strongly convex, differentiable, real valued function on D, i.e. a function whose Hessian is everywhere positive definite, and denote by $F(z, \bar{z})$ the strongly plurisubharmonic function on the tube $M = \frac{1}{2}D \oplus i\boldsymbol{R}^n$ defined by

$$F(z, \bar{z}) = f(z_1 + \bar{z}_1, z_2 + \bar{z}_2, \cdots, z_n + \bar{z}_n). \tag{1}$$

Consider in M the Kähler metric defined by the Levi invariant of F, i.e.

AMS (MOS) subject classifications (1970). Primary 53C25, 35J60, 32A07, 32F05.

*Supported in part by NSF grant 29258.

(2) $$ds^2 = 2g_{\alpha\bar{\beta}}(z, \bar{z})\, dz^\alpha d\bar{z}^\beta,$$

where

(3) $$g_{\alpha\bar{\beta}}(z, z) = \frac{\partial^2 F(z, \bar{z})}{\partial z_\alpha\, \partial \bar{z}_\beta} = \left.\frac{\partial^2 f(x)}{\partial x_\alpha\, \partial x_\beta}\right|_{x=z+\bar{z}};$$

its volume element is $dv = i^n \det(g_{\alpha\bar{\beta}})\, dz_1 \wedge d\bar{z}_1 \wedge \cdots \wedge dz_n \wedge d\bar{z}_n$. If the convex function $f(x)$ satisfies, for instance, the differential equation

(4) $$\det\left(\frac{\partial^2 f}{\partial x_i\, \partial x_j}\right) = 1,$$

already considered elsewhere **[6]**, **[8]**, then

$$g(z, \bar{z}) = \det(g_{\alpha\bar{\beta}}(z, \bar{z})) = 1,$$

and consequently the Ricci tensor **[2]**

$$R_{\alpha\beta} = -\frac{\partial^2}{\partial z_\alpha\, \partial \bar{z}_\beta}(\log g(z, \bar{z}))$$

vanishes identically. On the other hand, the full Riemann curvature tensor can be computed from the following formula, letting $(K^{\alpha\bar{\mu}}(z, \bar{z}))$ denote the transposed inverse matrix function of $(g_{\alpha\bar{\beta}}(z, \bar{z}))$ (cf. **[2]**, deduce from Lemma 5):

$$R^\alpha{}_{\beta\gamma\bar{\delta}} = -\frac{\partial}{\partial \bar{z}_\delta}\left(K^{\alpha\bar{\mu}}(z, \bar{z})\frac{\partial^3 F(z, \bar{z})}{\partial z_\beta \partial z_\gamma\, \partial \bar{z}_\mu}\right).$$

It is easy to verify that $R^\alpha{}_{\beta\gamma\bar{\delta}}$ vanishes identically, in view of (2), if and only if all third order derivatives of $f(x)$ vanish identically, i.e. if $f(x)$ is a quadratic polynomial. Thus the tubular domain M with the Kähler metric (3) is a Ricci flat manifold that is not locally flat, except in the case where $f(x)$ is one of the trivial solutions of (4).

It has already been reported that the Ricci flat Kähler metrics obtained in M by this method described above can never be complete **[6]** nor can they be defined in all of $\boldsymbol{C}^n$ **[8]** except for the trivial solutions. However we can give a simple example of a nontrivial solution $f(x)$ of (4) in $\boldsymbol{R}^n - \{0\}$ yielding a metric in $M = \boldsymbol{C}^n - i\boldsymbol{R}^n$ $(n \geq 2)$, namely

$$f(x) = \int_0^{|x|} (c + r^n)^{1/n}\, dr, \quad |x| = \left(\sum_{\alpha=1}^n (x_\alpha)^2\right)^{1/2}, \quad c > 0.$$

Other solutions can be obtained by direct sums of nontrivial and at most one trivial solution in lower dimensional spaces. For $n = 2$ there are several known representations of solutions of (4) (cf. **[1**, p. 216**]**).

One can similarly construct, more generally, Kähler metrics in a tubular domain $M = \frac{1}{2}D \oplus i\boldsymbol{R}^n$ satisfying an Einstein equation

$$R_{\alpha\bar{\beta}} = K g_{\alpha\bar{\beta}} \qquad (K = \text{constant})$$

by letting $f(x)$ be a solution of the equation

(5) $$\det\left(\frac{\partial^2 f(x)}{\partial x_i \partial x_j}\right) = e^{-Kf(x)},$$

in the place of (4). In this case it is clear from Myers' theorem [7] that, for $K > 0$, no solution of (5) can yield a complete metric in M. The case $K < 0$, however, seems to be quite different. For example, if D is a convex domain containing no complete straight line and $M = \frac{1}{2}D \oplus i\boldsymbol{R}^n$, in addition is a homogeneous complex manifold, then M is a Cartesian product of irreducible, homogeneous, Bergman manifolds. For each of these irreducible factors the Bergman metric satisfies Einstein's equation with $K < 0$; consequently the product of a suitable renormalization of these metrics satisfies Einstein's equation in M; thus M admits an Einstein metric, invariant under a transitive group of holomorphic transformations, so that M is complete with respect to this metric. This fact and others lead to the following conjecture.

CONJECTURE. Let D be an open convex domain in $\boldsymbol{R}^n$ containing no complete straight line. Then there exists in D, for any $K < 0$, a unique, convex function $f : D \to \boldsymbol{R}$ satisfying (5), approaching $+\infty$ uniformly in each bounded portion of the boundary, and bounded from below by $-c \log |x|$ for some $c > 0$ as $x \to \infty$ if D is unbounded.

Assuming, for the moment, the conjecture to be true for one value of K, say K_1, yielding a solution $f_1(x)$ of (5) with the asserted properties, one has a solution $f(x)$ for any other negative value of K by setting

$$f(x) = \frac{1}{K}(K_1 f_1(x) - n \log (K_1/K)).$$

The uniqueness of f with the properties required can be proved by means of the strong maximum principle [11] for the difference of two solutions of (5) with $K < 0$; similarly one should prove that the boundary condition on f is sufficient to show that the resulting Kähler metric on M makes the tubular domain complete.

We shall now construct an explicit example of a solution of equation (5) in a particular bounded, convex domain D, yielding on $M = \frac{1}{2}D + i\boldsymbol{R}^n$ an Einstein-Kähler metric with $K = -1$, with the properties that M is complete with respect to this metric and the metric is not locally homogeneous. To our best recollection, this is the first known example of a complete Einstein metric that is not locally homogeneous (incomplete examples were given explicitly in $\boldsymbol{R}^+ \times \boldsymbol{R}^3$ by T. J. Willmore [9], [10]).

The convex domain D is an open ball in $\boldsymbol{R}^n$ of unspecified radius $r_1 > 0$ and center at the origin. We seek a solution of (5) in D, that is a convex function diverging uniformly to $+\infty$ at the boundary. The uniqueness of the solution $f(x)$ and the symmetry of D under the orthogonal group $O(n)$ imply that $f(x)$ can be expressed as a radial function, $f(x) = y(r)$, $r = (\sum_{\alpha=1}^n x_\alpha^2)^{1/2}$, where y satisfies the ordinary differential equation (taking $K = -1$, without loss of generality)

(6) $$(y'/r)^{n-1} y'' = e^y.$$

Regularity at the origin requires the following initial conditions

$$y'(0) = 0, \qquad y''(0) = e^{y(0)/n}, \tag{7}$$

whence one obtains by recursion a solution expressed as a formal power series with respect to r^2. In order to prove that the functions $y(r)$ that are solutions of (6) and (7) actually generate an Einstein metric with all the asserted properties, one has to study the solution with some detail.

Replacing $y(r)$ by $y_\lambda(r) = y(\lambda r) + 2n \log \lambda$ for any $\lambda > 0$ yields a one-parameter family of solutions from one solution, showing the effect of different initial values $y(0)$ on the solution globally. It is also reasonable, in looking for comparison functions, to seek them with the same type of logarithmic homogeneity property.

DEFINITION. In view of equations (6), (7) we define a function $z(r)$ to be a supersolution (respectively, a subsolution) in an interval $-a < r < a$ for any $a > 0$ if $z'(0) = 0$ and, for all r in the interval, $z''(r) \geqq 0$ and $e^{-z(r)}(z'(r)/r)^{n-1} z''(r) \leqq 1$ (respectively $\geqq 1$). A solution of (6), (7) is the same thing as a simultaneous sub- and supersolution.

The following two elementary functions will be used as comparison functions in any open interval $(-a, a)$

$$z_l(r) = z_l(a, r) = -(n+1) \log (1 - r^2/a^2) + n \log((2n+2)/a^2)$$

and

$$z_u(r) = z_u(a, r) = (n+1) \log \sec \pi r/2a + n \log (n+1)\pi^2/4a^2.$$

One verifies that z_l and z_u are respectively a subsolution and a supersolution. In fact both z_l and z_u are differentiable and convex in the interval $(-a, a)$, and

$$e^{-z_l}(z_l'/r)^{n-1} z_l'' = 1 + r^2/a^2 \geqq 1,$$

$$e^{-z_u}(z_u'/r)^{n-1} z_u'' = ((2a/\pi r)[\sin(\pi r/2a)]^{n-1}) \leqq 1.$$

We apply now the maximum principle [**11**, pp. 6–7] as follows: if $g_1(r)$ and $g_2(r)$ are respectively a subsolution and a supersolution of (6), (7) in a common open interval, then $g_1(r) - g_2(r)$ cannot achieve a nonnegative relative maximum value at any interior point, unless they coincide in the whole interval (and are therefore a solution).

Let $y(r)$ be the solution of (6), (7) with initial value $y(0)$. We can show that $y(r)$ becomes $+\infty$ at the end points of an interval $(-A, A)$, where

$$(2(n+1))^{1/2} e^{-y(0)/2n} \leqq A \leqq \frac{\pi}{2}(n+1)^{1/2} e^{-y(0)/2n}.$$

In fact, if the upper bound were violated, choose a constant a strictly between the asserted upper bound and A and consider $y(r) - z_u(a, r)$: this function becomes $-\infty$ as $r \to a$, so that the difference would achieve an absolute maximum value at some interior point; but this maximum value would be positive, since $y(0) > z_u(a, 0)$, contradicting the maximum principle. The lower bound is proved similarly by considering $z_l(a, r) - y(r)$ for suitable a.

Now let a be the exact upper bound of the domain of regularity of $y(r)$. The same arguments can be used to show that, for $-a < r < a$, $z_l(a, r) \leqq y(r) \leqq z_u(a, r)$.

In order to evaluate the sharpness of the above estimate, one verifies that, in the interval $|r| < a$, the difference function $z_u(a, r) - z_l(a, r)$ is the bounded, positive, even, convex function

$$(n + 1) \log \frac{a^2 - r^2}{a^2 \cos(\pi r/2a)} + n \log \frac{\pi^2}{8}$$

taking its minimum value $n \log(\pi^2/8)$ at $r = 0$ and approaching the upper bound $n \log(\pi^2/8) + (n + 1) \log(4/\pi)$ as $r \to \pm a$.

We estimate next the asymptotic behavior of the derivatives of y by comparing them with the corresponding ones of z_u and z_l. Denote by $v(r)$, $w_u(r)$ and $w_l(r)$ respectively the functions $n^{-1} y'^n(r)$, $n^{-1} z_u'^n(r)$, $n^{-1} z_l'^n(r)$. Then one verifies directly that, in the interval $0 < r < a$,

$$\frac{v''(r)}{v'(r)} - \frac{n-1}{r} - (nv(r))^{1/n} = 0,$$

$$\frac{w_u''(r)}{w_u'(r)} - \frac{n-1}{r} - (nw_u(r))^{1/n} = (n-1)\left(\frac{\pi}{2a} \cot \frac{\pi r}{2a} - \frac{1}{r}\right) \leqq 0, \tag{8}$$

$$\frac{w_l''(r)}{w_l'(r)} - \frac{n-1}{r} - (nw_l(r))^{1/n} = \frac{2r}{a^2 + r^2} \geqq 0,$$

where the allowable functions are known to be strictly monotone increasing in the semiclosed interval $0 \leqq r < a$, vanishing of order n at $r = 0$; v is a solution of the equation derived from (6), w_u and w_l are respectively a super- and a subsolution. The conditions for the maximum principle are again valid for the differences $v - w_u$ and $w_l - v$, so that neither of these two difference functions can achieve local, positive valued maximum value at any interior point; the boundary conditions then imply that

$$w_l(r) \leqq v(r) \leqq w_u(r) \qquad (0 \leqq r < a)$$

and the equation (8) itself implies that $v(r)$ is strictly convex. As a corollary we have the following estimates for $y'(r)$:

$$z_l'(r) = \frac{2(n+1)r}{a^2 - r^2} \leqq y'(r) \leqq z_u'(r) = \frac{(n+1)\pi}{2a} \tan \frac{\pi r}{2a},$$

where the gap between the upper and lower bounds is again uniformly bounded

$$0 \leqq z_u'(r) - z_l'(r) = (n+1)\left(\frac{\pi}{2a} \tan \frac{\pi r}{2a} - \frac{2r}{a^2 - r^2}\right) \leqq n + 1 \quad (0 \leqq r < a).$$

Having estimated both upper and lower bounds for $y(r)$ and $y'(r)$, we obtain immediate bounds for $y''(r)$ by reading (6), namely $y'' = e^y(r/y')^{n-1}$ whence

$$\exp\{z_l (r/z_u')^{n-1}\} \leqq y'' \leqq \exp\{z_u (r/z_l')^{n-1}\}$$

In particular, we see that

$$y''(r) \geqq \exp\{z_l(r)(z_u'(r)/r)^{-n+1}\} = \frac{2^{2n-1}(n+1)\,a^{n+1}r^{n-1}\cot^{n-1}(\pi r/2a)}{\pi^{n-1}(a^2-r^2)^{n+1}},$$

the lower bound being asymptotic to $c^2/(a^2 - r^2)^2$ for some positive constant c as $r \to a$. An important consequence of this fact is that, as $r \to a$, the following improper integral diverges logarithmically,

$$(9) \qquad \int_0^r (y''(\rho))^{1/2}\, d\rho \geqq \frac{c}{a} \operatorname{arc\,tanh} \frac{r}{a}.$$

The solution $y(r)$ of (6), (7), which becomes infinite as $r \to a$ gives rise to the radial function $f(x)$ in the n-dimensional ball D of radius a in R^n, by setting

$$f(x) = y(r), \qquad r = \Bigl(\sum_{\alpha=1}^{n} (x_\alpha)^2\Bigr)^{1/2};$$

the function $f(x)$ is smooth in the interior, diverges uniformly to $+\infty$ at the boundary, is convex in the interior and satisfies the partial differential equation (5) for $K = -1$. Consequently the Kähler metric defined in the tube domain $M = \frac{1}{2}D + iR^n$ by (3) is an Einstein metric.

One verifies now without difficulty that M is complete with respect to this metric. In fact if we map M into the strip $[0, a] \times [0, \infty]$ by mapping the point $\frac{1}{2}(x + it) \in M$ $(x \in D,\, t \in R^n)$ to the point $(r, \rho) = (|x|, |t|)$ we obtain the following lower bound for the metric (3)

$$ds^2 = \frac{\partial^2 f(x)}{\partial x_\alpha\, \partial x_\beta}(dx^\alpha dx^\beta + dt^\alpha dt^\beta) \geqq y''(|x|)(dr^2 + d\rho^2) \geqq y''(r)\, dr^2.$$

Now suppose that we have a path Γ in M of finite length A, originating at the origin and parametrized by its arc length s $(0 \leqq s < A)$. Because of (9) we see that for each s the point $\frac{1}{2}(x_s + it_s)$ must satisfy

$$|x_s| \leqq a \tanh(as/c) < a \tanh(aA/c) < a$$

for the same constant c appearing in (9) so that x_s remains in a compact subdomain $D' \subset D$ defined by $|x| \leqq a' < a$. Let $\delta^2 = \inf_{0 \leqq r \leqq a'} y''(r)$; then one sees that the oscillation of $|t|$ along Γ is bounded by A/δ from above, so that the range of Γ is in a compact subdomain of M, proving the completeness.

Finally we show that, for $n \geqq 2$, the Einstein-Kähler metric just constructed in the tube $M = \frac{1}{2}D + iR^n$ where D is a euclidean n-ball is not locally homogeneous. For this purpose we carried on a long, tedious calculation on the Kählerian invariants of the metric.

We recall that the manifold M is the open submanifold of $\boldsymbol{C}^n$ defined by

$$M = \Bigl\{z = (z_1, \cdots, z_n)\, t\, \boldsymbol{C}^n \Bigm| \sum_{\alpha=1}^{n} (z_\alpha + \bar{z}_\alpha)^2 < a^2\Bigr\};$$

we have in M the plurisubharmonic potential defined by

$$F(z, \bar{z}) = f(x) = y(r) \qquad \left(x = (x_1, \cdots, x_n);\ x_\alpha = z_\alpha + \bar{z}_\alpha;\ r^2 = \sum_\alpha x_\alpha^2\right),$$

where $y(r)$ is the solution of (6), (7) which is regular in the interior and unbounded at the end points of the interval $(-a, a)$.

The metric given by (3) is then reduced to the expression

$$g_{\alpha\bar{\beta}}(z, \bar{z}) = \frac{y'(r)}{r}\delta_{\alpha\beta} + \left(y'' - \frac{y'}{r}\frac{x_\alpha x_\beta}{r^2}\right)$$

when $r > 0$, and to its limit value $y''(0)\,\delta_{\alpha\beta}$ at $r = 0$. The components $R^\lambda{}_{\alpha\beta}{}^\mu(z, \bar{z})$ can be calculated formally for $r > 0$, using the formula

$$R^\lambda{}_{\alpha\beta}{}^\mu = -\,g^{\mu\bar{\delta}}(z, \bar{z})\frac{\partial}{\partial \bar{z}_\delta}\left(g^{\lambda\bar{\gamma}}\frac{\partial^2 g_{\alpha\bar{\gamma}}}{\partial z_\beta}\right)$$

with the result

$$\begin{aligned} R^\lambda{}_{\alpha\beta}{}^\mu = {} & \left(\frac{1}{ry'} - \frac{y''}{y'^2}\right)(\delta^\lambda{}_\alpha\delta^\mu{}_\beta + \delta^\lambda{}_\beta\delta^\mu{}_\alpha) + \left(-\frac{1}{ry'} + \frac{1}{r^2 y''}\right)\delta^{\lambda\mu}\delta_{\alpha\beta} \\ & + \left(-\frac{1}{r^4 y''} - \frac{1}{r^3 y'} - \frac{y'''}{r^2 y' y''} + \frac{2y''}{r^2 y'^2}\right) \\ & \cdot(\delta^\lambda{}_\alpha x_\beta x_\mu + \delta^\lambda{}_\beta x_\alpha x_\mu + \delta^\mu{}_\alpha x_\beta x_\lambda + \delta^\mu{}_\beta x_\alpha x_\lambda + \delta^{\lambda\mu} x_\alpha x_\beta) \\ & + \left(\frac{1}{r^4 y''} + \frac{1}{r^3 y'} - \frac{2y'}{r^5 y''^2} - \frac{y' y'''}{r^4 y''^3}\right)\delta_{\alpha\beta} x_\lambda x_\mu \\ & + \left(-\frac{y''''}{r^4 y''^2} + \frac{y'''^2}{r^4 y''^3} + \frac{5y'''}{r^4 y' y''} - \frac{8y''}{r^4 y'^2}\right. \\ & \qquad \left. + \frac{2y'}{r^7 y''^2} + \frac{y' y'''}{r^6 y''^3} + \frac{3}{r^5 y'} + \frac{3}{r^6 y''}\right) x_\lambda x_\mu x_\alpha x_\beta. \end{aligned}$$

We interpret the curvature tensor at each point as a linear, selfadjoint endomorphism of the bundle $S^2(T^{(1,0)})$ of symmetric products of the holomorphic tangent bundle $T^{1,0}$. Owing to the rotational symmetry of M, it is clear that the following 2-dimensional (over $\boldsymbol{C}$) subspace of this bundle is invariant under the rotational group of isometries and hence under the curvature action: the space spanned by

$$\sum_{\alpha=1}^{n} \frac{\partial}{\partial z_\alpha} \otimes \frac{\partial}{\partial z_\alpha} \quad \text{and} \quad \sum_{\alpha,\beta=1}^{n} x_\alpha x_\beta \frac{\partial}{\partial z_\alpha} \otimes \frac{\partial}{\partial z_\beta}.$$

If M with this metric were locally homogeneous, then all eigenvalues of the linear action of the curvature tensor would be constant throughout M, and in particular the symmetric functions of the eigenvalues restricted to any invariant subspace, such as the one just exhibited. One can calculate now that the trace of the corresponding 2×2 matrix is

$$-\frac{y''''}{y''^2} + \frac{y'''^2}{y''^3} - \frac{2y''}{y'^2} + \frac{n-1}{r^2 y''} - \frac{n-3}{ry'},$$

and one readily verifies that the system of two equations consisting of equation (6) and the above expression equated to a constant is strictly overdetermined and inconsistent for $n \geqq 2$.

The example just exhibited and the estimates on its asymptotic behavior near the boundary will be probably of some value in establishing existence proofs of Kähler-Einstein metrics in other tube domains $M = \frac{1}{2}D + iR^n$, where D is any smoothly bounded, strongly convex domain in R^n.

REFERENCES

1. W. Blaschke, *Differentialgeometrie*. II. *Affine Differentialgeometrie*, Julius Springer, Berlin, 1923.

2. S. Bochner, *Vector fields and Ricci curvature*, Bull. Amer. Math. Soc. **52** (1946), 776–797. MR 8, 230.

3. S. Bochner and W. T. Martin, *Several complex variables*, Princeton Math. Ser., vol. 10, Princeton Univ. Press, Princeton, N.J., 1948. MR **10**, 366.

4. E. Calabi, *On Kähler manifolds with vanishing canonical class*, Algebraic Geometry and Topology (A Sympos. in Honor of S. Lefschetz), Princeton Univ. Press, Princeton, N.J., 1957, pp. 78–98. MR **19**, 62.

5. ———, *The space of Kähler metrics*, Proc. Internat. Congress Math., Amsterdam, 1954, pp. 206–207.

6. ———, *Improper affine hyperspheres of convex type and a generalization of a theorem by K. Jörgens*, Michigan Math. J. **5** (1958), 105–126. MR **21** #5219.

7. S. B. Myers, *Riemannian manifolds with positive mean curvature*, Duke Math. J. **8** (1941), 401–404. MR **3**, 18.

8. A.V. Pogorelov, *On the improper convex affine hyperspheres*, Geometriae Dedicata **1** (1972), 33–46.

9. T. J. Willmore, *An introduction to differential geometry*, Clarendon Press, Oxford, 1959, p. 238. MR **28** #2482.

10. ———, *On compact Riemannian manifolds with zero Ricci curvature*, Proc. Edinburgh Math. Soc. (2) **10** (1956), 131–133. MR **17**, 783.

11. M.H. Protter and H.F. Weinberger, *Maximum principle in differential equations*, Prentice-Hall, Englewood Cliffs, N. J., 1967.

UNIVERSITY OF PENNSYLVANIA

Proceedings of Symposia in Pure Mathematics
Volume 27, 1975

GENERALIZATIONS OF THE SCHWARZ-AHLFORS LEMMA TO QUASICONFORMAL HARMONIC MAPPINGS

S. I. GOLDBERG AND T. ISHIHARA*

Generalizations of the Schwarz-Ahlfors lemma to holomorphic mappings of hermitian manifolds have been given during the past five years notably by S. S. Chern and S. Kobayashi. On the other hand, an extension to real differentiable mappings was first given by Kiernan by considering harmonic K-quasiconformal mappings of Riemann surfaces. This was subsequently extended by T. Ishihara to the case where both the domain and target are n-dimensional Riemannian manifolds. More recently, the volume decreasing property of a different class of real harmonic mappings was studied by Chern and Goldberg (*On the volume decreasing property of a class of real harmonic mappings*, Amer. J. Math., to appear). In this paper, an investigation of the distance and volume decreasing properties of harmonic K-quasiconformal mappings of Riemannian manifolds is made and Ishihara's main results are improved and generalized. When the spaces are not of the same dimension corresponding results on intermediate volume elements are also obtained. An extension of Liouville's theorem and the little Picard theorem is also given.

AMS (MOS) subject classifications (1970). Primary 53A99, 53C99, 30A60.

*For a summary of the results, see *Harmonic quasiconformal mappings of Riemannion manifolds*, Bull. Amer. Math. Soc. **80** (1974), 562–566. The details are to appear in the Amer. J. Math. with this title.

Proceedings of Symposia in Pure Mathematics
Volume 27, 1975

HOLOMORPHIC MAPPINGS TO GRASSMANNIANS OF LINES

MARK L. GREEN*

The classical Picard theorem states that a nonconstant holomorphic map $f: \boldsymbol{C} \to \boldsymbol{P}_1$ (i.e. a meromorphic function) omits at most two points from its image. Picard also showed that a nonconstant holomorphic map $f: \boldsymbol{C} \to T$ to a surface of genus 1 omits no points at all, and that all holomorphic maps from $\boldsymbol{C}$ to a Riemann surface of genus $\geqq 2$ are constant. We may summarize this by:

PICARD'S THEOREM IN DIMENSION ONE. *Let $f: \boldsymbol{C} \to V - D$ be a holomorphic map, V a compact Riemann surface, D a finite collection of points. Then f is constant if $C_1(L_D) > \chi(V)$ or equivalently $C_1(K_V) + C_1(L_D) > 0$, where $\chi(V)$ is the Euler number of V, K_V is the canonical bundle of V, and L_D is the (positive) line bundle associated to the divisor D.*

This differential-geometric reformulation of the theorem is not merely a combinatorial sleight of hand, for there exist proofs of this theorem based on the Gauss-Bonnet formula. Such methods were carried further by Carlson and Griffiths [1], who proved:

CARLSON-GRIFFITHS PICARD THEOREM FOR EQUIDIMENSIONAL MAPS. *Let $f: \boldsymbol{C}^n \to V_n - D$ be a holomorphic map, where V is an n-dimensional Kähler manifold, D a hypersurface with normal crossings. If $C_1(K_V) + C_1(L_D)$ is a positive (1, 1) form, then f is degenerate in the sense that its Jacobian determinant vanishes identically.*

This is a remarkably general theorem, but the requirement that the domain and range have the same dimension is a severe restriction. For example, in the theory

---

AMS (MOS) subject classifications (1970). Primary 30A70, 32H25; Secondary 32H20.
*Partially supported by NSF grant GP-16651.

of minimal surfaces, holomorphic maps from the complex line or the unit disc to Grassmannians are encountered. The general Picard theorem which should be true, restricting ourselves to algebraic varieties, is:

CONJECTURE. *A holomorphic map* $f: \boldsymbol{C}^m \to V_n - D$, *where* V *is an n-dimensional algebraic variety,* D *a hypersurface with normal crossings,* m *arbitrary, must have image lying in a lower dimensional algebraic subvariety of* V *if* $C_1(K_V) + C_1(L_D)$ *is a positive* (1, 1) *form.*

When the conclusion of the theorem is satisfied, we will say that f is *algebraically degenerate.*

REMARK. Were this Conjecture established, one should also attempt the corresponding big Picard theorems and, under stronger hypotheses, Schottky-Landau theorems and hyperbolicity.

The simplest unknown cases of the conjecture are for maps to $\boldsymbol{P}_2 - D$, for D a curve of degree at least four with ordinary double points, and for maps to a surface of general type. In the first case, the result is known only for D the union of three curves in a linear pencil or the union of four hypersurface section curves (see Green [**6**]), for D a Brieskorn curve $\{z_0^{a_0} + z_1^{a_1} + z_2^{a_2} = 0\}$ where $\sum_i(1/a_i) < 1/2$ (see Toda [**7**] and Fujimoto [**3**]), and for D the dual curve of an algebraic plane curve of degree $\geqq 3$ with ordinary multiple points (see Green [**4**], note that here D will have worse singularities than those hypothesized in the Conjecture). There are partial results for D the union of a conic and two lines (see Green [**5**]). For surfaces, examples where a Picard theorem is known are scarcer—the product of two curves one of which has genus $\geqq 2$, surfaces whose universal cover is a bounded domain, and the family of hyperbolic hypersurfaces in $\boldsymbol{P}_3$ constructed in Green [**4**].

The new result I would like to discuss in this paper deals with holomorphic maps to Grassmannians of lines $G(2, n + 1)$, the latter denoting the lines in $\boldsymbol{P}_n$; it is an algebraic variety of dimension $2(n - 1)$. There are two families of hypersurfaces in $G(2, n + 1)$ which arise most naturally from geometric situations. There is the set of lines intersecting a fixed $(n - 2)$-dimensional subvariety of $\boldsymbol{P}_n$, of which the Schubert cycles of top dimension are a special case (i.e. when the subvariety is linear). Given a hypersurface S in $\boldsymbol{P}_n$, we can define a hypersurface T_S of $G(2, n + 1)$ in the second natural family to be the set of lines meeting S in fewer than degree(S) distinct points. If S is nonsingular, then T_S is the set of tangent lines to S.

THEOREM. *Let* $S \subseteq \boldsymbol{P}_n$ *be an irreducible hypersurface of degree* $\geqq 4$. *Then every holomorphic map* $f: \boldsymbol{C}^m \to G(2, n + 1)$ *either meets* T_S *or is algebraically degenerate.*

Before undertaking the proof, we state a theorem which gives a stronger conclusion for a generic class of S which we can describe explicitly. We will say a hypersurface S in $\boldsymbol{P}_n$ has *generic multiple tangents* if the dimension of the set of lines either intersecting S in at most two distinct points or lying completely in S is the same as it would be for a generic hypersurface of the same degree d as S (we will show later that this dimension is $\dim G(2, n + 1) - d + 2$).

We then have

THEOREM. *Let $S \subseteq \boldsymbol{P}_n$ be a hypersurface with generic multiple tangents. Then every holomorphic map $f: \boldsymbol{C}^m \to G(2, n+1)$ either meets T_S or has image lying in an algebraic subvariety of $G(2, n+1)$ of codimension* $\deg(S) - 3$.

REMARK. In the proof, it will emerge that these subvarieties can be given explicitly as the set of lines whose intersections with S, viewed as a set of deg S points on $\boldsymbol{P}_1$, are all equivalent to each other under automorphisms of $\boldsymbol{P}_1$.

PROOF. The proofs of the two theorems will proceed simultaneously. Assume the image of f does not meet T_S. Then the line $f(z)$ meets S in exactly $d = \deg(S)$ distinct points $a_1(z), \cdots, a_d(z)$. Locally, these are holomorphic functions of z because they are distinct roots of a polynomial equation with holomorphic coefficients. However, as $\boldsymbol{C}^m$ is simply connected, we can analytically continue to get global maps $a_1, \cdots, a_d$ from $\boldsymbol{C}^m$ to $\boldsymbol{P}_n$ (i.e. we can globally "label" the d intersections of our variable line with S). As $a_1, \cdots, a_d$ are collinear, we can take the cross-ratio

$$\frac{(a_{i_1} - a_{i_2})(a_{i_3} - a_{i_4})}{(a_{i_1} - a_{i_3})(a_{i_2} - a_{i_4})}$$

of any four, getting a map $\boldsymbol{C}^m \to \boldsymbol{P}_1 - \{0, 1, \infty\}$ as not two of the a_i ever coincide. This map must be constant by the classical Picard theorem. Thus, the cross-ratios of all 4-tuples of intersection points of the variable line with S are constant. By standard facts about the cross-ratio, this is the same as saying any two intersection sets $f(z_1) \cdot S, f(z_2) \cdot S$ are related by an automorphism of $\boldsymbol{P}_1$ (a projectivity). We conclude, dropping the labels on the points, that the image of f lies in the variety

$$V = \{l \in G(2, n+1) \mid l \cdot S \text{ is projectively equivalent to } f(0) \cdot S\}.$$

This is clearly an algebraic subvariety of $G(2, n+1)$, so all that remains to prove both theorems is to investigate its dimension. We do this via three lemmas.

LEMMA 1. *$\bar{V} \circ T_S \subseteq M_S$, where M_S is the set of lines intersecting S in at most two distinct points, or lying entirely in S, and $\bar{V}$ is the closure of V.*

PROOF. Since all the cross-ratios of 4-tuples of points in $f(0) \cdot S$ are different from $0, 1, \infty$ and as the cross-ratio of four points is one of these forbidden values when two points coincide and there are no further coincidences, whereas the cross-ratio is undefined when there are two or less distinct points in the r-tuple, the lines in $\bar{V} \circ T_S$ either are contained in S or must intersect S in a set of points so that all cross-ratios involve either four distinct points or else at most two distinct points. If l is tangent to S, this is possible only if the number of distinct points in $l \cdot S$ is two or less.

LEMMA 2. $\dim \bar{V} \leqq \dim M_S + 1$ *if S is not a union of hyperplanes.*

PROOF. Were we in $\boldsymbol{P}_n$ instead of a Grassmannian, this would be clear, as T_S is a hypersurface is S is not a union of hyperplanes. However, T_S is homologous to a multiple of a hyperplane section of $G(2, n+1)$ because the homology group in (real) codimension 2 is one dimensional (see Chern [2]).

This implies that $\bar{V} \circ T_S \neq 0$ and $\dim (\bar{V} \circ T_S) \geqq \dim \bar{V} - 1$. The result now follows from Lemma 1.

LEMMA 3. *For S irreducible and* $\deg(S) \geqq 4$, *we have* $\dim M_S \leqq \dim G(2, n+1) - 2$. *If S has generic multiple tangents, then* $\dim M_S \leqq \dim G(2, n+1) - \deg(S) + 2$.

PROOF. To see the first assertion, assume $\dim M_S \geqq \dim G(2, n+1) - 1$, so $\dim M_S = \dim T_S$. As S is irreducible a generic 2-plane section of S is an irreducible plane curve D of degree $\geqq 2$. Our assumption implies all the tangents to D are multiple tangents or inflexional tangents. This cannot be the case, for it dramatically contradicts the Plucker formulas.

To prove the second assertion, it suffices to show that for a generic surface S of degree d,

$$\dim M_S \leqq \dim G(2,n+1) - d + 2.$$

Consider the variety

$$W = \{(S, l) | S \text{ a hypersurface in } \boldsymbol{P}_n, l \in M_S\}$$

with the projection maps

$$\begin{array}{c} W \to G(2,n+1) \\ \downarrow \\ \{\text{hypersurfaces in } \boldsymbol{P}_n\}. \end{array}$$

Clearly $\dim W \geqq \dim \{\text{hypersurfaces in } \boldsymbol{P}_n\} + \dim M_S$ for S a generic hypersurface. The desired inequality for $\dim M_S$ is equivalent to showing the fibre $W_L = \{S | S \text{ a hypersurface in } \boldsymbol{P}_n, L \in M_S\}$, L fixed, has codimension at least $d-2$ in the variety of all hypersurfaces in $\boldsymbol{P}_n$. If $S = \{P(Z_0, \cdots, Z_n) = 0\}$ and L is the line $\{Z_2 = 0, Z_3 = 0, \cdots, Z_n = 0\}$, then the condition that L be tangent to S at points $a_1, \cdots, a_r$ to orders $k_1, \cdots, k_r$ respectively is equivalent to $P(Z_0, Z_1, 0, \cdots, 0)$ vanishing at $a_1, \cdots, a_r$ to orders $k_1, \cdots, k_r$. This imposes $k_1 + \cdots + k_r$ conditions on P, which are independent as the zeros of a polynomial in one variable can be chosen independently. The condition that $L \cdot S$ contain at most r points thus imposes $d - r$ independent conditions, where $d = \deg S$. Taking $r = 2$, we get the desired number of conditions on S if $L \in M_S$ and $L \not\subset S$. If $L \subset S$, we get the condition that $P(Z_0, Z_1) \equiv 0$, which imposes $d + 1$ independent conditions. We conclude W_L has codimension $d - 2$ in the variety of all hypersurfaces in $\boldsymbol{P}_n$, and from this, by our previous argument, the lemma follows.

We now return to the proof of the two theorems. It was shown that the image of a holomorphic map $f\colon \boldsymbol{C}^m \to G(2, n+1) - T_S$ lies in the variety

$$V = \{l \in G(2, n+1) | l \cdot S \text{ is projectively equivalent to } f(0) \cdot S\}.$$

By Lemma 2, $\dim V \leqq \dim M_S + 1$. By Lemma 3, if S is irreducible and $\deg(S) \geqq 4$, we have $\dim M_S \leqq \dim G(2, n+1) - 2$. So V has codimension at least one; hence f is algebraically degenerate.

If S has generic multiple tangents, then again by Lemma 3 we have $\dim M_S \leqq \dim G(2, n + 1) - d + 2$. So $\dim V \leqq \dim G(2, n + 1) - d + 3$, which proves the conclusion of the second theorem.

A corollary of particular interest is

COROLLARY. *Any holomorphic map* $f\colon \boldsymbol{C}^m \to G(2, n + 1) - T_S$, *where* S *is a hypersurface in* $\boldsymbol{P}_n$ *with generic multiple tangents and* $\deg(S) \geqq \dim G(2, n + 1) + 3$, *must be constant.*

By using the metric of constant negative curvature on $\boldsymbol{P}_1 - \{0, 1, \infty\}$, it is possible to impose a metric of holomorphic sectional curvatures $\leqq -1$ on $G(2, n + 1) - T_S$ when the hypotheses of the foregoing Corollary are satisfied. In particular, this implies

THEOREM. *If* S *is a hypersurface in* $\boldsymbol{P}_n$ *with generic multiple tangents (in particular, if* S *is a generic hypersurface) and* $\deg(S) \geqq \dim G(2, n + 1) + 3$, *then* $G(2, n + 1) - T_S$ *is hyperbolic.*

BIBLIOGRAPHY

1. James Carlson and Phillip Griffiths, *A defect relation for equidimensional holomorphic mappings between algebraic varieties*, Ann. of Math. (2) **95** (1972), 557–584.

2. S.S. Chern, *Complex manifolds without potential theory*, Van Nostrand Math. Studies, no. 15, Van Nostrand, Princeton, N.J., 1967. MR **37** #940.

3. H. Fujimoto, *On meromorphic maps into the complex projective space*, Marcel Dekker, New York, 1974, 119–132.

4. M. Green, *The complement of the dual of a plane curve and some new hyperbolic manifolds*, Proc. Tulane Conference in Value Distribution Theory (to appear).

5. ———, *On the functional equation* $f^2 = e^{2\phi_1} + e^{2\phi_2} + e^{2\phi_3}$ *and a new Picard theorem*, Trans. Amer. Math. Soc. **195** (1974), 223–230.

6. ———, *Some Picard theorems for holomorphic maps to algebraic varieties*, Amer. J. Math. (to appear).

7. N. Toda, *On the functional equation* $\sum_{i=0}^{p} a_i f_i^{n_i} = 1$, Tôhoku Math. J. (2) **23** (1971), 289–299. MR **45** #551.

UNIVERSITY OF CALIFORNIA, BERKELEY

Proceedings of Symposia in Pure Mathematics
Volume 27, 1975

SOME FUNCTION-THEORETIC PROPERTIES OF NONCOMPACT KÄHLER MANIFOLDS

R. E. GREENE AND H. WU*

The purpose of the following discussion is to present, with some indications of proof, a number of theorems concerning the influence of the geometric properties of noncompact Kähler manifolds on the function-theoretic properties of such manifolds. Those theorems not otherwise ascribed were originally stated in the authors' announcements [**4**, I, II, and III]. The discussion here is divided into four parts: §1, some geometric conditions under which a Kähler manifold is necessarily a Stein manifold; §2, some results on the nonexistence of L^p holomorphic functions and forms on noncompact Kähler manifolds satisfying certain curvature restrictions; §3, conditions under which a Kähler manifold necessarily fails to admit bounded nonconstant holomorphic functions; and §4, some function-theoretic properties of Kähler manifolds of nonnegative curvature.

1. Stein manifolds. If M is a complex submanifold of some complex euclidean space $\boldsymbol{C}^N$, then the restriction to M of a holomorphic function (or form) on $\boldsymbol{C}^N$ will be holomorphic on M. Thus in this case there exists on M an abundance, in a general sense, of such functions and forms so that M is especially interesting function-theoretically.

DEFINITION. A complex manifold M is a *Stein manifold* if there exist an integer N and a holomorphic proper embedding $F : M \to \boldsymbol{C}^N$. (Here and throughout,

AMS (MOS) subject classifications (1970). Primary 53C55, 32C10, 32E10; Secondary 32F05, 31C10.

*The research of both authors was partially supported by the National Science Foundation: grants GP-27576 (first author) and GP-34785 (second author).

a mapping $F : M_1 \to M_2$ of one manifold to another is said to be *proper* if for every compact subset K of M_2, $F^{-1}(K)$ is a compact subset of M_1.)

Any open Riemann surface is a Stein manifold (the Behnke-Stein theorem); but if the complex dimension of a complex manifold M is greater than 1, then M has to satisfy very restrictive conditions in order to be a Stein manifold. For example, if M is a domain in $\boldsymbol{C}^N$, then M is a Stein manifold if and only if M is holomorphically convex.

The definition given of a Stein manifold is not easily verified directly on the basis of differential geometric assumptions. For instance, if M is a Stein manifold then M admits a complete Kähler metric, namely the metric induced on M from that of $\boldsymbol{C}^N$ by a proper holomorphic embedding $F : M \to \boldsymbol{C}^N$; but not every complete noncompact Kähler manifold is a Stein manifold. For example, $P_n \boldsymbol{C} \times \boldsymbol{C}^m$, n, $m \geqq 1$, is not a Stein manifold because it has compact complex submanifolds. However, a well-known theorem of Grauert does provide a necessary and sufficient condition for a complex manifold to be a Stein manifold, and this condition is closely related to the geometry of the manifold:

A C^2 function $f : M \to \boldsymbol{R}$ on a complex manifold M is called *strictly plurisubharmonic* if the Hermitian quadratic form L_f defined in a holomorphic local coordinate system $(z_1, \cdots, z_n)$ by

$$L_f = \sum_{i,j=1}^{n} \frac{\partial^2 f}{\partial z_i \, \partial \bar{z}_j} dz_i \otimes d\bar{z}_j$$

is positive definite (the form L_f is coordinate-choice independent and so everywhere defined and C^∞ on M, as a straightforward computation shows).

THEOREM (GRAUERT). *A necessary and sufficient condition for a complex manifold M to be a Stein manifold is that there be a C^∞ nonnegative strictly plurisubharmonic proper function from M to $\boldsymbol{R}$.*

The necessity of the condition is easily verified: if $F : M \to \boldsymbol{C}^N$ is a proper holomorphic embedding and $D : \boldsymbol{C}^N \to \boldsymbol{R}$ is defined by $D((z_1, \cdots, z_N)) = \sum_{i=1}^N |z_i|^2$ then $D \cdot F$ is a function of the required sort. For the (much more difficult) proof of the sufficiency, see [3] or [8].

A relationship between the question of the existence of C^∞ strictly plurisubharmonic functions and geometric considerations appears in the following observation:

LEMMA 1. *A C^∞ function on a Kähler manifold M is strictly plurisubharmonic if its second derivative relative to arc length along every geodesic is positive.*

To prove this lemma, let q be any point of M and $(z_1, \cdots, z_n)$ a holomorphic normal coordinate system at q. Such a coordinate system exists because M is a Kähler manifold. Then

$$L_f\left(\left.\frac{\partial}{\partial z_1}\right|_q, \left.\frac{\partial}{\partial \bar{z}_1}\right|_q\right) = \left.\frac{\partial^2 f}{\partial z_1 \, \partial \bar{z}_1}\right|_q = \frac{1}{4}\left(\left.\frac{\partial^2 f}{\partial x_1^2}\right|_q + \left.\frac{\partial^2 f}{\partial y_1^2}\right|_q\right).$$

Because $\partial^2 f / \partial x_1^2 |_q$ and $\partial^2 f / \partial y_1^2 |_q$ are second derivatives relative to arc length along geodesics, they are positive by hypothesis so that

$$L_f\left(\frac{\partial}{\partial z_1}\bigg|_q, \frac{\bar\partial}{\partial \bar z_1}\bigg|_q\right) > 0.$$

The lemma now follows from the fact that the direction of $\partial / \partial z_1$ in $TM_q \otimes C$ could have been chosen arbitrarily.

On a complete, simply connected Riemannian manifold of curvature $\leqq 0$, the square of the distance from a fixed point has positive second derivative relative to arc length along geodesics [**1**]. In fact, the value of such a second derivative is always at least 2. Combining these results with Lemma 1 provides a proof of the following theorem (cf. [**15**]). Some additional estimates of the Levi form of various functions of ρ, which estimates involve the complex structure of M more closely, are given in [**4**].

THEOREM 1. *If M is a complete, simply connected Kähler manifold of everywhere nonpositive sectional curvature and $\rho : M \to R$ is the Riemannian distance function from a fixed point $p \in M$, then $\rho^2 : M \to R$ is a nonnegative C^∞ strictly plurisubharmonic proper function on M and so M is a Stein manifold. In fact, $L_{\rho^2} \geqq 2g$ where g is the Hermitian extension of the Riemannian metric of M to the complexification of the tangent spaces of M.*

The situation regarding noncompact Kähler manifolds of positive curvature is much more complicated, because no C^∞ proper function satisfying the hypothesis of Lemma 1 seems to arise naturally in this case. However, a geometrically significant continuous proper function does arise: Cheeger and Gromoll [**2**], following work of Gromoll and Meyer [**7**], have shown that there exists on a complete connected noncompact Riemannian manifold of nonnegative curvature a continuous function $\tau : M \to R$ which is proper and is geodesically convex in the sense that τ restricted to any geodesic in M is a convex function of arc length (in the usual one-variable sense of convexity). In [**5**], the present authors developed a smoothing technique to use in approximating geodesically convex functions. One result obtainable from this technique is the following lemma (see [**5**, §4]):

LEMMA 2. *If K is a compact subset of a Kähler manifold M, if $\tau : U \to R$ is a geodesically convex function on a neighborhood U of K, and if $\varphi : U \to R$ is a C^∞ strictly plurisubharmonic function on U, then given $\varepsilon > 0$ there exists a C^∞ function τ_ε defined in some neighborhood of K such that $| \tau_\varepsilon (q) - \tau (q) | < \varepsilon$ for all $q \in K$ and $\tau_\varepsilon + \varphi$ is (C^∞ and) strictly plurisubharmonic in some neighborhood of K.*

If M is a complete connected noncompact Kähler manifold of positive sectional curvature, then M is (real) diffeomorphic to euclidean space [**7**]. It can be shown then using de Rham's theorem that there exists a C^∞ strictly plurisubharmonic function $\varphi : M \to R$, φ not being required to be proper. The standard functional analysis technique for solution of the $\bar\partial$ problem on pseudoconvex domains in euclidean space (see, e.g. [**8**]) can be generalized to the Kähler manifold setting in such a way that the curvature of the manifold occurs explicitly in the relevant estimates. The following theorem can then be proved using this generalization together with Lemma 2 and the existence of such functions φ.

THEOREM 2. *If M is a complete, connected, noncompact, Kähler manifold of everywhere positive sectional curvature, then M is a Stein manifold.*

In fact, a stronger version of this result holds in which to infer that M is a Stein manifold, the sectional curvature of M need only be assumed to be everywhere nonnegative provided the additional assumptions are made that the Ricci curvature of M is everywhere positive and that the canonical bundle of M is topological trivial [**4**, III]. A related result is that if M is a complete connected noncompact Hermitian manifold and if there is a compact subset of M outside of which the metric of M is a Kähler metric of positive sectional curvature then M is obtainable from a Stein space by blowing up a finite number of points.

2. Nonexistence of L^p functions and forms. Once it is known that some complex manifold M is a Stein manifold so that many nonconstant holomorphic functions and nonzero holomorphic forms exist on M, it is natural to inquire whether such functions or forms can have finite L^2 or, more generally, L^p norms. In the case where M is a complete simply connected Kähler manifold of nonpositive sectional curvature, this question can be approached by using the fact that the exponential map $\exp_q : TM_q \to M$, $q \in M$, is a diffeomorphism with special properties in this case, e.g. exd_q is distance nondecreasing. In particular, if $S(r)$ = the geodesic sphere of radius $r > 0$ in M about q, then $S(r)$ is a compact C^∞ submanifold of codimension 1, and the distance nondecreasing property of $\exp_q$ implies that if $A(r)$ = the $(n-1)$-dimensional measure of $S(r)$ then $A(r) \geqq \sigma_{2n-1} r^{2n-1}$, where σ_{2n-1} = the area of the unit $(2n-1)$-sphere in R^{2n} and $2n = \dim_R M$. More generally, it can be shown that if $f: M \to R$ is a nonnegative C^∞ subharmonic function (f subharmonic means $\Delta f \geqq 0$ everywhere) on M then

$$f(q) \leqq \frac{1}{\sigma_{2n-1} r^{2n-1}} \int_{S(r)} f \omega_r$$

where ω_r = the volume form of the induced metric on $S(r)$. (This result is actually a Riemannian geometry statement proper, in which the complex structure of M plays no role; see [**4**, I].) This estimate implies the first part of the following theorem, once it is shown that if f is a holomorphic function then $|f|^p$, $1 \leqq p < +\infty$, can be approximated uniformly on compact sets by (nonnegative) subharmonic C^∞ functions. The functions $(|f|^2 + \varepsilon)^{p/2}$, $\varepsilon > 0$, provide the required approximations [**6**]. The second part is established by similar methods, using comparison with the constant negative holomorphic sectional curvature case rather than the euclidean space case.

THEOREM 3. *If M is a complete, simply connected Kähler manifold of nonpositive sectional curvature then:*

(a) *no holomorphic function $\not\equiv 0$ is in L^p, $1 \leqq p < +\infty$; in fact, if $f: M \to \boldsymbol{C}$*

is a holomorphic function then, for $1 \leqq p < +\infty$,

$$\int_{S(r)} |f|^p \, \omega_r \geqq |f(q)|^p \, \sigma_{2n-1} r^{2n-1};$$

and

(b) *if the sectional curvature of* M *is everywhere* $\leqq -c^2 < 0$ *and* $f: M \to C$ *is a holomorphic function then, for* $1 \leqq p < +\infty$,

$$\int_{S(r)} |f|^p \, \omega_r \geqq D_f \exp\{(2n-1)^{1/2} cr\}$$

where D_f *is independent of* r *and may be taken to be positive if* $f(q) \neq 0$.

A similar analysis of the corresponding integrals on manifolds of nonnegative curvature is hampered by the lack of any family of C^∞ submanifolds corresponding to the family $S(r)$. Once again, however, the proper geodesically convex function $\tau: M \to R$ constructed in **[2]** can be usefully employed: The function τ is Lipschitz continuous on all of M with Lipschitz constant 1, and although τ is not in general even once differentiable, it is once differentiable almost everywhere. Moreover, at those points at which the gradient grad τ of τ does exist, $\|\text{grad}\, \tau\| = 1$. These similarities to the distance function ρ in the nonpositive curvature case suggest analysis of the integrals of nonnegative subharmonic functions over M using the level sets $\{q \in M \mid \tau(q) = t\}$. An essential feature of this analysis is again the smoothing technique of **[5]**. A proof of the following theorem is given in **[6]**:

THEOREM 4. *Let* M *be a complete, connected, noncompact Riemannian manifold with curvature nonnegative outside some compact subset. Suppose that* $f: M \to R$ *is the limit uniformly on compact subsets of* M *of some sequence of* C^∞ *subharmonic functions on* M. *Then if* $f \geqq 0$ *but not* $\equiv 0$ *there exist constants* $A > 0$ *and* t_0 *such that*

$$\int_{\{q \in M \mid \tau(q) \leqq t\}} f \, d\mu \geqq A(t - t_0),$$

where μ *is the Lebesgue measure induced on* M *from the Riemannian metric of* M.

In the following applications of Theorem 4, it will be possible to construct the required C^∞ approximations of f directly. However, for the sake of generality it should be noted that a function $f: M \to R$ is the limit uniformly on compact subsets of M of a sequence of C^∞ subharmonic functions if and only if f is a continuous function which is subharmonic in the sense of the following definition:

DEFINITION. A continuous function $f: M \to R$ is *subharmonic* if for every domain D in M with compact closure and C^∞ boundary and every harmonic function $h: D \to R$ which has a continuous extension $\bar{h}: \bar{D} \to R$ satisfying $f(p) \leqq \bar{h}(p)$ for all $p \in \bar{D} - D$, the inequality $f(p) \leqq h(p)$ is satisfied for all $p \in D$.

Any function which is a limit uniformly on compact subsets of M of a sequence of C^∞ subharmonic functions is subharmonic in the sense of this extended definition. This fact follows easily from the standard arguments used in the eucilidean space case. To establish the converse requires approximation on compact sets of continuous subharmonic functions by C^∞ ones (cf. [**6**]). Such approximations can be obtained by convolution smoothing in the euclidean space case (e.g., [**8**, p. 20]), but on general Riemannian manifolds this method does not apply. However, the required approximations do exist in the general case: On an arbitrary Riemannian manifold M, such approximations can be obtained by using the function f to be approximated as the initial condition for the heat equation. The authors are indebted to P. Malliavin for pointing out that the required approximations can be obtained in this way.

Noting once again that if $f: M \to \boldsymbol{C}$ is a holomorphic function on a Kähler manifold then the functions $(|f|^2 + \varepsilon)^{p/2}$, $\varepsilon > 0$, are C^∞ nonnegative and subharmonic and that they converge uniformly on compact sets to $|f|^p$ as $\varepsilon \to 0^+$, one obtains the following consequence of Theorem 4:

COROLLARY. *If M is a complete, connected, noncompact Kähler manifold with sectional curvature nonnegative outside a compact set, then no holomorphic function $\not\equiv 0$ on M is in $L^p(M)$ for any p satisfying $1 \leqq p < +\infty$.*

A suitable computation of the Laplacian can be used to show that if α is a holomorphic $(q, 0)$ form, $q \geqq 1$, on a Kähler manifold of nonnegative Ricci curvature then $\langle \alpha, \alpha \rangle$ is a (C^∞) subharmonic function (the curvature of the manifold occurs explicitly in this computation) and again for any p satisfying $1 \leqq p < +\infty$ the functions $(\langle \alpha, \alpha \rangle + \varepsilon)^{p/2}$ provide approximations of $\|\alpha\|^p$ by C^∞ subharmonic functions. The next theorem then follows from Theorem 4:

THEOREM 5. *If M is a complete, connected, noncompact Kähler manifold of everywhere nonnegative sectional curvature, then no holomorphic $(q, 0)$ form $\not\equiv 0$ is in $L^p_{(q,0)}(M)$ for any p satisfying $1 \leqq p < +\infty$.*

In the case of $L^2_{(q,0)}(M)$, the following considerably stronger result is available [**4**, II].

THEOREM 6. *If M is a complete Kähler manifold of positive scalar curvature then no holomorphic $(n, 0)$ form $\not\equiv 0$ is in $L^2_{(n,0)}(M)$ $(n = \dim_{\boldsymbol{C}} M)$. More generally, if for some positive integer $k \leqq n$ the eigenvalues $r_1, \cdots, r_n$ of the Ricci tensor satisfy*

$$r_{i_1} + \cdots + r_{i_k} > 0 \quad \textit{for all } i_1 < \cdots < i_k,$$

then no holomorphic $(k, 0)$ form $\not\equiv 0$ is in $L^2_{(k,0)}(M)$.

COROLLARY. *If D is a domain in $\boldsymbol{C}^n$ which admits a complete Kähler metric of positive scalar curvature, then D necessarily has infinite Lebesgue measure.*

The corollary follows immediately from application of the theorem to the form $dz_1 \wedge \cdots \wedge dz_n$ on D. The proof of the theorem uses L^2 estimates of the type developed in the functional analysis of the $\bar{\partial}$ problem.

3. Bounded holomorphic functions. The contrast between the abundance of bounded nonconstant holomorphic functions on the unit disc in $\boldsymbol{C}$, which admits a Kähler metric of constant negative curvature, and the absence of such functions on $\boldsymbol{C}$, which has the property that the curvature of a Kähler metric of nonpositive curvature must be arbitrarily close to 0 somewhere outside any given compact set [**4**, I], suggests that some relationship exists between the absence of bounded nonconstant holomorphic functions and the existence of a 0 limit for the curvature at infinity. The following theorem gives exactly such a relationship:

THEOREM 7. *Let M be a complete, simply connected Kähler manifold of everywhere nonpositive sectional curvature. Suppose that for some point $q \in M$ and some positive constants C and ε,*

$$|\textit{sectional curvature}(p)| \leqq C(d(q, p))^{-2-\varepsilon}$$

for all $p \in M$, where d is the distance function determined by the Kähler metric. Then no nonconstant holomorphic function on M is bounded.

This result fails if ε is not required to be positive: if $n \geqq 3$ then the Kähler metric $(1 - z\bar{z})^{-n}\, dz d\bar{z}$ on the unit disc is complete and satisfies $|\text{curvature}(z)| \leqq C(d(0, z))^{-2}$ (0 = origin in $\boldsymbol{C}$).

To prove Theorem 7, one first notes that on a Kähler manifold a holomorphic function is harmonic and so its real and imaginary parts are harmonic. Hence, it suffices to prove that under the hypotheses given the manifold M has no bounded nonconstant harmonic real-valued functions. Since M is complete and simply connected and has nonpositive curvature, the mapping $\exp_q : TM_q \to M$ is a diffeomorphism. Pick a euclidean coordinate system on TM_q and let $(x_1, \cdots, x_{2n})$ be the corresponding C^∞ coordinate system on M. Now a real function $f : M \to R$ is harmonic if and only if

$$\sum_{i,j=1}^{2n} \frac{\partial}{\partial x_i}\left(g^{1/2} g^{ij} \frac{\partial f}{\partial x_j}\right) = 0$$

at every point of M. A theorem of Moser [**11**] asserts that this equation, which may be considered as an equation on R^{2n}, has no bounded nonconstant solution if the $2n \times 2n$ matrix whose ij entry is $g^{1/2}g^{ij}$ has eigenvalues which are bounded above and which are bounded below by a positive number. In the case at hand, it is easy to see that this condition is satisfied if the distance-nondecreasing map $\exp_q$ is a quasi-isometry, i.e. if there exists a constant A such that the $\exp_q$ images of two points of distance λ from each other in TM_q are of distance at most $A\lambda$ in M (their distance is at least λ because $\exp_q$ is distance-nondecreasing). To establish that $\exp_q$ has this property, one constructs a manifold of nonpositive curvature on which the exponential map is a quasi-isometry and to which M can be compared via the Rauch comparison theorem to prove the quasi-isometry property for M, thus completing the proof of Theorem 7.

In an as yet unpublished work, S.-T. Yau has established that, in contrast to the conditions required in the negative curvature case, on no complete Kähler manifold of positive sectional curvature are there any bounded nonconstant holomorphic functions. In fact, if M is any complete Riemannian manifold of nonnegative Ricci curvature whose sectional curvatures are bounded below then no nonconstant harmonic function on M is bounded. This result is related to Theorem 4 as follows:

Let M be a complete connected noncompact Riemannian manifold of everywere nonnegative sectional curvature. Yau's result implies that M has no bounded nonconstant harmonic functions. Then a generalization to Riemannian manifolds by Sario, Schiffer, and Glasner (see [**12**, p. 405]) of a theorem of Virtanen [**14**] on Riemann surfaces implies that any nonconstant harmonic function $f : M \to R$ has infinite Dirichlet norm, i.e. $\int_M \langle df, df \rangle \, d\mu = +\infty$. This property of M can be established directly from Theorem 4: Under the hypotheses on M indicated, for any harmonic 1-form α on M, $\langle \alpha, \alpha \rangle$ is subharmonic. If f is a harmonic function, then df is a harmonic 1-form since $\Delta(df) = d(\Delta f) \equiv 0$ when f is harmonic. Thus Theorem 4 applies to $\langle \alpha, \alpha \rangle$ so that $\int_M \langle df, df \rangle \, d\mu = +\infty$.

4. Kähler manifolds of nonnegative curvature. The techniques used to prove Theorem 2 can be utilized to prove cohomology vanishing theorems for Kähler manifolds which are not necessarily Stein manifolds:

THEOREM 8. *Let M be a complete Kähler manifold with everywhere positive Ricci curvature and everywhere nonnegative sectional curvature. If K is the canonical bundle of M and L is a holomorphic line bundle such that $L \otimes K^* > 0$, then $H^p(M, \mathcal{O}(L)) = 0$ for all $p \geqq 1$. (Here K^* denotes the dual of K and $L \otimes K^* > 0$ means that the line bundle $L \otimes K^*$ is positive in the sense that there is some Hermitian metric on $L \otimes K^*$ of positive curvature.)*

On a Kähler manifold satisfying the hypotheses of this theorem, the dual K^* of the canonical bundle K ($K =$ the nth exterior power of the holomorphic cotangent bundle, where $n = \dim_C M$) is a positive bundle. By generalizing the method of proof of the Kodaira embedding theorem [**10**], it can be shown that a sufficiently high tensor power of the bundle K^* has enough sections that the quotients of these sections provide the meromorphic functions called for in the following theorem:

THEOREM 9. *If M is a complete Kähler manifold with everywhere positive Ricci curvature and everywhere nonnegative sectional curvature, then there are nonconstant meromorphic functions on M. In fact, given a compact set $K \subseteq M$, there exists a positive integer N and a meromorphic mapping $\varphi : M \to P_N\mathbf{C}$ such that $\varphi | K$ is a holomorphic embedding.*

In connection with this theorem, it should be noted that not every noncompact Kähler manifold has nonconstant meromorphic functions: Siegel [**13**] has shown that certain complex tori have no nonconstant meromorphic functions. The complement of a single point in such a torus is then a noncompact Kähler manifold without nonconstant meromorphic functions, since such a meromorphic function would necessarily extend to be a nonconstant meromorphic function on the torus itself.

References

1. R. L. Bishop and B. O'Neill, *Manifolds of negative curvature*, Trans. Amer. Math. Soc. **145** (1969), 1–49.

2. J. Cheeger and D. Gromoll, *On the structure of complete manifolds of nonnegative curvature*, Ann. of Math. (2) **96** (1972), 413–443.

3. H. Grauert, *On Levi's problem and the imbedding of real-analytic manifolds*, Ann. of Math. (2) **68** (1958), 460–472. MR **20** #5299.

4. R. E. Greene and H. Wu, *Curvature and complex analysis*: I, Bull. Amer. Math. Soc. **77** (1971), 1045–1049; II, Bull. Amer. Math. Soc. **78** (1972), 866–870; III, Bull. Amer. Math. Soc. **79** (1973), 606–608.

5. ———, *On the subharmonicity and plurisubharmonicity of geodesically convex functions*, Indiana Univ. Math. J. **22** (1973), 641–653.

6. ———, *Integrals of subharmonic functions on manifolds of nonnegative curvature*, Invent. Math. (to appear).

7. D. Gromoll and W. Meyer, *On complete open manifolds of positive curvature*, Ann. of Math. (2) **90** (1969), 75–90.

8. L. Hörmander, *An introduction to complex analysis in several variables*, Van Nostrand, Princeton, N. J., 1966.

9. S. Kobayashi and H. Wu, *On holomorphic sections of certain Hermitian vector bundles*, Math. Ann. **189** (1970), 1–4.

10. K. Kodaira, *On Kähler varieties of restricted type (an intrinsic characterization of algebraic varieties)*, Ann. of Math. (2) **60** (1954), 28–48. MR **16**, 952.

11. J. Moser, *On Harnack's theorem for elliptic differential equations*, Comm. Pure Appl. Math. **14** (1961), 577–591. MR **28** #2356.

12. L. Sario and M. Nakai, *Classification theory of Riemann surfaces*, Springer-Verlag, New York, 1970.

13. C. L. Siegel, *Analytic functions of several complex variables*, Institute for Advanced Study, Princeton, N. J., 1950, pp. 104–106. MR **11**, 651.

14. K. I. Virtanen, *Über die Existenz von beschränkten harmonischen Funktionen auf offenen Riemannschen Flächen*, Ann. Acad. Sci. Fenn. Ser. A. I. Math.-Phys. No. 75 (1950), 8 pp. MR **12**, 403.

15. H. Wu, *Negatively curved Kähler manifolds*, Notices Amer. Math. Soc. **14** (1967), 515. Abstract #67T–327.

UNIVERSITY OF CALIFORNIA, LOS ANGELES

UNIVERSITY OF CALIFORNIA, BERKELEY

Proceedings of Symposia in Pure Mathematics
Volume 27, 1975

DIFFERENTIAL GEOMETRY AND COMPLEX ANALYSIS*

PHILLIP A. GRIFFITHS

0. Introduction. The theme of this paper will be negative curvature and complex analysis. The basic reasons underlying this theme may be schematically expressed as follows:

negative curvature ⇒ suitable functions are plurisubharmonic,
plurisubharmonic functions ⇒ consequences in function theory.

At the risk of oversimplification and omission, the theory as it presently exists may be outlined in the following way:

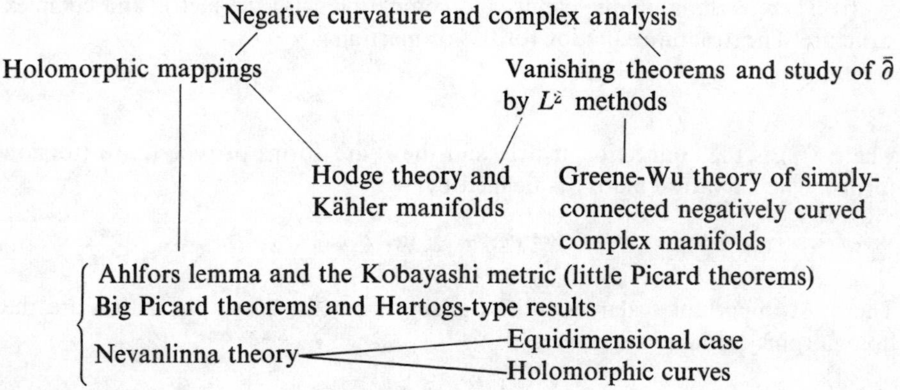

AMS (MOS) subject classifications (1970). Primary 30A70, 32H15, 32H20, 32H25; Secondary 14E99, 14N99, 14M99, 32J25, 53B35, 53C55, 53C65, 53C30.

*General lecture given at the Institute.

The reader may note the special position of Hodge theory and its subsequent applications to algebraic geometry in the outline. It is here that occurs what to me are the most interesting applications of the philosophy of negative curvature in all of its facets—both the linear and geometric theories appear, but time will not permit exploration of this subject (cf. **[G-S]** for heuristic discussion and further references). Similarly we will be unable to discuss the recent work of Greene and Wu, and for this we refer to their recent announcements in the *Bulletin of the American Mathematical Society*.

The plan for this paper is

1. Ahlfors lemma and applications.
2. Kobayashi metric and volume forms.
3. Nevanlinna theory—the equidimensional case.
4. Nevanlinna theory—holomorphic curves.

Generally speaking, I shall attempt to give the relevant definitions, some representative proofs, but shall omit all straightforward computations. The background references are **[C]**, **[K]**, **[Co-G]**, and **[G-K]** (letters refer to the bibliography at the end).

1. Ahlfors lemma and applications.

A. *Hermitian metrics.* Let M be a complex manifold with holomorphic coordinates $z_1,\cdots,z_m$. A Hermitian metric on M is given locally by

$$ds^2 = \sum h_{ij}\, dz_i\, d\bar{z}_j = \sum \varphi_i \bar{\varphi}_i$$

where (h_{ij}) is a C^∞ positive definite Hermitian matrix, and the φ_i are C^∞ (1, 0) forms which diagonalize the metric. Given such a metric, there are two basic consequences:

(i) There exists a unique connection compatible with the metric and complex structure. The structure equation for this connection is

$$d\varphi_i = \sum_j \varphi_{ij} \wedge \varphi_j + t_i \qquad (\varphi_{ij} + \bar{\varphi}_{ji} = 0)$$

where (φ_{ij}) is the connection matrix and the t_i are forms of type (2, 0) (torsion forms). The curvature matrix is defined by

$$\Phi_{ij} = d\varphi_{ij} - \sum_k \varphi_{ik} \wedge \varphi_{kj}.$$

The most important scalar quantities arising from the curvature matrix are the holomorphic sectional curvatures

$$\Phi(\xi) = \frac{1}{|\xi|^2}\left\{\sum \Phi_{ijkl}\, \xi^i \bar{\xi}^j \xi^k \bar{\xi}^l\right\}$$

determined by the (1, 0) vector ξ. The complex manifold M is said to be *negatively curved* in case there exists a Hermitian metric all of whose holomorphic sectional curvatures satisfy

$$\Phi(\xi) \leqq -A < 0$$

for some positive constant A. Multiplying the metric by $1/A$ allows us to assume that $A = 1$, and this will always be done.

(ii) If $S \subset M$ is a complex submanifold with induced metric, then we may choose the φ_i such that $\varphi_{s+1} = \cdots = \varphi_m = 0$ along S. Using the index range $1 \leqq \alpha, \beta \leqq s$ and the obvious notations, the curvature matrix for S is given by a formula

$$\Phi(S)_{\alpha\beta} = \Phi(M)_{\alpha\beta} - \sum_{\nu} A_{\alpha\nu} \wedge \bar{A}_{\nu\beta}$$

where $(A_{\alpha\nu})$ is a matrix of (1, 0) forms. In particular, if ξ is tangent to S, then $\Phi(S, \xi) \leqq \Phi(M, \xi)$, so that S is negatively curved in case M is. This principle that *curvatures decrease on complex submanifolds* is of fundamental importance, and perhaps may be explained as reflecting the ellipticity of the $\bar{\partial}$-operator. Indeed, in $\boldsymbol{C} \times \boldsymbol{C}$ the graph $(z, f(z))$ of a holomorphic function has negative curvature exactly because of $\partial f/\partial \bar{z} = 0$.

In case M is a Riemann surface, a Hermitian metric is $ds^2 = h\, dz\, d\bar{z}$ with associated (1, 1) form $\Omega = \frac{1}{2}(-1)^{1/2} h\, dz \wedge d\bar{z}$. The curvature matrix is a global (1, 1) form Φ and $\Phi = K \cdot \Omega$ where

$$K = -\frac{c}{h} \frac{\partial^2 \log h}{\partial z \partial \bar{z}} \qquad (c > 0)$$

is the Gaussian curvature. Thus

$$K \leqq 0 \Leftrightarrow \log h \text{ is subharmonic}$$

which

(a) using (ii) above shows that negative curvature has to do with plurisubharmonic functions, and

(b) marks the first appearance of the logarithm function, which is ubiquitous in the theory.

EXAMPLES.

(i) The Poincaré metric,

$$\begin{aligned} \pi(\rho) &= \frac{c\rho^2\, dz\, d\bar{z}}{(\rho^2 - |z|^2)^2} \qquad \text{on } \Delta_\rho = \{|z| < \rho\}, \\ &= c_1 \frac{dx\, dy}{y^2} \qquad \text{on } H = \{z = x + iy : y > 0\}, \end{aligned}$$

has constant Gaussian curvature $K = -1$ for suitable constants c, c_1.

(ii) The punctured disc $\Delta^* = \{0 < |t| < 1\}$ has universal covering H, and the induced Poincaré metric is

$$\pi(\Delta^*) = \frac{c_2\, dt\, d\bar{t}}{|t|^2(\log 1/|t|^2)^2}.$$

Two trivial but important properties of this metric deal with the circles $\gamma(\rho) = \{|t| = \rho\}$ and concentric punctured discs $\Delta^*(\rho) = \{0 < |t| < \rho\}$, and these are

(a) the length $l(\gamma(\rho)) \to 0$ as $\rho \to 0$,
(b) the area $\int_{\Delta^*(\rho)} \Omega < \infty$ for $\rho < 1$.
Both properties are immediate from the usual picture

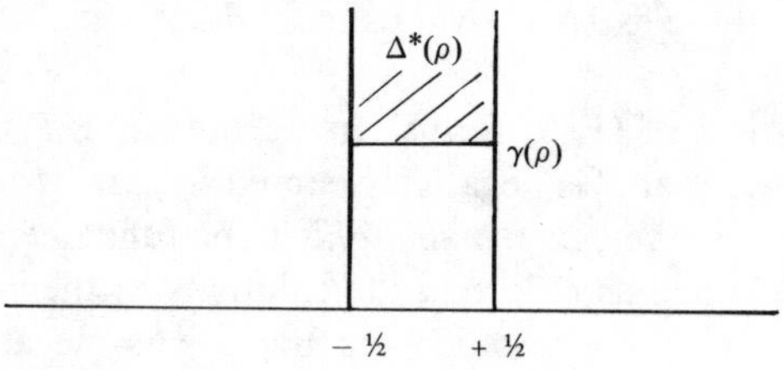

of the fundamental domain for $H \to \exp \Delta^*$ and the formula given above for the Poincaré metric on H.

(iii) On the Riemann sphere $\boldsymbol{P}^1 = \boldsymbol{C} \cup \{\infty\}$, the metric

$$\pi = \frac{c\, dz\, d\bar{z}}{(1 + |z|^2)^2}$$

has Gaussian curvature $K = +1$. Given a point $a \neq \infty$, we set

$$\sigma(a) = \frac{|a|^2}{1 + |a|^2}, \qquad \rho(a) = \sigma(a)(\log \mu/\sigma(a))^2$$

where $\mu > 1$ is constant. If a, b, c are distinct finite points, we set

$$\pi(a, b, c) = \left[\frac{1}{\rho(a)\rho(b)\rho(c)}\right] \pi.$$

At each point a, b, c this metric has asymptotically the same singularity as has the Poincaré metric $\pi(\Delta^*)$ at $t = 0$, and a computation shows that the Gaussian curvature of $\pi(a, b, c)$ is $\leqq -1$ for a suitable choice of μ. The fact that $\boldsymbol{P}^1 - \{a, b, c\}$ is negatively curved was traditionally deduced from the uniformization theorem, but the above elementary procedure of writing down negatively curved metrics globalizing the singularity of the Poincaré metric $\pi(\Delta^*)$ at $t = 0$ will work in situations where there is no uniformization.

B. *Ahlfors generalization of the Schwarz lemma.* A *pseudo-metric* on a Riemann surface is the same as a metric except that the coefficient function is allowed to vanish at isolated points. If $f : \Delta_\rho \to M$ is a nonconstant holomorphic mapping into a Hermitian manifold M, then $f^*(ds^2)$ is a pseudo-metric.

Let $h\, dz\, d\bar{z}$ be a pseudo-metric on Δ_ρ with Gaussian curvature K and $\pi(\rho) = h(\rho)\, dz\, d\bar{z}$ the Poincaré metric with constant Gaussian curvature $K_\rho \equiv -1$.

AHLFORS LEMMA. *If* $K \leqq -1$, *then* $h \leqq h(\rho)$.

PROOF. It will suffice to prove the case $\rho = 1$. For this we write $h = u(\sigma)h(\sigma)$ $(\sigma \leqq 1)$ and observe that

(i) $\lim_{\sigma\to 1} u(\sigma)(z) = u(1)(z)$,

(ii) $\lim_{|z|\to\sigma} u(\sigma)(z) = 0$ $(\sigma < 1)$.

Because of (i) it will suffice to show that $u(\sigma) \leqq 1$ for $\sigma < 1$, while (ii) implies that, for $\sigma < 1$, $u(\sigma)$ has an interior maximum at some point $z_0 = z(\sigma)$. By the maximum principle

$$0 \geqq \frac{\partial^2 \log u(\sigma)(z_0)}{\partial z \partial \bar{z}} = -K(z_0)h(z_0) + K_\sigma u(\sigma)(z_0)$$

which, using $K_\sigma \equiv -1$ and $K \leqq -1$, implies that

$$h(\sigma)(z_0) \geqq h(z_0),$$

or equivalently

$$u(\sigma)(z_0) \leqq 1. \qquad \text{Q.E.D.}$$

COROLLARY. *If M is a negatively curved complex manifold, then a holomorphic mapping $f: \Delta \to M$ is distance decreasing relative to the Poincaré metric on Δ and given metric on M.*

Applying this corollary to a holomorphic mapping $f: \Delta \to \Delta$, we find that

$$\frac{|f'(z)|}{1 - |f(z)|^2} \leqq \frac{1}{1 - |z|^2},$$

which is the intrinsic form of the Schwarz lemma due to Pick.

C. *Some applications of the Ahlfors lemma.*

SCHOTTKY-LANDAU THEOREM. *Let a, b, c be distinct points on the Riemann sphere $\boldsymbol{P}^1$ and $f: \Delta_r \to \boldsymbol{P}^1 - \{a, b, c\}$ a holomorphic mapping with $f'(0) \neq 0$. Then $r \leqq R(a, b, c, f(0), f'(0))$.*

PROOF. If $\pi(a, b, c)$ is the metric on $\boldsymbol{P}^1 - \{a, b, c\}$ constructed in example (iii) above, then the Ahlfors lemma applied to $f^*\pi(a, b, c) = h\, dz\, d\bar{z}$ gives

$$h(z) \leqq \pi(r)(z) \Rightarrow h(0) \leqq \frac{c}{r^2}$$

$$\Rightarrow r \leqq \left(\frac{c}{h(0)}\right)^{1/2} = R(a, b, c, f(0), f'(0)).$$

COROLLARY (LITTLE PICARD THEOREM). *A holomorphic mapping $f: \boldsymbol{C} \to \boldsymbol{P}^1 - \{a, b, c\}$ is constant.*

In general, we shall say that if a complex manifold M has the property that, for any holomorphic mapping $f: \Delta_r \to M$ with $f'(0) \neq 0$, the radius satisfies $r \leqq R(M, f(0), f'(0))$, then M has the *Schottky-Landau property*. The Ahlfors lemma implies that any negatively curved complex manifold has the Schottky-Landau property.

A second application of the Ahlfors lemma is the following extension result:

KWACK'S THEOREM. *If M is a compact negatively curved complex manifold, then any holomorphic mapping $f : \Delta^* \to M$ has a removable singularity at the origin.*

PROOF. Using the compactness of M and passing to subsequences when necessary, we may assume that any sequence of points $\{z_n\}$ tending to zero in Δ^* has images $w_n = f(z_n)$ tending to some point w_0. We must prove that w_0 is the same for any such sequence $\{z_n\}$.

To begin with, we observe from the distance decreasing property of f and the first property mentioned above of the Poincaré metric on Δ^* that the circles $\gamma(z_n)$ passing through z_n have images also tending to w_0. Let $P \subset\subset P'$ be concentric polycylindrical coordinate systems around w_0 in M. By what was just said, an annular ring A_n around the circle $\gamma(z_n)$ may be assumed to be mapped into P.

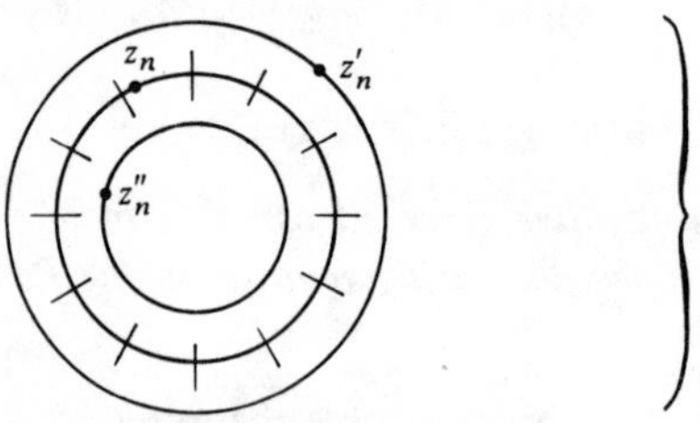

If our result were false, then in trying to let A_n become as large as possible and still be mapped into P we will encounter points z'_n and z''_n on the outer and inner boundaries of A_n whose images $w'_n = f(z'_n)$ and $w''_n = f(z''_n)$ lie on the boundary of P. Passing again to subsequences, we may assume that $w'_n \to w'_0$ and $w''_n \to w''_0$.

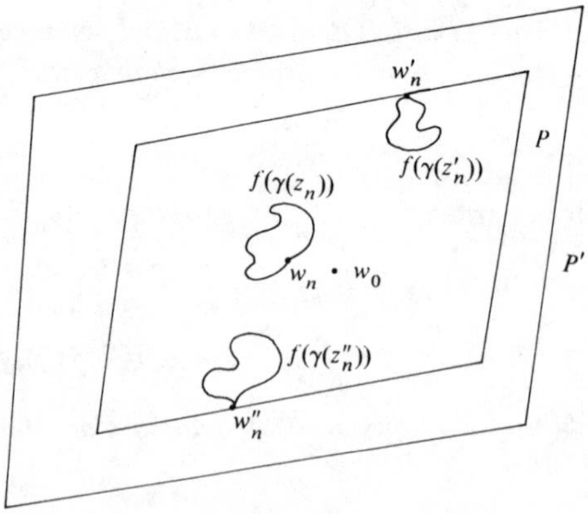

Let g be a holomorphic function on P' which vanishes at w_n ($n \gg 0$) but not at w'_n or w''_n, and set $h = g \circ f$. By the argument principle, the number of zeroes of h in A_n is given by

$$\frac{1}{2\pi}\int_{\gamma(z_n')} d \arg h - \frac{1}{2\pi}\int_{\gamma(z_n'')} d \arg h = 0.$$

On the other hand, $h(z_n) = 0$ by construction. Q.E.D.

COROLLARY (RIEMANN EXTENSION THEOREM). *A holomorphic mapping $f : \Delta^* \to \Delta$ has a removable singularity at the origin.*

PROOF. Let $\Gamma \subset \mathrm{Aut}(\Delta)$ be a properly discontinuous group operating without fixed points and having compact quotient.

FUNDAMENTAL DOMAIN FOR Γ

Applying Kwack's theorem to the composed mapping

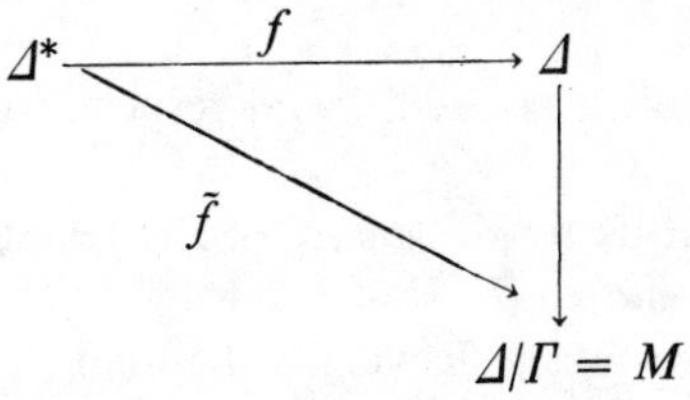

gives an extension of $\tilde{f}$, and hence one for f. Q.E.D.

It is amusing to compare the geometric argument just given with the usual analytic proof utilizing Laurent series.

2. The Kobayashi metric and volume forms.

A. *The Kobayashi metric.* Let M be a complex manifold and $T_x(M)$ the holomorphic tangent space at a point $x \in M$. For each vector $\xi \in T_x(M)$ we will, following Royden, define the *Kobayashi length* $F(x, \xi)$ as follows:

Let $\mathscr{F} = \mathscr{F}(x, \xi)$ be the class of all holomorphic mappings $f : \Delta_r \to M$ which satisfy $f(0) = x, f_*(\partial/\partial z) = \xi$.

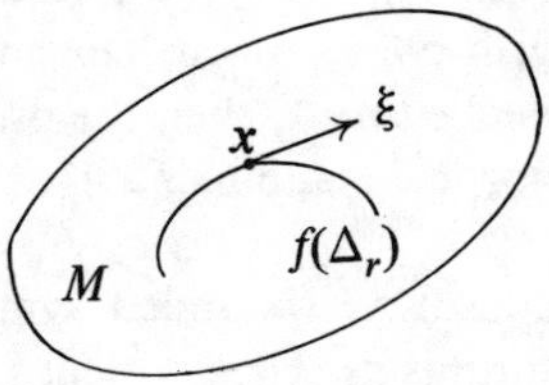

Then we set

$$F(x, \xi) = \inf_{f \in \mathscr{F}} \frac{1}{r}.$$

Royden has proved that $F(x, \xi)$ is semicontinuous on the tangent bundle $T(M)$.

Consequently, if $\gamma : [0, 1] \to M$ is a piecewise smooth path, then the integral

$$l(\gamma) = \int_0^1 F(\gamma(t), \gamma'(t)^{(1,0)})\, dt$$

exists and allows us to define the *Kobayashi pseudo-distance*

$$\rho(x, y) \qquad (x, y \in M)$$

by the usual procedure of minimizing the lengths of paths joining x and y.

The manifold M is *hyperbolic* in case $\rho(x, y)$ is a distance. Manifolds satisfying the Schottky-Landau property are hyperbolic—in particular, this is the case if M is negatively curved. Especially noteworthy are manifolds which are complete hyperbolic.

The Kobayashi metric is intrinsically defined on any complex manifold, and has many pleasant properties of which we wish to mention two:

(i) holomorphic mappings are distance decreasing relative to the Kobayashi metric; and

(ii) the metric ρ "behaves well" with respect to products, covering spaces, submanifolds, etc.

Time will not permit us to give many of the results concerning the Kobayashi metric, for which we refer to **[K]**, but we do want to call attention to two recent pretty applications. The first is Royden's theorem **[R]** that the Kobayashi metric equals the Teichmüller metric on the Teichmüller space. The proof is noteworthy for the insight which it gives into both the Teichmüller space and Kobayashi metric.

The second application, which will be discussed in some detail, is a recent Big Picard type result due to Borel in a form proved by Kobayashi and Ochiai (cf. **[K-O]** and **[K-K]**). We begin with the observation that, because of the distance decreasing property of ρ, Kwack's theorem discussed above holds with the same proof in case M is a compact hyperbolic manifold. A significant generalization occurs when M is an open set in a compact, complex space N. Then we say that M is *hyperbolically embedded* in case: (i) M is hyperbolic, and (ii) if $\{x_n\}$, $\{y_n\}$ are sequences of points in M with $x_n \to x$, $y_n \to y$, and $\rho(x_n, y_n) \to 0$, then $x = y$. Intuitively, ρ should distinguish points on the closure $\bar{M}$.

If M is hyperbolically embedded in N, then the proof of Kwack's theorem still applies to prove that a holomorphic mapping $f : \Delta^* \to M$ extends to a holomorphic mapping $f : \Delta \to N$.

To apply this result, we consider a bounded symmetric domain $D \subset \boldsymbol{C}^n$ and arithmetic group Γ of automorphisms. The quotient space $M = D/\Gamma$ is a negatively curved, quasi-projective algebraic variety admitting the Baily-Borel compactification $N = \overline{D/\Gamma}$. Kobayashi and Ochiai proved that M is hyperbolically embedded in N, thus proving

BOREL'S THEOREM. *A holomorphic mapping $f : \Delta^* \to D/\Gamma$ extends to $f : \Delta \to \overline{D/\Gamma}$.*

A special case of Borel's theorem is when $M = \mathbf{P}^1 - \{a, b, c\}$ and $N = \mathbf{P}^1$. In this situation the fact that M is hyperbolically embedded in N is obvious from the metric written down in the first lecture, and the above result is the usual Big Picard Theorem.

B. *Volume forms.* Thus far our discussion has centered around holomorphic mappings where the domain is one-dimensional. In the general several variables case, the first situation to study is the *equidimensional case* of a holomorphic mapping $f : D \to M$ between complex manifolds of the same dimension, and where f is assumed to be *nondegenerate* in the sense that the Jacobian determinant $J(f)$ is not identically zero. In this situation volume forms will play the analogous role of metrics in the one-dimensional case, with the Ricci form being the analogue of the curvature.

A *volume form* on a complex manifold M is given by a positive $C^\infty(n, n)$ form Ω. Locally $\Omega = h(z)\,\Phi(z)$ where h is a positive C^∞ function and

$$\Phi(z) = \prod_{\nu=1}^{m} (((-1)^{1/2}/2)\, dz_\nu \wedge d\bar{z}_\nu)$$

is the Euclidean volume form. A *pseudo-volume form* is the same, except that locally $h = |g|^2 h_0$ where h_0 is positive and g is a holomorphic function. If $f : D \to M$ is an equidimensional, nondegenerate holomorphic mapping and Ω is a volume form on M, then $f^*\Omega$ is a pseudo-volume form on D.

Associated to a pseudo-volume form Ω is its *Ricci form* Ric Ω, a global $C^\infty(1, 1)$ form defined locally by

$$\operatorname{Ric} \Omega = dd^c \log h$$

where $d^c = ((-1)^{1/2}/4\pi)(\bar{\partial} - \partial)$. (The operator $dd^c = ((-1)^{1/2}/2\pi)\partial\bar{\partial}$ is intrinsically defined by the complex structure, and plays in several variables the analogous role to the Laplacian in one variable.) Ricci forms are functorial in the sense that $f^*(\operatorname{Ric} \Omega) = \operatorname{Ric}(f^*\Omega)$.

The conditions

$$(*) \qquad \operatorname{Ric} \Omega > 0, \qquad (\operatorname{Ric} \Omega)^n = \underbrace{\operatorname{Ric} \Omega \wedge \cdots \wedge \operatorname{Ric} \Omega}_{n} \geqq \Omega$$

will play the analogous role to the Gaussian curvature condition $K \leqq -1$ in the one variable case.

EXAMPLES. (i) When M is a Riemann surface with Hermitian metric $h\, dz\, d\bar{z}$, the associated (1, 1) form $\Omega = ((-1)^{1/2}/2)\, h\, dz \wedge d\bar{z}$ is a volume form and

$$\operatorname{Ric} \Omega = c(-K)\Omega$$

where $c > 0$ is a constant and K is the Gaussian curvature. Our signs have been chosen so as to keep as many as possible of them positive during the discussion.

(ii) On the unit ball $B \subset \mathbf{C}^n$ or unit polycylinder $P \subset \mathbf{C}^n$ there is a unique volume form Π, the *Poincaré-Bergman volume form*, which is invariant under the biholomorphic automorphism group and which satisfies

$$\operatorname{Ric} \Pi > 0, \qquad (\operatorname{Ric} \Pi)^n = \Pi.$$

(iii) On the complex projective space $\boldsymbol{P}^n$ with homogeneous coordinates $Z = [Z_0, \cdots, Z_n]$, the differential form $\phi = dd^c \log \|Z\|^2$ is the (1, 1) form associated to the Fubini-Study Kähler metric on $\boldsymbol{P}^n$. The volume form $\Psi = \phi^n$ satisfies $\operatorname{Ric} \Psi = -(n+1)\phi$. (To check the signs, recall that the usual metric on $\boldsymbol{P}^1$ has positive curvature, hence negative Ricci form.)

Hyperplanes A in $\boldsymbol{P}^n$ are given by linear equations

$$\langle A, Z \rangle = A_0 Z_0 + \cdots + A_n Z_n = 0,$$

and we set (cf. § 1)

$$\sigma(A) = \frac{|\langle A, Z \rangle|^2}{\|A\|^2 \|Z\|^2}, \qquad \rho(A) = \sigma(A)\left(\log \frac{\mu}{\sigma(A)}\right)^2.$$

Given $n+2$ hyperplanes $\{A_\nu\}$ in general position, i.e. no $n+1$ are linearly dependent, define

$$\Omega(A_\nu) = \left[\prod_{\nu=1}^{n+2} \frac{1}{\rho(A_\nu)}\right] \Psi.$$

This is a volume form on $\boldsymbol{P}^n - \bigcup_{\nu=1}^{n+2} A_\nu$ having singularities along the A_ν of the same character as those of the Poincaré volume form on $(\Delta^*)^k \times \Delta^{n-k}$. For suitable choice of constant μ, one checks directly that

$$\operatorname{Ric} \Omega(A_\nu) > 0, \qquad \operatorname{Ric} \Omega(A_\nu)^n \geqq \Omega(A_\nu).$$

In the case $n = 1$, our construction reduces to the singular metric $\pi(a, b, c)$ on $\boldsymbol{P}^1$ given in § 1. The condition of $n+2$ hyperplanes comes from the $n+1$ factor in $\operatorname{Ric} \Psi = -(n+1)\phi$. Analogues of $\Omega(A_\nu)$ exist on general smooth projective varieties with the anticanonical divisor playing the role of the "$n+1$" in the present case—cf. **[Ca-G]** for further discussion.

C. *The Ahlfors lemma for volume forms and applications.* Let D be either the unit ball $B \subset \boldsymbol{C}^n$ or unit polycylinder $P \subset \boldsymbol{C}^n$ with Poincaré-Bergman volume form Π, and let Ω be a pseudo-volume form on D.

LEMMA (CHERN-KOBAYASHI). *If Ω satisfies the conditions* (∗) *above, then* $\Omega \leqq \Pi$.

PROOF. Writing $\Omega = u \cdot \Pi$, the proof is almost exactly the same as that of the Ahlfors lemma, where the only new step uses the Hadamard inequality

$$[\operatorname{Trace}(h_{ij})/n] \geqq [\det(h_{ij})]^{1/n}$$

for a positive Hermitian matrix (h_{ij}).

As an application of the Ahlfors lemma for volume forms, we let B_r be the ball of radius r in $\boldsymbol{C}^n$, $\{A_\nu\}$ a set of $n+2$ hyperplanes in general position in $\boldsymbol{P}^n$, and $f: B_r \to \boldsymbol{P}^n - \bigcup_{\nu=1}^{n+2} A_\nu$ a holomorphic mapping with Jacobian determinant $Jf(0) \neq 0$. Then the same proof as that of the Schottky-Landau theorem above leads to the

COROLLARY. *Under the above conditions,* $r \leqq R(\{A_\nu\}, f(0), Jf(0))$.

In particular, an entire holomorphic mapping $f : \boldsymbol{C}^n \to \boldsymbol{P}^n - \bigcup_\nu A_\nu$ is necessarily degenerate, a result due to A. Bloch (1926), and which has been recently rediscovered by Fujimoto and Green.

3. Equidimensional Nevanlinna theory.

A. *General philosophy.* In § 2 we saw that a nondegenerate holomorphic mapping $f : \boldsymbol{C}^n \to \boldsymbol{P}^n$ must meet at least one of $n + 2$ hyperplanes $\{A_\nu\}$ in general position. More precisely, given $f(0)$ and $Jf(0)$, there is a largest r such that the ball $B_r = \{z \in \boldsymbol{C}^n : \|z\| \leqq r\}$ of radius r can miss $f^{-1}(A_1 + \cdots + A_{n+2})$. Applying the same reasoning to balls centered around other points in $\boldsymbol{C}^n$, we arrive in principle at a lower bound on the size of $f^{-1}(A_1 + \cdots + A_{n+2})$.

Nevanlinna theory is a precise and far reaching quantitative study of the size of $f^{-1}(A_1 + \cdots + A_{n+2})$. The *First Main Theorem* (= F.M.T.) gives an upper bound on the magnitude of $f^{-1}(A)$ for any hyperplane A. The *Second Main Theorem* (= S.M.T.) gives a lower bound on $f^{-1}(A_1 + \cdots + A_{n+2})$, and when played off against one another these two estimates yield the famous *defect relation* of Rolf Nevanlinna.

In one complex variable, what is being studied are the solutions to the equation

$$f(z) = a \qquad (z \in \boldsymbol{C},\ a \in \boldsymbol{P}^1) \tag{3.1}$$

where f is an entire meromorphic function (hence the synonym *value distribution theory* for the subject). The size of $f^{-1}(a)$ in this case means the number $n(a, r)$ of solutions to (3.1) in the disc $|z| \leqq r$. The F.M.T. bounds $n(a, r)$ from above by the *order function* $T(f, r)$, an increasing convex function of $\log r$ which plays for general meromorphic functions a role analogous to the degree of a rational function or maximum modulus of an entire holomorphic function. The S.M.T. gives a lower bound of approximately the sort

$$n(a, r) + n(b, r) + n(c, r) \geqq T(f, r). \tag{3.2}$$

The F.M.T. may be viewed as a noncompact version of the *Wirtinger theorem*, which says that the area of an algebraic curve $C \subset \boldsymbol{P}^2$ is equal to the intersection number of C with any line. The lower bound (3.2) is proved as follows: Given $f : \boldsymbol{C} \to \boldsymbol{P}^1$ and the metric $\pi(a, b, c)$ on $\boldsymbol{P}^1 - \{a, b, c\}$, set $f^*\pi(a, b, c) = h\, dz\, d\bar{z}$. Then Picard's theorem says that h must have some singularities, and the S.M.T. gives a formula for the size of the singular set of h in terms of $T(f, r)$.

In these two remaining lectures we shall discuss in more detail how the theory works for nondegenerate, equidimensional holomorphic mappings $f : \boldsymbol{C}^n \to \boldsymbol{P}^n$ and nondegenerate holomorphic curves. When coupled with a standard discussion regarding line bundles and divisors on complex manifolds, the equidimensional theory goes through whenever $\boldsymbol{P}^n$ is replaced by an arbitrary algebraic variety with the anticanonical divisor replacing the $n + 1$ hyperplanes in general position, but the situation regarding holomorphic curves in general algebraic varieties is still pretty much open.

B. *The order function and F.M.T.* Let $f : \boldsymbol{C} \to \boldsymbol{P}^n$ be a holomorphic mapping and $\psi = dd^c \log \|Z\|^2$ the standard Kähler form on $\boldsymbol{P}^n$. Wirtinger's theorem suggests that the quantity $t(f, r) = \int_{\Delta_r} f^*\psi$ should be related to the number $n(A, r)$ of points of intersection of the analytic curve $f(\Delta_r)$ with a hyperplane A in $\boldsymbol{P}^n$. This is made precise by *Crofton's formula*

$$t(f, r) = \int_A n(A, r)\Psi(A)$$

expressing the area of the piece of analytic curve $f(\Delta_r)$ as the average number of points of intersection with hyperplanes $A \in \boldsymbol{P}^{n*}$, the dual projective space.

For reasons arising from Jensen's theorem, and ultimately related to twice integrating the operator dd^c, the growth of f is more conveniently measured by the *order function*

$$T(f, r) = \int_0^r t(f, \rho)\, d\rho/\rho.$$

For an entire holomorphic function $f: \boldsymbol{C} \to \boldsymbol{C} \subset \boldsymbol{P}^1$, it is an easy consequence of Crofton's formula and the Poisson-Jensen formula that the order function $T(f, r)$ is essentially the maximum of $\log |f|$ in Δ_r.

To define the order function for a holomorphic mapping $f: \boldsymbol{C}^m \to \boldsymbol{P}^n$, we shall restrict f to the lines through the origin and use the 1-dimensional order function just introduced. Although not the most natural definition of the order function, this is the quickest and will suffice for our purposes.

In $\boldsymbol{C}^m$, we let $\boldsymbol{P}^{m-1}$ be the projective space of lines ξ passing through the origin and $\Psi(\xi) = (dd^c \log \|z\|^2)^{m-1}$ the canonical density on $\boldsymbol{P}^{m-1}$.

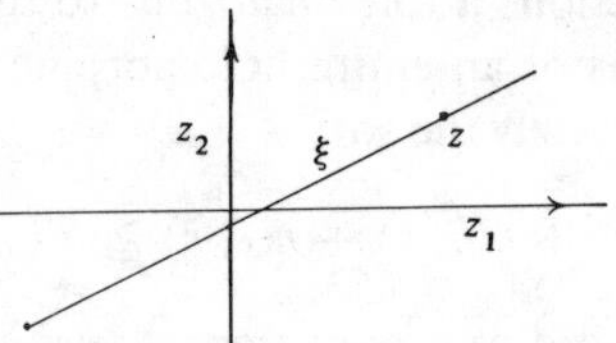

Given $f: \boldsymbol{C}^m \to \boldsymbol{P}^n$, for each line ξ we may restrict f to ξ and then define the order function $T(f, r, \xi)$ as above. Using this, the order function for f is given by

$$T(f, r) = \int_\xi T(f, r, \xi)\Psi(\xi).$$

We will now describe how one measures the size of an analytic hypersurface $V \subset \boldsymbol{C}^m$; for simplicity, we shall always assume that $0 \notin V$. Denoting by $V[r] = V \cap B_r$ that part of V in the ball of radius r, we set

$$n(V, r) = \frac{\operatorname{vol} V[r]}{r^{2m-2}}, \qquad N(V, r) = \int_0^r n(V, \rho)\frac{d\rho}{\rho}.$$

For each line $\xi \in \boldsymbol{P}^{m-1}$, the intersection $V_\xi = V \cap \xi$ is a discrete set of points in ξ, and a variant of Crofton's formula is

(i) $$n(V, r) = \int_{\xi} n(V_{\xi}, r)\Psi(\xi).$$

Moreover, results of Lelong and Stoll (cf. [G-K]) show that

(ii) $$V \text{ is algebraic of degree } d \Leftrightarrow n(V, r) \leqq d.$$

Because of (i) and (ii) it seems reasonable to use the *counting function* $N(V, r)$ to measure the growth of the analytic hypersurface V.

Now let $f : \boldsymbol{C}^m \to \boldsymbol{P}^n$ be a holomorphic mapping which is nondegenerate in the sense that the image does not lie in a hyperplane. For each such hyperplane A, the inverse image $A_f = f^{-1}(A)$ is an analytic hypersurface in $\boldsymbol{C}^m$, and we set

$$n(A, r) = n(A_f, r), \qquad N(A, r) = N(A_f, r).$$

Crofton's formula and (i) above imply that

(3.3) $$T(f, r) = \int_A N(A, r)\Psi(A).$$

The F.M.T. expresses the relation between $T(f, r)$ and $N(A, r)$ for a *particular* hyperplane A. To state this formula we recall the function

$$\sigma(A) = |\langle A, Z\rangle|^2/\|A\|^2\|Z\|^2$$

on $\boldsymbol{P}^n$, and we denote by Λ the unique closed $2m - 1$ on $\boldsymbol{C}^m - \{0\}$ which is invariant under unitary transformations and satisfies $\int_{\|z\|=r} \Lambda = 1$ for all radii r. The F.M.T. is the formula

(F.M.T) $$N(A, r) + \int_{\|z\|=r} \log \frac{1}{\sigma(A)} \Lambda = T(f, r) + O(1, A).$$

This equation is proved quite easily by integrating twice the Poincaré equation of currents

(3.4) $$dd^c \log \frac{1}{\sigma(A)} = f^* \psi - A_f,$$

and, as mentioned previously, should be viewed as a noncompact form of Wirtinger's theorem. Since $\sigma(A) \leqq 1$, a corollary of the F.M.T. is the famous *Nevanlinna inequality*

(3.5) $$N(A, r) \leqq T(f, r) + O(1, A).$$

The reader may wish to compare (3.3) and (3.5) .For an entire meromorphic function $f(z)$, (3.5) bounds the number of zeroes of f in the disc $|z| \leqq r$ by the maximum modulus of f.

C. *The S.M.T. and defect relation.* Let $f : \boldsymbol{C}^n \to \boldsymbol{P}^n$ be an equidimensional holomorphic mapping whose Jacobian determinant Jf is not identically zero. This implies that $f(\boldsymbol{C}^n)$ cannot lie in a hyperplane. We want to measure how much this image meets a set $\{A_\nu\}$ of $n + 2$ hyperplanes in general position.

Let $\Omega(A_\nu)$ be the volume form on $\boldsymbol{P}^n$ with singularities on $A_1 + \cdots + A_{n+2}$

which was constructed in the second section. The pull-back $f^*\Omega(A_\nu) = \Omega_f(A_\nu)$ is a singular pseudo-volume form on C^n, having "zeroes" along the ramification divisor $R = \{Jf = 0\}$ and "poles" on $f^{-1}(A_1 + \cdots + A_{n+2})$.

It was proved in § 2 that $f^{-1}(A_1 + \cdots + A_{n+2})$ must be nonempty. To measure the size of this analytic hypersurface, we write $\Omega_f(A_\nu) = h\Phi$ where $\Phi = \prod_{j=1}^n((-1)^{1/2}/2)\, dz_j \wedge d\bar{z}_j$ is the Euclidean volume form. The function h is non-negative with zeroes on R and poles on $f^{-1}(A_1 + \cdots + A_{n+2})$. Considering the locally L^1-function $\log h$ as a distribution, we arrive at the equation of currents (cf. (3.4))

$$dd^c \log h + \sum_\nu f^{-1}(A_\nu) = R + f^* \operatorname{Ric} \Omega(A_\nu). \tag{3.6}$$

Integrating (3.6) twice yields the

$$\text{(S.M.T)}\qquad \begin{aligned} &\int_{\|z\|=r} \log h \cdot \Lambda + \sum_\nu N(A_\nu, r) \\ &\quad = N(R, r) + \int_0^r \left\{\int_{B_\rho} \operatorname{Ric} \Omega_f(A_\nu) \wedge \phi^{n-1}\right\} \frac{d\rho}{\rho} \end{aligned}$$

where for simplicity we have assumed that $h(0) = 1$.

To see better how the S.M.T. leads to a lower bound on $\sum_\nu N(A_\nu, r)$, we shall restrict to the case $n = 1$, although the final estimates (3.12) below are the same in the general case. In fact, the general situation is done in basically the same way, the only new ingredient being the use of the Hadamard inequality as in the extension of the Ahlfors lemma to volume forms.

In the case $n = 1$, $\operatorname{Ric} \Omega_f(A_\nu) \geqq \Omega_f(A_\nu)$ by the condition $(*)$ on negative curvature. Since $N(R, r) \geqq 0$, the S.M.T. implies the estimate

$$\int_0^r \left(\int_{\Delta_\rho} h\, dz\, d\bar{z}\right) \frac{d\rho}{\rho} \leqq \sum_\nu N(A_\nu, r) + \frac{1}{2\pi} \int_{|z|=r} \log h\, d\theta. \tag{3.7}$$

At this point we use the notation

$$T^\#(r) = \int_0^r \left(\int_{\Delta_\rho} h\, dz\, d\bar{z}\right) \frac{d\rho}{\rho}.$$

If we use the ubiquitous *concavity of the logarithm*

$$\frac{1}{2\pi} \int \log h\, d\theta \leqq \log\left(\frac{1}{2\pi} \int h\, d\theta\right)$$

and obvious computation

$$\frac{1}{2\pi} \int_{|z|=r} h\, d\theta = \frac{1}{r^2} \frac{d^2 T^\#(r)}{(d \log r)^2}$$

in (3.7), we arrive at the inequality ($r \geqq 1$)

$$T^\#(r) \leqq \sum_\nu N(A_\nu, r) + \log\left[\frac{d^2 T^\#(r)}{(d \log r)^2}\right]. \tag{3.8}$$

If there were no derivatives in the last term on the R.H.S. of (3.8), then we would obviously have a lower bound on $\sum_\nu N(A_\nu, r)$. Even so, a clever but simple calculus argument gives

$$\frac{d^2T^{\#}(r)}{(d\log r)^2} \leqq [T^{\#}(r)]^{1+\delta} \quad \|,$$

where the notation "$\|$" means that the stated inequality holds outside an exceptional open set E satisfying $\int_E dr/r < +\infty$. Combining this inequality with (3.8) gives, for any $\varepsilon > 0$, the lower bound estimate

$$(3.10) \qquad (1-\varepsilon)T^{\#}(r) \leqq \sum_\nu N(A_\nu, r) \quad \|$$

where the exceptional intervals depend of course on the particular ε chosen.

Now in principle we are done. The inequality (3.10) gives a lower bound on the size of $f^{-1}(A_1 + \cdots + A_{n+2})$. To obtain the defect relation, one first proves that

$$(3.11) \qquad T^{\#}(r) = T(f, r) + (\text{negligible terms})$$

by explicitly taking into account the form of $\Omega(A_\nu)$. Combining this with (3.5) and (3.10) gives the simultaneous ineqaulities

$$(3.12) \qquad \begin{aligned} N(A_\nu, r) &\leqq T(f, r) + O(1), \\ \sum N(A_\nu, r) &\geqq (1-\varepsilon)\, T(f, r) \quad \|. \end{aligned}$$

Using the first inequality in (3.12), we may define the *Nevanlinna defect*

$$\delta(A_\nu) = 1 - \limsup_{r\to\infty} \frac{N(A_\nu, r)}{T(f, r)}$$

with the properties that

$$\begin{gathered} 0 \leqq \delta(A_\nu) \leqq 1, \\ \delta(A_\nu) = 1 \text{ if } f \text{ omits the hyperplane } A_\nu. \end{gathered}$$

Using the second inequality in (3.12),

$$\begin{aligned} \sum_\nu \delta(A_\nu) &\leqq (n+2) - \limsup_{n\to\infty} \left\{\sum N(A_\nu, r)/T(f, r)\right\} \\ &\leqq n + 1 + \varepsilon, \end{aligned}$$

which yields the *defect relation*

$$\sum_\nu \delta(A_\nu) \leqq n + 1$$

in its usual form.

4. Holomorphic curves and some open problems.

A. *Statement of the theorems of Ahlfors and Bloch.* A holomorphic mapping $f: \Delta_r \to \boldsymbol{P}^n$ will be called a holomorphic curve. We say that the curve is *entire* in case $r = +\infty$, and *nondegenerate* in case the image does not lie in a linear hyperplane. A classical theorem of E. Borel (1896) states that a nondegenerate entire holomorphic curve must meet at least one of $n + 2$ hyperplanes $\{A_\nu\}$ in general position. For $n = 1$ this reduces to the usual Picard theorem. Following a preliminary attempt by H. and J. Weyl, the corresponding defect relation

$$\sum_{\nu} \delta(A_\nu) \leqq n + 1 \tag{4.0}$$

was proved by Ahlfors (1941). An analogue of the Schottky-Landau theorem for $f : \Delta_r \to \boldsymbol{P}^n - (A_1 + \cdots + A_{n+2})$ was proved by A. Bloch (1926). We shall discuss how the method of negative curvature leads to a proof of Ahlfors' theorem in a similar fashion to the equidimensional case treated in the last lecture. Such an approach has thus far failed to yield the Bloch theorem, and there are some nice open problems in this area.

For simplicity we shall usually restrict ourselves to the case $n = 2$ of holomorphic curves in the projective plane. Our $n + 2$ hyperplanes in general position then become four lines A_1, A_2, A_3, A_4 spanning a quadrilateral Q.

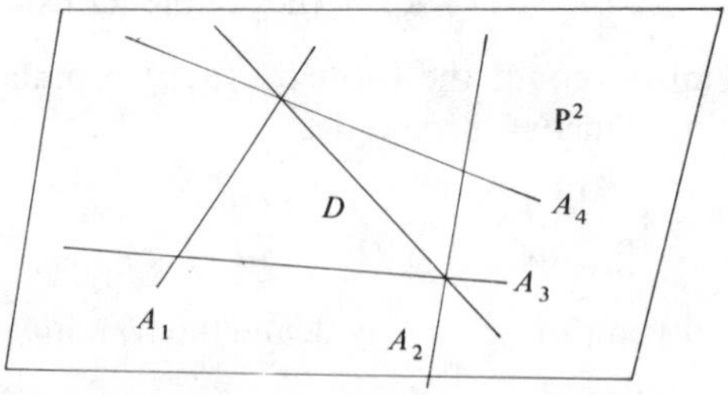

The diagonals D of this quadrilateral give lines $\boldsymbol{P}^1$ meeting Q in 2 points. Thus $\boldsymbol{P}^1 - \boldsymbol{P}^1 \cap Q \cong \boldsymbol{C}^*$, and in this way we find nonconstant but necessarily degenerate maps $f : \boldsymbol{C} \to \boldsymbol{P}^2 - Q$. The proof of Borel's theorem shows that any such f must map into one of the diagonals.

Recalling the Kobayashi metric $F(x, \xi)$ from § 2, Bloch's theorem is:

$F(x, \xi) > 0$ *unless* x *is on a diagonal* D *and* ξ *is tangent to* D.

This is a beautiful result, and the only proof of which I am aware is the highly nontransparent one given originally by Bloch. Finding a more conceptual argument for this theorem is an open problem for the theory.

Returning to the general case of a nondegenerate entire holomorphic curve $f : \boldsymbol{C} \to \boldsymbol{P}^n$, the order function $T(f, r)$ and counting function $N(A, r)$ have been defined and satisfy the Nevanlinna inequality $N(A, r) \leqq T(f, r) + C$. As before we may define the *defect*

$$\delta(A) = 1 - \limsup_{r \to \infty} \frac{N(A, r)}{T(f, r)}$$

with the properties that

$$0 \leqq \delta(A) \leqq 1, \qquad \delta(A) = 1 \quad \text{if } f(\boldsymbol{C}) \text{ misses } A.$$

The Ahlfors defect relation (4.1) is a quantitative refinement of the Borel theorem. This result concerns a 1-dimensional curve in a high-dimensional space, and the interpolation between dimensions 1 and n is accomplished via the *osculating curves* associated to the holomorphic curve. We shall see how these osculating curves arise naturally when one attempts to use the method of negative curvature which proved successful in the equidimensional case.

B. *Negative curvature and the Ahlfors theorem.* Let $f: \boldsymbol{C} \to \boldsymbol{P}^2$ be a nondegenerate entire holomorphic mapping given by a homogeneous coordinate vector $Z(t) = [z_0(t), z_1(t), z_2(t)]$ $(t \in \boldsymbol{C})$. Suppose that $\{A_\nu\}$ is a set of N lines in general position, A is a generic line, and recall the notations

$$\psi_0 = dd^c \log \|Z(t)\|^2 \quad (= f^* \text{ (standard Kähler form on } \boldsymbol{P}^2)),$$
$$\rho_0(A) = \frac{|\langle Z(t), A\rangle|^2}{\|Z(t)\|^2 \|A\|^2} \quad (= 0 \Leftrightarrow Z(t) \text{ lies on the line } A).$$

Motivated by the equidimensional case and explicit expression

$$\frac{c\, dz\, d\bar{z}}{|z|^2(\log(1/|z|^2))^2}$$

for the Poincaré metric on the punctured disc, we are prompted to consider the singular metric

$$\omega_0 = \left\{\prod_\nu \frac{1}{\rho_0(A_\nu)[\log(\mu/\rho_0(A_\nu))]^2}\right\}\psi_0.$$

In computing the Ricci forms of this and all future metrics, we shall ignore the $[\log(\mu/\rho_0(A_\nu))]^{-2}$ terms, the reason being that these terms essentially always help to make curvatures more negative—especially around the singular points. To assist in computing Ric ω_0, we shall also use the following comments:

(i) $\operatorname{Ric}(u\psi) = dd^c \log u + \operatorname{Ric} \psi$ where u is a positive function and ψ is a positive (1, 1) form;

(ii) $\operatorname{Ric} \psi_0 \geqq -2\, \psi_0$, since the holomorphic sectional curvatures of $\boldsymbol{P}^n$ are all equal to $+2$, and curvatures decrease on complex submanifolds; and

(iii) $dd^c \log(1/\rho_0(A)) = \psi_0$, by definition.

Using (i)—(iii) and ignoring the $[\log(\mu/\rho_0(A_\nu))]^{-2}$ terms, we find that

$$\operatorname{Ric} \omega_0 \geqq (N-3)\, \psi_0. \tag{4.1}$$

Consequently, for $N \geqq 3$ the metric ω_0 has negative curvature.

However, this curvature cannot be bounded away from zero for the following reason:

Given on the punctured disc $\Delta^* = \{0 < |z| < 1\}$ a metric $h\, dz\, d\bar{z}$ with Gaussian curvature $K \leqq -1$, then $\int_{|z| \leqq 1-\varepsilon} h\, dz\, d\bar{z} < \infty$ by the Ahlfors lemma and second property of the Poincaré metric on Δ^*. Now, on the other hand, the holomorphic curve may, at some point t_0, be unramified so that $\psi_0(t_0) \neq 0$, but have arbitrarily high order of contact with one of the lines, say A_1. Then the denominator $\rho(A_1)\, [\log(\mu/\rho(A_1))]^2$ in ω_0 becomes zero to arbitrarily high order and so $\int_{|t-t_0|<\delta} \omega^0 = +\infty$.

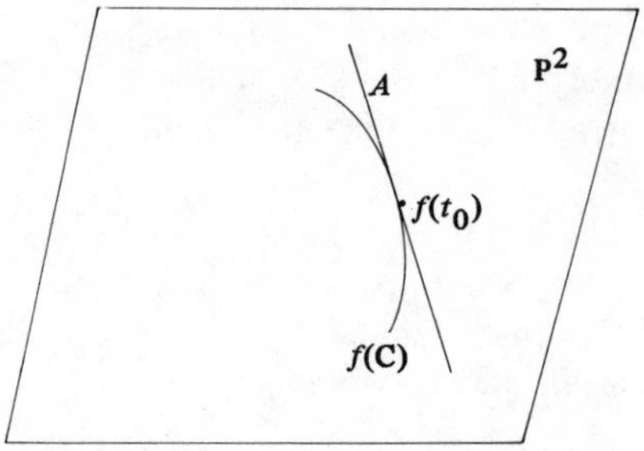

This suggests that we additionally consider the functions

$$\rho_1(A) = \frac{\|Z(t) \wedge Z'(t), A\|^2}{\|Z(t) \wedge Z'(t)\|^2 \|A\|^2},$$

which vanish at points t_0 where $f(t_0)$ meets A tangentially. Using the notations

$$\psi_1(A) = dd^c \log\|Z(t) \wedge Z'(t), A\|^2$$
$$\psi_1 = dd^c \log\|Z(t) \wedge Z'(t)\|^2,$$

the singular metric

$$\omega_1 = \left\{\prod_\nu \frac{\rho_1(A_\nu)}{\rho_0(A_\nu)[\log(\mu/\rho_0(A_\nu))^2]}\right\} \psi_0$$

is always integrable. The Ricci form

(4.2) $$\text{Ric } \omega_1 = N\psi_0 + \text{Ric } \psi_0 - N\psi_1 + \sum_\nu \psi_1(A_\nu),$$

where we continue to ignore the $[\log(\mu/\rho_0(A_\nu))]^{-2}$ terms. To compensate for the term $- N\psi_1$, we are thus prompted to consider a second metric

$$\omega_2 = \left\{\prod_\nu \frac{1}{\rho_1(A_\nu)\,[\log(\mu/\rho_1(A_\nu))]^2}\right\} \psi_1.$$

Using (4.2),

(4.3) $$\text{Ric } \omega_1 + \text{Ric } \omega_2 \geqq N\psi_0 + \text{Ric } \psi_0 + \text{Ric } \psi_1.$$

In order to conclude that ω_1 and ω_2 form, so to speak, a *negatively curved pair of metrics*—thereby forcing a defect relation as before—it is necessary to relate Ric ψ_0 and Ric ψ_1.

Now $\psi_0 = dd^c \log \|Z(t)\|^2$ is the pull-back of the standard Kähler metric on $\boldsymbol{P}^2$ under the given mapping f. Similarly, $\psi_1 = dd^c \log \|Z(t) \wedge Z'(t)\|^2$ is the pull-back of the standard Kähler metric on the dual projective space $\boldsymbol{P}^{2*}$ of lines in $\boldsymbol{P}^2$ under the *dual curve mapping* $f^* : \boldsymbol{C} \to \boldsymbol{P}^{2*}$, which is given by

$$f^*(t) = \text{tangent line to } f(\boldsymbol{C}) \text{ at } f(t).$$

In classical algebraic geometry, the relation between degree (C) and degree (C^*), for an algebraic curve C and its dual C^* is provided by the *Plücker formulae*, whose

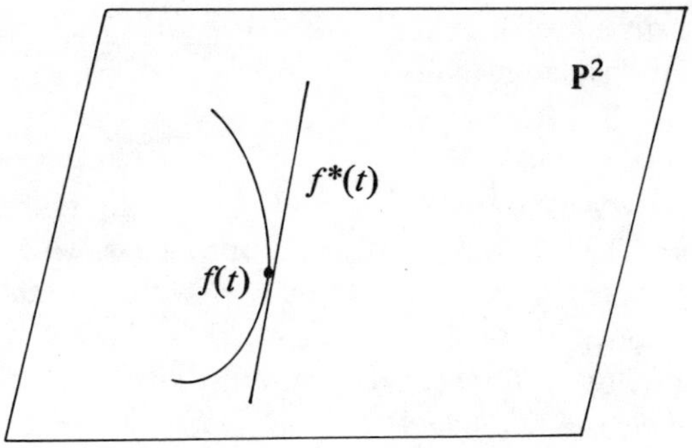

extension to holomorphic curves was given by H. and J. Weyl. For plane curves the relevant relations are

(4.4) $$\operatorname{Ric}\psi_0 = -2\psi_0 + \psi_1, \qquad \operatorname{Ric}\psi_1 = -2\psi_1 + \psi_0.$$

Plugging these into (4.3) gives

(4.5) $$\operatorname{Ric}\omega_1 + \operatorname{Ric}\omega_2 \geqq (N-1)\psi_0 - \psi_1.$$

On the other hand, following the same procedure as in the equidimensional case, the first equation in (4.4) gives

$$\int_0^r\left(\int_{\Delta_\rho}\psi_1\right)\frac{d\rho}{\rho} < 2\int_0^r\left(\int_{\Delta_\rho}\psi_0\right)\frac{d\rho}{\rho} + C\log\left[\int_0^r\left(\int_{\Delta_\rho}\psi_0\right)\frac{d\rho}{\rho}\right]\|,$$

which we shall write as

(4.6) $$\psi_1 < (2+\delta)\psi_0 \;\|.$$

Combining (4.5) and (4.6) yields

(4.7) $$\operatorname{Ric}\omega_1 + \operatorname{Ric}\omega_2 \geqq (N-3-\delta)\,\psi_0 \;\|.$$

Thus, for $N \geqq 4$, the pair of metrics ω_1, ω_2 taken together has negative curvature and both metrics are integrable. In fact, one easily calculates the curvature is bounded away from zero except at intersection points $A_i \cap A_j$.

To circumvent this final difficulty, Mike Cowen introduced Hölder exponents with the following conclusion: Setting

$$\varphi_1 = c_1\left\{\prod_\nu \frac{\rho_1(A_\nu)}{\rho_0(A_\nu)[\log(\mu/\rho_0(A_\nu))]^2}\right\}^{1/2}\psi_0,$$

$$\varphi_2 = c_2\left\{\prod_\nu \frac{1}{\rho_1(A_\nu)[\log(\mu/\rho_1(A_\nu))]}\right\}\psi_1,$$

for suitable choice of constants c_1, c_2, μ,

(4.8) $$2\operatorname{Ric}\varphi_1 + \operatorname{Ric}\varphi_2 \geqq (N-3-\varepsilon)\psi_0 + (\varphi_1+\varphi_2)\;\|.$$

Since the singular divisor of 2 Ric φ_1 + Ric φ_2 is just $\sum_\nu f^{-1}(A_\nu)$, (4.8) may be rewritten in distributional language as

$$\sum_\nu f^{-1}(A_\nu) \geqq (N - 3 - \varepsilon)\, \psi_0 \,\|,$$

which then leads to the Ahlfors defect relation as in the equidimensional case.

C. *Some problems*. The following collection of problems, which to the best of my knowledge are still open, deal for the most part with the relationship between holomorphic mappings and the Kobayashi metric and algebraic geometry. The basic underlying question is *to understand in terms of the algebro-geometric properties of a variety V the possible holomorphic mappings $f : D \to V$ from open sets D in $\boldsymbol{C}^k$ into V*, generalizing as far as possible the understanding of the case when V is an algebraic curve obtained through the classical uniformization theorem.

(i) Let M be a complex Hermitian manifold. The holomorphic tangent spaces are denoted by $T_x(M)$, and $\zeta^{1,0}$, $\zeta^{0,1}$ denote the projections of a tangent vector ζ into $T(M)$, $\overline{T(M)}$ respectively. A complex manifold M' is said to be *ε-quasi-conformally equivalent to M* if there is a diffeomorphism $f : M' \to M$ such that $\|f_*(\xi)^{0,1}\|/\|f_*(\xi)\| < \varepsilon$ for every $\xi \in T_x(M')$.

PROBLEM. Suppose that M is a compact, complex manifold which is hyperbolic in the sense of Kobayashi. Then for sufficiently small ε, is any complex manifold M' which is ε-quasi-conformally equivalent to M necessarily hyperbolic?

It seems to me quite likely that the answer to this question is yes. If so, then given an analytic family $\{M_t\}_{t\in\Delta}$ of compact, complex manifolds when M_0 is hyperbolic, the M_t would be hyperbolic for sufficiently small t.

(ii) The next two problems deal with complex manifolds (usually algebraic varieties) M which are *hyperbolic on a Zariski open set* in the sense that the Kobayashi length $F(x, \xi) > 0$ unless x lies on a subvariety S and ξ is tangent to S there. This definition is prompted by the Bloch theorem concerning $\boldsymbol{P}^2 - \{4$ lines in general position$\}$. Because of his result and similar examples of Mark Green [G], it seems that being hyperbolic on a Zariski open set may be more fruitful in studying algebraic varieties than the requirement of strict hyperbolicity. In any case, being hyperbolic on a Zariski open set is invariant under *birational transformations*, thus affording some additional flexibility.

PROBLEM. Let $\{M_t\}_{t\in\Delta}$ be an analytic family of compact, complex manifolds where M_t is hyperbolic on a Zariski open set. In particular, the M_t may be hyperbolic. Then is M_0 hyperbolic on a Zariski open set?

(iii) Recall that an algebraic surface M is of *general type* if the graded canonical ring $\bigoplus_n H^0(M, \mathcal{O}(nK))$ has transcendence degree two. Examples are nonsingular surfaces of degree $\geqq 5$ in $\boldsymbol{P}^3$.

PROBLEM. Is an algebraic surface of general type hyperbolic on a Zariski open set? Is the complement $\boldsymbol{P}^2 - C$ of a smooth plane curve of degree $\geqq 5$ hyperbolic on a Zariski open set? Is a general smooth surface of degree $\geqq 5$ in $\boldsymbol{P}^3$ hyperbolic?

An affirmative answer to problems (i) and (ii) should allow one to obtain information on at least the last part of this problem by checking a few special surfaces and the applying techniques from deformation theory.

(iv) There is a notion of *measure hyperbolic* due to Pelles (cf. [E]) which bears the same relation to volume forms Ω satisfying (∗) in §2 as does hyperbolic to negatively curved metrics. In particular, any surface of general type is measure hyperbolic.

PROBLEM. If M is an algebraic surface which is measure hyperbolic, then is M of general type?

By looking at the classification of algebraic surfaces, one sees that this problem is equivalent to showing that *a K3 surface M is not measure hyperbolic* (elliptic surfaces are never measure hyperbolic). Letting $P(r_1, r_2) = \{(z_1, z_2) : |z_1| < r_1, |z_2| < r_2\}$, roughly speaking one must construct nondegenerate holomorphic maps $f : P(r_1, r_2) \to M$ where the product $r_1 r_2 \to \infty$, which is a uniformization type of question. In any case, a dense set of $K3$'s is *not* measure hyperbolic.

(v) Let M be a compact algebraic variety, and $D \subset M$ a divisor with normal crossings such that $M - D$ is complete (in a suitable sense) hyperbolic on a Zariski open set.

(EXAMPLE. $M = \boldsymbol{P}^n$ and $D = (n + 2)$ hyperplanes in general position.)

PROBLEM. Can one find a lower bound on the size of $f^{-1}(D)$ for a nondegenerate entire holomorphic curve $f : \boldsymbol{C} \to M$?

If so, then in order to prove defect relations one might not need to construct negatively curved metrics so explicitly as has been the case thus far. Solving this problem would probably necessitate obtaining information on the following two questions:

(vi) PROBLEM. Let M be a compact algebraic variety and $D \subset M$ a divisor with normal crossings such that $M - D$ is complete hyperbolic. Then can one estimate the Kobayashi length $F(x, \xi)$ for $M - D$ as x tends to D?

In this connection, we should like to point out that A. Sommese (Princeton thesis, 1973) has shown that given a *complete*, negatively curved ds^2 on $M - D$, then this metric is, in a suitable sense, asymptotic to the Poincaré metric on Δ^* when one approaches D from a normal direction.

(vii) PROBLEM. In what, if any, sense is the Kobayashi metric $F(x, \xi)$ of a hyperbolic manifold M negatively curved? Specifically, given a holomorphic mapping $f : \Delta \to M$ and setting $h(z) = F(f(z), f_*(\partial/\partial z))$, then can one say anything about $\Delta \log h$ (taken in the distributional sense)?

(viii) PROBLEM. Let M be a compact, complex manifold and $D \subset M$ a divisor such that $M - D$ is complete hyperbolic. Then does any holomorphic mapping $f : \Delta^* \to M - D$ extend to $f : \Delta \to M$? If $M - D$ is only assumed to be complete hyperbolic on a Zariski open set but f is taken to be nondegenerate, then does the same extension theorem hold?

EXAMPLES. $\boldsymbol{P}^2 - \{5 \text{ lines in general position}\}$ and $\boldsymbol{P}^2 - \{4 \text{ lines in general position}\}$.

BIBLIOGRAPHY

This bibliography is not intended to be complete, but rather to give references for background material together with a few references for the topics discussed in the lectures through which additional sources can be found.

The basic references for general information about complex manifolds are

[C] S. S. Chern, *Complex manifolds without potential theory*, Van Nostrand Math. Studies, no. 15, Van Nostrand, Princeton, N. J., 1967. MR **37** #940.

[W] R. Wells, *Differential analysis on real and complex manifalds*, Prentice-Hall, Engelwood Cliffs, N.J., 1972.

Additional material on complex manifolds and a discussion of volume forms and the Kobayashi metric are given in

[K] S. Kobayashi, *Hyperbolic manifolds and holomorphic mappings*, Pure and Appl. Math., no. 2, Dekker, New York, 1970. MR **43** #3503.

Related references concerning the Kobayashi metric and Big Picard theorems are

[R] H. Royden, *Remarks on the Kobayashi metric*, these PROCEEDINGS

[K-O] S. Kobayashi and T. Ochiai, *Satake compactification and the great Picard theorem*, J. Math. Soc. Japan **23** (1971), 340–350. MR **45** #5418.

[K-K] P. Kiernan and S. Kobayashi, *Satake compactification and extension of holomorphic mappings*, Invent. Math. **16** (1972), 237–248.

Equidimensional Nevanlinna theory is discussed in

[Ca-G] J. Carlson and P. Griffiths, *A defect relation for equidimensional holomorphic mappings between algebraic varieties*, Ann. of Math. (2) **95** (1972), 557–584.

[G-K] P. Griffiths and J. King, *Nevanlinna theory and holomorphic mappings between algebraic varieties*, Acta Math. **130** (1973), 145–220.

The second reference contains a general discussion of the formalism of Nevanlinna theory and gives some applications to Big Picard theorems including a proof of the Borel theorem discussed in lecture 2. The theory of holomorphic curves together with an extensive historical bibliography is given in

[Co-G] M. Cowen and P. Griffiths, *Holomorphic curves and metrics of negative curvature*, J. Analyse (to appear)

Holomorphic curves in some more general algebraic varieties are discussed in

[G] M. Green, *Some Picard theorems for holomorphic maps to algebraic varieties*, Thesis, Princeton, University, Princeton, N. J., 1972.

An expository account of the application of Nevanlinna theory to Hodge theory and algebraic geometry is given in

[G-S] P. Griffiths and W. Schmid, *Variation of Hodge structure (a discussion of recent results and some methods of proof)*, Proc. Tata Institute Conference on Discrete Groups and Moduli, Bombay (1973). (to appear).

Finally, a discussion of the extension of the Kobayashi metric to intermediate volumes is contained in

[E] D. Eisenman (Pelles), *Intrinsic measures on complex manifolds and holomorphic mappings*, Mem. Amer. Math. Soc. No. 96 (1970). MR **41** #3807.

HARVARD UNIVERSITY

Proceedings of Symposia in Pure Mathematics
Volume 27, 1975

ON THE CURVATURE OF RATIONAL SURFACES

NIGEL HITCHIN

1. Introduction. Among the differential-geometric vanishing theorems for Kähler manifolds we have the following:

(a) If a Kähler manifold X has positive *scalar curvature,* the plurigenera P_m vanish for $m > 0$.

(b) If X has positive *holomorphic sectional curvature*, the fundamental group π_1 is trivial and $P_m = 0$ for $m > 0$.

(c) If X has positive *Ricci curvature*, the dimension of the space of holomorphic p-forms $h^{p,0}$ is zero for $p > 0$, $\pi_1 = 1$ and $P_m = 0$ for $m > 0$.

We notice now that the objects which vanish are not only invariants of the complex structure but also *birational* invariants. In particular, they are all zero for rational algebraic varieties. This leads one to conjecture that rational varieties are characterized by admitting a Kähler metric with some positivity of curvature which will force the vanishing of one or more of these invariants.

For curves this is clearly true—a one-dimensional Kähler manifold with positive curvature is biholomorphically equivalent to $\boldsymbol{P}^1$ by any of the above arguments, and conversely $\boldsymbol{P}^1$ admits a Kähler metric of positive curvature. For regular surfaces (i.e., the first Betti number $b_1 = 0$) with positive scalar curvature, vanishing theorem (a) above together with Kōdaira's classification implies rationality. The question we are concerned with is the converse. Do all rational surfaces admit a Kähler metric of positive scalar curvature? We prove the following:

THEOREM. *Almost all rational surfaces admit a Hodge metric of positive scalar curvature.*

AMS (MOS) subject classifications (1970). Primary 53C55.

To be specific, we prove the theorem for surfaces obtained by blowing up k distinct points on $\boldsymbol{P}^2$ or one of the surfaces F_n ($\boldsymbol{P}^1$ bundles over $\boldsymbol{P}^1$). Every rational surface is a deformation of such a surface and they are in a certain sense generic.

We first consider a class of Hodge metrics on the minimal models and show in particular that they have positive scalar curvature. Then we start blowing up. If we start with a Hodge manifold X and blow up a point p to obtain $\hat{X}$, then we can introduce in a natural way a Hodge metric on $\hat{X}$ (this is the metric used by Kōdaira in the proof of his theorem that any Hodge manifold is algebraic). In dimension > 2, the metric on $\hat{X}$ can be made to have positive scalar curvature if the metric on X has, but in dimension 2 we need a positivity condition involving the Ricci curvature and the holomorphic sectional curvature at p. For successive blow ups this restricts the choice of p and accounts for the restriction on rational surfaces in the theorem. Fortunately the condition is satisfied for all points on the minimal models.

We may remark that the Hodge metrics we introduce on the surfaces F_n all have positive holomorphic sectional curvature, although for $n > 1$ the F_n do not admit Kähler metrics of positive Ricci curvature. The metric on F_1 does in fact have positive Ricci curvature. We also point out (cf. the Calabi conjecture) that a surface with first Chern class $c_1 > 0$ admits a Hodge metric of positive *scalar* curvature.

2. Rational surfaces. We review here some basic facts about rational surfaces and refer to **[10]** for details.

A rational surface is obtained from $\boldsymbol{P}^2$ by blowing up and blowing down. To blow up a point p on a complex manifold X^n we take local complex coordinates in a neighborhood $N(p)$ and identify $N(p)$ with a neighborhood U of $0 \in \boldsymbol{C}^n$. We then form the manifold $\hat{U} \subset \boldsymbol{C}^n \times \boldsymbol{P}^{n-1}$ defined by $\{(x, y) \in \boldsymbol{C}^n \times \boldsymbol{P}^{n-1}: x \in y \;\&\; x \in U\}$ where $x \in y$ means x lies on the line through 0 defined by y. We have a projection $\pi\colon \hat{U} \to U$ where $\pi^{-1}(0) \cong \boldsymbol{P}^{n-1}$ and π maps $\hat{U} - \boldsymbol{P}^{n-1} \to U - \{0\}$ biholomorphically. Using the identification we replace $N(p) \subset X$ by $(N(p))^\wedge$ and form a new complex manifold $\hat{X}$ where the point p has been "blown up"—replaced by a $\boldsymbol{P}^{n-1}$. For surfaces, p is replaced by a $\boldsymbol{P}^1$ with self-intersection -1, and conversely, given such a nonsingular rational curve on a surface X, it may be "blown down" to a point, i.e., $X = \hat{Y}$ for some Y.

We have the projection map $\pi : \hat{X} \to X$ and we denote by E the exceptional curve $\pi^{-1}(p)$. The first Chern classes of $\hat{X}$ and X are then related by the following:

$$c_1(\hat{X}) = \pi^* c_1(X) - [E] \tag{2.1}$$

where $[E]$ denotes the cohomology class defined by the exceptional curve. If C is a nonsingular curve passing through p, then C lifts to a nonsingular curve $\hat{C}$ where the self-intersection numbers are related by

$$\hat{C}^2 = C^2 - 1. \tag{2.2}$$

If p is a point of multiplicity m of C, then the cohomology classes of $\hat{C}$ and C are related by

$$[\hat{C}] = \pi^*[C] - m[E]. \tag{2.3}$$

A surface which cannot be blown down any further, i.e., which has no rational curve of self-intersection -1, is called a minimal model. The minimal models of rational surfaces are the surfaces $\boldsymbol{P}^2$ and F_n $(n \neq 1)$ where $F_n = \boldsymbol{P}(H^n \oplus 1)$, the projective bundle associated to the vector bundle $H^n \oplus 1$ over $\boldsymbol{P}^1$, H being the line bundle defined by a hyperplane section. The manifolds F_{2n}—the Hirzebruch surfaces—are all diffeomorphic to $S^2 \times S^2$.

Topologically, blowing up a point on a surface is equivalent to attaching $\boldsymbol{C}P^2$ with opposite orientation ($\overline{\boldsymbol{C}P}^2$) and so any rational surface is diffeomorphic to $S^2 \times S^2$ or $\boldsymbol{C}P^2 \# n\overline{\boldsymbol{C}P}^2$ $(n \geqq 0)$. As an example of blowing up and down, $\boldsymbol{P}^1 \times \boldsymbol{P}^1$ is obtained from $\boldsymbol{P}^2$ by blowing up two points and blowing down the line joining them.

3. Curvature and rationality. Let g be a Kähler metric and $\{e_\alpha\}$ a local unitary frame field. Let $\{\theta_\alpha\}$ be the dual field of (1, 0) forms. Then the curvature form $\Theta_{\alpha\beta}$ may be written as

$$\Theta_{\alpha\beta} = R_{\alpha\beta\gamma\delta}\,\theta_\gamma \wedge \bar{\theta}_\delta$$

using the summation convention. $R_{\alpha\beta\gamma\delta}$ satisfies the following symmetry conditions:

$$R_{\alpha\beta\gamma\delta} = \bar{R}_{\beta\alpha\delta\gamma}, \qquad R_{\alpha\beta\gamma\delta} = R_{\gamma\beta\alpha\delta} = R_{\alpha\delta\gamma\beta}.$$

If $\xi = \xi_\alpha e_\alpha$ is a tangent vector, then the *holomorphic sectional curvature* K in the direction ξ is defined by

$$K(\xi) = 2R_{\alpha\beta\gamma\delta}\,\xi_\alpha \bar{\xi}_\beta \xi_\gamma \bar{\xi}_\delta/(\xi_\alpha \bar{\xi}_\alpha)^2;$$

the *Ricci curvature* in the direction ξ is defined by

$$\mathrm{Ric}(\xi) = R_{\alpha\alpha\gamma\delta}\,\xi_\gamma \bar{\xi}_\delta/(\xi_\alpha \bar{\xi}_\alpha);$$

the *scalar curvature* is defined by

$$R = R_{\alpha\alpha\beta\beta}.$$

By the symmetries above, all these quantities are real. Note that if $\mathrm{Ric} > 0$ (i.e., for all ξ) then $R > 0$. It also follows from a lemma of Berger **[1]** that $K > 0$ implies $R > 0$. From this we obtain from the vanishing theorem of Kobayashi and Wu **[4]** the vanishing of holomorphic sections in (a), (b) and (c) of the introduction. The vanishing of the fundamental group is due to Tsukamoto **[12]** and Kobayashi **[3]**.

We now give a curvature criterion for rationality of surfaces:

PROPOSITION (3.1). *Let X be a regular Kähler surface with positive scalar curvature. Then X is rational.*

PROOF. By the vanishing theorem of Kobayashi and Wu, if X has positive scalar curvature, then the pluricanonical bundle K^m $(m > 0)$ has no holomorphic sections, i.e., the plurigenera $P_m = \dim H^0(X, \mathcal{O}(K^m))$ vanish. We now use a theorem of Kōdaira [7]—if X is a complex surface with b_1 even and $P_1 = 0$, then X is algebraic. Since our X is regular, $b_1 = 0$ and so X is algebraic, and furthermore the arithmetic genus $(= 1 - b_1/2 + P_1)$ is 1. Now the theorem of Castelnuovo (see [10]) states that if X is algebraic with arithmetic genus 1 and $P_2 = 0$, then X is rational. Hence X is rational.

With what sort of positivity of curvature can we attempt to characterize rational surfaces—scalar curvature, holomorphic sectional curvature or Ricci curvature? We see next that positive Ricci curvature is much too strong a property to be carried by all rational surfaces.

If X has positive Ricci curvature, then in particular the first Chern class c_1 is positive. This property puts a heavy restriction on X:

PROPOSITION (3.2). *Let X be a complex surface. Then $c_1(X) > 0$ iff $X \cong \mathbf{P}^1 \times \mathbf{P}^1$ or is obtained from $\mathbf{P}^2$ by blowing up $k \leqq 8$ distinct points such that no three are collinear, no six of them lie on a conic, and no eight of them lie on a cubic with one of them a double point. For $k \leqq 6$ these are the Del Pezzo surfaces.*

PROOF. If $c_1 > 0$ then X is a Hodge manifold and hence algebraic, since c_1 is represented by a (1, 1) form coming from an integral class. Also, by Kōdaira's vanishing theorem $h^{1,0}$ $(= \dim H^1(X, \mathcal{O}(K)))$ and P_m $(= \dim H^0(X, \mathcal{O}(mK)))$ are zero, and hence by Castelnuovo's theorem X is rational.

Now by Nakai's criterion [9], $c_1 > 0$ iff

$$c_1^2 > 0 \tag{3.3}$$

and

$$c_1 \cdot [C] > 0 \tag{3.4}$$

for any curve C.

Let $\pi: \hat{Y} \to Y$ be a blowing down and suppose $c_1(\hat{Y}) > 0$. Then if C is a curve in Y,

$$\begin{aligned} c_1 \cdot [C] = \pi^* c_1 \cdot \pi^*[C] &= c_1(\hat{Y}) \cdot \pi^*[C] && \text{from (2.1)} \\ &= c_1(\hat{Y}) \cdot ([\hat{C}] + m[E]) && \text{from (2.3)} \\ &> 0 && \text{from (3.4).} \end{aligned}$$

Also, from (2.1), $c_1(Y)^2 = c_1(\hat{Y})^2 + 1 > 0$. Hence if $c_1(\hat{Y}) > 0$, so is $c_1(Y)$. Thus if we blow down X to a minimal model M then $c_1(M) > 0$.

If D is a nonsingular curve of genus g on an algebraic surface, then

$$- c_1 \cdot [D] + [D]^2 = 2(g - 1). \tag{3.5}$$

In particular, if D is rational $(g = 0)$ and $c_1 > 0$, then

$$[D]^2 > - 2. \tag{3.6}$$

Now the minimal model F_n has a nonsingular rational curve of self-intersection $-n$ (the section at infinity of the bundle $\boldsymbol{P}(H^n \oplus 1)$); hence from (3.6), F_n cannot have positive c_1 for $n > 1$. Thus the minimal model M must be $\boldsymbol{P}^2$ or $\boldsymbol{P}^1 \times \boldsymbol{P}^1$. Another consequence of (3.6) is that X must be obtained by blowing up distinct points on a minimal model, since blowing up a point on an exceptional curve E gives us from (2.2) a rational curve of self-intersection -2. Since blowing up a point on $\boldsymbol{P}^1 \times \boldsymbol{P}^1$ is equivalent to blowing up two points on $\boldsymbol{P}^2$, we may say that $X \cong \boldsymbol{P}^1 \times \boldsymbol{P}^1$ or is obtained from $\boldsymbol{P}^2$ by blowing up k distinct points. From (2.1), $c_1(X)^2 = 9 - k$; so, from (3.3), $k \leqq 8$.

Now let Y be a surface obtained by blowing up $k \leqq 8$ points $\{p_i\}$ on $\boldsymbol{P}^2$ and such that $c_1(Y) \not> 0$. By (3.4) there must be a connected curve C in Y such that $c_1 \cdot [C] \leqq 0$. Project C onto a curve D in $\boldsymbol{P}^2$ where D is of degree d and has multiplicity m_i at p_i. Then

$$0 \leqq g \leqq \tfrac{1}{2}(d-1)(d-2) - \textstyle\sum \tfrac{1}{2} m_i(m_i - 1). \tag{3.7}$$

From (2.1) and (2.3) we get

$$\begin{gathered} 0 \geqq c_1 \cdot [C] = (\pi^* c_1 - \textstyle\sum [E_i])(\pi^*[D] - \sum m_i [E_i]) = 3d - \sum m_i, \\ \textstyle\sum m_i \geqq 3d. \end{gathered} \tag{3.8}$$

Put $a = \sum m_i$; then, since $m_i > 0$ and $k \leqq 8$,

$$8 \textstyle\sum m_i^2 \geqq k \sum m_i^2 \geqq (\sum m_i)^2 = a^2. \tag{3.9}$$

Substituting in (3.7) we get

$$a^2/8 - a \leqq (d-1)(d-2),$$

i.e., $a \leqq 4 + (8(d^2 - 3d + 4))^{1/2}$. From (3.8) this gives

$$(3d-4)^2 \leqq 8(d^2 - 3d + 4),$$

i.e., $d \leqq 4$.

If $d = 1$, then from (3.7), $m_i = 1$ and from (3.8), $k \geqq 3$.

If $d = 2$, $m_i = 1$ and $k \geqq 6$.

If $d = 3$, then from (3.7), $m_i = 1$ or 2, with at most one $m_i = 2$. If all m_i's are 1, then (3.8) gives a contradiction, so we must have $k = 8$, $m_1 = 2$ and $m_2 = \cdots = m_7 = 1$.

If $d = 4$, we get equality throughout and in (3.9), $k = 8$, $m_1 = \cdots = m_8 = m$, but this gives a contradiction in (3.7).

Thus the cases $d = 1$, 2 and 3 above exhaust the possibilities for Y and give Proposition (3.2).

Note that Calabi's conjecture would imply that all of the surfaces in Proposition (3.2) admit Kähler metrics with Ric > 0. The standard metrics on $\boldsymbol{P}^2$ and $\boldsymbol{P}^1 \times \boldsymbol{P}^1$ have Ric > 0—in the next section we see that F_1 ($\cong \boldsymbol{P}^2$ with a point blown up) admits a Hodge metric with Ric > 0.

Having considered positive Ricci curvature, we are left with the holomorphic sectional curvature K and the scalar curvature R. The metrics in the next section all have $K > 0$ but in § 5 we can only construct metrics with $R > 0$.

4. The surfaces F_n. Let E be a hermitian vector bundle over a Kähler manifold X, then we can define a Kähler metric on $\boldsymbol{P}(E)$ as follows. Let $\pi: E - \{0\} \to X$ be the projection, where $\{0\}$ denotes the zero section and define a form $\tilde{\Phi}$ on $E - \{0\}$ by:

$$\tilde{\Phi} = \pi^* \omega + is\, \partial\bar{\partial} \log \|w\|^2 \tag{4.1}$$

where ω is the fundamental form of the Kähler metric on X, $w \in E - \{0\}$ and $s > 0$. This projects to a closed form Φ on $\boldsymbol{P}(E)$ which is positive definite if s is chosen small enough, i.e., a Kähler form.

We use this construction for $F_n \cong \boldsymbol{P}(H^n \oplus 1)$. We take the standard metric on the base space $\boldsymbol{P}^1$, and since H^2 is the tangent bundle of $\boldsymbol{P}^1$ we get a natural hermitian metric on the vector bundle $H^n \oplus 1$. If z_1 is an inhomogeneous coordinate on $\boldsymbol{P}^1$, then dz_1 is a local section of H^{-2} and so we can locally represent $w \in H^n \oplus 1$ by $w = (z_1, w_1(dz_1)^{-n/2}, w_2)$. The metric on $\boldsymbol{P}^1$ is given by $dz_1\, d\bar{z}_1/(1 + z_1\bar{z}_1)^2$; hence the fibre metric in $H^n \oplus 1$ is

$$\|w\|^2 = w_1 \bar{w}_1 (1 + z_1\bar{z}_1)^n + w_2 \bar{w}_2. \tag{4.2}$$

If we take a local inhomogeneous coordinate $z_2 = w_2/w_1$ then the form Φ from (4.1) becomes

$$\Phi = i\partial\bar{\partial}[\log(1+z_1\bar{z}_1) + s \log((1 + z_1\bar{z}_1)^n + z_2\bar{z}_2)]. \tag{4.3}$$

This is the metric for which we must calculate the curvature, but first we make some remarks.

1. $H_2(F_n, \boldsymbol{Q})$ is generated by a fibre $z_1 = 0$ and the section $z_2 = 0$. It is easy to see from (4.3) that the induced metric on the fibre has fundamental form $is\, dz_2 \wedge d\bar{z}_2/(1 + z_2\bar{z}_2)^2$ and on the section $i(1 + ns)dz_1 \wedge d\bar{z}_1/(1 + z_1\bar{z}_1)^2$. Hence if we take s to be rational and multiply by N/π for some large integer N, Φ will take integer values on $H_2(F_n, \boldsymbol{Z})$ and thus we have a *Hodge* metric.

2. $SU(2)$ acts on $\boldsymbol{P}^1$ as isometries of the standard metric and lifts to an action on $H^n \oplus 1$ preserving the fibre metric. Hence our metric (4.1) is invariant under the action of $SU(2)$. Since the group acts transitively on $\boldsymbol{P}^1$, we need only compute the curvature along one fibre, i.e., at $z_1 = 0$. This simplifies the calculation somewhat.

Now let $f = (1 + z_1\bar{z}_1)^n + z_2\bar{z}_2$ and $F = \log f$. To compute the curvature of (4.3) we have to know the derivatives up to order 4 of F at $z_1 = 0$. It helps if we observe the following:

$$\begin{gathered} \partial^2 f/\partial z_1\, \partial z_2 = \partial^2 f/\partial \bar{z}_1\, \partial z_2 = 0 \quad \text{for all } z_1, \\ \frac{\partial^2 f}{\partial z_1\, \partial \bar{z}_1} = n, \qquad \frac{\partial^4 f}{\partial z_1\, \partial \bar{z}_1\, \partial z_1\, \partial \bar{z}_1} = 2n(n-1), \\ \frac{\partial f}{\partial z_1} = 0, \qquad \frac{\partial^3 f}{\partial z_1\, \partial z_1\, \partial \bar{z}_1} = 0, \end{gathered} \tag{4.4}$$

at $z_1 = 0$.

Let ∂_i denote a partial derivation, then

$$\partial_1 \partial_2 F = \partial_1 (\partial_2 f / f) = \partial_1 \partial_2 f / f - \partial_1 f \partial_2 f / f^2.$$

Hence, at $z_1 = 0$,

$$(4.5) \qquad \frac{\partial^2 F}{\partial z_1 \partial \bar{z}_1} = \frac{n}{(1 + z_2 \bar{z}_2)}, \quad \frac{\partial^2 F}{\partial z_2 \partial \bar{z}_2} = \frac{1}{(1 + z_2 \bar{z}_2)^2}, \quad \frac{\partial^2 F}{\partial z_1 \partial \bar{z}_2} = 0.$$

$\partial_1 \partial_2 \partial_3 F = \partial_1 \partial_2 \partial_3 f / f - \partial_2 \partial_3 f \partial_1 f / f^2$— etc., $+ 2\partial_1 f \partial_2 f \partial_3 f / f^3$. Hence at $z_1 = 0$,

$$(4.6) \qquad \begin{aligned} &\frac{\partial^3 F}{\partial \bar{z}_1 \partial z_1 \partial z_1} = 0, && \frac{\partial^3 F}{\partial \bar{z}_2 \partial z_1 \partial z_1} = 0, \\ &\frac{\partial^3 F}{\partial \bar{z}_1 \partial z_1 \partial z_2} = \frac{-n\bar{z}_2}{(1 + z_2 \bar{z}_2)^2}, && \frac{\partial^3 F}{\partial \bar{z}_2 \partial z_1 \partial z_2} = 0, \\ &\frac{\partial^3 F}{\partial \bar{z}_1 \partial z_2 \partial z_2} = 0, && \frac{\partial^3 F}{\partial \bar{z}_2 \partial z_2 \partial z_2} = \frac{-2\bar{z}_2}{(1 + z_2 \bar{z}_2)^3}. \end{aligned}$$

$$\begin{aligned} \partial_1 \partial_2 \partial_3 \partial_4 F = {} & \partial_1 \partial_2 \partial_3 \partial_4 f / f - \partial_2 \partial_3 \partial_4 f \partial_1 f / f^2 - \cdots - \partial_3 \partial_4 f \partial_1 \partial_2 f / f^2 \\ & - \cdots + 2\partial_1 \partial_2 f \partial_3 f \partial_4 f / f^3 + \cdots - 6\partial_1 f \partial_2 f \partial_3 f \partial_4 f / f^4. \end{aligned}$$

At $z_1 = 0$,

$$(4.7) \qquad \begin{aligned} \frac{\partial^4 F}{\partial z_1 \partial \bar{z}_1 \partial z_1 \partial \bar{z}_1} &= \frac{2n(n-1)}{(1 + z_2 \bar{z}_2)} + \frac{2n^2}{(1 + z_2 \bar{z}_2)^2}, \\ \frac{\partial^4 F}{\partial z_1 \partial \bar{z}_1 \partial z_2 \partial \bar{z}_2} &= \frac{-n}{(1 + z_2 \bar{z}_2)^2} + \frac{2n}{(1 + z_2 \bar{z}_2)^3}, \\ \frac{\partial^4 F}{\partial z_2 \partial \bar{z}_2 \partial z_2 \partial \bar{z}_2} &= \frac{2(2z_2 \bar{z}_2 - 1)}{(1 + z_2 \bar{z}_2)^4}, \end{aligned}$$

and the other terms of this form vanish.

Now if $\Phi = i\partial\bar{\partial}G = ig_{\alpha\beta}\, dz_\alpha \wedge d\bar{z}_\beta$ and $g^{\alpha\beta}$ is the inverse of the matrix $g_{\alpha\beta}$, the curvature tensor as defined in §3 is given by the following:

$$(4.8) \qquad 2R_{\alpha\beta\gamma\delta} = g^{\alpha\rho} g^{\gamma\sigma} \left[g^{\varepsilon\tau} \frac{\partial^3 G}{\partial \bar{z}_\varepsilon \partial z_\rho \partial z_\sigma} \cdot \frac{\partial^3 G}{\partial z_\tau \partial \bar{z}_\beta \partial \bar{z}_\delta} - \frac{\partial^4 G}{\partial z_\rho \partial \bar{z}_\beta \partial z_\sigma \partial \bar{z}_\delta} \right]$$

In our case $G = \log(1 + z_1 \bar{z}_1) + sF$ and from (4.5) at $z_1 = 0$, the matrix $g^{\alpha\beta}$ is:

$$\begin{bmatrix} (1 + z_2 \bar{z}_2)/(1 + z_2 \bar{z}_2 + ns) & 0 \\ 0 & (1 + z_2 \bar{z}_2)^2/s \end{bmatrix};$$

so from (4.5)—(4.8) we can compute the curvature. We find

$$2R_{1111} = \frac{2}{(1 + z_2 \bar{z}_2 + ns)^2} [(1 + z_2 \bar{z}_2)^2 + ns(1 - (n-1) z_2 \bar{z}_2)],$$

$$2R_{1122} = \frac{n}{(1 + z_2 \bar{z}_2 + ns)^2} [z^2 \bar{z}_2^2 - 1 + ns(2z_2 \bar{z}_2 - 1)], \quad 2R_{2222} = 2/s.$$

All the other terms (except those obtained from the above by symmetry) are zero. Let us put $a = 2R_{1111}$, $b = 2R_{1122}$, $c = 2R_{2222}$. We have the following inequalities:

$$4 > a > 2((1 + ns)^{-2} - n(n-1)s), \quad |b| < n, \quad c = 2/s. \tag{4.9}$$

Let $\xi = \xi_1 e_1 + \xi_2 e_2$ be a unit tangent vector and $\{e_1, e_2\}$ the unitary basis relative to which the components of the curvature tensor are a, b, c and put $u = \xi_1 \bar{\xi}_1$, $v = \xi_2 \bar{\xi}_2$; then the holomorphic sectional curvature becomes a quadratic form:

$$K(\xi) = au^2 + 4buv + cv^2. \tag{4.10}$$

Now $c > 0$ and from (4.9), $a > 0$ if $s(1 + ns)^2 < 1/n(n-1)$ (if $n = 1$ a is positive anyway). The discriminant of (4.10) is

$$\begin{aligned} 16b^2 - 4ac &< 16(n^2 - (1 + ns)^{-2}/s + n(n-1)) \\ &< 0 \quad \text{if } s(1 + ns)^2 < 1/n(2n-1). \end{aligned}$$

Hence if we choose s sufficiently small to satisfy this condition, the holomorphic sectional curvature $K > 0$, and even at the point at infinity of the fibre since the bounds in (4.9) are independent of z_2.

The eigenvalues of the Ricci form $\mathrm{Ric}(\xi)$ are $(a + b)/2$, $(b + c)/2$; thus if $n = 1$ we get, from (4.9),

$$a + b > 2(1 + s)^{-2} - 1, \qquad b + c > 2s^{-1} - 1.$$

Hence if we choose s small enough ($< \frac{1}{3}$ for example) $\mathrm{Ric} > 0$. In other words F_1 possesses a Hodge metric with positive Ricci curvature. We thus obtain the following.

PROPOSITION (4.11). *The surfaces F_n admit Hodge metrics of positive holomorphic sectional curvature. For $n = 0, 1$, F_n even has positive Ricci curvature.*

5. Blowing up. If X is a Hodge manifold and we blow up a point $p \in X$ to get $\hat{X}$ then in **[5]**, **[8]** Kōdaira defines a Hodge metric on $\hat{X}$. We are going to investigate the scalar curvature of this metric.

Let U be a neighborhood of $0 \in \boldsymbol{C}^n$. Take neighborhoods $U'' \subset U' \subset U$ and a C^∞ function f such that $f \equiv 1$ on U'' and $f \equiv 0$ outside U'. Then consider the form $\tilde{\theta}$ on $U \times \boldsymbol{C}^n - \{0\}$ defined by

$$\tilde{\theta} = i\partial\bar{\partial}(p_1^* f p_2^* \log \| \ \|^2) \tag{5.1}$$

where p_1, p_2 are the two projections. $\tilde{\theta}$ projects down to a form on $U \times \boldsymbol{P}^{n-1}$ and we take the induced form on $\hat{U} \subset U \times \boldsymbol{P}^{n-1}$ (see §2). Using local coordinates we identify $\hat{U}$ with $N(E)$ $(=\hat{N}(p))$ a neighborhood of the exceptional variety E and extend the form by 0 to obtain a form θ globally defined on $\hat{X}$. If ω is the fundamental form of the Hodge metric on X, then we set

(5.2) $$\hat{\omega}_t = \pi^*\omega + t\theta.$$

If $t > 0$ is sufficiently small and $= q/\pi$ where q is rational then $\hat{\omega}_t$ defines a Hodge metric on $\hat{X}$ after multiplication by some large positive integer.

Suppose ω has positive scalar curvature; then we want to show that $\hat{\omega}_t$ has positive scalar curvature for small enough t. The idea of the proof is the following: we take a neighborhood $N''(p)$ on which the function $f \equiv 1$ and show that under certain conditions on ω, $\hat{\omega}_t$ has positive scalar curvature in $\pi^{-1}(N''(p)) = N''(E)$ for $t < t_1$. On $\hat{X} - N''(E)$, $\hat{\omega}_0 = \pi^*\omega$ is a well-defined metric and thus if R_t is the sealar curvature of $\hat{\omega}_t$, then as $t \to 0$, $R_t \to R_0$ since $\hat{\omega}_t$ is a C^∞ function of t. Now $R_0 > 0$ everywhere and since $\hat{X} - N''(E)$ is compact we can find a t_2 such that, for $t < t_2$, $\hat{\omega}_t$ has positive scalar curvature outside $N''(E)$. Hence, if $t < \min(t_1, t_2)$, $\hat{\omega}_t$ has everywhere positive scalar curvature.

Let us consider then $\hat{\omega}_t$ in $N''(E)$. If R_t is bounded below on $N''(E) - E$, by $k > 0$ it is positive on $N''(E)$ so we can consider

$$N''(E) - E \underset{\pi}{\overset{\cong}{\rightarrow}} N''(p) - \{p\} \cong U'' - \{0\},$$

i.e., we need only consider a metric on the punctured ball. Using π and the local coordinates $\{z_\alpha\}$ we can write the metric (5.2) on $U'' - \{0\}$ as:

(5.3) $$\hat{\omega} = \omega + it\partial\bar{\partial}\log\|z\|^2, \qquad \hat{g}_{\alpha\beta} = g_{\alpha\beta} + \frac{t}{r^2}\left[\delta_{\alpha\beta} - \frac{\bar{z}_\alpha z_\beta}{r^2}\right].$$

The metric $\hat{g}$ clearly depends on the choice of local coordinates—we choose geodesic coordinates for the Kähler metric g. Relative to these coordinates we can write a Taylor series for g in terms of the curvature and its derivatives at 0:

(5.4) $$g_{\alpha\beta} = \delta_{\alpha\beta} - 2R_{\alpha\beta\gamma\delta}\, z_\gamma \bar{z}_\delta + \cdots.$$

In particular, $(g_{\alpha\beta} - \delta_{\alpha\beta})/r^2 = C_{\alpha\beta}$ is bounded in U''. Notice that as a matrix $\bar{z}_\alpha z_\beta/r^2 = P_{\alpha\beta}$ where P is a unitary projection onto the vector with components $\{z_\alpha\}$. We may now write (5.3) as:

(5.5) $$\hat{g} = \frac{r^2 + t}{r^2}\left[1 - \frac{tP}{r^2 + t} + \frac{r^4 C}{r^2 + t}\right]$$

(5.6) $$= \frac{r^2 + t}{r^2}\left[1 - \frac{tP}{r^2 + t}\right]\left[1 + \frac{tr^2 PC}{r^2 + t} + \frac{r^4 C}{r^2 + t}\right].$$

Recall that the scalar curvature of $\hat{g}$ is given by:

(5.7) $$\hat{R} = -\tfrac{1}{2}\hat{g}^{\beta\alpha}\,\partial^2(\log\det\hat{g})/\partial z_\alpha\,\partial\bar{z}_\beta$$

so we have to compute $\hat{g}^{-1}$ and $\det\hat{g}$. From (5.6), using the fact that $P^2 = P$ and $t/r^2 + t < 1$,

(5.8) $$\hat{g}^{-1} = \left[1 + \frac{r^4 C}{r^2 + t} + \frac{tr^2 PC}{r^2 + t}\right]^{-1}\left[\frac{r^2}{r^2 + t} + \frac{tP}{r^2 + t}\right].$$

Note that

$$\left\| \frac{r^4 C}{r^2+t} + \frac{tr^2 PC}{r^2+t} \right\| \leqq r^2 \|C\| \tag{5.9}$$

and $r^2/r^2+t<1$, $t/r^2+t<1$ and so from (5.8), $\hat{g}^{-1}$ is bounded in $r^2<\varepsilon$ for some ε independent of t.

$$\det \hat{g} = \left[\frac{r^2+t}{r^2}\right]^{n-1} \det\left[1+\frac{r^4 C}{r^2+t}+\frac{tr^2 PC}{r^2+t}\right]$$

from (5.6), so taking the logarithm,

$$\begin{aligned}\log\det\hat{g} &= (n-1)(\log(r^2+t)-\log r^2)\\ &\quad + \log\det\left[1+\frac{tr^2 PC}{r^2+t}+\frac{r^4 C}{r^2+t}\right].\end{aligned} \tag{5.10}$$

Now consider the last term in (5.10):

$$H = \log\det\left(1+[tr^4 PC + r^6 C]/r^2(r^2+t)\right).$$

LEMMA (5.11).

$$\frac{\partial^2 H}{\partial z_\alpha \partial \bar{z}_\beta} = \frac{-2\partial^2}{\partial z_\alpha \partial \bar{z}_\beta}\left[\frac{1}{r^2(r^2+t)}(tR_{\rho\sigma\gamma\delta} z_\rho \bar{z}_\sigma z_\gamma \bar{z}_\delta + r^4 R_{\gamma\delta} z_\gamma \bar{z}_\delta)\right] + O(r^2)$$

where by $f \in O(r^n)$ *we mean* $|f| < Ar^n$ *as* $r \to 0$ *where* A *is independent of* t, *and* $R_{\rho\sigma\gamma\delta}$, $R_{\gamma\delta}$ *denote the curvature tensor and Ricci curvature tensor of* g *at* 0.

PROOF. Put

$$G = \det\left(1+\frac{1}{r^2(r^2+t)}[tr^4 PC + r^6 C]\right);$$

then from (5.9), $G = 1 + O(r^2)$. Put $f = 1/r^2(r^2+t)$; then

$$\frac{\partial f}{\partial z_\alpha} = \frac{-\bar{z}_\alpha(2r^2+t)}{r^4(r^2+t)^2},$$

$$\frac{\partial^2 f}{\partial z_\alpha \partial \bar{z}_\beta} = \frac{-\delta_{\alpha\beta}(2r^2+t)}{r^4(r^2+t)^2} + \frac{2\bar{z}_\alpha z_\beta}{r^6(r^2+t)^3}(3r^4+3r^2 t+t^2).$$

Hence

$$\left\|\frac{\partial f}{\partial z_\alpha}\right\| < \frac{2}{r^3(r^2+t)}, \qquad \left\|\frac{\partial^2 f}{\partial z_\alpha \partial \bar{z}_\beta}\right\| < \frac{8}{r^4(r^2+t)}.$$

Now, if we expand the determinant G we shall get

$$\begin{aligned}G &= 1 + \operatorname{trace}(tr^4 PC + r^6 C)/r^2(r^2+t) + \cdots\\ &\quad + \text{terms of the form } f^{m+n} t^m r^{4n} a_1 \cdots a_m b_1 \cdots b_n \quad \text{with } m+n>1,\end{aligned}$$

where the a_i are entries in $r^4 PC$ and hence C^∞ of $O(r^4)$ and the b_i in $r^2 C$ hence C^∞ of $O(r^2)$. By the inequalities above, one differentiation reduces the order in r of each term by 1 and so the extra terms are of order

$$O\left(\frac{t^m\, r^{2m+4n-2}}{(r^2+t)^{m+n}}\right) = O(r^{2m+2n-2});$$

hence for $m + n > 1$, $\in O(r^2)$. Similarly $\partial G/\partial z_\alpha \in O(r)$ and $\partial^2 G/\partial z_\alpha\, \partial \bar{z}_\beta$ is bounded. Now

$$\frac{\partial^2 H}{\partial z_\alpha\, \partial \bar{z}_\beta} = \frac{1}{G}\frac{\partial^2 G}{\partial z_\alpha\, \partial \bar{z}_\beta} - \frac{1}{G^2}\frac{\partial G}{\partial z_\alpha}\frac{\partial G}{\partial \bar{z}_\beta}$$

and so, using the estimates for the order of G and its derivatives,

$$\frac{\partial^2 H}{\partial z_\alpha\, \partial \bar{z}_\beta} = \frac{\partial^2}{\partial z_\alpha\, \partial \bar{z}_\beta}(\mathrm{trace}(tr^4 PC + r^6 C)/r^2(r^2+t)) + O(r^2).$$

From the Taylor series (5.4), $\mathrm{trace}(r^4 PC) = -2R_{\rho\sigma\gamma\delta}\, z_\rho \bar{z}_\sigma z_\gamma \bar{z}_\delta + O(r^6)$ and $\mathrm{trace}(r^2 C) = -2R_{\gamma\delta}\, z_\gamma \bar{z}_\delta + O(r^4)$. Using the estimates again we obtain the lemma.

We return to (5.10) and now see that

$$\frac{\partial^2 \log}{\partial z_\alpha\, \partial \bar{z}_\beta} \det \hat{g} = (n-1)\left[\frac{\delta_{\alpha\beta}}{(r^2+t)} - \frac{\bar{z}_\alpha z_\beta}{(r^2+t)^2} - \frac{\delta_{\alpha\beta}}{r^2} + \frac{\bar{z}_\alpha z_\beta}{r^4}\right]$$

(5.12)

$$\cdot \frac{-2\partial^2}{\partial z_\alpha\, \partial \bar{z}_\beta}\left[\frac{1}{r^2(r^2+t)}(tR_{\rho\sigma\gamma\delta}\, z_\rho \bar{z}_\sigma z_\gamma \bar{z}_\delta + r^4 R_{\gamma\delta}\, z_\gamma \bar{z}_\delta)\right] + O(r^2).$$

Also, from (5.8)

$$\hat{g}^{-1} = \left(\frac{r^2}{r^2+t} + \frac{tP}{r^2+t} - \frac{r^6 C}{(r^2+t)^2} - \frac{tr^4 PC}{(r^2+t)^2}\right.$$

(5.13)

$$\left. - \frac{r^4 t}{(r^2+t)^2} CP - \frac{t^2 r^2}{(r^2+t)^2} PCP\right) + O(r^4).$$

We can also replace $C_{\alpha\beta}$ in (5.13) by $\tilde{C}_{\alpha\beta} = -2R_{\alpha\beta\gamma\delta}\, z_\gamma \bar{z}_\delta / r^2$ from the Taylor series (5.4) and then using the formula (5.7) for scalar curvature we can write

(5.14) $$\hat{R} = R_0 + R_1 + O(r^2)$$

where R_0 is a term independent of the curvature of g at 0 and R_1 is linear in the curvature at 0, hence a linear combination of the scalar curvature R, the Ricci curvature in the direction $\{z_\alpha\}(\mathrm{Ric}(z))$ and the holomorphic sectional curvature in that direction $K(z)$.

LEMMA (5.15). $R_0 = (n-1)(n-2)t/2(r^2+t)^2$.

PROOF. From (5.12) and (5.13) we have to take the trace:

$$-\tfrac{1}{2}\mathrm{trace}\left[\frac{r^2}{r^2+t} + \frac{tP}{r^2+t}\right](n-1)\left[\frac{1}{(r^2+t)^2} - \frac{r^2 P}{(r^2+t)^2} - \frac{1}{r^2} + \frac{P}{r^2}\right]$$
$$= (n-1)(n-2)t/2(r^2+t)^2.$$

LEMMA (5.16).

$$R_1 = \frac{1}{2(r^2+t)^3}[2r^4(r^2+t)\,R + 4tr^2(3r^2+4t)\mathrm{Ric}(z) + t(t^2 - 5r^2t - 4r^4)K(z)].$$

PROOF. The expansion of $\hat{g}^{-1}$ in (5.13) contributes one term a_1 to R_1 and the two curvature expressions in (5.12) contribute a_2 and a_3, the rest. We first consider the $\hat{g}^{-1}$ contribution:

$$\text{(1)}\quad a_1 = \frac{-\,\mathrm{trace}}{2(r^2+t)^2}\,[r^6\tilde{C} + tr^4 P\tilde{C} + tr^4\tilde{C}P + t^2r^2P\tilde{C}P] \times (n-1)\left[\frac{1}{(r^2+t)} - \frac{r^2P}{(r^2+t)^2} - \frac{1}{r^2} + \frac{P}{r^2}\right].$$

Since

$$\mathrm{trace}(P\tilde{C}P^2) = \mathrm{trace}(P\tilde{C}P) = \mathrm{trace}(P\tilde{C}) = \mathrm{trace}(\tilde{C}P) = -K(z)$$

and

$$\mathrm{trace}\,\tilde{C} = -2\mathrm{Ric}(z),$$

we obtain

$$\text{(2)}\quad \begin{aligned} a_1 &= \frac{(n-1)}{2(r^2+t)^3}[-2tr^4\,\mathrm{Ric}(z) + (2r^2+t)tr^2K(z)],\\ a_2 &= \left[\frac{r^2\delta_{\alpha\beta}}{r^2+t} + \frac{tz_\alpha\bar{z}_\beta}{r^2(r^2+t)}\right]\frac{\partial^2}{\partial z_\alpha\,\partial\bar{z}_\beta}\left[\frac{tu}{r^2(r^2+t)}\right] \end{aligned}$$

where $u = R_{\rho\sigma\gamma\delta}\,z_\rho\,\bar{z}_\sigma\,z_\gamma\,\bar{z}_\delta$. Now

$$\partial u/\partial z_\alpha = (R_{\alpha\sigma\gamma\delta} + R_{\gamma\sigma\alpha\delta})\,\bar{z}_\sigma\,z_\gamma\,\bar{z}_\delta,$$
$$\partial^2u/\partial z_\alpha\,\partial\bar{z}_\beta = (R_{\alpha\beta\gamma\delta} + R_{\alpha\delta\gamma\beta} + R_{\gamma\beta\alpha\delta} + R_{\gamma\delta\alpha\beta})\,z_\gamma\,\bar{z}_\delta.$$

Using this and the expressions for $\partial f/\partial z_\alpha$ and $\partial^2 f/\partial z_\alpha\,\partial\bar{z}_\beta$ in Lemma (5.11) ,we compute a_2 using the formula:

$$\frac{\partial^2}{\partial z_\alpha\,\partial\bar{z}_\beta}(fu) = \frac{\partial^2 f}{\partial z_\alpha\,\partial\bar{z}_\beta}u + \frac{\partial f}{\partial z_\alpha}\frac{\partial u}{\partial\bar{z}_\beta} + \frac{\partial f}{\partial\bar{z}_\beta}\frac{\partial u}{\partial z_\alpha} + f\frac{\partial^2 u}{\partial z_\alpha\,\partial\bar{z}_\beta}.$$

We obtain

$$\text{(3)}\quad \begin{aligned} a_2 &= \frac{4tr^2\,\mathrm{Ric}(z)}{(r^2+t)^2} + \frac{tK(z)}{2(r^2+t)^3}(-nr^2(2r^2+t) - 2r^4 - 4r^2t + t^2),\\ a_3 &= \left[\frac{r^2}{r^2+t}\delta_{\alpha\beta} + \frac{tz_\alpha\bar{z}_\beta}{r^2(r^2+t)}\right]\cdot\frac{\partial^2}{\partial z_\alpha\,\partial\bar{z}_\beta}\left[\frac{v}{r^2(r^2+t)}\right] \end{aligned}$$

where

$$v = r^4R_{\gamma\delta}\,z_\gamma\,\bar{z}_\delta\,,$$

$$\frac{\partial v}{\partial z_\alpha} = 2r^2\,\bar{z}_\alpha\,R_{\gamma\delta}\,z_\gamma\,\bar{z}_\delta + r^4R_{\alpha\delta}\,\bar{z}_\delta$$

$$\frac{\partial^2 v}{\partial z_\alpha\,\partial\bar{z}_\beta} = 2z_\beta\,\bar{z}_\alpha\,R_{\gamma\delta}\,z_\gamma\,\bar{z}_\delta + 2r^2(\delta_{\alpha\beta}\,R_{\gamma\delta}\,z_\gamma\,\bar{z}_\delta + \bar{z}_\alpha\,R_{\gamma\beta}\,z_\gamma + z_\beta\,R_{\alpha\delta}\,\bar{z}_\delta) + r^4R_{\alpha\beta}$$

and computing as above, we obtain

$$a_3 = \frac{r^4 R}{(r^2+t)^2} + \frac{r^2 t \mathrm{Ric}(z)}{(r^2+t)^3}((n+1)r^2 + 4t).$$

Taking $R_1 = a_1 + a_2 + a_3$ we finally get the expression in the lemma.

Now we have $\hat{R} = R_0 + R_1 + O(r^2)$ and so in order for $\hat{g}$ to have positive scalar curvature we need $R_0 + R_1$ to be bounded below by a positive constant k independent of t.

PROPOSITION (5.17). *If $n > 2$ and $R > \lambda > 0$ then there exists $t_1 > 0$ such that for $t < t_1$, $R_0 + R_1 > \lambda$.*

PROOF. Consider $P = 2(r^2+t)^3(R_0 + R_1) - 2\lambda(r^2+t)^3$. From (5.15) and (5.16),

$$\begin{aligned} P = {} & r^6(2R - 2\lambda) + r^4 t(2R + 12\mathrm{Ric}(z) - 4K(z) - 3\lambda) \\ & + r^2((n-1)(n-2)t + (16\mathrm{Ric}(z) - 5K(z) - 3\lambda)t^2) \\ & + (n-1)(n-2)t^2 + K(z)t^3 - \lambda t^3. \end{aligned}$$

Regard P as a cubic polynomial in r^2. If we choose t small enough that $(n-1)\cdot(n-2) + (K(z) - \lambda)t > 0$, then $P(0) > 0$. If t is small enough that $(n-1)(n-2) + (16\mathrm{Ric}(z) - 5K(z) - 3\lambda)t > 0$ then $P'(0) > 0$. Now if, furthermore,

$$\begin{aligned} & t^2(2R + 12\,\mathrm{Ric}(z) - 4K(z) - 3\lambda)^2 \\ & \qquad < 3(2R - 2\lambda)[(n-1)(n-2)t + (16\mathrm{Ric}(z) - 5K(z) - 3\lambda)t^2] \end{aligned}$$

(which is clearly possible if t is sufficiently small) then P' does not change sign and so P' is positive everywhere and $P(r^2) > 0$.

Hence $R_0 + R_1 > \lambda$.

COROLLARY (5.18). *Let X be a compact Kähler manifold of dimension > 2 with everywhere positive scalar curvature. Suppose we blow up any point $p \in X$ to obtain $\hat{X}$; then $\hat{X}$ admits a Kähler metric of positive scalar curvature.*

PROOF. $\hat{R} = R_0 + R_1 + O(r^2)$. From (5.17), $R_0 + R_1 > \lambda > 0$ so we can find ε independent of t such that for $r^2 < \varepsilon$, $\hat{R} > 0$ for $t < t_1$. We now go through the construction at the beginning of this section, taking $N''(p) = \{z \in U : r^2 < \varepsilon\}$.

We are primarily interested in the case $n = 2$ and then $R_0 = 0$ and so we are thrown back onto the positivity of R_1. In particular, if we take a fixed $w \in \boldsymbol{C}^2$, then $\lim_{s\to 0} \hat{R}(sw) = \lim_{s\to 0} R_1(sw) = K(w)/2$. Hence the scalar curvature of $\hat{g}$ at a point on the exceptional curve E corresponding to the direction w is $K(w)/2$. We need much more than $K > 0$ however, in order to extend the metric over $\hat{X}$ with positive scalar curvature—we need the condition $R_1 > \lambda > 0$.

As an example, take $X = \boldsymbol{P}^2$ with curvature tensor given by $R_{\alpha\beta\gamma\delta} = \delta_{\alpha\beta}\delta_{\gamma\delta} + \delta_{\alpha\delta}\delta_{\gamma\beta}$ then $K(z) = 4$, $\mathrm{Ric}(z) = 3$, $R = 6$ and so

$$R_1 = (6r^6 + 16r^4 t + 14r^2 t^2 + 2t^3)/(r^2+t)^3 > 2.$$

Hence we can blow up any point on P^2 and get positive scalar curvature. Also, since the construction of the metric is local, we can blow up any number of points. We turn now to the surfaces F_n and see if $R_1 > \lambda > 0$.

PROPOSITION (5.19). *Let $g(s)$ be one of the metrics constructed on F_n in §4. Then for sufficiently small s, $R_1 > \lambda > 0$.*

PROOF. As in (4.10) we take $u = \bar{z}_1 z_1$, $v = z_2 \bar{z}_2$ and express the various curvatures as follows:

$$K(z) = (au^2 + 4buv + cv^2)/(u + v)^2,$$
$$2\mathrm{Ric}(z) = ((a + b)u + (b + c)v)/(u + v),$$
$$2R = a + 2b + c.$$

We saw in §4 that for sufficiently small s we can make $a > 0$, so take λ such that $a > 2\lambda > 0$. Now consider

$$\begin{aligned} Q &= [(r^2 + t)^3 R_1 - \lambda(r^2 + t)^3](u + v)^2 \\ &= u^2(r^6(a + 2b + c - \lambda) + r^4 t(3a + 8b + c - 3\lambda) \\ &\qquad + r^2 t^2(3a + 8b - 3\lambda) + t^3(a - \lambda)) \\ &\quad + 2uv(r^6(a + 2b + c - \lambda) + r^4 t(4a + 4c - 3\lambda) \\ &\qquad + r^2 t^2(4a - 2b + 4c - 3\lambda) + t^3(2b - \lambda)) \\ &\quad + v^2(r^6(a + 2b + c - \lambda) + r^4 t(a + 8b + 3c - 3\lambda) \\ &\qquad + r^2 t^2(3c + 8b - 3\lambda) + t^3(c - \lambda)) \\ &= Eu^2 + 2Fuv + Gv^2. \end{aligned}$$

From (4.9), $|a|$, $|b|$ and $|\lambda|$ are bounded above by constants independent of s. Hence since $c = 2/s$ we can make all the coefficients of G positive by taking $s < s_1$ for some $s_1 > 0$. We write E as follows:

$$E = cr^4(r^2 + t) + (a + 2b - \lambda)r^6 + (3a + 8b - 3\lambda)r^2 t(r^2 + t) + t^3(a - \lambda).$$

Now since $a - \lambda > \lambda > 0$, there exists $\varepsilon > 0$ independent of s such that, if $r^2 < t\varepsilon$, $E - cr^4(r^2 + t) > 0$. If $r^2 \geqq t\varepsilon$, then

$$\begin{aligned} &|(a + 2b - \lambda) r^6 + (3a + 8b - 3\lambda) r^2 t(r^2 + t)| \\ &\qquad < (4 + 2n + \lambda) r^6 + (12 + 8n + 3\lambda) r^6/\varepsilon(1 + \varepsilon). \end{aligned}$$

Hence, by taking s sufficiently small, $s < s_2$, we can make

$$cr^4(r^2 + t) > cr^6 > |(a + 2b - \lambda) r^6 + (3a + 8b - 3\lambda) r^2 t(r^2 + t)|$$

and so make $E > 0$ for all r^2, t.

Consider now F. We can clearly make all the coefficients except $(2b - \lambda)$ positive by taking $s < s_3 < \min(s_1, s_2)$. If $F > 0$ then since $E > 0$ and $G > 0$, $Q > 0$; so suppose $F < 0$. Then

$$0 > F > t^3(2b - \lambda). \tag{5.20}$$

Now replacing λ by $3\lambda/2$ in the argument for the positivity of E, we see that $E > (r^2 + t)^3\, \lambda/2 > \lambda t^3/2 > 0$. Also, $G > t^3(c - \lambda) > 0$. Hence from (5.20),

$$F^2 - EG < t^6[(2b - \lambda)^2 - (c - \lambda)\lambda/2].$$

Since $|b| < n$ we can thus make $F^2 - EG < 0$ by taking $s < s_4$, in which case $Q > 0$ and the lemma is proved.

COROLLARY (5.21). *Let X be obtained by blowing up k distinct points on $\boldsymbol{P}^2$ or F_n. Then X admits a Hodge metric of positive scalar curvature.*

In the next section we shall see that such surfaces are in a deformational sense generic.

We finally remark that from Proposition (3.2) a surface with $c_1 > 0$ is obtained by blowing up $k \leqq 8$ distinct points on $\boldsymbol{P}^2$ or is equal to $\boldsymbol{P}^1 \times \boldsymbol{P}^1$. Hence from (5.21) we have the following: *If X is a surface with $c_1 > 0$, then X admits a Hodge metric of positive scalar curvature.*

6. Deformations. In this final section we consider the precise sense in which "almost all" is used in the main theorem.

By a family of complex structures we mean a complex manifold Z together with a holomorphic map $f:Z \to B$ onto a complex manifold B such that Z is a differentiable fibre bundle over B. If X is biholomorphically equivalent to $f^{-1}(u)$ for some $u \in B$ then Z is a family of deformations of X. A family $Z \to B$ is semi-universal if given any other family $Z_0 \to B_0$ there exists a not necessarily unique holomorphic map $g:B_0 \to B$ such that Z_0 is isomorphic to the induced family g^*Z.

PROPOSITION (6.1). *Let X be a rational surface. Then there exists a family $Z \to B$ of deformations of X such that*

(a) *$Z \to B$ is semi-universal relative to local deformations of X,*

(b) *there exists an open dense set $U \subset B$ such that the fibres over U are obtained by blowing up k distinct points on a minimal model.*

It is in this sense that we mean that the sufaces of Corollary (5.21) are generic.

PROOF. Let $f:W \to A$ be a family of deformations of a surface Y. Then we can form a family of deformations $V(W) \to W$ where the fibre over $w \in f^{-1}(u) \subset W$ is just $f^{-1}(u)$ with the point w blown up—we take $V(W)$ equal to the fibre product $W \times_A W$ with the diagonal blown up. The projection onto W is inherited from the projection onto the first factor in $W \times_A W$. If we iterate the process we get a family $V^m(W) \to V^{m-1}(W)$ and clearly any surface obtained from a fibre of $W \to A$ by successively blowing up m points is contained in this family. Also, those surfaces obtained by blowing up a point on an exceptional curve produced by the previous $m - 1$ blow ups form a subspace of codimension at least one; hence there is an open set $U \subset V^{m-1}(W)$ such that the fibres over U are obtained by blowing up m distinct points on the fibres of W.

We apply this set-up by taking W to be a trivial deformation for the minimal models $\boldsymbol{P}^2$ and $\boldsymbol{P}^1 \times \boldsymbol{P}^1$ and for F_n, $n > 1$, the semi-universal family with $A = \boldsymbol{C}^{n-1}$

as constructed by Suwa **[11]**, **[8]**. Hence, if the rational surface X is obtained by blowing up k times from a minimal model we take $Z = V^k(W)$ in the proposition.

We have to show that this family is semi-universal for small deformations. Let $f: Z \to B$ be a family of deformations of $X = f^{-1}(u)$ and suppose X contains an exceptional curve of self-intersection -1. Such a curve is stable under small deformations (Kodaira **[6]**) and (Iitaka **[2]**) there exists a neighborhood N of u over which the exceptional curves can be unformly blown down. In other words, if we restrict Z to N we can blow down Z to Z' where Z' defines a family of deformations of X with its exceptional curve blown down and $N \subset Z'$ induces the family Z from $V(Z') \to Z'$. Continuing in this way, we can induce Z over some neighborhood N' from $V^k(Z'') \to Z''$ where Z'' is a family of deformations of a minimal model. Since W is semi-universal this is induced from W and hence $Z|_{N'}$ is induced from $V^k(W)$.

References

1. M. Berger, *Sur les variétés d'Einstein compactes*, C.R. IIIe Réunion du Groupement des Math. d'Expression Latine (Namur, 1965), Librarie Universitaire, Louvain, 1966, pp. 35–55. MR **38** #6502.

2. S. Iitaka, *Deformations of compact complex surfaces*, Global Analysis, Papers in Honor of K. Kodaira, Univ. of Tokyo Press, Tokyo; Princeton Univ. Press, Princeton, N.J., 1969, pp. 267–272. MR **40** #8086.

3. S. Kobayashi, *On Compact Kähler manifolds with positive definite Ricci tensor*, Ann. of Math. (2) **74** (1961), 570–574. MR **24** #A2922.

4. S. Kobayashi and H. H. Wu, *On holomorphic sections of certain Hermitian vector bundles*, Math. Ann. **189** (1970), 1–4. MR **42** #5281.

5. K. Kodaira, *On Kähler varieties of restricted type (an intrinsic characterization of algebraic varieties)*, Ann. of Math. (2) **60** (1954), 28–48. MR **16**, 952.

6. ———, *On stability of compact submanifolds of complex manifolds*, Amer. J. Math. **85** (1963), 79–94. MR **27** #3002.

7. ———, *On the structure of compact complex analytic surfaces*. I, Amer. J. Math. **86** (1964), 751–798. MR **32** #4708.

8. J. Morrow and K. Kodaira, *Complex manifolds*, Holt, Rinehart and Winston, New York, 1971. MR **46** #2080.

9. Y. Nakai, *Non-degenerate divisors on an algebraic surface*, J. Sci. Hiroshima Univ. Ser. A, **24** (1960), 1–6. MR **25** #78.

10. I. R. Šafarevič, et al., *Algebraic surfaces*, Trudy Mat. Inst. Steklov. **75** (1965) = Proc. Steklov Inst. Math. **75** (1965). MR **32** #7557.

11. T. Suwa, *Stratification of local moduli spaces of Hirzebruch manifolds* (to appear).

12. Y. Tsukamoto, *On Kählerian manifolds with positive holomorphic sectional curvature*, Proc. Japan Acad. **33** (1957), 333–335. MR **19**, 880.

Institute for Advanced Study

Proceedings of Symposia in Pure Mathematics
Volume 27, 1975

HOLOMORPHIC EXTENSION FOR NONGENERIC *CR*-SUBMANIFOLDS

L. R. HUNT AND R. O. WELLS, JR.*

1. Introduction. A classical theorem of Hartogs asserts that every holomorphic function in at least two complex variables cannot have nonremovable singularities. In particular, each function which is holomorphic on a neighborhood of the unit sphere S^{2n-1} in $\boldsymbol{C}^n$, $n \geqq 2$, extends to a holomorphic function on the closed unit ball. Thus these holomorphic functions are extended over a submanifold in $\boldsymbol{C}^n$ of one higher real dimension than that of the sphere. We say that S^{2n-1} is *extendible* to the closed unit ball. We are interested in the problem of holomorphic extension of submanifolds of $\boldsymbol{C}^n$ in the following setting. Let M be a real k-dimensional C^∞ submanifold of some open set of $\boldsymbol{C}^n$, $n \geqq 2$. Given a point $p \in M$, under what conditions is it possible to extend all holomorphic functions on some fixed neighborhood of p in M to holomorphic functions on a submanifold of one higher real dimension? In other words, when is M locally extendible to a higher dimensional submanifold? Stein [8] proved that if M is a real hypersurface in $\boldsymbol{C}^n$, and if the Levi form on M at p is nonvanishing, then M is locally extendible to an open set; this is a local version of the Hartogs phenomenon (cf. also [5], [9], and [13]).

Suppose that M is a differentiable submanifold of some open set in $\boldsymbol{C}^n$, and that each point on M near a point $p \in M$ has the same number of holomorphic (complex) tangent vectors (which are now called *CR*-tangent vectors, see § 2). In this case we say that M is *locally CR* at p, and if the number of holomorphic tangent vectors to M at p is minimal, we say that M is *generic* at p (and it follows that M is generic near p). In the generic case, Bishop [1] devised an important method for constructing analytic discs with boundaries on M and which, depending on some

AMS (MOS) subject classifications (1970). Primary 32D10, 32F99.

*Research supported by NSF GP-190011 at Rice University.

parameters, shrink to the point p. If the Levi form on M at p is nonvanishing, it was shown in **[2]** and **[10]** that all holomorphic functions on M extend to holomorphic functions on these analytic discs, and that these analytic discs contain a submanifold of dimension $(k + 1)$. In other words, a generic submanifold with a nonvanishing Levi form at a point is locally extendible to one higher dimension near that point. For a detailed survey of the holomorphic extension problem and the related holomorphic approximation problem see **[13]**, which gives more background information, bibliography, and examples than we present here.

The Levi form is well defined in the locally CR-case, and the purpose of this paper is to extend the holomorphic extension theorem mentioned above (**[2]** and **[10]**) to this more delicate case. To do this we show that a locally CR-submanifold of $\mathbf{C}^n$ can be realized as a graph over a generic submanifold in a lower dimensional complex Euclidean space, in which the functions defining the graph are CR-functions (i.e., satisfy the tangential Cauchy-Riemann equations). Using Bishop's family of analytic discs for the generic submanifold, and applying a theorem of R. Nirenberg on extensions of CR-functions **[7]**, we construct a family of analytic discs with boundaries on our given CR-submanifold which effect the desired holomorphic extension.

In § 2 we prove a basic lemma concerning a relation between CR-submanifolds and graphs of CR-functions. In § 3 we prove the holomorphic extension theorem for locally CR-submanifolds, discussed above. As an application of this result, we are able to verify the conjecture made in **[13]** (cf. **[2]**, **[10]**, **[12]**) which asserts that a CR-submanifold M is locally holomorphically convex at p if and only if the Levi form on M vanishes identically near p.

2. *CR*-submanifolds. Let M be a real k-dimensional C^∞ submanifold of an open set in $\mathbf{C}^n$, $n \geqq 2$. Suppose $T(M)$ is the real tangent bundle to M with fibre $T_p(M)$ for $p \in M$. Letting J: $T(\mathbf{C}^n) \to T(\mathbf{C}^n)$ be the almost complex tensor for $\mathbf{C}^n$, we define

$$H_p(M) = T_p(M) \cap JT_p(M),$$

the vector space of *CR-tangent vectors* to M at p. Thus $H_p(M)$ is the maximal complex subspace of $T_p(\mathbf{C}^n)$ which is contained in $T_p(M)$. It is well known that

$$\max(k - n, 0) \leqq \dim_{\mathbf{C}} H_p(M) \leqq k/2.$$

We call $\dim_{\mathbf{C}} H_p(M)$ the CR-dim(M) at p. If CR-dim(M) is constant near p, then M is called a *locally CR-submanifold* at p. M is called *generic* if CR-dim(M) at p is minimal, i.e., equal to $\max(k - n, 0)$. If M is generic at p then M is generic near p, as is easy to check. If M is not generic at $p \in M$, then we let

$$d = CR\text{-dim}(M) - \max(k - n, 0)$$

(at p) be the *exceptionality* at p, and we will say that p is an exceptional point of order d. Thus a locally CR-submanifold defined near p is either generic or has a positive exceptionality, constant near p. Note that for a locally CR-submanifold M at p, the vector spaces $H_{p'}(M)$ form a subbundle of $T(M)$ for p' near p.

Assume without loss of generality that M is a *CR*-submanifold defined in an open set in $\boldsymbol{C}^n$, i.e., *CR*-dim(M) is constant on M. Since we are only studying local questions this is valid. There are however global obstructions that a submanifold be *CR* (cf. [**11**]). A function $f \in C^1(M)$ is a *CR-function* at $p \in M$ if $\bar{X}f(p') = 0$, for p' near p and X any local section of $H(M)$ near p. This is equivalent to saying that $\bar{\partial}_M f = 0$, where

$$\mathscr{E}(M) \xrightarrow{\bar{\partial}_M} \mathscr{E}^{0,1}(M) \xrightarrow{\bar{\partial}_M} \mathscr{E}^{0,2}$$

is the *CR*-complex on M (cf. [**2**], [**13**]). Moreover, f is a *CR*-function on M if f is a *CR*-function at each point in M.

Let M be a *CR*-submanifold of an open set in $\boldsymbol{C}^n$ of real dimension $k > n$, of *CR*-dimension $m \geqq k - n$, and let $d = m - (k - n)$ be the exceptionality of M. In other words each point of M is an exceptional point of order d, and this exceptionality measures the extent to which M is not generic. We suppose for simplicity that the origin $0 \in M$, and we shall do all of our local analysis near this distinguished point (without any loss of generality). The submanifold M can be parametrized in the following manner. There is a neighborhood U of 0 in $\boldsymbol{R}^l \times \boldsymbol{C}^m$, where $l + 2m = k$, and functions

$$h : U \longrightarrow \boldsymbol{R}^l, \qquad g : U \longrightarrow \boldsymbol{C}^d,$$

where $h(0) = g(0) = dh(0) = dg(0)$. Let $(t_1, \cdots, t_l, w_1, \cdots, w_m) = (t, w) \in \boldsymbol{R}^l \times \boldsymbol{C}^m$ be coordinates in U, and let $(z_1, \cdots, z_n)$ be coordinates in $\boldsymbol{C}^n$. Letting $\boldsymbol{C}^n = \boldsymbol{C}^l \times \boldsymbol{C}^m \times \boldsymbol{C}^d$, we then parametrize M by

$$\tau : U \longrightarrow \boldsymbol{C}^n, \qquad \tau(t, w) = (t + ih(t, w), w, g(t, w)). \tag{2.1}$$

Thus M is locally defined by $\tau(U)$, and for convenience we set $M = \tau(U)$. By the implicit function theorem any *CR*-submanifold M of real dimension k and of *CR*-dimension m and of exceptionality d has such a local representation near any of its points (cf. [**1**]).

Associated with τ is the mapping $\sigma : U \to \boldsymbol{C}^{n-d}$ given by

$$\sigma(t, w) = (t + ih(t, w), w), \tag{2.2}$$

and we let $N = \sigma(U)$. Then one easily checks that N is a generic *CR*-submanifold defined near the origin in $\boldsymbol{C}^{n-d}$. Thus we have the diagram

$$\begin{array}{ccc} U & \xrightarrow{\ \tau\ } & M \subset \boldsymbol{C}^n \\ & \searrow^{\sigma} & \downarrow \pi \\ & & N \subset \boldsymbol{C}^{n-d} \end{array} \tag{2.3}$$

where π is the natural projection on the first $(n - d)$ coordinates of $\boldsymbol{C}^n$. We call N the *associated generic submanifold* to M. Now σ is a diffeomorphism onto its image and thus the function

$$\gamma : N \longrightarrow C^d$$

given by $\gamma(z) = g \circ \sigma^{-1}(z)$, $z \in N$, is a well-defined vector-valued function on $N \subset C^{n-d}$. Moreover the graph of γ in $C^{n-d} \times C^d = C^n$ is precisely the submanifold M. We now have the following elementary but important proposition.[1]

PROPOSITION 1. *Let M be a differentiable submanifold parameterized as in* (2.1). *Then M has CR-dimension m and hence exceptionality d if and only if $\gamma_1, \cdots, \gamma_d$ are CR-functions on the projection $N = \pi(M)$ in C^{n-d}.*

PROOF. Consider the differentiable mapping $\gamma : N \to C^d$, and thus $d\gamma : T(N) \to T(C^d)$. Now we have the natural CR-splitting (cf. [2], [10], [13])

$$T(N) = T^{1,0}(N) \oplus T^{0,1}(N) \oplus Y(N),$$

where $H(N) = T^{1,0}(N)$, and $Y(N)$ is a real subbundle of real rank l. Thus we see that

$$d\gamma \mid_{H(N)} : H(N) \longrightarrow T(C^d)$$

is a real linear mapping which decomposes into $\partial_N \gamma + \bar{\partial}_N \gamma$, where ∂_N and $\bar{\partial}_N$ are the C-linear and C-conjugate-linear components of the real mapping $d\gamma|_{H(N)}$. Thus $d\gamma|_{H(N)}$ will be C-linear if and only if $\bar{\partial}_N \gamma$ vanishes identically, and these are the tangential Cauchy-Riemann equations on N. Moreover, the graph of $d\gamma|_{H(N)}$ over $H(N)$ in $T(M)$ will be a C-linear subbundle of $T(M)$ of m complex dimensions, since $d\gamma|_{H(N)}$ is C-linear. Thus, we see that the maximal complex subspace of $T_p(M)$, for each $p \in M$, contains at least m dimensions. Now let p be any point on M. Suppose $T_p(M)$ contained a complex subspace L of greater than m complex dimensions. Then since $d\pi(T_p(M)) = T_{\pi(p)}(N)$, and $d\pi$ is C-linear and injective when restricted to $T_p(M)$, it follows that $d\pi(L)$ would be a C-linear subspace of $T_{\pi(p)}(N)$ of greater than m dimensions, which contradicts the hypothesis. q.e.d.

3. Holomorphic extension. Let M be a CR-submanifold parametrized as in (2.1) and let N be the associated generic submanifold as given by (2.2). On any CR-submanifold there is a well-defined *Levi form* (cf. [2], [10], or [13]). In particular, the Levi form on M at $p \in M$ is a quadratic form

$$L_p(M) : H_p(M) \times H_p(M) \longrightarrow T_p(M) \otimes C/(H_p(M) \oplus \bar{H}_p(M))$$

which reflects the lack of integrability of the subbundle $H(M) \subset T(M)$. For our purposes, we recall that for $p = 0$, $L_0(M) \neq 0$ is equivalent to, after a complex-linear change of coordinates in C^n and C^m,

$$\partial^2 h_1(0,0)/\partial w_1 \, \partial \bar{w}_1 \neq 0,$$

where $h = (h_1, \cdots, h_l)$ and $w = (w_1, \cdots, w_m) \in C^m$.

We can now state our principal result.

---

[1]This was pointed out to us in the case where N is a hypersurface by R. Harvey (cf. [4]).

THEOREM 3.1. *Let M be a C^∞ k-dimensional CR-submanifold of CR-dimension m of an open set in $\boldsymbol{C}^n$, where $k > n$, and suppose $L_p(M) \neq 0$, for $p \in M$. Given a compact neighborhood K of p in M, then there exists a $(k + 1)$-dimensional connected topological submanifold with boundary $\tilde{M}$, such that $\partial\tilde{M} \cap K \neq \emptyset$, and such that K is extendible to $\tilde{M} \cup K$.*

REMARK. In other words M is locally extendible at the point p to a topological submanifold of one higher dimension (cf. **[10]**, **[12]**).

Before we prove Theorem 3.1 we want to indicate a principal consequence. A compact set $K \subset \boldsymbol{C}^n$ is *holomorphically convex* if the *spectrum* (maximal ideal space) of the topological algebra $\mathcal{O}(K)$ coincides, under the Gelfand mapping with K (cf. **[4]**). A closed set $S \subset \boldsymbol{C}^n$ is *locally holomorphically convex at* $p \in S$ if there is a fundamental neighborhood system (in S) at p consisting of holomorphically convex compact sets. A *holomorphic set* in $\boldsymbol{C}^n$ is a closed set which has a fundamental neighborhood system in $\boldsymbol{C}^n$ consisting of open Stein submanifolds (domains of holomorphy). A closed set $S \subset \boldsymbol{C}^n$ is a *locally holomorphic set* at $p \in S$ if there is a fundamental neighborhood system in S consisting of holomorphic sets.

We now have the following characterization of locally holomorphic and locally holomorphically convex *CR*-submanifolds in terms of the basic geometric invariant, the Levi form (cf. the survey papers **[12]** and **[13]**).

THEOREM 3.2. *Let M be a C^∞ CR-submanifold of an open set in $\boldsymbol{C}^n$. Then the following are equivalent*:

(a) *M is locally holomorphic at $p \in M$.*

(b) *M is locally holomorphically convex at $p \in M$.*

(c) *The Levi form $L_x(M)$ vanishes for each x in a neighborhood of p.*

REMARK. That (a) and (c) are equivalent was shown directly in **[6]** without recourse to an extension theorem (unavailable at the time) of the type given in Theorem 3.1.

PROOF. This theorem has been proved in the case that M was generic at p, i.e. $m = k - r$ (cf. **[2]**, **[10]**). In particular, (c) $\Rightarrow$ (a) $\Rightarrow$ (b) is proven in detail in **[10]**. To see that (b) $\Rightarrow$ (c) we assume that M is locally holomorphically convex at $p \in M$, and that the Levi form does not vanish identically near p. Then there is a compact holomorphically convex set K which is a neighborhood (in M) of p, and there is an interior point (in M) $q \in K$ so that $L_q(M) \neq 0$. By Theorem 3.1, there exists a topological manifold $\tilde{M}$ of dimension $(k + 1)$, depending on K, such that the restriction mapping

$$\mathcal{O}(\tilde{M} \cup K) \xrightarrow{\ r\ } \mathcal{O}(K)$$

is surjective. Since r is also injective, it follows that the spectrums $\mathcal{S}(\mathcal{O}(\tilde{M} \cup K))$ and $\mathcal{S}(\mathcal{O}(K))$ coincide. Then points in $\tilde{M} - K$, which is not empty, are elements of $\mathcal{S}(\mathcal{O}(K))$, thus contradicting the holomorphic convexity of M. q.e.d.

PROOF OF THEOREM 3.1. We assume we have the parametrizations (2.1) and (2.2) and that p is taken to be the origin, as before. Let L_c be the linear subspace of $\boldsymbol{C}^n$ obtained by setting $z_{l+2} = c_2, \cdots, z_{l+m} = c_m$, for fixed $(c_2, \cdots, c_m) \in \boldsymbol{C}^{m-1}$. Thus $L_c \cong \boldsymbol{C}^{n-m+1}$, and we let

$$M_c = L_c \cap M, \qquad N_c = \pi(L_c) \cap N,$$

where π is the projection $(z_1, \cdots, z_n) \to (z_1, \cdots, z_{n-d})$. We assume throughout that $d > 0$, since Theorem 3.1 is known in the generic case ([2], [10]). Thus we have a diagram (cf. (2.3))

$$\begin{array}{ccc} U_c & \xrightarrow{\ \tau_c\ } & M_c \subset L_c \cong \boldsymbol{C}^{n-m+1} \\ & \searrow^{\sigma_c} & \downarrow \pi \\ & & N_c \subset \pi(L_c) \cong \boldsymbol{C}^{n-d-m+1} \end{array}$$

by fixing $w_2 = c_2, \cdots, w_m = c_m$ in the parameter space U and restricting the mappings τ and σ. Thus M_c and N_c are CR-submanifolds of CR-dimension $= 1$, in a lower dimensional complex Euclidean space, and we still have $L_0(M) \neq 0$, $L_0(N) \neq 0$. We are now in a position to utilize deep results due to Bishop [1] and R. Nirenberg [7] respectively.

Namely, by the work of Bishop (as amplified in [2] and [10]), there is a $\hat{k} - 1$ parameter family of analytic discs (where we set $\hat{k} = k - 2(m - 1)$, i.e., $\hat{k} = \dim_{\boldsymbol{R}} U_c$),

$$F_c : I^{\hat{k}-1} \times \bar{\Delta} \longrightarrow \pi(L_c),$$

where $I = [0, 1]$ and $\Delta = \{\zeta \in \boldsymbol{C} : |\zeta| < 1\}$. Moreover F has the following properties:

(a) $F_c|_{\operatorname{int}(I^{\hat{k}-1} \times \bar{\Delta})}$ is a C^1 embedding (cf. [2]),
(b) the image of $F_c|_{I^{\hat{k}-1} \times \partial\Delta}$ is contained in N_c,
(c) the mapping F_c depends continuously on the parameter c,
(d) $F_c(s, \zeta)$ is holomorphic in $\zeta \in \Delta$ for fixed $s \in I^{\hat{k}-1}$,
(e) there is one point $s_0 \in I^{\hat{k}-1}$ so that $F(s_0 \times \bar{\Delta})$ is a single point on N_c (a degenerate analytic disc).

We call the family of analytic discs with the above properties a *Bishop family* of analytic discs associated with the generic submanifold N_c. By part (a) we see that the image of $I^{\hat{k}-1} \times \bar{\Delta}$ is a submanifold of dimension $\hat{k} + 1$. We notice that if we consider the Bishop family above for each c in a neighborhood V of $0 \in \boldsymbol{C}^{m-1}$, we obtain a family of analytic discs, the image of which is a C^1-submanifold of $\boldsymbol{C}^{n-d}$ of real dimension $(k + 1)$.[2]

Now R. Nirenberg has shown in [7] that any CR-function on N_c has a continuous extension to the image of the Bishop family F_c, which is holomorphic on the interior of a fixed analytic disc (i.e., on $F_c(s \times \Delta)$ for fixed $s \in I^{\hat{k}-1}$). Namely, the

[2]We could have required F to be a C^r mapping, in all parameters, for any fixed $r > 0$, but it is unknown if F can be made to be C^∞.

hypotheses of Nirenberg's general CR-extension theorem are satisfied in the special case where CR-dim $= 1$ and the Levi form is nonvanishing. Now let $\Gamma = (\Gamma_1, \cdots, \Gamma_d)$ be the extension to the Bishop family F_c given by Nirenberg of the particular CR-functions $\gamma = (\gamma_1, \cdots, \gamma_d)$ given by Proposition 1.1. Here we are restricting γ to N_c, and of course, M_c is the graph over N_c of the functions $\gamma|_{N_c}$. Let

$$F : I^{\hat{k}-1} \times \bar{\Delta} \times V \longrightarrow \boldsymbol{C}^{n-d}$$

be the *full* family of Bishop discs whose boundaries lie on the generic submanifold N (recall that V is a neighborhood of 0 in $\boldsymbol{C}^{m-1}$ in which the parameter c varies and thus $F_c = F|_{I^{\hat{k}-1} \times \bar{\Delta} \times \{c\}}$). Letting $P = I^{\hat{k}-1} \times V$, we have F: $P \times \bar{\Delta} \to \boldsymbol{C}^{n-d}$ and $F(p_0 \times \bar{\Delta})$ is a point on N, a degenerate analytic disc. Moreover, $F(P \times \bar{\Delta})$ is a C^1 submanifold of $\boldsymbol{C}^{n-d}$ of dimension $(k + 1)$ with boundary. Then the *graph* of the functions Γ defined on $F(P \times \bar{\Delta})$ will be a topological submanifold of $\boldsymbol{C}^n$ of dimension $(k + 1)$ with boundary (we know only that the extension Γ of γ are continuous functions, according to [7]). Then there is defined via the graph of Γ a mapping

$$\tilde{F} : P \times \bar{\Delta} \longrightarrow \boldsymbol{C}^n$$

given by the composition

$$\tilde{F}(p, \zeta) = (F(p, \zeta), \Gamma(F(p, \zeta)))$$

and since Γ is holomorphic on the interiors of the analytic discs, it follows that F is a family of analytic discs which degenerates to a single point. Since one may require that the image of the boundaries of the discs lie in any arbitrarily given neighborhood in M of 0, the theorem then follows from the extension theorem for such families of analytic discs [9] as in the generic case (cf. [2], [10]). q.e.d.

REFERENCES

1. E. Bishop, *Differentiable manifolds in complex Euclidean space*, Duke Math. J. **32** (1965), 1–21.MR **34** #369.

2. S. J. Greenfield, *Cauchy-Riemann equations in several variables*, Ann. Scuola Norm. Sup. Pisa (3) **22**(1968), 275–314. MR **38** #6097.

3. F. Reese Harvey and H. Blaine Lawson, Jr., *Boundaries of complex-analytic varieties*, Bull. Amer. Math. Soc. **80** (1974), 180–183.

4. F. R. Harvey and R. O. Wells, Jr., *Compact holomorphically convex subsets of a Stein manifold*, Trans. Amer. Math. Soc. **136** (1969), 509–516. MR **38** #3470.

5. L. Hörmander, *An introduction to complex analysis in several variables*, Van Nostrand, Princeton, N.J., 1966. MR **34** #2933.

6. L. R. Hunt, *Locally holomorphic sets and the Levi form*, Pacific J. Math. **42** (1972), 681–688.

7. R. Nirenberg, *On the H. Lewy extension phenomenon*, Trans. Amer. Math. Soc. **168** (1972), 337–356. MR **46** #392.

8. K. Stein, *Zur theorie der Funktionen mehrerer komplexer Veränderlichen. Die Regularitätshüllen niederdimensionaler Mannigfaltigkeiten*, Math. Ann. **114** (1937), 543–569.

9. R. O. Wells, Jr., *On the local holomorphic hull of a real submanifold in several complex variables*, Comm. Pure Appl. Math. **19** (1966), 145–165. MR **33** #5948.

10. ———, *Holomorphic hulls and holomorphic convexity of differentiable submanifolds*, Trans. Amer. Math. Soc. **132** (1968), 245–262. MR **36** #5392.

11. R. O. Wells, *Compact real submanifolds of a complex manifold with nondegenerate holomorphic tangent bundles,* Math. Ann. **179** (1969), 123–129. MR **38** #6104.

12. ———, *Holomorphic hulls and holomorphic convexity*, Complex Analysis (Proc. Conf. Rice Univ., Houston, Tex., 1967), Rice Univ. Studies, **54** (1968), No.4, 75–84. MR **39** #3029.

13. ———, *Function theory on differentiable submanifolds*, Contributions to analysis, A collection of papers dedicated to Lipman Bers, Academic Press, New York, 1974, pp. 407–441.

Texas Tech University

Rice University

Proceedings of Symposia in Pure Mathematics
Volume 27, 1975

HOLOMORPHIC EXTENSION THEOREMS

PETER KIERNAN

The paper[1] discusses some recent extension theorems in several complex variables which generalized the big Picard theorem. The emphasis is on holomorphic extension theorems and the main tool for obtaining the results is the Kobayashi pseudo-distance defined on a complex space.

AMS (MOS) subject classifications (1970). Primary 32—02, 32H20, 32H25.

[1]Published by Marcel Dekker in the Proceedings of the Tulane Conference on Value Distribution Theory (1973).

Proceedings of Symposia in Pure Mathematics
Volume 27, 1975

RESIDUES AND CHERN CLASSES

JAMES R. KING

In this note we will discuss the notion of the residue of a form in several complex variables. Then we will investigate the relationship between intersection theory and the cup product in cohomology using residues. The greater part of these results will appear shortly in [3] and [4], so we will content ourselves with a brief outline of these papers and then go on to discuss their application to Lefschetz fixed point formulas and other residue formulas.

1. Residues. Let $D \subset V$ be an $n - 1$ dimensional complex submanifold of the complex n-manifold V. A C^∞ r-form η on $V - D$ is said to be *semimeromorphic* with a pole of first order along D if for any holomorphic function t which defines D locally, $\eta = dt/t \wedge \psi + \varphi$, where ψ and φ are C^∞ forms in the domain of t. We say that t defines D locally if t is defined on an open subset of V and vanishes to first order along D and if $D = \{t = 0\}$ in the domain of t. If $i : D \to V$ denotes the inclusion map, the *residue* of η is a C^∞ form defined locally by $2\pi(-1)^{1/2}i^*\psi$. This definition differs from the usual definition in one complex variable by a factor of $2\pi(-1)^{1/2}$; but this will simplify the formulas for high codimension.

This procedure yields a well-defined form on D, independent of t, which is of type $(p - 1, q)$ if η is of type (p, q). See [6] and [3] for details.

A more geometric interpretation of the residue is given if we introduce $S_\varepsilon(D)$, the normal ε-sphere bundle. This is the boundary of the ε-neighborhood of D obtained by imbedding the normal bundle of D as a neighborhood of D in V. Then we also have the equation:

$$\operatorname{res} \eta = \lim_{\varepsilon \to 0} \operatorname{Fib} \int_{S_\varepsilon(D)} \eta.$$

AMS (MOS) subject classifications (1970). Primary 14C15, 14C30, 32A25, 32C30, 32C35, 32J25; Secondary 35N15, 55B45.

This means that for fixed ε we integrate out the fiber coordinate of η to obtain a form of degree $r - 1$. Then the limit of these forms on D is taken in the C^∞ topology.

Both of these procedures generalize the idea of taking a residue that one has in one complex variable. The second approach makes sense for submanifolds of higher codimension. If $W \subset V$ is a submanifold of complex codimension d, we may still take the normal ε-sphere bundle $S_\varepsilon(W)$ and write $\operatorname{res} \eta = \lim_{\varepsilon\to 0} \operatorname{Fib}\int_{S_\varepsilon(W)} \eta$ to define an $r - 2d + 1$ form on W. One question arises at once: for what η is such a limit defined? We will dodge this question, although reasonable sufficient conditions can be given (cf. [**1**]), and will reduce this case to the earlier situation of codimension one by blowing up along W.

If $\sigma : \hat{V} \to V$ is the monoidal transform of V along W, the restriction of σ to $\sigma^{-1}(V - W)$ is a biholomorphism. Therefore, σ^* induces a bijection of C^∞ forms on $V - W$ to C^∞ forms on $\hat{V} - \hat{W}$, where $\hat{W} = \sigma^{-1}(W)$. $\hat{W}$ is a submanifold of codimension 1 in V and is biholomorphic to the projectivized normal bundle of W in V. Furthermore, $\sigma^{-1}(S_\varepsilon(\hat{W})) = S_\varepsilon(\hat{W})$, so we can factor the residue map for forms on $V - W$ into two steps. First integrate out the fiber coordinate of the circle bundle $S_\varepsilon(W) \to \hat{W}$ to get $\operatorname{res} \sigma^*\eta$. Then since $\hat{W}$ is $\boldsymbol{P}(N_W)$, the projective normal bundle, $\operatorname{res} \eta$ equals the fiber integral $\operatorname{Fib} \int_{\boldsymbol{P}(N_W)} \operatorname{res} \sigma^*\eta$.

Since integration over the fiber is a continuous mapping of forms in the C^∞ topology, this is the same as the previous definition if $\operatorname{res} \sigma^*\eta$ exists. It has the advantage that a sufficient condition for the existence of $\operatorname{res} \sigma^*\eta$ is already known:

DEFINITION. A C^∞ form η on $V - W$ is a residue form of *kernel type* along W if $\sigma^*\eta$ is semimeromorphic along $\hat{W}$ with a pole of order one, $\sigma : \hat{V} \to V$ being the monoidal transform of V along W. The *refined residue* of η is $\operatorname{res} \sigma^*\eta$, a C^∞ form on W. The *residue* of η is the form on W given by the fiber integral $\operatorname{Fib} \int_{\boldsymbol{P}(N_W)} \operatorname{res} \sigma^*\eta$, recalling that $\hat{W} = \boldsymbol{P}(N_W)$.

The mapping $\eta \to \operatorname{res} \eta$ is a linear mapping from the (p, q) forms on V of kernel type along W to the $(p - d, q - d + 1)$ forms on W. If η is d- or $\bar{\partial}$-closed, so is $\operatorname{res} \eta$ and $\operatorname{res} \sigma^*\eta$; in fact this is true if $d\eta$ or $\bar{\partial}\eta$ is C^∞ on all of V, since res commutes with these operators up to sign.

2. Existence of residue forms. It is easy to see by a partition of unity that for any C^∞ form φ on $\hat{W}$, there is a residue form η such that $\operatorname{res} \sigma^*\eta = \varphi$. Therefore, given a C^∞ form ψ on W, to show that there exists an η such that $\operatorname{res} \eta = \psi$ requires that we produce a form on $\hat{W}$ whose fiber integral is ψ.

If p: $\hat{W} = \boldsymbol{P}(N_W) \to W$ is the projection map defined by the restriction of σ, we will write $p_*\varphi$ for the fiber integral $\operatorname{Fib} \int_{\boldsymbol{P}(N_W)} \varphi$ for any form φ on $\hat{W}$.

If $|\ \ |$ is the length given by a hermitian metric on N_W, we define the forms $\mu = (1/2\pi i)\partial \log|\tau|^2$ and $\omega = \bar{\partial}\mu$ on the total space of N_W. These are forms of kernel type on N_W with poles along the zero section. The form ω is the pull-back of a form on $\boldsymbol{P}(N_W)$ which we also denote by ω. This form restricts to the Kähler form of the Fubini-Study metric on each fiber, so $p_*\, \omega^{d-1} = 1$. Therefore, if we set $\tilde{\psi} = p^*\psi \wedge \omega^{d-1}$, $p_*\tilde{\psi} = \psi$.

It is more interesting to ask, for reasons that we will see below, whether a form φ on W is the residue of a form η such that $d\eta$ is C^∞ on all of V. We can also ask whether a specified $\tilde{\varphi}$ on W is res $\sigma^*\eta$ for such a form η (that is, $d\eta$ is C^∞ on V, not merely that $d\sigma^*\eta$ is C^∞ on $\hat{V}$). Since res $d\eta = -d$ res $\eta = 0$ in this case, such φ or $\tilde{\varphi}$ must be d-closed.

EXISTENCE THEOREM FOR RESIDUES. *If φ is a* (i) *d-closed or* (ii) *$\bar{\partial}$-closed r-form on W, there is a residue form η on V such that res $\eta = \varphi$ and* (i) *$d\eta$ or* (ii) *$\bar{\partial}\eta$ is C^∞ on V. Also, given any hermitian metric in the normal bundle N_W, η can be chosen so that* res $\sigma^*\eta = p^*\varphi \wedge Q(\omega)$, *where $Q(\omega) = \omega^{d-1} + p^*c_1 \wedge \omega^{d-2} + \cdots + p^*c_{d-1}$, the form c_i being the ith Chern form of N_W with respect to the given metric. Furthermore, in case* (i), *if φ has Hodge filtration a, η can be chosen with Hodge filtration $a + d$. In case* (ii), *if φ is of type (a, b), then η can be found of type $(a + d, b + d - 1)$.*

PROOF. A proof is given by [4] using the results of [3]. The method is to construct a fine resolution of the constant sheaf $\boldsymbol{C}_V$ or the sheaf Ω_V^a using residue forms.

The appearance of the Chern forms is explained by looking at the infinitesimal case: we assume that $V = N_W$, the total space of the normal bundle, and $W = Z$, the zero section. On N_W the following fundamental equation of forms holds: $-p^*c_d = \omega \wedge Q(\omega)$. See [1] for a proof. If we set $\eta = \mu \wedge Q(\omega)$, $d\eta = \bar{\partial}\eta = -p^*c_d$ is C^∞ on N_W; res $\eta = 1$ on Z and res $\sigma^*\eta = Q(\omega)$ on $\boldsymbol{P}(N_W)$. This is the prototype for the existence theorem.

On the other hand, the form $\zeta = \mu \wedge \omega^{d-1}$ has refined residue ω^{d-1} and residue 1, but $d\zeta = \omega^d$ is not C^∞. In the discussion of intersection in the "wrong" dimension, we will see that in general there is an obstruction to finding a form η with $d\eta$ C^∞ and with refined residue ω^{d-1}.

The interest of this existence theorem lies in the following fact:

PROPOSITION. *If η is a residue form of kernel type on V along W such that $d\eta$ is C^∞ on all of V, the form $-d\eta$ is cohomologous to i_* of the de Rham class of* res η, *where $i_*\colon H^*(W) \to H^*(V)$ is the Gysin homomorphism associated with the inclusion $i\colon W \to V$.*

REMARK. By definition the Gysin map is defined as the adjoint of i^* under the cup product pairing. If φ is a closed form on W and ψ is a closed form on V, the class of $i_*\varphi$ is the cohomology class such that

$$(i_*\varphi \cup \psi)[V] = \int_W \varphi \wedge i^*\psi.$$

In particular we can view $i_*\varphi$ as a functional (or current) on all of the forms on V by setting $i_*\varphi(\psi) = \int_W \varphi \wedge i^*\psi$. We also will write $W \wedge \varphi$ to denote the current $i_*\varphi$.

If res $\eta = 1$, this proposition says that W is the Poincaré dual of $-d\eta$.

PROOF. For any closed form ψ on V,

$$\int_V d\eta \wedge \psi = \lim_{\varepsilon\to 0}\int_{V-B_\varepsilon} d(\eta \wedge \psi) = -\lim_{\varepsilon\to 0}\int_{S_\varepsilon} \eta \wedge \psi = -\int_W \operatorname{res}\eta \wedge \psi.$$

If we consider $[\eta]$ to be the current, or functional, on the forms on V given by $[\eta](\psi) = \int_V \eta \wedge \psi$ and $d[\eta](\psi)$ to be $(-1)^{\deg \eta+1} [\eta](d\psi)$, then the proposition may be written as

$$W \wedge \operatorname{res} \eta = d[\eta] - [d\eta].$$

If we compute de Rham cohomology with currents instead of C^∞ forms, this is a statement that $W \wedge \operatorname{res} \eta$ and $-d\eta$ are cohomologous. This equation may be extended to the case where W is a subvariety with singularities. Any locally L^1 form on W will be called res η if it satisfies the equation above. Thus the residue measures the failure of Stokes' theorem to hold for a form η with singularities.

3. Residues and intersections. For simplicity let us restrict our attention to the case where W is a complex submanifold of V and η is a residue form of kernel type with res $\eta = 1$ on W. Suppose X is any pure dimensional subvariety of V satisfying

$$(*) \qquad \dim W + \dim X = \dim V + \dim W \cap X.$$

Then the restriction of η to X is a locally L^1 form on X which is C^∞ on $X - X \cap W$. (See [3, Proposition 5.3].)

Let $\tilde{\eta}$ be the restriction of η to X. At points which are regular points of both X and $X \cap W$, $\tilde{\eta}$ is a residue form on X along $X \cap W$. In any case we may define

$$W \cap X \wedge \operatorname{res} \tilde{\eta} = d[\tilde{\eta}] - [d\tilde{\eta}].$$

From the point of view of V, we may write

$$W \cap X \wedge \operatorname{res} \tilde{\eta} = d[X \wedge \eta] - [X \wedge d\eta].$$

It is proved in [3, Theorem 6.3] that if res $\eta = 1$ on W, res $\tilde{\eta}$ is the intersection multiplicity, $m(W, X)$, of W and X, at least or the regular locus of $W \cap X$. Thus if we define the current $W \cdot X$ by $W \cdot X(\psi) = \int_{W \cap X} m(W, X)\psi$ for compactly supported forms ψ on V, we have

$$W \cdot X = d[X \wedge \eta] - [X \wedge d\eta],$$

i.e., $W \cdot X(\psi) = \int_X (\eta \wedge d\psi - d\eta \wedge \psi)$. A corollary to this is

PROPOSITION. *If η is a residue form with* res $\eta = 1$ *on W and if $d\eta$ is C^∞ on V, then $W \cdot X$ is cohomologous to $- X \wedge d\eta$, provided $(*)$ holds.*

COROLLARY. *If W_1 and W_2 are complex submanifolds, and η_i, $i = 1, 2$, is a residue form with residue 1 on W_i, then provided $d\eta_i$ is C^∞ on V and $(*)$ holds, the current W_1, W_2 is cohomologous to $d\eta_1 \wedge d\eta_2$.*

EXAMPLE. If E is a k-dimensional holomorphic vector bundle over V and τ is a holomorphic section, the Chern form $c_k(E)$ is cohomologous to the zero set $T = \{\tau = 0\}$ (counted with multiplicity), provided that the codimension of T is k.

PROOF. On the total space of E, $c_k(E) = - d\eta$, where $\eta = \mu \wedge (\omega^{k-1} + \cdots + c_{k-1}(E)) = \mu \wedge Q(\omega)$.

As a special case of the example, if $D \subset V$ is a divisor, D is cohomologous to the

Chern form of the associated line bundle L_D. In particular, if $V = \boldsymbol{P}^n$ and H^k is a k-dimensional linear subspace, then H^{n-1} is cohomologous to the Kähler form of the Fubini-Study metric $\nu = (i/2\pi)\partial\bar\partial \log \sum_{i=0}^{n} |z_i|^2$. Then by the corollary, $H^k = H_1 \cdot H_2 \cdot \cdots \cdot H_{n-k}$ is cohomologous to ν^{n-k}. Thus for any subvariety X of $\boldsymbol{P}^n$ of dimension $n-k$, $X \cdot H^k$ is cohomologous to $X \wedge \nu^{n-k}$. Applying both currents to the function 1, we see that $\int_X \nu^{n-k}$ equals the intersection number of X and H^k, which by definition is the degree of X, n_X. Therefore, since ν^k generates $H^{2k}(\boldsymbol{P}^n)$, X is cohomologous to $n_X \nu^k$. If W is a submanifold of $\boldsymbol{P}^n$, then by the existence theorem for residues and the ordinary de Rham theorem, there is a residue form ζ_W such that res ζ_W is 1 on W and $-d\zeta_W = n_W \nu^{n-k}$, where dim $W = k$. Consequently, the following currents are cohomologous: $W \cdot X$, $W \wedge n_X \nu^k$, and $n_W\, n_X\, \nu^n$. Applying these currents to the function 1, it follows that the intersection number of W and X equals the product of the degrees; this is Bezout's theorem. The case where W has singularities is proved by intersecting $W \times X$ with the diagonal in $\boldsymbol{P}^n \times \boldsymbol{P}^n$. There will be more about the diagonal in the following sections.

4. Intersections in the 'wrong' dimension. Suppose that W is a submanifold and X is a subvariety of V, but

$$\dim (W \cap X) = \dim W + \dim X - \dim V + e,$$

where $e > 0$ (we assume that all subvarieties are pure dimensional, but the general case may be treated by looking at each component separately). Then if η is a residue form with res $\eta = 1$ on W and with $d\eta$ a C^∞ form on V, the expression

$$R = d[X \wedge \eta] - [X \wedge d\eta]$$

still defines a current supported on $W \cap X$, provided that no component of X is contained properly in W. This uses the fact that η is L^1 on X, despite the fact that $e > 0$; this is proved in [3, Proposition 5.3].

By definition R represents the cohomology class of $-X \wedge d\eta$, which is the cup product of X and W. In fact frequently there is a locally L^1 form φ on $W \cap X$ such that $R = W \cap X \wedge \varphi$. If $\sigma : \hat{V} \to V$ is the monoidal transform of V along W, we define $\hat{X} \subset \hat{V}$ to be the closure of $\sigma^{-1}(X - W)$. Then, by definition,

$$R = \sigma_*(d[\hat{X} \wedge \sigma^*\eta] - [\hat{X} \wedge d\sigma^*\eta]) = p_* \operatorname{res} \sigma^*\eta,$$

where $p \colon \hat{X} \cap \hat{W} \to X \cap W$. If the image of every irreducible component of $\hat{X} \cap \hat{W}$ contains an open set in $X \cap W$, then this push forward is a locally L^1 form φ on $W \cap X$; cf. Proposition 5.1 of [3].

In case the intersection is *nondegenerate* we may compute φ. By this we mean that W, X, and $W \cap X$ are submanifolds and the intersection of the tangent spaces $T_{W,x} \cap T_{X,x} = T_{W \cap X,x}$ for all $x \in W \cap X$. In this case $\hat{X} \cdot \hat{W} = \boldsymbol{P}(N')$, where N' is the normal bundle of $W \cap X$ in X. Therefore

$$\varphi = \sum_{i=0}^{d-1} c_i(N_W) \wedge p_*\, \omega^{d-i+1},$$

and since the restriction of ω to $\hat{X} \cap \hat{W}$ is just the ω-form for $\boldsymbol{P}(N')$, $p_*\omega^j$ can be computed in terms of the Chern classes of N'. The relation

$$0 = \omega^{d-e} + p^* c_1(N') \wedge \omega^{d-e-1} + \cdots + p^* c_{d-e}(N')$$

implies that the formal power series of forms $p_*(1/(1 - t\omega)) = 1/c_t(N')$, where $c_t(N') = \sum_{i=0}^{d-e} c_i(N')t^i$. Therefore, φ is the coefficient of t^e in the series $c_t(N_W)/c_t(N')$. This proves

PROPOSITION. *If W and X intersect nondegenerately, then the current $W \cap X \wedge \varphi$ represents the cup product of the classes of W and X, where φ is the coefficient of t^e in $c_t(N_W)/c_t(N')$. N' is the normal bundle of $W \cap X$ in X.*

EXAMPLE (BOTT). If τ is a holomorphic section of a holomorphic k-vector bundle E over V, then if τ is nondegenerate on the $(n - k + e)$-dimensional zero set $T = \{\tau = 0\}$, then $c_k(E)$ is cohomologous to the coefficient of t^e, in T, $c_t(E)/c_t(N_T)$.

EXAMPLE. If f: $V \to V$ is a holomorphic mapping of an n-dimensional compact complex manifold, the (topological) Lefschetz number $L(f) = \sum_{q=0}^{n}(-1)^q \cdot \operatorname{trace} f_q^*$, where f_q^*: $H^q(V, \boldsymbol{C}) \to H^q(V, \boldsymbol{C})$. In terms of differential forms, if $\{\varphi_i\}$ is a basis for $H^*(V, \boldsymbol{C})$, and $\{\psi_j\}$ is a dual basis (under the Poincaré pairing), then

$$L(f) = \int_V \Big(\sum_i (-1)^{\deg \varphi_i} f^* \varphi_i \wedge \psi_i\Big).$$

But if we set $\Phi = \sum(-1)^{\deg\varphi_i} p_1^* \varphi_i \wedge p_2^* \psi_i$ and $\Gamma =$ graph of f in $V \times V$, then $L(f) = \int_\Gamma \Phi$. It is formal to check that Φ is cohomologous to Δ (check out on closed forms of the type $p_1^* \varphi_i \wedge p_2^* \psi_j$). Therefore, $L(f) = \Gamma \cdot \Delta(1)$ if $\Gamma \cdot \Delta$ exists. By the above, if S is the fixed point set of f and f is nondegenerate along S, $L(f) = \int_S c_t(T_V)/c_t(N_S)$, taking the coefficient of $t^{\dim S}$.

5. Final remarks. The original motivation for this work was an aspect of the theory that perhaps has not been stressed enough in the preceding section. In the theory of algebraic cycles, one has the Abel-Jacobi mapping of algebraic cycles homologous to zero into the intermediate Jacobian varieties. To prove certain functorial properties of these mappings it was necessary to show that $W \cdot X$ and $S(-d\eta)$ are cohomologous with the Hodge filtration preserved (cf. [2] and [5]). In the C^∞ case this is not hard because one can perturb X, but in the holomorphic case it is necessary to do intersection theory without budging X. This is the approach outlined in § 3 and proved in [3].

The results in § 4 have the curious flavor that they are topological results with Hodge filtration. A natural question is: what about the Dolbeault cohomology? The results of § 4 hold for the Dolbeault cohomology as well. However the $\bar{\partial}$-Lefschetz fixed point theorem does not follow from the results given, but it is related to them. For this, one must find a residue form η for the diagonal in $V \times V$ and split it into parts η_p, $p = 0, \cdots, n$, corresponding to the (p, q) forms on the second factor. The residue formula for isolated fixed points has been derived by Tong [8], Toledo [7], and L. and R. Sibner [9]; but the theory for higher dimensional fixed point sets, analogous to that of § 4, involves higher order residues. If this were worked out, it should imply the Riemann-Roch theorem for complex manifolds.

Bibliography

1. R. Bott, *A residue formula for holomorphic vector fields*, J. Differential Geometry **1** (1967), 311–330.

2. P. Griffiths, *Some results on algebraic cycles on algebraic manifolds*, Proc. Bombay Colloq. on Algebraic Geometry, Oxford Univ. Press, London, 1968.

3. J. King, *Global residues and intersections on a complex manifold*, Trans. Amer. Math. Soc. **192** (1974), 163–200.

4. ———, *Refined residues, Chern forms, and intersections*, Value Distribution Theory, part A, Dekker, New York, 1974.

5. ———, *The Abel-Jacobi map for families of Kähler manifolds* (to appear).

6. J. Leray, *Le calcul différentiel et intégral sur une variété analytique complexe* (*Problème de Cauchy*. III), Bull. Soc. Math. France **87** (1959), 81–180. MR **23** #A3281.

7. D. Toledo, *On the Atiyah-Bott formula for isolated fixed points*, J. Differential Geometry **8** (1973), 401-436.

8. Y. -L. L. Tong, *de Rham's integrals and Lefschetz fixed formula for d coholomogy*, Bull. Amer. Math. Soc. **78** (1972), 420–422.

9. L. Sibner and R. Sibner, ···, Contributions to Analysis, A collection of papers dedicated to Lipman Bers, Academic Press, New York, 1974.

Massachusetts Institute of Technology

Proceedings of Symposia in Pure Mathematics
Volume 27, 1975

SOME CLASSICAL THEOREMS FOR HOLOMORPHIC MAPPINGS INTO HYPERBOLIC MANIFOLDS

MYUNG H. KWACK

Griffiths [1] generalized Landau-Schottky's theorem, Bloch's theorem and Koebe's distortion theorem for a class of holomorphic mappings into complete canonical algebraic manifolds. Wu [3] proved Bloch's theorem for a class of holomorphic mappings into a complete hermitian manifold whose holomorphic sectional curvatures are bounded from above by a negative constant.

In this paper we would like to point out that similar theorems can be proved for a class of holomorphic mappings into hermitian manifolds which are complete hyperbolic. With a slight change in assumption, analogues of the above theorems can also be proved for a class of holomorphic mappings into hyperbolic manifolds. Proofs are similar to the ones given in [1] and [3].

1. Mappings into hermitian manifolds. Let $f : M \to N$ be a holomorphic mapping from a connected complex manifold to a complex manifold. All manifolds will have complex dimension n unless otherwise stated. Let $0 \in M$ and $x_0 \in N$ with $f(0) = x_0$. Using local coordinates $z_1, \cdots, z_n$ at 0 and $w_1, \cdots, w_n$ at x_0, f may be written locally as

$$w_i = f_i(z_1, \cdots, z_n), \qquad i = 1, \cdots, n.$$

We will write $\det(\partial f_i/\partial z_k)$ for the Jacobian determinant of f in terms of the local coordinates $z_1, \cdots, z_n$ and $w_1, \cdots, w_n$. Now we state Landau-Schottky's theorem (Theorem 1) and Koebe's theorem (Theorem 2).

AMS (MOS) subject classifications (1970). Primary 32H20; Secondary 32A10, 32A30.
Key words and phrases. Hyperbolic manifolds, homomorphic functions, Bloch's theorem.

THEOREM 1. *Let N be a complete hermitian manifold whose holomorphic sectional curvatures are bounded from above by a negative constant. Let $B(r)$ be the ball $\{z = (z_1, \cdots, z_n) : \|z\| < r\}$ of radius r in C^n where $\|z\|^2 = \sum_{i=1}^n |z_i|^2$. We let $o = (0, \cdots, 0) \in B(r)$, $x_0 \in N$ and $a > 0$. Then for any holomorphic mapping $f : B(r) \to N$ with $f(o) = x_0$ and $|\det \partial f_i/\partial z_k(o)| \geqq a$, we have that $r \leqq R$, where $R = R(x_0, a)$ is a constant independent of f.*

To state our next theorem, let ds_N^2 denote the hermitian metric on a complex manifold N. For a holomorphic mapping f: $B \to N$ of the unit ball $B = \{z = (z_1, \cdots, z_n): \sum_{i=1}^n |z_i|^2 < 1\}$ into N, we consider the differential f_*: $T_0(B) \to T_{f(o)}(N)$ where $T_p(X)$ denotes the tangent space of X at p. If $e_1, \cdots, e_n$ is a unitary basis for the tangent space, we write

$$f_*(\partial/\partial z_1) \wedge \cdots \wedge f_*(\partial/\partial z_n) = Jf(o)\,(e_1 \wedge \cdots \wedge e_n)$$

so that $Jf(o)$ is the length of $f_*(\partial/\partial z_1) \wedge \cdots \wedge f_*(\partial/\partial z_n)$.

THEOREM 2. *Let $f : B \to N$ be a holomorphic mapping of the unit ball B in C^n into a complete hermitian manifold N whose holomorphic sectional curvatures are bounded from above by a negative constant. Let $o = (0, \cdots, 0) \in B$, $x_0 \in N$ and $a > 0$ and assume that f satisfies the conditions $f(o) = x_0$ and $|Jf(o)| \geqq a$. Let $d_N(p, q)$ be the distance function on N coming from the hermitian metric ds_N^2. Then there exist r_0 with $0 < r_0 \leqq 1$ and a strictly increasing upper-semicontinuous function $u(r)$ defined for $0 \leqq r \leqq r_0$ such that*

$$u(\|z\|) \leqq d_N(x_0, f(z))$$

for all $z \in B(r_0) = \{z \in C^n : \|z\| < r_0\}$ and all holomorphic mappings as above.

We will quote two theorems which will be used in the proof.

THEOREM A (KOBAYASHI [2]). *Let N be a complete hermitian manifold whose holomorphic sectional curvatures are bounded from above by a negative constant. Then N is a complete hyperbolic manifold.*

THEOREM B (KOBAYASHI [2]). *Let M be a complex manifold and N a complete hyperbolic manifold. Then the set F of holomorphic mappings from M into N is locally compact with respect to the compact-open topology. For a point p of M and a compact subset K of N, the subset $F(p, K) = \{f \in F : f(p) \in K\}$ of F is compact.*

In particular every sequence $\{f^k\}$ of holomorphic mappings of the ball $B(r)$ of radius r in C^n into a complete hyperbolic manifold N with $f^k(o) = x_0$ contains a subsequence which converges uniformly on every compact subset of $B(r)$.

LEMMA 1. *Let $\{f^k\}$ be a sequence of holomorphic mappings from the unit ball B of C^n to a hyperbolic manifold which converges uniformly to f on compact subsets of B. Further assume that $f^k(o) = x_0$ for all k. Then* $\det(\partial f_i^k/\partial z_j(o))$ *converges to* $\det(\partial f_i/\partial z_j(o))$, *where $f^k = (f_1^k, \cdots, f_n^k)$ is the local expression of f^k in terms of local coordinates $z_1, \cdots, z_n$ at $o \in B$ and $w_1, \cdots, w_n$ at $x_0 \in N$.*

PROOF. Let U denote a local coordinate neighborhood of x_0 with the local coordinates $w_1, \cdots, w_n$. There exist positive numbers r_0, r_1 and r_2 such that

(i) $0 < r_0 < r_0 + r_1 < r_2$, and

(ii) $\{p \in N : d_N(p, x_0) < r_2\} \subseteq U$, where d_N denotes the hyperbolic distance on N [2] (or the hermitian distance as in Theorem 2).

Let t_0 with $0 < t_0 \leqq 1$ satisfy $f(B(t_0)) \subset \{p \in N: d_N(x_0, p) < r_0\}$. There exists an integer K such that, for all $k \geqq K$ and for all $x \in B(t_0)$, $d_N(f^k(x), f(x)) \leqq r_1$. It follows that

$$\begin{aligned} d_N(f^k(x), x_0) &\leqq d_N(f^k(x), f(x)) + d_N(f(x), x_0) \\ &\leqq r_1 + r_0 < r_2 \end{aligned}$$

and $f^k(\overline{B(t_0)}) \subset U$ for all $k \geqq K$. Since uniform convergence of holomorphic functions defined in a domain in C^n entails uniform convergence of the corresponding partial derivatives of all orders, we have

$$\lim_{k\to\infty} \det(\partial f_j^k/\partial z_j(o)) = \det(\partial f_i/\partial z_j(o)).$$

PROOF OF THEOREM 1. If the theorem is false, then there exist a sequence of numbers $\{r_k\}$ tending to infinity and a sequence of holomorphic mappings $f^k : B(r_k) \to N$ with $f^k(o) = x_0$ and $|\det(\partial f_j^k/\partial z_i(o))| \geqq a$, where $f^k = (f_1^k, \cdots, f_n^k)$ is the local expression of f^k in terms of the local coordinates $z_1, \cdots, z_n$ of C^n and $w_1, \cdots, w_n$ at x_0. Define $h^k : B \to N$ of the unit ball B in C^n into N by $h^k(z) = f^k(r_k z)$. Then $h^k(o) = x_0$ and

$$|\det(\partial h_j^k/\partial z_i(o))| = |\det(\partial f_j^k/\partial z_i(o))| \cdot r_k^n \geqq a r_k^n.$$

By Theorem B, there is a subsequence which converges uniformly on compact subsets of B to a holomorphic mapping $h : B \to N$. But this contradicts that $|\det(\partial h_j^k/\partial z_i(o))| \geqq a r_k^n$ as r_k tends to infinity. We remark that the condition on the determinant in the assumption may be replaced by a condition on the Jacobian.

Wu [3] proved the following theorem.

THEOREM C. *Let N be a complete hermitian manifold whose holomorphic sectional curvatures are bounded from above by a negative constant. Let F_a denote the set of all holomorphic mappings $f : B \to N$ of the unit ball about o of C^n into N such that $f(o) = x_0$ and $|Jf(o)| \geqq a > 0$. Then there exists a positive number r, $r < 1$, depending only on a such that, for all $f \in F$, f is univalent on the open ball $B(r)$ of radius r about o of C^n.*

Using Theorems B and C, Professor Griffiths' arguments carry over to yield Theorem 2.

PROOF OF THEOREM 2. We will use the notation F_a to denote the set of mappings satisfying the conditions stated in Theorem 2. If $f \in F_a$, we define

$$u_f(r) = \inf_{\|z\|=r} d_N(x_0, f(z)), \qquad u(r) = \inf_{f\in F_a} u_f(r).$$

It follows immediately that $u(\|z\|) \leqq d_N(x_0, f(z))$ for $0 \leqq \|z\| \leqq 1$. By Theorem

C there exists a constant r_1 with $0 < r_1 < 1$ that such every mapping $f \in F_a$ is univalent on $B(r_1)$. This will show that u_f is continuous and strictly increasing on the interval $0 \leqq r \leqq r_1$. We will show that $u(r)$ is strictly increasing on the interval $[0, r]$. Let $0 \leqq r < r' \leqq r_1$. Let $f^k \in F_a$ and $z_k \in B$ with $\|z_k\| = r'$ be such that

$$\lim_{k\to\infty} d_N(x_0, f^k(z)) = u(r').$$

By passing to a subsequence if necessary, we may assume that $\{f^k\}$ converges uniformly on compact subsets of $B(r_1)$ to a univalent holomorphic mapping $g : B(r_1) \to N$. If we set $\lim_{k\to\infty} z_k = z$, then

$$\|z\| = r' \quad \text{and} \quad u(r) \leqq d_N(x_0, f^k(w)) < d_N(x_0, f^k(z_k))$$

where $\|w\| = r$. Passing to the limit as $k \to \infty$,

$$u(r) = d_N(x_0, g(w)) \leqq d_N(x_0, g(z)) = u(r').$$

Since g is one-to-one, $u(r) < u(r')$.

2. Mappings into hyperbolic manifolds. In this section we given Landau-Schottky's theorem, Bloch's theorem and Koebe's distortion theorem for holomorphic mappings into hyperbolic manifolds (see [2] for the definition).

THEOREM 3 (LANDAU-SCHOTTKY). *Let N be a hyperbolic manifold and $B(r)$ the ball $\{z = (z_1, \cdots, z_n): \sum_{k=1}^n |z_k|^2 < r^2\}$ of radius r in $\mathbf{C}^n$. Let $o = (0, \cdots, 0) \in B(r)$, $x_0 \in N$ and $a > 0$. Then for any holomorphic mapping $f: B(r) \to N$ with $f(o) = x_0$ and $|\det(\partial f_k/\partial z_j(o))| \geqq a$, where $f = (f_1, \cdots, f_n)$ is the local expression of f with respect to the local coordinates $z_1, \cdots, z_n$ of $\mathbf{C}^n$ and $w_1, \cdots, w_n$ of x_0, we have that $r \leqq R$ where $R = R(x_0, a)$ is a constant independent of f.*

Proof of Theorem 3 is similar to the proof of Theorem 1 using the following Lemma 2.

LEMMA 2. *Let $\{f^k\}$ be a sequence of holomorphic mappings of the unit ball B of $\mathbf{C}^n$ into a hyperbolic manifold N such that $f^k(o) = x_0$ for fixed $x_0 \in N$. Then there exists a positive number t, $t < 1$, such that on $B(t)$ a subsequence of $\{f^k\}$ converges uniformly to a holomorphic mapping f of $B(t)$ to N.*

PROOF. Let U be a local coordinate neighborhood of x_0. There exists a positive number t' such that $\{p \in N : d_N(x_0, p) < t'\}$ is contained in U where d_N denotes the hyperbolic distance of N. Since for all k $d_N(f^k(p), f^k(q)) \leqq d_B(p, q)$ where d_B is the hyperbolic distance of B, $f^k(\{p \in B: d_B(o, p) < t'\}) \subset U$. By Theorem B there exists a positive number t, $t \leqq t'$, such that a subsequence of $\{f^k\}$ converges uniformly on $B(t)$ to a holomorphic mapping f of $B(t)$ to N.

Let N be a hyperbolic manifold with the hyperbolic distance d_N and B the unit ball about o in $\mathbf{C}^n$. A univalent ball $\Delta(p, r) = \{q \in N : d_N(p, q) < r\}$ for a holomorphic mapping $f : B \to N$ is by definition a ball of N such that f maps some open set in B biholomorphically onto $\Delta(p, r)$.

THEOREM 4 (BLOCH). *Let N be a hyperbolic manifold and B the unit ball about o in $\mathbf{C}^n$. Let $x_0 \in N$ and $a > 0$. Then for any holomorphic mapping $f: B \to N$ with $f(o) = x_0$ and $|\det(\partial f_k/\partial z_j(o))| \geqq a$, where $f = (f_1, \cdots, f_n)$ is the local expression of f with respect to the local coordinates $z_1, \cdots, z_n$ of $\mathbf{C}^n$ and $w_1, \cdots, w_n$ of x_0, there exists a constant $r = r(x_0, a)$ independent of f such that $\Delta(x_0, r)$ is a univalent ball for all mappings f as above.*

PROOF. Let F denote the set of all mappings $f: B \to N$ satisfying the conditions stated in the theorem. Let U be the local coordinate neighborhood of x_0 with the local coordinates $w_1, \cdots, w_n$. Given $f \in F$ we let $r(f)$ denote the radius of the maximal univalent ball $\Delta(x_0, r(f))$ for f. If the theorem is false, there exists a sequence $\{f^k\}$ $\subset F$ such that $\lim_{k\to\infty} r(f^k) = 0$. By taking a subsequence if necessary, we may assume that $r(f^k) \leqq r'$ for all k where r' is a positive number such that $\Delta(x_0, r')$ $\subset U$. We may further assume by Lemma 2 that the sequence $\{f^k\}$ converges uniformly to a mapping $f \in F$ on compact subsets of $B(t)$ for some $0 < t \leqq 1$. As shown in the proof of Lemma 1, there is t_0, $0 < t_0 \leqq t$, such that

$$f^k(B(t_0)) \subset \Delta(x, r') \quad \text{for all } k, \text{ and}$$
$$f(B(t_0)) \subset \Delta(x_0, r').$$

Since $f \in F$, there is a univalent ball $\Delta = \Delta(x_0, r)$ of radius r for f such that (i) $0 < r \leqq r'$, and (ii) if A is the open set which contains o and which f maps univalently onto Δ, then $\bar{A} \subset B(t_0)$.

Let Δ_k be the maximal univalent ball of radius $r(f^k)$ about x_0 for f^k. Let A_k be the open set in B which contains o and which f^k maps univalently onto Δ_k. Let ε be any positive number with $0 < 2\varepsilon < r$. There exists an integer K such that

(i) $d_N(f(z), f^k(z)) < \varepsilon$ for all $z \in \bar{A}$ and $k \geqq K$, and

(ii) $r(f^k) < r - 2\varepsilon$ for all $k \geqq K$.

Then $\bar{A}_k \subset A$ for all $k \geqq K$. For if not, there is a point z_k in $\bar{A}_k \cap \partial A$ for all $k \geqq K$ and we have

$$\begin{aligned} d_N(f(z_k), x_0) &\leqq d_N(f(z_k), f^k(z_k)) + d_N(f^k(z_k), x_0) \\ &\leqq \varepsilon + r - 2\varepsilon = r - \varepsilon. \end{aligned}$$

On the other hand $d_N(f(z_k), x_0) = r$. This is a contradiction. Therefore $\bar{A}_k \subset A$ for all $k \geqq K$.

Let $k \geqq K$. Since Δ_k is the maximal univalent ball for f^k, ∂A_k must contain a point p_k at which $|\det(\partial f_j^k/\partial z_i(p_k))| = 0$ where the determinants are calculated in terms of the local coordinates of U. It follows that if p is an accumulation point of p_k, $|\det(\partial f_j/\partial z_i(p))| = 0$. Since $p \in A$ and f is univalent on A, this is a contradiction. This proves the theorem.

THEOREM 5. *Let N be a hyperbolic manifold and B the unit ball about o in $\mathbf{C}^n$. Let $x_0 \in N$ and $a > 0$. Then for any holomorphic mapping $f: B \to N$ with $f(o) = x_0$ and $|\det(\partial f_j/\partial z_i(o))| \geqq a$, where $f = (f_1, \cdots, f_n)$ is the local expression of f with respect to the local coordinates $z_1, \cdots, z_n$ of $\mathbf{C}^n$ and $w_1, \cdots, w_n$ of x_0, there exists a* *positive constant $R = R(x_0, a)$ such that the restriction of f to $B(R)$ is one-to-one for all mappings as above.*

PROOF. By Theorem 4 there is a positive constant r such that for $f \in F$, where F denotes the set of all mappings $f: B \to N$ satisfying the conditions stated in the theorem,

(i) $\Delta(x_0, r)$ is a univalent ball for f and

(ii) if A_f is the open set containing o which f maps univalently onto $\Delta(x_0, r)$, then $\bar{A}_f$ is compact.

We will show that $\inf_{f \in F} \operatorname{dist}(o, A_f) > 0$, where $\operatorname{dist}(o, A_f)$ is the euclidean distance from o to the set A_f. If this is false, there exists a sequence $\{f^k\} \subset F$ such that $\operatorname{dist}(o, A_{f^k})$ converges to 0 as $k \to \infty$. Using Lemma 1 and Lemma 2, we may assume, by taking a subsequence if necessary, that $\{f^k\}$ converges to a mapping $f: B(t) \to N$ uniformly on compact subsets of $B(t)$ for some t, $0 < t \leqq 1$. Let $p_k \in \bar{A}_k = \bar{A}_{f^k}$ satisfy

$$\| p_k - o \| = \operatorname{dist}(o, A_k).$$

It follows that p_k converges to 0 and $\lim_{k\to\infty} d_N(f(p_k), x_0) = 0$. Since $d_N(f^k(p_k), x_0) \leqq d_N(f^k(p_k), f(p_k)) + d_N(f(p_k), x_0)$, it can be shown that

$$\lim_{k\to\infty} d_N(f^k(p_k), x_0) = 0.$$

On the other hand $d_N(f^k(p_k), x_0) = r$. This proves the theorem.

THEOREM 6 (KOEBE). *Let N be a hyperbolic manifold and B the unit ball about o in $\mathbf{C}^n$. Let $x_0 \in N$ and $a > 0$. Consider a holomorphic mapping $f: B \to N$ such that $f(o) = x_0$ and $|\det(\partial f_j / \partial z_i(o))| \geqq a$, where $f = (f_1, \cdots, f_n)$ is the local expression of f with respect to the local coordinates $z_1, \cdots, z_n$ of $\mathbf{C}^n$ and $w_1, \cdots, w_n$ of x_0. Then there is a positive constant $r_0 = r_0(x_0, a)$, $r_0 \leqq 1$, and a strictly increasing upper-semi-continuous function $u(r)$ defined for $0 \leqq r \leqq r_0$ such that $u(r) \leqq d_N(x_0, f(z))$ for all $z \in B(r_0)$ and for all f as above.*

Proof of Theorem 6 is the same as the proof of Theorem 2 using Theorems 4 and 5.

REFERENCES

1. P. Griffiths, *Holomorphic mappings into canonical algebraic varieties*, Ann. of Math. (2) **93** (1971), 439–458.
2. S. Kobayashi, *Hyperbolic manifolds and holomorphic mappings*, Dekker, New York, 1970.
3. H. Wu, *Normal families of holomorhphic mappings*, Acta Math. **119** (1967), 193–233.

HOWARD UNIVERSITY

Proceedings of Symposia in Pure Mathematics
Volume 27, 1975

SOME RESULTS IN f-STRUCTURES MOTIVATED BY THE COUSIN PROBLEM

RICHARD S. MILLMAN*

1. Introduction. In his lecture to the International Congress of Mathematicians, 1950, H. Cartan [**1**] gave a formulation, due to Weil, of the second Cousin problem. The formulation is as follows: Given a holomorphic principal fiber bundle ξ which, when considered as a real (C^∞) bundle, is the trivial (or product) bundle, how can we decide whether ξ is also trivial as a holomorphic bundle? By viewing this problem differential geometrically, the author was able to give a sufficient condition for holomorphic triviality which is both necessary and sufficient when the structure group is solvable [**4**]. The key to this method is to notice that if ξ has base space M and structure group G then the total space of ξ must be diffeomorphic to $M \times G$ and must have a complex structure J such that $\pi : M \times G \to M$ and the action of G on $M \times G$ are both holomorphic. To compute J we regarded J as an almost complex structure whose Nijenhuis torsion vanished. It is this last version which generalizes in a natural way to f-structures. We will define a certain kind of f-structure (which we call a Cousin structure) on $M \times G$ (where M is an f-manifold and G is a Lie group with f-structure) which generalizes the above situation. This will also generalize the standard embedding of a complex manifold M into the almost compact manifold $M \times R$.

Let M^n be a smooth (C^∞) n-manifold, $T_m M$ be the tangent space at $m \in M$ and let $\chi(M)$ be the set of all smooth vector fields on M. An *f-structure* on M is a tensor, f_M, of type (1, 1) on M such that (1) $f_M^3(X) + f_M(X) = 0$ for all $X \in \chi(M)$; and (2)

AMS (MOS) subject classifications (1970). Primary 53C15; Secondary 32M05.

*This research supported in part by a Summer Research Award at Southern Illinois University at Carbondale.

f_M has constant nullity on M. An *f-manifold* is a manifold M together with an f-structure. This notion has been studied by Yano and Ishihara [9] among others. If $\gamma : M_1 \to M_2$ then $(\gamma_m)_*$ or $\dot{\gamma}_m$ will denote the differential of γ at $m \in M$. We may occasionally omit the point m if there is no danger of confusion. If M_1 (resp. M_2) is an f-manifold with f-structure f_1 (resp. f_2) then γ is an f-map if $f_2\, \gamma_*(X) = \gamma_*(f_1 X)$ for all $X \in \chi(M)$. An f-structure is integrable if about each point there is a coordinate system in which f has the constant components

$$f = \begin{bmatrix} 0 & -I_p & 0 \\ I_p & 0 & 0 \\ 0 & 0 & 0 \end{bmatrix}$$

where I_p is the $p \times p$ identity matrix (p is one-half the rank of f). In [2], Ishihara and Yano show that the vanishing of the Nijenhuis torsion tensor of f,

$$N(X, Y) = [fX, fY] - f[fX, Y] - f[X, fY] + f^2[X, Y]$$

where $X, Y \in \chi(M)$ is equivalent to the integrability of f. The product of M_1 and M_2 is the f-manifold $M_1 \veebar M_2$ where the f-structure on $M_1 \times M_2$ is defined by $f((X_1, X_2)) = (f_1 X_1, f_2 X_2)$ for $X_1 \in \chi(M_1)$, $X_2 \in \chi(M_2)$. We shall assume for the remainder of the paper that M has an f-structure f_M and the Lie group G has an f-structure f_G. If f is an f-structure on $M \times G$ such that $\pi : M \times G \to M$ (which is projection) and $\alpha : (M \times G) \veebar G \to M \times G$ (which is right action of G on $M \times G$) are f-maps then f is called a *Cousin structure* on $M \times G$.

Let R_λ be right multiplication by $\lambda \in G$. Let $\Lambda^p(M, \hat{G})$ be the space of $\hat{G}$-valued p-forms on M (where $\hat{G}$ is the Lie algebra of G) and for $\omega \in \Lambda^1(M, \hat{G})$ define l: $M \times G \to R$ as

$$l(m, \lambda) = \dim \{(A, B) \in T_{m,\lambda}(M \times G) \,|\, f_M A = 0 \text{ and } f_G B = -\dot{R}_\lambda \omega(A)\}.$$

We say that $\omega \in \Lambda^1(M, \hat{G})$ is an *admissible* 1-*form* if l is a constant function and if for all $m \in M$ and $A \in T_m M$,

$$f_G^2 \omega(A) + f_G \omega(f_M A) + \omega(f_M^2 A) + \omega(A) = 0.$$

The set of admissible forms will be denoted by $\mathfrak{A}$.

f_G is *bi-invariant* if both R_λ (right multiplication by $\lambda \in G$) and L_λ (left multiplication by $\lambda \in G$) are f-maps.

THEOREM A. *$M \times G$ admits a Cousin structure if and only if f_G is bi-invariant. Furthermore, if there is a Cousin structure on $M \times G$, then there is a one-to-one correspondence between Cousin structures on $M \times G$ and admissible 1-forms given as follows: If $\omega \in \mathfrak{A}$ then define f^ω to be the f-structure on $M \times G$ given by: For $m \in M$, $\lambda \in G$, $A \in T_m M$, $B \in T_\lambda G$,*

$$f^\omega_{m,\lambda}(A, B) = (f_M A, f_G B + (R_\lambda)_* \omega(A)).$$

The proof of this appears in [5]. We remark that if both f_M and f_G are complex structures then $\omega \in \mathfrak{A}$ if and only if ω is of type (0, 1). In [5], we also obtain information about the integrability of f^ω.

2. f-Lie groups [7]. Theorem A implies that we should understand bi-invariant f-structures on Lie groups (or more simply *f-Lie groups*). There is, of course, the obvious categorical motivation for this. We shall present two methods of attacking the problem of f-Lie groups. The first method makes use of the Lie algebra of an f-Lie group. The second makes use of a Calabi-Eckmann-Morimoto type construction. The proof of all assertions in this section is contained in [7].

Let $(\ker f)_m = \{X \in T_m M \mid f_m(X) = 0\}$ for $m \in M$ and let $(\operatorname{im} f)_m = \{X \in T_m M \mid X = f_m Y \text{ for some } Y \in T_m M\}$. The rank of f is $\dim(\operatorname{im} f)_m = r$ and the nullity of f is $\dim(\ker f)_m = n_0$. If $r = 0$ then f is called *trivial.* Let $\hat{G}$ be the Lie algebra of G and $g \in G$, $X \in \hat{G}$. As usual we define ad $g : G \to G$ by ad $g(x) = gxg^{-1}$ and Ad $X : \hat{G} \to \hat{G}$ by Ad $X(Y) = [X, Y]$. An f-structure in *bi-invariant* if both left and right multiplication are f-maps. An easy computation shows

PROPOSITION 1. *If f is a bi-invariant f-structure on a Lie group, then $f([X, Y]) = [f(X), Y]$ for all $X, Y \in \hat{G}$.*

The proof of the following corollary is immediate since from Proposition 1 the Nijenhuis torsion of a bi-invariant f-structure must vanish for all left-invariant vector fields on G.

COROLLARY. *A bi-invariant f-structure on a Lie group is intergrable.*

In [7] we also prove

THEOREM B. *Every f-Lie group is covered by the product of a complex Lie group and a Lie group with trivial f-structure.*

It is not true that every f-Lie group is the product of a complex Lie group and a Lie group with trivial f-structure as the following example shows. Let $G = \boldsymbol{C} \times R$ where $\boldsymbol{C}$ is the complex line (considered as a complex manifold) and R is the real line with trivial f-structure. If $D = \{(n + in, n) \mid n \text{ is an integer}\}$, then G/D is an f-Lie group which is not the product of a complex Lie group and an f-Lie group with trivial f-structure. (G/D is of course diffeomorphic to $\boldsymbol{C} \times S^1$ but the f-structure on G/D is not the product f-structure of $\boldsymbol{C} \times S^1$.)

In [6], the author gave a construction which gave to the product of f-manifolds an almost complex structure which generalized the construction of Morimoto [8] on the product of almost contact manifolds. We call this the Calabi-Eckmann-Morimoto construction on the product of f-manifolds. If G is an f-Lie group we shall construct a complex structure on $G \times G$. We refer to [6] for details. Let $\{X_1, \cdots, X_{n_0}\}$ be a global basis for $\ker f$. Since $T_g G = (\ker f)_g \oplus (\operatorname{im} f)_g$ we may define the differential forms η_i $(i = 1, \cdots, n_0)$ on G by $(\eta_i)_m(X) = a_i(m)$ where $X_m = \sum a_i(m) X_i(m) + \tilde{X}_m$ and $\tilde{X}_m \in (\operatorname{im} f)_m$ for $m \in G$.

LEMMA 1. *If $m, g \in G$ then $L_g^* \eta_m = \eta_{gm}$. We say that the f-Lie group G is an f-contact Lie group if $[X_i, X_j] = 0$ for all $1 \leqq i < j \leqq n_0$.*

LEMMA 2. *If G is an f-contact Lie group then $(R_g)_* X_j = X_j$ and $R_g^* \eta_j = \eta_j$ for all $1 \leqq j \leqq n_0$ and $g \in G$.*

If $p, q \in G$ we define an almost complex structure on $G \times G$ by

$$J_{p,q}(X, Y) = \left(f(x) - \sum \eta_i(Y)X_i(p), f(Y) + \sum_i \eta_i(X)X_i(q)\right)$$

where $X \in T_p G$ and $Y \in T_q G$. Using Lemmas 1 and 2 we can prove

THEOREM C. *Every compact connected f-contact Lie group is isomorphic to a torus.*

The requirement that the f-Lie group G be an f-contact Lie group is necessary in Theorem C as the following example shows: Let T^k be a complex torus of complex dimension k. If SO(n) is the usual group of special orthogonal matrices with the trivial f-structure then $G = \mathrm{SO}(n) \times T^k$ is a compact, connected f-Lie group of rank $2k$ and dimension $2k + n(n-1)/2$.

BIBLIOGRAPHY

1. H. Cartan, *Problèmes globaux dans la théorie des fonctions analytiques de plusieurs variables complexes*, Proc. Internat. Congress of Mathematicians (Cambridge, Mass., 1950), vol. 1, Amer. Math. Soc., Providence, R.I., 1952, pp. 152–164. MR **13**, 548.

2. S. Ishihara and K. Yano, *On integrability conditions of a structure f satisfying* $f^3 + f = 0$, Quart. J. Math. Oxford Ser. (2) **15** (1964), 217–222. MR **29** #3991.

3. S. Kobayashi and K. Nomizu, *Foundations of differential geometry*. Vol. II, Interscience Tracts in Pure and Appl. Math., no. 15, vol. 2, Interscience, New York, 1969. MR **38** #6501.

4. R. Millman, *Complex structures on real product bundles with applications to differential geometry*, Trans. Amer. Math. Soc. **166** (1972), 71–99. MR **46** #2086.

5. ———, *On certain types of manifolds with f-structure*, J. Math. Soc. Japan **26** (1974), 83–91.

6. ———, *f-structures with parallelizable kernel on manifolds*, J. Differential Geometry **9** (1974) (to appear).

7. ———, *Groups in the category of f-manifolds*, Fund. Math. (to appear).

8. A. Morimoto, *On normal almost contact structures*, J. Math. Soc. Japan **15** (1963), 420–436. MR **29** #548.

9. K. Yano and S. Ishihara, *Structure defined by f satisfying* $f^3 + f = 0$, Proc. U.S.-Japan Seminar in Differential Geometry (Kyoto, 1965), Nippon Hyoronsha, Tokyo, 1966, pp. 153–166. MR **36** #815.

SOUTHERN ILLINOIS UNIVERSITY AT CARBONDALE

Proceedings of Symposia in Pure Mathematics
Volume 27, 1975

HOLOMORPHIC EQUIVALENCE AND NORMAL FORMS OF HYPERSURFACES

JÜRGEN MOSER*

In the complex space C^{n+1} $(n \geqq 1)$ we consider a real hypersurface Σ of real codimension 1 and investigate the local properties of Σ under biholomorphic transformations of C^{n+1}. If p is a point of Σ and Σ' another such hypersurface and $p' \in \Sigma'$ we call Σ and Σ' equivalent at the points p, p' respectively, if there exists a biholomorphic transformation ϕ near p taking p and Σ into p' and Σ'. Our aim is to find necessary and sufficient conditions for equivalence of two such hypersurfaces. For $n = 0$ the problem concerns curves under conformal mappings and is trivial, at least if one deals with real analytic curves. For $n \geqq 1$ pseudo-convexity, or more generally, the signature of the Levi forms are clearly invariant. However, for $n \geqq 1$ there are many more invariant properties as we will show.

For $n = 1$ this problem was studied in two profound papers [**2**] by E. Cartan after this question had been raised already by H. Poincaré in 1907 (see [**1**]). Cartan's approach consisted in setting up structure equations for a set of differential forms which gave rise to a connection and invariant curvature forms. In a more recent paper [**3**], N. Tanaka attempted a generalization of Cartan's work for $n \geqq 2$. However, his results are restricted to hypersurfaces which he called regular,[1] and this constitutes in fact a severe limitation.

To attack this equivalence problem we subject Σ to holomorphic mappings ϕ taking p into the origin and Σ into a normal form. It turns out that this normal form osculates a hyperquadric Σ_0 to high order and, moreover, this osculation takes place along distinguished curves. This family of curves satisfies a system of

AMS (MOS) subject classifications (1970). Primary 32C05, 32H19.

*Partially supported by grant AFOSR-71-2055.

[1]See [**3**, p. 407].

second order differential equations on Σ and is holomorphically invariantly associated with Σ.

We formulate our results first for real analytic hypersurfaces. Let z^α ($\alpha = 1,2,\cdots, n+1$) denote the coordinates in C^{n+1}; we single out

$$z^{n+1} = w = u + iv,$$

and set $z = (z^1, z^2, \cdots, z^n)$. After an appropriate linear coordinate change Σ can be written as

$$v = F(z, \bar{z}, u) \tag{1}$$

where F is real analytic and vanishes with its first derivatives at the origin. The basic assumption for the following is that the Levi form

$$\langle z, \bar{z}\rangle = \sum_{\alpha,\bar{\beta}=1}^{n} g_{\alpha\bar{\beta}}\, z^\alpha z^{\bar{\beta}}, \qquad g_{\alpha\bar{\beta}} = \frac{\partial^2 F}{\partial z^\alpha\, \partial z^{\bar{\beta}}}$$

is nondegenerate at the origin.

We subject (1) to a holomorphic mapping

$$\phi\begin{pmatrix} z \\ w \end{pmatrix} \to \begin{pmatrix} f(z, w) \\ g(z, w) \end{pmatrix} \tag{2}$$

which leaves the origin fixed and has nonvanishing Jacobian there. We will try to choose ϕ so that the resulting hypersurface osculates the hyperquadric Σ_0, $v = \langle z, \bar{z}\rangle$ to a high degree. Therefore it is important to know first the mappings preserving Σ_0 and we list some of the main properties of Σ_0 needed.

(i) Σ_0 is a Lie group with the group operation

$$(z_1, w_1) \oplus (z_2, w_2) = (z_1 + z_2, w_1 + w_2 + 2i\langle z_1, \bar{z}_2\rangle)$$

for $(z_k, w_k) \in \Sigma_0$ ($k = 1,2$). This gives rise to linear mappings

$$\begin{pmatrix} z \\ w \end{pmatrix} \to \begin{pmatrix} z + b \\ w + 2i\langle \bar{b}, z\rangle + s + i\langle b, \bar{b}\rangle \end{pmatrix}, \tag{3}$$

where $b \in C^n$, $s \in R^1$ if we write $b = z_2$; $s + i\langle b, \bar{b}\rangle = w_2$.

(ii) The isotropy group of Σ_0 consisting of the holomorphic mappings leaving Σ_0 and the origin fixed consists of fractional linear mappings of the form

$$\begin{pmatrix} z \\ w \end{pmatrix} = \begin{pmatrix} C(z + aw)\delta^{-1} \\ pw\delta^{-1} \end{pmatrix} \tag{4}$$

where

$$\delta = 1 - 2i\langle z, \bar{a}\rangle - (r + i\langle a, \bar{a}\rangle)w; \qquad \langle Cz, C\bar{z}\rangle = p\langle z, \bar{z}\rangle,$$
$$0 \neq p \in R^1, r \in R^1, a \in C^n.$$

This group G_0 has $(n+1)^2 + 1$ real dimensions. The full group G of holomorphic mappings preserving Σ_0 is generated by (3) and (4). To describe these groups appropriately we use projective coordinates. On Σ_0 we can describe a family of curves, which is invariant under G, by intersecting Σ_0 with all complex lines which intersect the complex tangent spaces of Σ_0 transversally. These curves, which are circles or lines, will be called chains.

In our aim to construct a normal form we will factor out mappings $\phi_0 \in G_0$. We may assume that F in (1) differs from $\langle z, \bar{z}\rangle$ only by cubic terms and therefore will consider only holomorphic mappings ψ which preserve the origin and this form of Σ. From (4) it is easily seen that any such holomorphic map ψ can be uniquely factored as $\psi = \phi \cdot \phi_0$ with $\phi_0 \in G_0$ and ϕ satisfying

$$(5) \qquad \phi(0) = 0, \quad d\phi(0) = \mathrm{id}, \quad \mathrm{Re}(\partial^2 g(0, 0)/\partial w^2) = 0$$

when written in the form (2). In the following we will impose this normalization (5) on ϕ.

To formulate the result we consider the space F of functions $F(z, \bar{z}, u)$, real analytic at the origin and write

$$F = \sum_{k,l=0}^{\infty} F_{kl}(z, \bar{z}, u)$$

where $F_{kl}(\lambda z, \mu\bar{z}, u) = \lambda^k \mu^l F_{kl}(z, \bar{z}, u)$. Thus F_{kl} are polynomials in z, $\bar{z}$ with analytic coefficients in u satisfying

$$\overline{F_{kl}(z, \bar{z}, u)} = F_{kl}(z, \bar{z}, \bar{u}).$$

We call (k, l) the type of F_{kl}.

Using the notation of tensor calculus, we define the contraction F_{kl}^* of F_{kl} with respect to the Hermitian from $\langle z, \bar{z}\rangle$: If $k, l \geqq 1$ and

$$F_{kl} = \sum a_{\alpha_1, \cdots, \alpha_k\, \bar{\beta}_1, \cdots, \bar{\beta}_l} z^{\alpha_1} \cdots z^{\alpha_k} z^{\bar{\beta}_1} \cdots z^{\bar{\beta}_l}$$

where we assume that the coefficients are symmetric with respect to permutations of the $\alpha_1, \cdots, \alpha_k$ and the $\bar{\beta}_1, \cdots, \bar{\beta}_l$, we set

$$F_{kl}^* = \sum b_{\alpha_1, \cdots, \alpha_{k-1}\, \bar{\beta}_1, \cdots, \bar{\beta}_{l-1}} z^{\alpha_1} \cdots z^{\alpha_{k-1}} z^{\bar{\beta}_1} \cdots z^{\bar{\beta}_{l-1}}$$

with $b_{\alpha_1, \cdots, \alpha_{k-1}\, \bar{\beta}_1, \cdots, \bar{\beta}_{l-1}} = \sum_{\alpha_k, \bar{\beta}_l} g^{\alpha_k \bar{\beta}_l} a_{\alpha_1, \cdots, \alpha_k\, \bar{\beta}_1, \cdots, \bar{\beta}_l}$.

We introduce the following decomposition of F: $\mathscr{F} = \mathscr{R} + \mathscr{N}$, where $\mathscr{R}$ consists of series of the types

$$R = \sum_{\min(k,l) \leqq 1} R_{kl} + G_{11}\langle z, \bar{z}\rangle + (G_{10} + G_{01})\langle z, \bar{z}\rangle^2 + G_{00}\langle z, \bar{z}\rangle^3,$$

G_{jm} being of type (j, m) and

$$\mathscr{N} = \{N \in \mathscr{F}, N_{kl} = 0\ \min(k, l) \leqq 1;\ N_{22}^* = N_{32}^{**} = N_{33}^{***} = 0\}.$$

One verifies that this is indeed a decomposition with $\mathscr{R} \cap \mathscr{F} = (0)$ and we can define the projection P of $\mathscr{F}$ into $\mathscr{R}$, so that $\mathscr{R}$ is the range and $\mathscr{N}$ the null space of P. We remark that $F \cdot \mathscr{N} \in \mathscr{N}$ for every $F \in \mathscr{F}$, i.e. $\mathscr{N}$ is an ideal in $\mathscr{F}$ under multiplication, a fact, which is needed in the convergence proof of the following theorem.

THEOREM. *If Σ is a nondegenerate real analytic hypersurface represented by* (1) *then there exists a unique holomorphic mapping ϕ of the form* (2) *satisfying* (5) *such that the transformed hypersurface has the form*

$$v = \langle z, \bar{z} \rangle + N(z, \bar{z}, u) \quad \textit{with } N \in \mathcal{N}.$$

We call this the normal form of Σ.

The same result holds for formal power series expansions and the uniqueness asserts, in particular, that the only formally holomorphic transformation preserving Σ_0 and the origin are those of G_0.

We note that $N(0, 0, u) = 0$ so that $\phi(\Sigma)$ and Σ_0 have the u-axis λ in common. Moreover,

$$0 \neq N_{kl} \in \mathcal{N} \quad \text{implies} \quad \begin{cases} k + l \geqq 6 & \text{for } n = 1, \\ k + l \geqq 4 & \text{for } n \geqq 2, \end{cases}$$

which shows that Σ can be osculated by the image $\phi^{-1}(\Sigma_0)$ of the hyperquadric to order 5 or 3, inclusively, for $n = 1$, $n \geqq 2$, respectively, and the osculation takes place along a curve $\phi^{-1}(\lambda)$. On the other hand the osculation is, in general, not closer. If, however, $N_{42} = 0$ for $n = 1$ or $N_{22} = 0$ for $n \geqq 2$ we speak of an 'umbilical point". The property of a point to be umbilical is actually independent of our normalization and therefore has an invariant meaning.

The set of curves $\phi^{-1}(\lambda)$, along which the osculation between Σ and $\phi^{-1}(\Sigma_0)$ takes place, are transversal to the complex tangent planes and are described by a system of second order differential equations. Through every point $p \in \Sigma$ and any real line in the real tangent space $T_p(\Sigma)$, but transversal to the complex tangent space, there exists a unique curve of this type. This system of second order differential equations is invariant under holomorphic transformation and is meaningful even if Σ admits only a finite number of derivatives. For Σ_0 these curves agree with the chains introduced above.

This result can be viewed as an analogue of Levi-Civita's parallelism in classical differential geometry. The above curves correspond to the geodesics. There is also an analogue to frames at each point $p \in \Sigma$, and these frames can be identified with the elements $\phi_0 \in G_0$. The study of this parallelism is still underway in joint work with S. S. Chern who developed an alternate intrinsic approach by generalizing Cartan's structure equations.

ADDED IN PROOF. Details and proofs for this note as well as Chern's derivation of the structure equations are included in the forthcoming paper: S. S. Chern and J. K. Moser, *Real hypersurfaces in complex manifolds*, Acta Math. (1974) (to appear).

REFERENCES

1. H. Poincaré, *Les fonctions analytiques de deux variables et la représentation conforme*, Rend. Circ. Mat. Palermo (1907), 185–220.

2. E. Cartan, *Sur la géométrie-pseudo conforme des hypersurfaces de deux variables complexes.* I, Ann. Mat. Pura Appl. (4) **11** (1932), 17–90; Oeuvres II, 2, 1231–1304; II, Ann. Scuola Norm. Sup. Pisa (2) **1** (1932), 333–354; Oeuvres III, 2, 1217–1238.

3. N. Tanaka, *On the pseudo-conformal geometry of hypersurfaces of the space of n complex variables*, J. Math. Soc. Japan **14** (1962), 397–429. MR **26** #3086.

COURANT INSTITUTE OF MATHEMATICAL SCIENCES

Proceedings of Symposia in Pure Mathematics
Volume 27, 1975

ON COMPACT KÄHLER MANIFOLDS WITH POSITIVE HOLOMORPHIC BISECTIONAL CURVATURE

TAKUSHIRO OCHIAI*

1. Introduction. Let (M, g) be a compact Kähler manifold of dimension m. If p and p' are planes in the tangent space each invariant with respect to the almost complex structure J, then the (*holomorphic*) *bisectional curvature* $H(p, p')$ is defined by

$$H(p\ p') = R(X, JX, Y, JY) \tag{1.1}$$

where R is the Riemann curvature tensor and X (resp. Y) is a unit vector in the plane p (resp. p'). The above notion has been introduced in Goldberg and Kobayashi [**1**]. When M is a complex projective space $P^m(\boldsymbol{C})$ and g is a Fubini-Study metric with constant holomorphic sectional curvature 1, $H(p, p')$ satisfies the inequality

$$\tfrac{1}{2} \leqq H(p, p') \leqq 1. \tag{1.2}$$

Thus $P^m(\boldsymbol{C})$ admits a Kähler metric with everywhere positive bisectional curvature. Up to date we do not know such an example of M other than $P^m(\boldsymbol{C})$. In fact we have rather substantial supporting evidence to suspect that the converse should be true. So we propose here the following.

CONJECTURE A. *If an m-dimensional compact complex manifold M admits a Kähler*

AMS (MOS) subject classifications (1970). Primary 53C55; Secondary 32C15.

Key words and phrases. Kähler manifolds, holomorphic bisectional curvature, ample bundles, ample tangent bundles.

*Supported partially by the Sakokai Foundation, Tokyo, Japan.

metric g with everywhere positive bisectional curvature, then M is biholomorphic to a complex projective space $P^m(\boldsymbol{C})$ of dimension m.

Today we can only verify Conjecture A for a few but interesting cases. The purpose of this note is to report what is known so far and what needs to be done. This note grows out of the joint work by Kobayashi and myself **[1]**, **[2]**, **[3]**, **[4]**. Before finishing this section we have to remark that Bianchi's identity implies

$$H(p, p') = R(X, Y, X, Y) + R(X, JY, X, JY) \tag{1.3}$$

so that $H(p, p')$ is essentially the sum of two sectional curvatures. It follows that everywhere positive *sectional curvature* implies everywhere positive *bisectional curvature.*

2. Let (M, g) be an m-dimensional compact Kähler manifold with everywhere positive bisectional curvature. We know the Ricci tensor S can be expressed as

$$S(Y, Y) = \sum_{i=1}^{m} R(X_i, JX_i, Y, JY) \tag{2.1}$$

where we choose X_i's so that $\{X_1, \cdots, X_m, JX_1, \cdots, JX_m\}$ is an orthonormal basis of the tangent space. Thus we have

FACT 1. *The Ricci tensor is positive definite everywhere.*

If we denote by $K(M)$ the canonical line bundle of M, Fact 1 implies that $K(M)$ is negative in the sense of Kodaira **[1]**. Hence Kodaira's vanishing theorem **[1]** implies

FACT 2. (i) *Any holomorphic k-form* $(1 \leqq k \leqq m)$ *is zero. In particular the arithmetic genus of M is* 1. (ii) *The pluri-genus* $p_l = \dim H^0(M, K(M)^l)$ *is zero for any* $l > 0$.

(i) is originally due to Bochner **[1]**. On the other hand the well-known theorem of Myers **[1]** tells us that Fact 1 implies the finiteness of the fundamental group $\pi_1(M)$ of M. If we observe the Riemann-Roch-Hirzebruch formula for the arithmetic genus (Hirzebruch **[1]**), from (i) of Fact 2 as well as the remark above, we will see

FACT 3. *M is simply connected.*

The above is due to Kobayashi **[1]**. Now Bishop and Goldberg **[1]** (cf. Goldberg and Kobayashi **[1]**) proved the following remarkable

FACT 4. *The second Betti number of M is* 1.

Bishop and Goldberg proved Fact 4 under the assumption of everywhere positivity of sectional curvature. Later Goldberg and Kobayashi generalized it to this case. It is Frankel **[1]** (cf. Goldberg and Kobayashi **[1]**) who proved the following geometrically interesting

FACT 5. *Let V and W be closed complex submanifolds of M such that* $\dim V + \dim W \geqq m$. *Then* $V \cap W \neq \emptyset$.

Finally it is almost unnecessary to note that from the celebrated theorem of Kodaira **[2]** Fact 1 implies

FACT 6. *M is an algebraic manifold.*

Now let us pause for a moment to digress to the special case $m = 1,2$. In the case $m = 1$, (i) of Fact 2 tells us that the genus of M is zero. Hence M should be biholomorphic to $P^1(\boldsymbol{C})$. In the case $m = 2$, first of all, Fact 4 implies M is minimal (i.e., M cannot be blown down.) The criterion of Castelnuovo (cf. Šafarevič **[1]**) asserts that a simply connected algebraic surface with $p_l = 0$ for $l > 0$ is rational. Thus from Fact 6 and (ii) of Fact 2 we see M is a rational surface. Then the classification of minimal rational surfaces (e.g., Šafarevič **[1]**) informs us that M should be biholomorphic either to $P^2(\boldsymbol{C})$ or to a ruled surface. Since a ruled surface is by definition fibred over a curve, the application of Fact 5 to different two fibres implies that M cannot be a ruled surface. In summing up we have

THEOREM 1. *In the case* $m = 1,2$, *Conjecture* A *is true.*

The above is due to Frankel **[1]** (see also Goldberg and Kobayashi **[1]**).

REMARK. Howard and Smyth **[1]** classified all the compact Kähler surfaces with everywhere nonnegative bisectional curvature. For complex dimension 2, it follows from their results that everywhere nonnegative bisectional curvature is equivalent to everywhere nonnegative sectional curvature.

3. We need to put ourselves in a more general setting. Thus let $\pi : E \to M$ be a holomorphic complex vector bundle over M of fibre dimension r. Let $E - 0$ denote the bundle space E minus its zero section. We denote by $P(E)$ the quotient $(E - 0)/\boldsymbol{C}^*$ by the multiplicative group $\boldsymbol{C}^*$ of nonzero complex numbers acting on $E - 0$. Clearly, $E - 0$ is a holomorphic principal bundle over $P(E)$ with group $\boldsymbol{C}^*$. Let $L(E)$ be the associated complex line bundle over $P(E)$. If M is a point, i.e., E is just a complex vector space, $P(E)$ is the projective space of E and $L(E)$ is the so-called tautological line bundle. In the case $r = 1$, $P(E)$ is identical to M and $L(E) = E$.

According to Hartshorne **[1]**,[1] a holomorphic complex vector bundle $\pi : E \to M$ is called an *ample bundle* if $L(E^*)$ is negative in the sense of Kodaira **[1]** where E^* is the dual bundle of E. The reason why this notation of ampleness is relevant to Conjecture A is given by

THEOREM 2. *If M satisfies the assumption in Conjecture* A, *then the holomorphic tangent bundle of M is ample.*

The above theorem is proved in Kobayashi and Ochiai **[1]** as well as in Griffiths **[1]**. At this moment we have no idea whether the converse of Theorem 2 is true or not. As explained later, the ampleness behaves rather nicely in algebraic operations of vector bundles. Thus it seems to be natural, as Hartshorne **[2]** has proposed, to set the following (formally stronger than Conjecture A).

---

[1]Accidentally we have adopted different notation from Hartshorne **[1]**. Our $P(E)$ is written as $P(E^*)$ in Hartshorne **[1]**. Thus our $P(E^*)$ corresponds to $P(E^{**}) = P(E)$ in his notation.

CONJECTURE B. *Let M be a compact complex manifold of dimension m. If the holomorphic tangent bundle of M is ample, then M is biholomorphic to $P^m(\boldsymbol{C})$.*

Let us list some of the fundamental properties of ample bundles proved by Hartshorne [**1**].

FACT 7. (i) *If E is ample, then any quotient bundle of E is also ample.*
(ii) *If $E = E_1 \oplus E_2$, then E is ample if and only if both E_1 and E_2 are ample.*
(iii) *If E_1 and E_2 are ample, then $E_1 \otimes E_2$ is also ample.*
(iv) *If E is an ample bundle of fibre dimension r, then the exterior powers $\wedge^k E$ $(1 \leqq k \leqq r)$ are ample.*
(v) *If E is ample and W is a closed complex submanifold of M, then the restriction of E to W is ample.*
(vi) *If E is ample, then E can be a holomorphically trivial bundle only if M is a point.*

From (iv) above we have

FACT 8. *Let M be as in Conjecture* B.
(i) *The canonical bundle of M is negative.*
(ii) *In particular M is an algebraic manifold.*

As in the proof of Theorem 1, we can prove (as is done in Hartshorne [**2**])

THEOREM 1′. *Conjecture* B *is true in the case $m = 1$ or 2.*

Suppose now there is a holomorphic fibration $\tau : M \to N$. We denote by F the bundle tangent along the fibre. Then we have the following short exact sequence

$$0 \to F \to T(M) \to \tau^*T(N) \to 0 \tag{3.1}$$

where $T(M)$ (resp. $T(N)$) is a holomorphic tangent bundle of M (resp. N) and $\tau^*T(N)$ is the pull-back of $T(N)$ with respect to τ. Now assume $T(M)$ is ample. From (i) of Fact 7 we see $\tau^* T(N)$ is also ample. Take any point $Z \in N$. From (v) of Fact 7, the restriction of $\tau^* T(N)$ to the fibre $\tau^{-1}(Z)$ over Z is again ample. Clearly the restricted bundle is holomorphically trivial. Thus from (vi) of Fact 7 we conclude that $\tau^{-1}(Z)$ is discrete.

FACT 9. *Let M satisfy the assumption in Conjecture* B. *If there exists a holomorphic fibration $\tau : M \to N$, then each fibre is discrete.*

The following has essentially appeared in Hartshorne [**2**].

PROPOSITION 1. *Let M be as in Conjecture* B. *Suppose $P^1(\boldsymbol{C})$ is holomorphically imbedded in M. Then $C_1(M)[P^1(\boldsymbol{C})] \geqq m + 1$ where $C_1(M)$ is the first Chern class of M.*

In fact, from (v) of Fact 7, the restriction $T(M)|_{P^1(C)}$ of $T(M)$ to $P^1(\boldsymbol{C})$ is ample. We define the normal bundle N of $P^1(\boldsymbol{C})$ by the short exact sequence

$$0 \to T(P^1(\boldsymbol{C})) \to T(M)|_{P^1(C)} \to N \to 0. \tag{3.2}$$

Then from (i) of Fact 7, N is ample. On the other hand, the well-known theorem of Grothendieck [**1**] asserts that every holomorphic vector bundle over $P^1(\boldsymbol{C})$ can be written holomorphically as a sum of line bundles. Thus we have

$$N = L_1 \oplus \cdots \oplus L_{m-1} \tag{3.3}$$

for some holomorphic line bundles L_j $(1 \leqq j \leqq m-1)$. The property (ii) of Fact 7 implies that each L_j $(1 \leqq j \leqq m-1)$ should be ample (i. e., positive). We denote by d_j the Chern class of L_j $(1 \leqq j \leqq m-1)$. Since L_j is positive we have $d_j[P^1(\boldsymbol{C})] \geqq 1$. From the standard fact on Chern classes we have

$$C_1(M)[P^1(\boldsymbol{C})] = C_1(P^1(\boldsymbol{C}))[P^1(\boldsymbol{C})] + \sum_{j=1}^{m-1} d_j[P^1(\boldsymbol{C})] \geqq 2 + (m-1) = m+1.$$

As a direct application of Fact 9 and Proposition 1 we will see that Conjecture B is true if we assume M is homogeneous. So let M be as in Conjecture B. Furthermore, assume the group Aut (M) of all holomorphic diffeomorphisms of M acts on M transitively. Then the classification theorem of compact homogeneous complex manifolds with Kähler metric due to Wang [**1**] and Borel and Remmert [**1**] asserts that such M should be written as

$$M = A \times M_1 \times \cdots \times M_k \tag{3.4}$$

where A is a complex torus and M_j $(1 \leqq j \leqq k)$ is a quotient space of a complex simple Lie group. Since the holomorphic tangent bundle of A is holomorphically trivial, (vi) of Fact 7 and Fact 9 imply that M should be of the form G/P where G is a complex simple Lie group. We know all the possible pairs of G and P. Furthermore, we know from Fact 9 that P is a maximal complex subgroup. Then we know that $H_2(M, \boldsymbol{Z}) = \boldsymbol{Z}$ and $P^1(\boldsymbol{C})$ is imbedded in M as a generator of $H_2(M, \boldsymbol{Z})$ (cf. Borel and Hirzebruch [**1**]). Since everything is computable, Proposition 1 now implies that the only possibility is when M is $P^m(\boldsymbol{C})$.

THEOREM 3. *Conjecture* B *is true in the case M is holomorphically homogeneous.*

For further details, see Kobayashi and Ochiai [**2**].

Now the following seems to be more or less well known.

PROPOSITION 2. *Let E and F be holomorphic complex vector bundles over M. Then for $0 \leqq j \leqq m$ we have*

$$H^j(M, F \otimes S^k(E^*)) = H^j(P(E), \rho^*F \otimes L(E)^{-k})$$

for $k = 0, 1, 2, \cdots$ where $S^k(E^*)$ denotes the kth symmetric tensor power of the dual E^* of E and $\rho^* F$ the pull-back bundle of F by the projection $\rho : P(E) \to M$.

For the proof see Kobayashi and Ochiai [**1**] or Griffiths [**1**]. We need to compute the canonical bundle $K(P(E))$ of $P(E)$. That is given by

PROPOSITION 3. *Let E be a holomorphic complex vector bundle over M with fibre dimension r; then*

$$K(P(E)) = L(E)^r \cdot \rho^*(K(M) \cdot \wedge^r E^*)$$

where E^* is dual to E, $K(M)$ is the canonical bundle of M and $\rho : P(E) \to M$ is the projection.

For the proof see Kobayashi and Ochiai **[1]**. The combination of Proposition 2 and Proposition 3 gives us the following

THEOREM 4. *Let M be as in Conjecture* B; *then we have $H^j(M, S^kT) = 0$ for $j \geqq 1$ and $k = 0, 1, 2, \cdots$, where T is the holomorphic tangent bundle of M.*

In fact from Proposition 3 we have $K(P(T^*)) = L(T^*)^m$. Since T is ample, $K(P(T^*))$ is negative. From Proposition 2 we have

$$H^j(M, S^kT) = H^j(P(T^*), L(T^*)^{-k}) \quad \text{for } j \geqq 0,\ k \geqq 0. \tag{3.5}$$

By Serre's duality we have

$$\begin{aligned} H^j(P(T^*), L(T^*)^{-k}) &= H^{2m-1-j}(P(T^*), K(P(T^*)) \cdot L(T^*)^k) \\ &= H^{2m-1-j}(P(T^*), L(T^*)^{m+k}). \end{aligned}$$

Since $L(T^*)$ is negative, Kodaira's vanishing theorem tells us that $H^{2m-1-j}(P(T^*), L(T^*)^{m+k}) = 0$ unless $j = 0$. Thus we see $H^j(M, S^k T) = 0$ for $j \geqq 1$ and $k \geqq 0$.

Suppose now M is obtained as a complete intersection of r hypersurfaces of degree $a_1, \cdots, a_r$ in $P^{m+r}(\boldsymbol{C})$ with $a_i \geqq 2$ and $m \geqq 2$. Using the Riemann-Roch-Hirzebruch formula carefully we can estimate dim $H^1(M, T)$. The result is stated in Kobayashi and Ochiai **[1]** as

PROPOSITION 4. *$H^1(M, T) \neq 0$ except for the case $r = 1$ and $a_1 = 2$ (i.e., the case where M is a quadratic in $P^{m+1}(\boldsymbol{C})$).*

We remark that a quadratic in $P^{m+1}(\boldsymbol{C})$ is holomorphically homogeneous. Thus combining Proposition 4 with Theorem 3, we obtain

THEOREM 5. *Conjecture* B *is true in the case M is obtained as a complete intersection of hypersurfaces in some complex projective space.*

Another good property of ample bundles is given in Bloch and Gieseker **[1]**.

FACT 10. *Let E be an ample bundle over M with fibre dimension r, and $W \subset M$ an analytic subspace of dimension $k \leqq r$. Then $C_k(E)[W] > 0$ where $C_k(E)$ is the kth Chern class of E.*

As a corollary to the above we have

FACT 11. *Let M be as in Conjecture* B. *Then the Euler number of M is positive.*

REMARK. Using the same type of argument as in Theorem 4, J. Le Potier **[1]** has shown that if E is an ample bundle over M of fibre dimension r, then $H^{p,q}(M, E) = 0$ for $p + q \geqq m + r$ where $H^{p,q}(M, E)$ is the cohomology of type (p, q) with the value in E. For further discussion of ample bundles see Griffiths **[1]**.

4. Hirzebruch and Kodaira [1] have given a characterization of the complex projective space. A similar characterization for the complex hyperquadratics has been given by Brieskorn [1]. We can give a slightly different characterization for them (cf. Kobayashi and Ochiai [3]). We state it here only in the form suited for the later purpose.

THEOREM 6. *Let M be an m-dimensional compact complex manifold with a positive line bundle F. If we have*

$$C_1(M) \geqq (m+1)\, C_1(F), \tag{4.1}$$

then M is biholomorphic to $P^m(\mathbf{C})$.

Similarly we obtain

THEOREM 7. *Let M be an m-dimensional compact complex manifold with a positive line bundle F. If we have*

$$C_1(M) = m\, C_1(F), \tag{4.2}$$

then M is biholomorphic to a hyperquadratic in $P^{m+1}(\mathbf{C})$.

The proof for Theorem 6, roughly speaking, goes as follows. First, using first Hilbert's polynomial (cf. Hirzebruch [1]) it is shown that $\dim H^0(M, F) = m + 1$ and $C_1(F)^m [M] = 1$. The latter implies F has no base point. Thus we have a well-defined holomorphic mapping from M into $P^m(C)$ by the use of F. It should be easy to see this mapping is biholomorphic.

PROPOSITION 5. *Let M be as in Conjecture* B. *Further assume $m = 3$. Then the complex dimension of the group* Aut(M) *of all holomorphic diffeomorphisms of M is at least* 7.

OUTLINE OF PROOF. Let C_j $(1 \leqq j \leqq 3)$ be the Chern classes of M, and g be the Chern class of $L(T^*)^{-1}$ where T^* denotes the dual to the holomorphic tangent bundle T of M. The well-known formula says

$$g^3 - C_1 g^2 + C_2 g - C_3 = 0 \tag{4.3}$$

under the identification $C_j = \tau^* C_j$, where $\tau \colon P(T^*) \to M$ is the projection. Integrating along the fibre, the formula (4.3) implies

$$g^5 [P(T^*)] = (C_1^3 - 2C_1 C_2 + C_3)\, [M]. \tag{4.4}$$

Since $L(T^*)$ is negative, we have $g^5 [P(T^*)] > 0$. Thus we have

$$(C_1^3 - 2C_1 C_2 + C_3)\, [M] > 0. \tag{4.5}$$

On the other hand the Riemann-Roch-Hirzebruch formula tells us

$$\sum_{j=0}^{3} (-1)^j \dim H^j(M, T) = \{\tfrac{1}{2}(C_1^3 - 2C_1 C_2 + C_3) + 5\, C_1 C_2/24\}[M], \tag{4.6}$$

and

$$\text{(4.7)} \qquad \text{the arithmetic genus} = C_1 C_2[M]/24.$$

From (i) of Fact 8 we know the arithmetic genus is 1. Thus from Theorem 4, (4.6), (4.5) and (4.7) we have

$$\text{(4.8)} \qquad \dim H^0(M, T) = \{\tfrac{1}{2}(C_1^3 - 2C_1 C_2 + C_3) + 5\, C_1 C_2/24\}\,[M] \geqq 6.$$

A more careful estimation gives us $\dim H^0(M, T) \geqq 7$ (see Kobayashi and Ochiai [4]).

THEOREM 8. *Let M be as in Conjecture* B. *Assume further* (i) $m = 3$, (ii) *the second Betti number is* 1, *and* (iii) *the group* Aut(M) *of all the holomorphic diffeomorphisms of M contains a compact subgroup K of real dimension* $\geqq 7$. *Then M is biholomorphic to* $P^3(M)$.

OUTLINE OF PROOF. Let α be the positive generator of the image of $H^2(M, \mathbf{Z})$ in $H^2(M, \mathbf{R})$. Then $C_1(M)$ can be written for some positive integer l as

$$\text{(4.9)} \qquad C_1(M) = l\alpha.$$

From Theorems 6 and 7 we may assume $l = 1$ or 2. Now it is well known that the upper bound for the real dimension of K is 15. From the classification theorem of compact Lie groups, we have only a few possibilities for K. Thus it is possible to examine in detail the possible orbit structure of K in M. By checking everything carefully we conclude that the case $l = 1$ or 2 cannot occur. For full details see Kobayashi and Ochiai [4].

At this stage we have to raise another

CONJECTURE C. *If M satisfies the assumption of Conjecture* A, *then there exists at least one Einstein-Kähler metric on M.*

Aubin [1] seems to have succeeded more or less in proving this conjecture. The theorem of Matsushima [1] asserts that if M admits an Einstein-Kähler metric, then the group Aut(M) is the complexification of a compact subgroup. Combining Proposition 5 and Theorem 8 with the remark above, we obtain

THEOREM 9. *Let M be a compact complex manifold of dimension* 3. *Assume M satisfies the assumption in Conjecture* A. *Then M is biholomorphic to* $P^3(M)$, *provided Conjecture* C *is true.*

Unfortunately, Theorem 8 or Theorem 9 is the farthest we can go for three-dimensional cases. In view of Aubin's paper [1], it appears to be not so difficult to prove Conjecture C. At this moment there is some gap in Theorem 7.1 of Kobayashi and Ochiai [4] which states that Conjecture A is true in the case $m = 3$. We hope we can fill that gap in the near future.

5. Berger [1] proved that an m-dimensional compact Kähler-Einstein manifold M with everywhere positive sectional curvature is isometric to $P^m(\mathbf{C})$ with a Fubini-

Study metric. Goldberg and Kobayashi [1] generalized this to the case of everywhere positive bisectional curvature. For a compact Kähler manifold with constant scalar curvature, the Ricci form is harmonic. From Fact 4 it follows that a compact Kähler manifold with constant scalar curvature and with everywhere positive bisectional curvature is an Einstein manifold. Hence,

FACT 12. *An m-dimensional compact Kähler manifold with constant scalar curvature and with everywhere positive bisectional curvature is holomorphically isometric to* $P^m(C)$ *with a Fubini-Study metric.*

Let (M, g) be a compact Kähler manifold of dimension m. As before we denote by J the canonically associated almost complex structure, and by R the Riemann curvature tensor. In addition to the notion of sectional curvature and holomorphic sectional curvature, we define the Kählerian sectional curvature of M as follows. Let p be a plane in the tangent space. Let X and Y be an orthonormal basis of p. Denote by $K(p)$ the sectional curvature of M, i.e.,

$$K(p) = R(X, Y, X, Y). \tag{5.1}$$

The *Kählerian sectional curvature* of the plane p is defined by

$$K^*(p) = \frac{4\,K(p)}{1 + 3g(X, JY)^2} \tag{5.2}$$

If p is J-invariant, $K^*(p)$ is nothing but the holomorphic sectional curvature of p.

For a Kähler manifold (M, g) we consider three kinds of pinchings. Let δ be a positive number such that $\delta \leqq 1$. We say (M, g) is *δ-pinched* if there is a constant A such that

$$\delta A \leqq K(p) \leqq A \quad \text{for all planes } p. \tag{5.3}$$

We say that (M, g) is *δ-Kähler pinched* if there is a constant A such that

$$\delta A \leqq K^*(p) \leqq A \quad \text{for all planes } p. \tag{5.4}$$

We say that M is *δ-holomorphically pinched* if there is a constant A such that

$$\delta A \leqq K^*(p) \leqq A \quad \text{for all planes invariant by } J. \tag{5.5}$$

Then the following is well known (cf. Kobayashi and Nomizu [1]).

FACT 13. *A compact Kähler manifold* (M, g) *of dimension m is holomorphically isometric with the complex projective space* $P^m(C)$ *with a Fubini-Study metric if any one of the following three conditions is satisfied:*

$$M \text{ is } \tfrac{1}{4}\text{-pinched}; \tag{5.6}$$

$$M \text{ is 1-holomorphically pinched}; \tag{5.7}$$

$$M \text{ is 1-Kähler pinched}. \tag{5.8}$$

Klingenberg [1] has proved the following

FACT 14. *A compact m-dimensional Kähler manifold* (M, g) *with Kähler pinching* $> 9/16$ *has the same homotopy type as the complex projective space* $P^m(\boldsymbol{C})$.

The following is also known.

FACT 15. *A compact m-dimensional Kähler manifold* (M, g) *with holomorphic pinching* $> 4/5$ *has the same homotopy type as the complex projective space* $P^m(\boldsymbol{C})$.

Howard [**1**] has proved the following

FACT 16. *There exists a sequence* δ_m *with* $0 < \delta_m < 1$ *such that any m-dimensional compact Kähler manifold* (M, g) *with* δ_m*-Kähler pinching is biholomorphic to* $P^m(\boldsymbol{C})$.

For further references of the implication of various pinching assumptions, see Kobayashi and Nomizu [**1**, Note 23]. See also Ruh's note in this conference.

BIBLIOGRAPHY

T. Aubin, [**1**] *Métriques riemanniennes et courbure*, J. Differential Geometry **4**(1970), 383–424.

M. Berger, [**1**] *Sur les variétés d'Einstein compactes*, C.R. IIIe Reunion Math. Expression latine, Namur, 1965, pp. 35–55.

R.L. Bishop, and S.I. Goldberg, [**1**] *On the second cohomology group of a Kähler manifold of positive curvature*, Proc. Amer. Math. Soc. **16** (1965), 119–122. MR **30** #2441.

S. Bochner, [**1**] *Curvature and Betti numbers*, Ann. of Math. (2) **49** (1948), 379–390. MR **9**, 618.

S. Bloch and D. Gieseker, [**1**] *The positivity of the Chern classes of an ample vector bundle*, Invent. Math. **12** (1971), 112–117.

A. Borel and F. Hirzebruch, [**1**] *Characteristic classes and homogeneous spaces*. I, Amer. J. Math. **80** (1958), 458–538. MR **21** #1586.

A. Borel and R. Remmert, [**1**] *Über kompakte homogene Kählershe Mannigfaltigkeiten*, Math. Ann. **145** (1961/62), 429–439. MR **26** #3088.

E. Brieskorn, [**1**] *Ein Satz über die komplexen Quadriken*, Math. Ann. **154** (1964), 184–193. MR **29** #5257.

T.T. Frankel, [**1**] *Manifolds with positive curvature*, Pacific J. Math. **11** (1961), 165–174. MR **23** #A600.

S.I. Goldberg and S. Kobayashi, [**1**] *On holomorphic bisectional curvature*, J. Differential Geometry **1** (1967), 225–233.

P.A. Griffiths, [**1**] *Hermitian differential geometry, Chern classes and positive vector bundles*, Global Analysis (Paper in honor of Kodaira), Princeton Univ. Press, Princeton, N.J., 1970.

A. Grothendieck, [**1**] *Sur la classification des fibrés holomorphes sur la sphère de Riemann*, Amer. J. Math. **79** (1957), 121–138. MR **19**, 315.

R. Hartshorne, [**1**] *Ample vector bundle*, Inst. Hautes Études Sci. Publ. Math. **29**(1966), 63–94.

[**2**] *Ample subvarieties of algebraic varieties*, Lecture Notes in Math., vol. 156, Springer-Verlag, Berlin and New York, 1970.

F. Hirzebruch, [**1**] *Topological methods in algebraic geometry*, Springer-Verlag, Berlin and New York, 1970.

F. Hirzebruch and K. Kodaira, [**1**] *On the complex projective spaces*, J. Math. Pures Appl. (9) **36** (1957), 201–216. MR **19**, 1077.

A. Howard, [**1**] *A remark on Kähler pinching*.

A. Howard and B. Smyth, [**1**] *On Kähler surfaces of non-negative curvature*, J. Differential Geometry **5** (1971), 491–502.

W. Klingenberg, [1] *Manifolds with restricted conjugate locus*, Ann. of Math. (2) **78** (1963), 527–547; II, ibid. **80** (1964), 330–339. MR **28** #2506.

S. Kobayashi, [1] *On compact Kähler manifolds with positive definite Ricci tensor*, Ann. of Math. (2) **74** (1961), 570–574. MR **24** #A2922.

S. Kobayashi and K. Nomizu, [1], *Foundations of differential geometry*, Interscience Tracts in Math., vol. 15, Interscience, New York, 1969.

S. Kobayashi and T. Ochiai, [1] *On complex manifolds with positive tangent bundle,* J. Math. Soc. Japan **22** (1970), 499–525.

[2] *Compact homogeneous complex manifolds with positive tangent bundle*, Differential Geometry in honor of K. Yano, Kinokuniya, Tokyo, 1972, pp. 221–232.

[3] *Characterizations of complex projective spaces and hyperquadrics*, J. Math. Kyoto Univ. **13** (1972), 31–47.

[4] *Three-dimensional compact Kähler manifolds with positive holomorphic bisectional curvature*, J. Math. Soc. Japan **24** (1972), 465–480.

K. Kodaira, [1] *On a differential-geometric method in the theory of analytic stacks*, Proc. Nat. Acad. Sci. U.S.A. **39** (1953), 1268–1273. MR **16**, 618.

[2] *On Kähler varieties of restricted type (An intrinsic characterization of algebraic varieties)*, Ann. of Math. (2) **60** (1954), 28–48. MR **16**, 952.

J. Le Potier, [1] *Théorèmes d'annulation en cohomologie*, C.R. Acad. Sci. Paris Sér A **276** 535–537.

Y. Matsushima, [1] *Sur la structure du groupe d'homéomorphismes analytiques d'une certaine variété kaehlérienne*, Nagoya Math. J. **11** (1957), 145–150. MR **20** #995.

S.B. Myers, [1] *Riemannian manifolds with positive mean curvature*, Duke Math. J. **8** (1941), 401–404. MR **3**, 18.

I. R. Šafarevič, et al., [1] *Algebraic surfaces*, Trudy Mat. Inst. Steklov. **75** (1965) = Proc. Steklov Inst. Math. **75** (1965). MR **32** #7557; **35** #6685.

H.C. Wang, [1] *Closed manifolds with homogeneous complex structure*, Amer. J. Math. **76** (1954), 1-32. MR **16**, 518.

OSAKA UNIVERSITY

PARTIAL DIFFERENTIAL EQUATIONS

Proceedings of Symposia in Pure Mathematics
Volume 27, 1975

ON THE SIZE OF A STABLE MINIMAL SURFACE IN R^3

J. L. BARBOSA AND M. DO CARMO

Let $M \subset R^3$ be a minimal surface. A *domain* $D \subset M$ is a connected open set in M such that its closure $\bar{D}$ is compact and its boundary ∂D is a finite union of piecewise differentiable curves. Each normal variation of D which leaves ∂D fixed determines a vector field $V = uN$, when N is a unit normal field and $u : \bar{D} \to R$ is a differentiable function vanishing on ∂D. The second derivative of the area function of the variation is, modulo a factor,

$$I(V, V) = \int_{\bar{D}} u(-u\Delta u + 2Ku)dM,$$

where Δ is the laplacian of the induced metric, K its gaussian curvature and dM its element of area. We say that D is *stable* if $I(V, V) > 0$, for all such V. This means that D is a minimum for the area function of all normal variations leaving ∂D fixed.

THEOREM. *Let $g : M \to S^2$ be the gauss map of $M \subset R^3$ and let $D \subset M$ be a domain. If area $g(D) < 2\pi$, D is stable.*

INSTITUTO DE MATEMATICA PURA E APLICADA

AMS (MOS) subject classifications (1970). Primary 49F10, 49F25; Secondary 53B25, 58E05.

Proceedings of Symposia in Pure Mathematics
Volume 27, 1975

GEOMETRY OF THE SPECTRUM. I*

M. BERGER

0. Introduction. These three lectures contain the following material: After an introduction (§ 1) to the spectrum of a riemannian manifold thru vibrating membranes, § 2 contains the basic definitions and material on a riemannian manifold (r.m.). Here, in contrast to Singer's lectures, *Geometry of the spectra.* II [**35**], we will only be interested in *the spectrum* of a r.m. M for real valued functions $M \to \boldsymbol{R}$. § 3 is devoted to the study of *the fundamental solution of the real heat equation* (FSRHE) on M, to the asymptotic expansion one can deduce from it and to applications. The computational part of the existence of the FSRHE will be presented in detail, first because it will be a key point in § 4, and second because the notion of asymptotic expansion (for more general spectra) will be widely used in Singer's lectures. In § 4, we will state and give the important points of proof of a result of Colin de Verdière ([**13**] or [**14**]) to the effect that, for a generic r.m., the spectrum determines the set of lengths of the periodic geodesics. In § 5, one finds more general results and applications.

After Colin de Verdière's result was known, Chazarain in [**10**] and [**11**] and Duistermaat in [**16**] found a more neat statement and proof of his result of §4, based on the deep Hörmander theory of Fourier integral operators. For the time being Colin de Verdière's proof is far more elementary but also far more technical and computational. So the reader might well find it rewarding to consult [**16**]. Moreover in [**16**] he will find two important results concerning the spectrum of a r.m. The first result (also due to Guillemin [**19**]) characterizes by their spectra the r.m.'s all of whose geodesics are periodic; the second states that if a r.m. does not have all periodic geodesics then its spectrum is distributed on the real line asymptotically as regularly as possible. See also [**39**].

AMS (MOS) subject classifications (1970). Primary 53C20; Secondary 58E10.
*General lecture given at the Institute.

Paradoxically, we will almost everywhere be careless with analysis (differentiability, convergence of series) for clarity's sake, the reason being that there are already quite a few notions and computations to understand and, for careful analysis, we refer the reader to the bibliography.

1. Prelude.

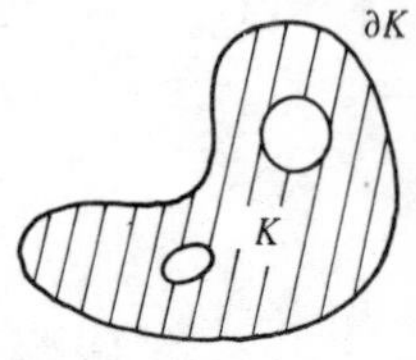

1.1. SPECTRUM OF A DOMAIN. Let $K \subset \boldsymbol{R}^d$ be a very regular bounded domain with nice boundary ∂K (for $d = 2$, one can think of K as a vibrating membrane, fixed along ∂K). The vibrations of K are the functions $F : K \times \boldsymbol{R} \ni (x, t) \mapsto F(x, t) \in \boldsymbol{R}$ with

1.1.1. $$\Delta F + \partial^2 F/\partial t^2 = 0 \quad \text{and} \quad F|_{\partial K \times \boldsymbol{R}} = 0$$

with $\Delta = -\sum_{i=1}^{d}(\partial^2/\partial x_i^2)$. The Stone-Weierstrass theorem permits us to separate the variables, i.e. to study only the F's of type $F(x, t) = f(x)\,\theta(t)$ with $f : K \to \boldsymbol{R}$, $\theta : \boldsymbol{R} \to \boldsymbol{R}$. For these F, 1.1.1 boils down to

$$\Delta f / f = -\theta''/\theta = \lambda$$

where λ has to be a constant, connected to the frequencies of our vibrations since $\theta'' + \lambda\theta = 0$. That is a good reason to be interested in the *spectrum* of K,

$$\mathrm{Spec}(K) = \{0 < \lambda_1 \leqq \lambda_2 \leqq \cdots\},$$

consisting of all λ's such that there exists an f with $f \neq 0$, $\Delta f = \lambda f$, $f|_{\partial K} = 0$; each λ is written in Spec(K) a number of times equal to its multiplicity, that is dim $\{f : \Delta f = \lambda f\}$.

1.2. EXAMPLES. Despite its simplicity and its physical background, Spec is almost an unknown person; in fact Spec(K) is known only for K a ball or a rectangular parallelopiped, and few other K. For example even the smallest eigenvalue λ_1 is not known for a general triangle, or an ellipse, etc.

1.3. QUESTIONS. A natural question is to study the map

$$\mathrm{Spec} : \{\text{bounded domains}\} \to \boldsymbol{R}^{\boldsymbol{N}}.$$

To determine the image seems to be too hard a question for the present time; the only solved analogous problem seems to be that of a vibrating string (see **[30]** for a recent account). Even the injectivity problem is completely open now; it is trivial that for any K and any isometry $\varphi : \boldsymbol{R}^d \circlearrowleft$, Spec ($\varphi(K)$) = Spec($K$) but essentially nothing is known for the converse question: Does Spec(K) = Spec(K') imply K isometric to K'? That question is often referred to as "Can one hear the shape of a drum?"; see **[22]**. A partial result is: Spec(K) = Spec(a ball) implies K is a ball of the same radius so isometric to that ball; this follows from 1.4 or 1.5 below; see **[6**, p. 121**]**. Another one can be found in the last line of p. 25 of **[23]**.

1.4. SOME ANSWERS. To be less greedy than asking to hear the complete shape of K, one can ask only to deduce from Spec(K) some information on the geometry of K, say for example Volume(K) (Lebesgue measure), Volume(∂K)

(i.e. $(d-1)$ dimensional measure of ∂K, which makes sense since ∂K is nice), $\int_{\partial K}$(mean curvature of ∂K) $d(\partial K)$, the fundamental group $\pi_1(K)$, the topology of K up to homeomorphism, the lengths of the billiard closed trajectories (or equivalently the light polygons) of K:

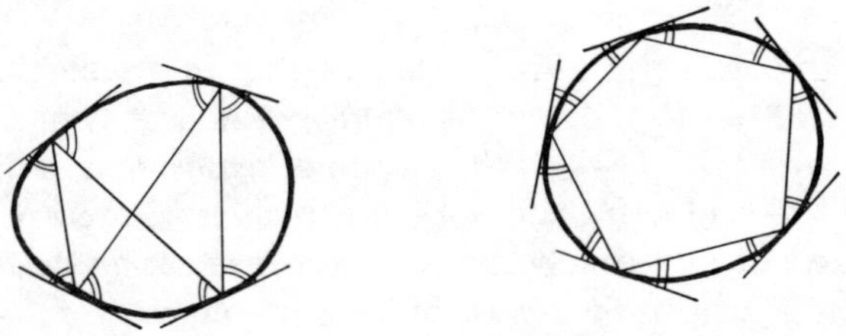

Present knowledge is the following (for the details, we refer to [27], [25] and [37]): From Spec(K) one can recover: Volume(K), Volume(∂K), $\int_{\partial K}$(mean curvature of ∂K) $d(\partial K)$; if $d=2$: $\int_{\partial K}\rho^2(\partial K)$ where ρ is the curvature of ∂K; and if $d=3$: $\int_{\partial K}(k_1-k_2)^2\, d(\partial K)$, where k_1, k_2 are the two principal curvatures of ∂K. When $d=2$, the Gauss-Bonnet formula implies that $\int_{\partial K}$ (curvature) $d(\partial K)$ gives us the Euler characteristic of K, i.e. its number of holes, so in that case K is known up to homeomorphism.

From [3]; 5.1.1 below and private talks it seems reasonable to expect that soon one will be able to recover the lengths of the light polygons from the spectrum.

1.5. ABOUT λ_1. There exist a lot of inequalities connecting λ_1, or more generally the k first eigenvalues of K, with the geometry of K; see [32].

1.6. See also nice new results in [21].

2. The spectrum of a riemannian manifold. A general reference for this section can be [8]; see also [38, pp. 254, 256].

2.1. RIEMANNIAN MANIFOLDS. In these lectures, by a *riemannian manifold* we mean exclusively a *compact connected without boundary* C^∞ manifold M endowed with a C^∞ riemannian tensor g; we use the notations (M, g) and $d = \dim M$. Two r.m.'s (M, g), (M', g') are said to be *isometric* if there exists a diffeomorphism $f: M \to M'$ such that $f^*g' = g$. Everything we will be interested in behaves nicely under isometry, and uniqueness problems are always up to isometry; so we define a *riemannian structure* as an equivalence class of r.m. under isometries. For example, when $d = 1$, the set of riemannian structures is just indexed by the total length L.

A r. m. carries a natural metric, *denoted by r* or by $\cdots$. It carries also geodesics and the associated exponential maps $\exp_m : T_mM \to M$, where $T_m M$ is the tangent space to M at m. The *injectivity radius* of M, denoted by ρ, is the biggest $\rho > 0$ such that $\exp_m$ is, for every $m \in M$, a diffeomorphism from the open euclidean ball of radius ρ in $T_m M$ onto its image (the existence of ρ is insured by the compactness of M).

A *periodic geodesic* of (M, g) is a geodesic $\gamma : \boldsymbol{R} \to M$ which is a periodic map and has to be distinguished from a geodesic loop $\gamma : [a, b] \to M$ for which only $\gamma(a) = \gamma(b)$ is required:

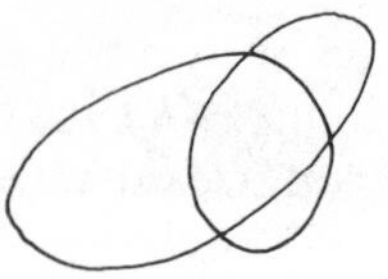

periodic geodesic

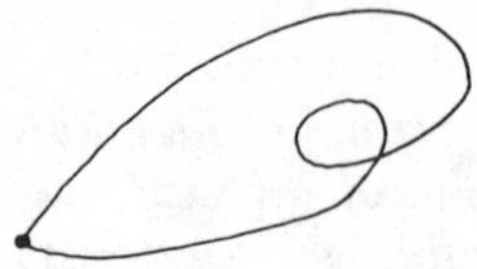

$\gamma(a) = \gamma(b)$

geodesic loop

Riemannian structures also carry algebraic tensor invariants, in particular: the *scalar curvature* $\tau : M \to \boldsymbol{R}$, the *Ricci curvature* ρ, a bilinear symmetric form on M, and its square norm $|\rho|^2 : M \to \boldsymbol{R}$, the *curvature tensor* R, a 4-linear form, and its square norm $|R|^2$ and also the covariant derivatives DR, $D^2R, \cdots$, of R.

2.2. THE LAPLACIAN AND THE SPECTRUM. A r.m. has a *canonical measure, denoted by* v or $v(m)$; its total mass is the *volume* of (M, g): Volume(M, g).

For us the basic object that (M, g) carries is the *laplacian* Δ, acting on the C^∞ functions $f : M \to \boldsymbol{R}$. Various definitions exist for Δ. Here are some of them:

2.2.1.1. df is a 1-form on M, so thru g-euclidean duality it yields a vector field $(df)^\#$, the gradient of g, whose divergence div $((df)^\#)$ is $-\Delta f$.

2.2.1.2. df being a 1-form, its covariant derivative Ddf is a 2-form; the trace, with respect to g, of Ddf again is $-\Delta f$.

2.2.1.3. The last definition can be read geometrically as follows: to compute Δf at $m \in M$, pick an orthonormal set of geodesics $\{\gamma_i\}_{i=1,\cdots,d}$, parametrized by arc length and passing through m at $t = 0$; then

$$(\Delta f)(m) = -\sum_{i=1}^{d} \frac{d^2(f \circ \gamma_i)}{dt^2}(0).$$

2.2.1.4. A complete computational formula in a coordinate chart $\{x_i\}$ is

$$\Delta f = -\frac{1}{g}\sum_{i,j} \frac{\partial(g\, g^{ij}\,(\partial f / \partial x_j))}{\partial x^i} \quad \text{with } g = \det(g_{ij}).$$

2.2.1.5. THE MORALIST DEFINITION. Δ is the only natural differential operator whose symbol is $-\|\cdot\|^2$ and with no zero-order term.

A basic formula is the one giving Δf when f depends only on the distance $r = \overline{m\cdot}$ to a fixed point m, i. e. there exists φ with $f = \varphi(\overline{m\ \cdot})$; then, inside the ball of center m and radius ρ one has

2.2.2.1. $$\Delta f = -\varphi'' - ((d-1)/r + \theta'/\theta)\,\varphi',$$

where $\cdot'$, $\cdot''$ mean the first and second derivatives with respect to r and where θ is a C^∞ function which is nothing but the ratio between the measure v and the Lebesgue measure of T_mM transferred to M via $\exp_m$.

The *spectrum* of (M, g), Spec(M, g), is the set of eigenvalues of Δ, i. e. the λ's such that there exists an $f \neq 0$ with $\Delta f = \lambda f$. We write

$$\operatorname{Spec}(M, g) = \{0 = \lambda_0 < \lambda_1 \leqq \lambda_2 \leqq \cdots\}$$

each λ being written a number of times equal to its multiplicity mul$(\lambda) = \dim\{f : \Delta f = \lambda f\}$; this multiplicity is finite because Δ is elliptic, a property which ensures also the discreteness of the spectrum. The spectrum does not start before 0 because Δ is positive, thanks to $\int_M f\Delta f\cdot v = \int_M |df|^2 v$; it starts really at 0 with multiplicity one, for $\Delta f = 0$ implies $df = 0$; hence f has to be constant and conversely.

2.2.2.2. Note that Spec(M, g) depends only on the riemannian *structure* of (M, g) (cf. 2.1).

2.3. EXAMPLES.

2.3.1. If $d = \dim M = 1$, and if the length of (M, g) is L (cf. 2.1), then the eigenvalues of Δ are $4\pi^2 L^{-2} n^2$, $n \in \boldsymbol{N}$, each of multiplicity 2 except 0. Hence, when $d=1$, the spectrum characterizes the riemannian structure and even λ_1 does!

2.3.2. The spectrum is completely known for (S^d, g_0), the d-dimensional *sphere* endowed with its canonical riemannian structure, for any d; also for the canonical structures on $P^d(\boldsymbol{R})$, $P^d(\boldsymbol{C})$ the *real* and *complex projective spaces*. The spectrum is, in some sense, also completely known for symmetric spaces and some homogeneous spaces, at least in theory (Casimir operators, etc.).

2.3.3. The spectrum is really fun for *flat tori*. We follow here **[29]**; a flat torus is $M = \boldsymbol{R}^d/\Lambda$, where Λ is any lattice of $\boldsymbol{R}^d$, endowed with g_0/Λ, g_0 being the canonical riemannian tensor of $\boldsymbol{R}^d$ which has no problems in going to the quotient since it is invariant by translations. $(\boldsymbol{R}^d/\Lambda, g_0/\Lambda)$, $(\boldsymbol{R}^d/\Lambda', g_0/\Lambda')$ are isometric as in 2.1 if and only if the lattices Λ, Λ' are related by an isometry of $\boldsymbol{R}^d$. Now it is not hard to see that if $\Lambda^* = \{\xi \in \boldsymbol{R}^d : (\xi \mid \eta) \in \boldsymbol{Z}\ \forall \eta \in \Lambda\}$ is the dual lattice of Λ, then the λ's of Spec$(\boldsymbol{R}^d/\Lambda, g_0/\Lambda)$ are the $4\pi^2\ \|\xi\|^2$ with ξ running thru Λ^*, the multiplicity of $4\pi^2\ \|\xi\|^2$ being the number of $\eta \in \Lambda^*$ such that $\|\xi\|^2 = \|\eta\|^2$. Now Witt discovered in 1941 two lattices Λ, Λ' in $\boldsymbol{R}^{16}$, not isometric but with always the same number of elements of any given norm. Hence, for the two corresponding flat tori:

2.3.4. $$\mathrm{Spec}(\boldsymbol{R}^d/\Lambda, g_0/\Lambda) = \mathrm{Spec}(\boldsymbol{R}^d/\Lambda', g_0/\Lambda')$$

but with $(\boldsymbol{R}^d/\Lambda, g_0/\Lambda)$, $(\boldsymbol{R}^d/\Lambda', g_0/\Lambda')$ not isometric. This is to be compared with 1.3; see also 3.5.4. Examples of flat tori with the same spectrum and not isometric are also known to exist for $n = 12$; they cannot happen for $n = 2$ and from 3 to 11 the question is open; see **[24]** for a recent reference.

2.4. QUESTIONS AND FEW ANSWERS.

2.4.1. As in 1.3 one can ask to study the map (see 2.1 and 2.3):

$$\mathrm{Spec}\colon \{\text{riemannian structures}\} \to \boldsymbol{R}^{\boldsymbol{N}}.$$

The problem of describing the image looks too formidable today, even to talk about it. From 2.3.4 we see that injectivity is false in general.

2.4.2. Noting the discrete character of the Milnor's type counterexamples, an open problem is to decide whether or not there exists a nontrivial family $(M, g(t))$ of riemannian structures with

$$\mathrm{Spec}(M, g(t)) = \mathrm{Spec}(M, g(0)) \text{ for every } t$$

(isospectral deformation). See some information in the introduction of **[26]**. Also we can think of some special cases of injectivity, as at the end of 1.3: does Spec$(M, g) = $ Spec(S^d, g_0) imply (M, g) isometric to (S^d, g)? See a partial answer in 3.5.3. Same question with (S^d, g_0) replaced by $(P^d(\boldsymbol{R}), g_0)$ or $(P^{d/2}(\boldsymbol{C}), g_0)$.

2.4.3. A less ambitious program is to deduce from Spec(M, g) some information on the geometry of (M, g): Volume (M, g), total scalar curvature $\int_M \tau v$ of (M, g), topology of M up to homeomorphism or only its Euler characteristic, its Betti numbers, fundamental group. See in 3.5 some partial answers to those questions, as corollaries of the asymptotic expansion. Comparing with 1.4, one can also hope to deduce from Spec(M, g) the lengths of the pending, for compact without boundary riemannian manifolds, of the light polygons of a bounded domain: those are the *periodic geodesics* and we will prove in § 4 that, at least generically, Spec(M, g) determines them.

2.4.4. Other questions can be:

2.4.4.1. Tell from Spec(M, g) whether or not (M, g) is a kähler manifold: see a partial answer in [**18**] (cf. 3.5.5).

2.4.4.2. Find inequalities, if possible sharp ones, involving λ_1 : see a recent account in [**7**].

2.4.5. As already said in § 0 we are interested in these lectures only in the spectrum of Δ acting on the functions $M \to \boldsymbol{R}$. But a spectrum exists for any elliptic operator on any vector bundle over M, for example for the canonical laplacian acting on exterior p-forms ($p = 1, \cdots, d$) on (M, g). For that, see Singer's lectures [**35**] and [**2**].

3. The real heat equation, the asymptotic expansion and applications. Recall (§ 0) that we pay little attention to analysis, estimates, etc. but on the other hand we need precise computations; for rigorous analysis, see [**8**, pp. 204–215]. In this section and the next one, (M, g) denotes always a r.m. (see 2.1).

3.1. The FSRHE. The (real) heat equation on (M, g) works with functions $G: M \times \boldsymbol{R}_+ \ni (m, t) \mapsto G(m, t) \in \boldsymbol{R}$ fulfilling the two conditions

$$\Delta G + \partial G / \partial t = 0, \qquad G(\cdot, 0) = f,$$

where Δ is the laplacian acting on m and $f: M \to \boldsymbol{R}$ is a given initial condition. That problem will be solved for any f as soon as one will know a FSRHE, defined as follows:

3.1.1. Definition. *A FSRHE on (M, g) is an F: $M \times M \times \boldsymbol{R}_+^* \to \boldsymbol{R}$ such that:*

(i) *F is C^0 on the three variables, C^2 on the two first and C^1 on the third, and then $\Delta_2 F + \partial F / \partial t = 0$, where Δ_2 is the laplacian on the second variable;*

(ii) *for every $m \in M$: $\lim_{t \to 0+} F(m, \cdot, t) = \delta_m$, where δ_m is the Dirac distribution at m.*

3.1.2. Theorem. *A FSRHE exists and is unique.*

The uniqueness can be seen in [**8**, pp. 205–206], and uses the formula 3.1.4 below. The existence will be proved in 3.3 below.

3.1.3. For every $\lambda \in$ Spec(M, g) of multiplicity mul(λ), let us choose an orthonormal set $\{\varphi_j\}$, $_{j=1, \cdots, \mathrm{mul}(\lambda)}$, of eigenfunctions of Δ for λ, i.e., $\langle \varphi_i, \varphi_j \rangle = \int_M \varphi_i \varphi_j v = \delta_{ij}$ (Kronecker symbol). When λ runs through Spec(M, g), the totality of the φ_j's, say $\{\varphi_i\}$, is called an *orthonormal set of eigenfunctions of Δ*. As for Fourier series, we have, for every $f: M \to \boldsymbol{R}: f = \sum_i \langle \varphi_i, f \rangle \varphi_i$ in the L^2-sense. And then:

3.1.4. PROPOSITION. *If $\{\varphi_i\}$ is an orthonormal set of eigenfunctions, then, for every m, n, t, the series*

$$\sum_i e^{-\lambda_i t} \varphi_i(m) \varphi_i(n)$$

converges and

$$F(m, n, t) = \sum_i e^{-\lambda_i t} \varphi_i(m) \varphi_i(n).$$

Modulo analysis, the proof will be done by the uniqueness of the FSRHE if we can show that (i) and (ii) of 3.1.1 holds for that series. Now

$$\left(\Delta_2 + \frac{\partial}{\partial t}\right)\left(\sum_i e^{-\lambda_i t} \varphi_i(m) \varphi_i(n)\right)$$
$$= \sum_i e^{-\lambda_i t} \varphi_i(m) (\Delta \varphi_i)(n) + \sum_i \frac{\partial}{\partial t}(e^{-\lambda_i t}) \varphi_i(m) \varphi_i(n) = 0$$

since $\Delta \varphi_i = \lambda_i \varphi_i$ and $(\partial/\partial t)(e^{-\lambda_i t}) = -\lambda_i e^{-\lambda_i t}$. For any $f: M \to \boldsymbol{R}$,

$$\lim_{t\to 0+} \int_M \sum_i e^{-\lambda_i t} \varphi_i \varphi_i(n) f(n) v(n) = \lim_{t \to 0+} \sum_i e^{-\lambda_i t} \langle \varphi_i, f \rangle \varphi_i$$
$$= \sum_i \langle \varphi_i, f \rangle \varphi_i = f$$

after 3.1.3. From the proposition follows the

3.1.5. COROLLARY. *For every $t > 0$ the series $\sum_i e^{-\lambda_i t}$ converges and $\sum_i e^{-\lambda_i t} = \int_M F(m, m, t) v(m)$.*

3.2. HEURISTIC CONSIDERATIONS AND COMPUTATIONS.

3.2.1. To construct the FSRHE we use a successive approximation method, but here the initial value has to be chosen carefully. We recall first that in $\boldsymbol{R}^d$ (though not compact) there exists a unique (under the condition to be decreasing at infinity) FSRHE given by

$$F_0(\cdot, \cdot, t) = (4\pi t)^{-d/2} e^{-r^2/4t}$$

where $r = \overline{mn}$ is the euclidean distance. The condition (ii) comes, after some work, from the value of the Gauss integral $\int_R e^{-x^2} dx$; the condition (i) comes from 2.2.2.1:

$$\left(\Delta_2 + \frac{\partial}{\partial t}\right)((4\pi t)^{-d/2} e^{-r^2/4t}) = (4\pi t)^{-d/2} e^{-r^2/4t}\left[-\left(-\frac{1}{2t} + \frac{r_2}{4t_2}\right)\right.$$
$$\left. - \left(\frac{d-1}{2} + \frac{\theta'}{\theta}\right)\left(-\frac{r}{2t}\right) - \frac{n}{2t} + \frac{r^2}{4t^2}\right] = 0$$

because $\theta = 1$ by the very definition of θ in 2.2.2.1.

3.2.2. Now the idea is: On a r.m. (M, g), for small t's, the FSRHE F differs little from the pull-back of F_0 by $\exp^{-1}$; as Kac says in **[22]**, the particles diffusing on M, if concentrated at m for $t = 0$, at the beginning do not realize that M is compact and not euclidean; they behave as if M were infinite and without curvature (a proof of that will in fact follow from the results below). So we are led to consider on $M \times M \times \boldsymbol{R}$ the function

$$G = (4\pi t)^{-d/2}\, e^{-r^2/4t}$$

where r is the distance on (M, g); note G is smooth when $\overline{mn} < \rho$ (see 2.1), i.e. we should work on the open submanifold $M_2' = \{(m, n) \in M \times M : \overline{mn} < \rho\}$. Everything being understood in M_2', we compute:

$$(\Delta_2 + \partial/\partial t)\, G = r/2t \cdot \theta'/\theta \cdot G \quad \text{from 2.2.2.1.}$$

So not only (i) is not O.K. but moreover the singularity at $t = 0$ of G is worse for $(\Delta_2 + \partial/\partial t)G$. If we want to replace G by $G \cdot u_0$, where $u_0 : M_2' \to \boldsymbol{R}$ is a function to be found, we have to compute the laplacian $\Delta(AB)$ of a product of two functions A, B: $M \to \boldsymbol{R}$; the formula is $\Delta(AB) = (\Delta A)\, B - 2(dA \mid dB) + A(\Delta B)$, where the middle term is the scalar product of the differentials of A, B respectively. Since in $G \cdot u_0$ the function G depends only on r, then $(dG \mid du_0) = G' \cdot u_0'$ (recall $\cdot'$ is the radial derivative); hence

$$\left(\Delta_2 + \frac{\partial}{\partial t}\right)(Gu_0) = G\left[\left(\frac{\theta'}{2\theta}\, u_0 - u_0'\right)\frac{r}{2t} + \Delta_2 u_0\right].$$

Then, if we can solve $(\theta'/2\theta)u_0 - u_0' = 0$, $(\Delta_2 + \partial/\partial t)\, G$ will have at $t = 0$ the same singularity as G; the solution is $u_0 = \theta^{1/2}$. Finally, for every $k \in \boldsymbol{N}$, we set:

$$S_k = G(u_0 + tu_1 + \cdots + t^k u_k),$$

where the u_i's are functions $M_2' \to \boldsymbol{R}$ to be found. Computation shows if the u_i's satisfy the conditions

3.2.3. $$u_i' + (\theta'/2\theta + i/r)\, u_i + r^{-1}\, \Delta_2 u_{i-1} = 0 \quad \text{for } i = 1, \cdots, k,$$

then

$$(\Delta_2 + \partial/\partial t)\, S_k = (4\pi t)^{-d/2} t^{k-d/2}\, e^{-r^2/4t}\, \Delta_2 u_k \cdot$$

The above 3.2.3 conditions are easily solved by induction and yield C^∞ functions on M_2'.

3.2.4. In particular, if $k > d/2$, then $(\Delta_2 + \partial/\partial t)S_k$ will be extendible to a continuous function for $t = 0$, and more generally to a C^s function if $k > d/2 + s$. But we have first to pass from M_2' to $M \times M$; to do this we introduce a C^∞ function $\eta : \boldsymbol{R} \to \boldsymbol{R}$ looking like

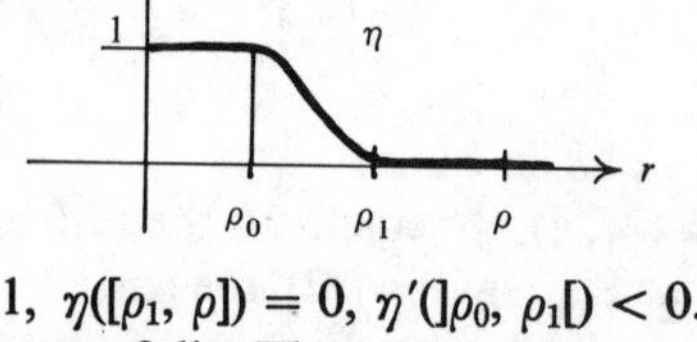

$0 < \rho_0 < \rho_1 < \rho$

i.e. $\eta([0, \rho_0]) = 1$, $\eta([\rho_1, \rho]) = 0$, $\eta'(]\rho_0, \rho_1[) < 0$. Note, later in 4.9.1, we will have to choose η more carefully. Then we set

3.2.5. $$H_k = (4\pi t)^{-d/2}\, \eta\, e^{-r^2/4t} \sum_{i=0}^{k} t^i u_i, \qquad K_k = (\Delta_2 + \partial/\partial t)\, H_k.$$

There are functions C^∞ on $M \times M \times \boldsymbol{R}^*_+$, and by the construction of η and 3.2.5 they satisfy the following:

3.2.6. PROPOSITION. *For any* k,

$$H_k = (4\pi t)^{-d/2}\, e^{-r^2/4t} \sum_{i=0}^{k} t^i U_i, \qquad K_k = (4\pi t)^{-d/2}\, e^{-r^2/4t} \sum_{i=-1}^{k} t^i V_i,$$

where the U_i*'s and the* V_i*'s are* C^∞ *functions on* $M \times M$ *with supports such that:*

$\operatorname{supp}(U_i)\ \{(m, n) \in M \times M : 0 \leqq \overline{mn} \leqq \rho_1\}$ *for every* $i = 0, 1, \cdots, k$,
$\operatorname{supp}(V_i)\ \{(m, n) \in M \times M : \rho_0 \leqq \overline{mn} \leqq \rho_1\}$ *for every* $i = -1, 0, 1, \cdots, k-1$,
$\operatorname{supp}(V_k)\ \{(m, n) \in M \times M : 0 \leqq \overline{mn} \leqq \rho_1\}$,
$V_{-1}(m, n) < 0$ *whenever* $\rho_0 < \overline{mn} < \rho_1$.

Moreover $\lim_{t \to 0+} H_k(m, \cdot, t) = \delta_m$.

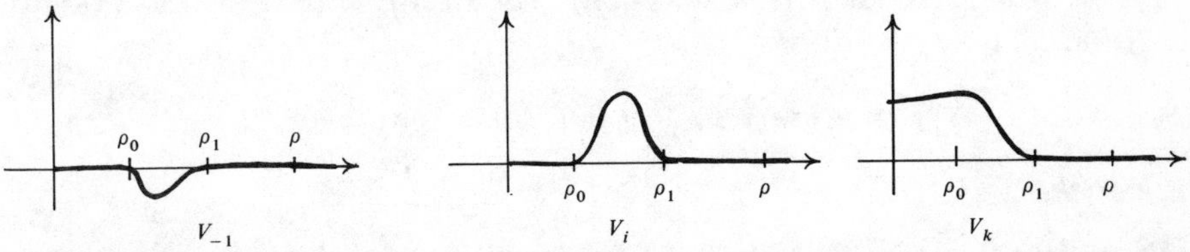

3.3. EXISTENCE OF THE FSRHE. Now we look for a FSRHE F as in the successive approximations $(I + e)^{-1} = \sum_n (-1)^n e^n$. The first step is to make a kind of algebra with the functions $M \times M \times \boldsymbol{R}^*_+ \to \boldsymbol{R}$. This is achieved by setting:

3.3.1. $$(A * B)(m, n, t) = \int_0^t \int_M A(m, q, \theta)\, B(q, n, t - \theta)\, v(q)\, d\theta;$$

the $\cdot * \cdot$ is associative and we set $A^{*n} = A * \cdots * A$ (n times) and $A^{*0} * B = B$ for any B. Then:

3.3.2. THEOREM. *For every* η *as above and every* $k > d/2$ *the series,* F,

$$\sum_{n=0}^{\infty} (-1)^n K_k^{*n} * H_k$$

converges and is a FSRHE, hence the FSRHE.

We omit the proof of convergence; the estimates are not too hard; see [**8**, p. 212]. We only check (i) and (ii) of 3.1.1; the axiom (ii) follows smoothly from

3.3.3. For every $k > d/2$, $F = H_k + O(t^{k+1-d/2})$.

To prove (i), we need first the following formula, valid for any A:

3.3.4. $$(\Delta_2 + \partial/\partial t)(A * H_k) = A + A * K_k.$$

For $\Delta_2(A * H_k) = A * (\Delta_2 H_k)$ by 3.3.1 while $(\partial/\partial t)(A * H_k)$ has two terms, one which comes from the t in $\int_0^t$, the other from the t in $H_k(q, n, t - \theta)$; the second yields $\partial H_k / \partial t$ while the first yields A thanks to the last assertion in 3.2.6.

Then 3.3.4 implies

$$\left(\Delta_2 + \frac{\partial}{\partial t}\right)\left(\sum_{n=0}^{\infty}(-1)^n K_k^{*n} * H_k\right) = \sum_{n=1}^{\infty}(-1)^n K_k^{*n} + \sum_{n=0}^{\infty}(-1)^n K_k^{*n} * K_k = 0.$$

3.4. THE ASYMPTOTIC EXPANSION. From 3.1.5, 3.2.6 and 3.3.3 we get, for every $k > d/2$,

3.4.1. $$\sum_i e^{-\lambda_i t} = \int_M H_k(m, m, 0)\, v(m) + O(t^{k+1-d/2}).$$

Hence, if we set for every i,

3.4.2. $$a_i = \int_M u_i(m, m)\, v(m),$$

we have the

3.4.3. THEOREM (ASYMPTOTIC EXPANSION). *For every r.m. there exist a_i's ($i = 0, 1, \cdots$) with*

$$\sum_i e^{-\lambda_i t} = (4\pi t)^{-d/2} \sum_{i=0}^{k} a_i t^i + O(t^{k+1-d/2})$$

for every k.

From the value of $u_0 = \theta^{1/2}$ in 3.2.2 follows $u_0(m, m) = 1$ for every m; hence

3.4.4. $$a_0 = \int_M v = \text{Volume}(M, g).$$

By their definition 3.4.3 or by 3.2.3, the $u_i(m, m)$ are universal with respect to riemannian structures, from which one deduces they are orthogonal polynomial invariants in the curvature tensor of (M, g) and its covariant derivatives. Using that, arguments on the degree of homogeneity, testing with standard manifolds and computations, one shows

3.4.5. (easily) $\quad a_1 = \frac{1}{6}\int_M \tau v \quad$ (already folklore in 1965);

3.4.6. (less easily) $\quad a_2 = \frac{1}{360}\int_M (2|R|^2 - 2|\rho|^2 + 5\tau^2)\, v \quad$ ([8], [27]);

3.4.7. (by hard computations, [33]) $\quad a_3 = \frac{1}{6!}\int_M f v,$

with

$$f = -\frac{142}{63}|d\tau|^2 - \frac{26}{63}|D\rho|^2 - \frac{1}{9}|DR|^2 + \frac{5}{9}\tau^3 - \frac{2}{3}\tau|\rho|^2$$
$$+ \frac{2}{3}\tau|R|^2 - \frac{4}{7}\rho_a^b \rho_b^c \rho_c^a + \frac{20}{63}\rho^{ab}\rho^{cd}R_{acbd} - \frac{8}{63}\rho^{uv}R_u{}^{abc}R_{vabc}$$
$$+ \frac{8}{21}R^{abcd}R_{ab}{}^{uv}R_{cduv} \quad \text{(Einstein summation convention)},$$

which, for $d = 2$, boils down to

$$a_3 = \frac{2}{7!}\int_M (4\tau^3 - 9\,|d\tau|^2)\,v.$$

The a_i' are zero for any $i > 0$ for a manifold with zero curvature tensor. Also the asymptotic expansion is explicitly and completely known for compact simply connected semisimple Lie groups; see **[9]**.

3.5. Applications of the Asymptotic Expansion.

3.5.1. From 3.4.3 and 3.4.4, one sees the spectrum gives us the dimension of M and the volume of (M, g).

3.5.2. When $d = 2$, the Gauss-Bonnet formula $\chi(M) = \frac{1}{2}\pi^{-1}\int_M \tau v$ for the characteristic of M, coupled with 3.4.5, shows that the spectrum (when $d = 2$) determines the characteristic of M; note that it does not determine M up to homeomorphism since, when χ is even, there are two topological types of compact surfaces with the same characteristic, one orientable, the other not. To recover the topology of M thru Spec(M, g) remains, even when $d = 2$, an open question. Note also that, although a_1 is a topological invariant when $d = 2$, a_2 is no longer a topological invariant when $d = 4$, because essentially the Gauss-Bonnet formula, when $d = 4$, reads

$$\chi(M) = \frac{1}{32\,\pi^2}\int_M (|R|^2 - 4\,|\rho|^2 + \tau^2)\,v.$$

A better reason, as was pointed out to us by Gilkey, is that $\chi(M \times N) = \chi(M)\,\chi(N)$, the time being $a_2(M \times N) = a_2(M)a_0(N) + a_1(M)a_1(N) + a_2(N)a_0(M)$ for a product metric.

3.5.3. Consider the *assertion* Σ_d: Spec(M, g) = Spec(S^d, g_0) implies (M, g) isometric to (S^d, g_0). From 3.4.4, 3.4.5 and 3.4.6 it follows (see **[8**, pp. 227 and 229**]**) that Σ_2 and Σ_3 are true. In **[36]** it is proved, using 3.4.7, that Σ_4, Σ_5 and Σ_6 are also true. It is an open question to decide if Σ_d is or is not valid for any d; compare with 1.3 and 3.5.4 just below. They are also uniqueness results for some standard manifolds other than spheres; see **[36]** for a recent account.

3.5.4. Let (see 2.4.5) Spec$^p(M, g)$ be the spectrum of the laplacian for p-exterior forms on M; among other uniqueness results, there is proved in **[31**, p. 281**]**, that Spec$^i(M, g)$ = Spec$^i(S^d, g_0)$ for $i = 0$ and $i = 1$ implies (M, g) isometric to (S^d, g_0). Note however that for Milnor's flat tori in 2.3.3, one has

$$\text{Spec}^i(\boldsymbol{R}^d/\Lambda, g_0/\Lambda) = \text{Spec}^i(\boldsymbol{R}^d/\Lambda', g_0/\Lambda') \quad \text{for every } i = 0, 1, \cdots, 16.$$

See also **[21]** and **[34]**.

3.5.5. If one introduces, in a complex riemannian manifold (M, g), besides Δ, the complex laplacian Δ^c, then it is proved in **[18]** that if Spec(M, g, Δ) = Spec$(M, g, 2\Delta^c)$ then the manifold is kähler. See **[18]** for other results of this type.

4. Spectrum and lengths of periodic geodesics. We have lots of definitions and notions to state before we are able to state the main result 4.5.1

4.1. Nondegenerate Critical Submanifolds. Given a C^∞ Hilbert-manifold

Ω and a C^∞ function $f : \Omega \to \boldsymbol{R}$, one has the notion of a *nondegenerate critical submanifold* W: W has to be a connected submanifold satisfying:

(i) every $w \in W$ is a critical point of f;

(ii) for every $w \in W$ there exists a direct sum of closed subspaces $T_W\Omega = T_W W + N_W$ such that the restriction $\mathrm{Hess}_W f|_{N_w}$ of the hessian of f at w to N_W is a nondegenerate quadratic form.

To such a nondegenerate critical submanifold we attach two integers: its dimension and its *index* j_W, which is the constant value of the index of $\mathrm{Hess}_W f|_{N_w}$ when w runs thru W.

4.2. Periodic Geodesics as Critical Points. Now let (M, g) be a r. m. and $\Omega(M)$ the set of absolutely continuous maps $\gamma : S^1 \to M$ whose speed $\dot{\gamma}$ is square-integrable (S^1 is the standard circle). One knows that $\Omega(M)$ has a natural Hilbert structure ([**17**] is a reference for the present section). The *energy function* $E : \Omega(M) \to \boldsymbol{R}$ is $E(\gamma) = \int_{S^1} |\dot{\gamma}(t)|^2\, dt$. In fact E is C^∞ and its critical points are precisely the periodic geodesics, including those of length zero. The critical value of a periodic geodesic is the square of its length.

We *denote by* $\mathscr{L}$ the set of the lengths of the periodic geodesics of (M, g); it is a closed set in $\boldsymbol{R}$, thanks to the compactness of M. A length $L \in \mathscr{L}$ is said to be *nondegenerate* if the set of all critical points of E of energy equal to L is a finite union of nondegenerate critical submanifolds; for every such manifold W, we introduce its *nullity* $n_W = \dim W - 1$ and its *index* j_W as in 4.1 when $L \neq 0$; when $L = 0$, $n_W = \dim W$. The motivation for this is that the index and nullity are nothing but the classical index and nullity of a periodic geodesic $\gamma \in W$ as defined by vector fields along γ. For example $L = 0$ is nondegenerate, with index zero and nullity this time equal to d, because the set of trivial geodesics is identified with M itself thru constant maps.

4.2.1. Definition. *A r.m. is said to be* $G^\cdot$ *if every* $L \in \mathscr{L}$ *is nondegenerate; it is said to be* $G^{\cdot\cdot}$ *if every* $L \in \mathscr{L}\backslash 0$ *is nondegenerate, of nullity* 0 *and if,* γ *being a periodic geodesic of length L, the set of critical points of energy* L^2 *consists exactly of the two submanifolds*

$$W_+ = \{t \mapsto \gamma(t + a) : a \in [0, 1[\} \text{ and } W_- = \{t \mapsto \gamma(a - t) : a \in [0, 1[\}.$$

4.2.2. *Note.* If (M, g) is $G^\cdot$ then $\mathscr{L}$ is necessarily discrete in $\boldsymbol{R}$; if it is $G^{\cdot\cdot}$ then all periodic geodesics are isolated and of lengths all distinct. In the $G^{\cdot\cdot}$ case we will denote by j_L the common index of the periodic geodesics which belong to L.

Note. From the Abraham bumpy metrics theorem [**1**] it follows that the properties $G^\cdot$ and $G^{\cdot\cdot}$ are generic.

4.3. Finite-Dimensional Approximations to $\Omega(M)$. To study the periodic geodesics one need not necessarily work with the infinite-dimensional $\Omega(M)$; it is enough to work with a tool designed by Morse; a general reference is [**28**, pp. 88–92]. Set

$$X_n = \left\{(u_0, u_1, \cdots, u_{n-1}) \in (]0, 1[)^n : \sum_{i=0}^{n-1} u_i = 1\right\},$$

$$M_0^n = \{(x_0, x_1, \cdots, x_{n-1}) \in M^n : i = 0, 1, \cdots, n-1 : \overline{x_i x_{i+1}} < \rho \text{ (and } x_n = x_0)\}.$$

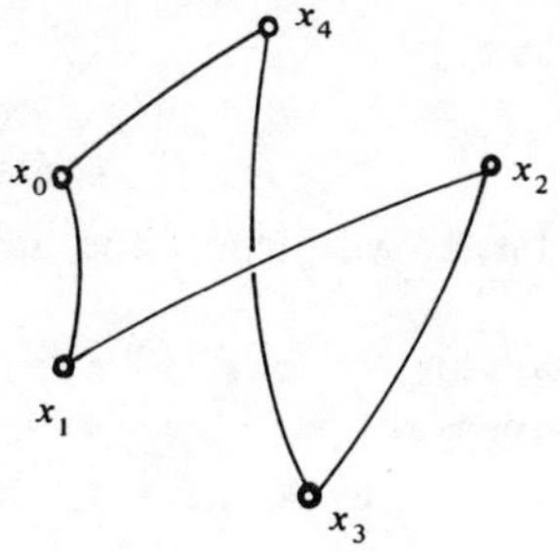

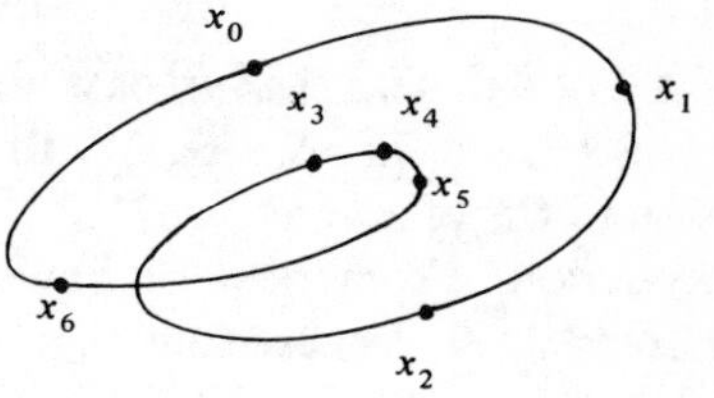

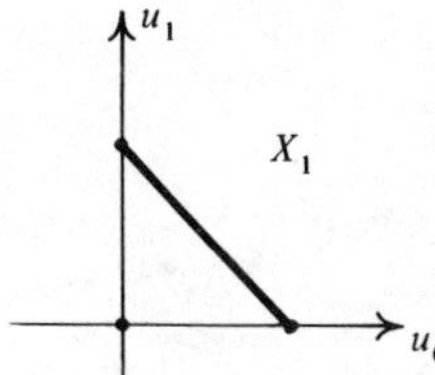

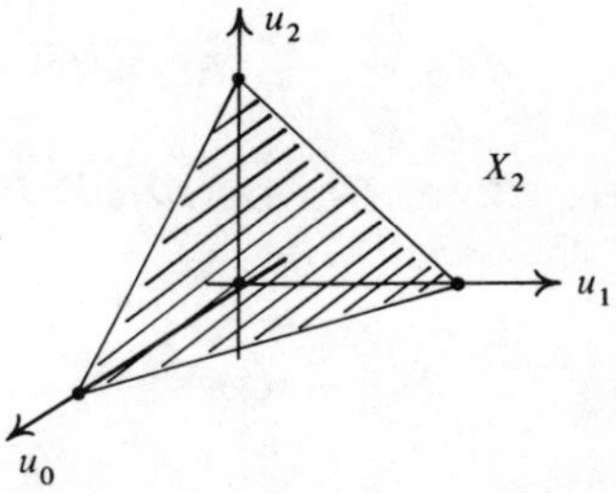

For $U \in X_n$ define an embedding $b_U : M_0^n \to \Omega(M)$ in the obvious way: $\gamma = b_U(x_0, \cdots, x_{n-1})$ is the broken closed geodesic, parametrized proportionally to arc length on every segment, such that

$\gamma(0) = x_0$,

$\gamma|_{[0,u_0]}$ is a geodesic from x_0 to x_1,

$\gamma|_{[u_0, u_0+u_1]}$ is a geodesic from x_1 to x_2,

...

$\gamma|_{[u_0+\cdots+u_{n-1}, 1]}$ is a geodesic from x_{n-1} to x_0.

These geodesics are unique by the definition of M_0^n and ρ in 2. 1, so b_U is well defined.

We now define

$$E_U = E \circ b_U : M_0^n \to \boldsymbol{R},$$
$$h = \mathrm{Id}_{X_n} \times b_U : M_0^n \times X_n \to \Omega(M) \times X_n,$$
$$\tilde{E} : M_0^n \times X_n \to \boldsymbol{R} \quad \text{by } \tilde{E}((x_i), U) = (E_U)((x_i)).$$

Note.

4.3.1. $$E_U((x_i)) = \sum_{i=0}^{n-1} \frac{\overline{x_i x_{i+1}}^2}{u_i}.$$

We have then the elementary

4.3.2 PROPOSITION. *The map h is an embedding and yields a bijection from the critical points of E onto the couples (γ, U) such that γ is a periodic geodesic of length $L < \rho/\mathrm{Max}(u_i)$. Moreover, if W is a nondegenerate critical submanifold of E of energy $L^2 < (n\rho)^2$, then $h^{-1}(W \times X_n)$ is a nondegenerate critical submanifold of E of nullity equal to $n_W + n$ and index equal to j_W.*

4.3.3. *Note.* For a critical point of E, necessarily:

$$\frac{\overline{x_0 x_1}}{u_0} = \frac{\overline{x_1 x_2}}{u_1} = \cdots = \frac{\overline{x_n x_0}}{u_{n-1}}.$$

4.4. A FOURIER-LIKE TRANSFORM.

4.4.1. Set $\boldsymbol{C}^+ = \{z \in \boldsymbol{C} : \mathrm{Re}(z) > 0\}$; for any real α, z^α will denote the unique extension to $\boldsymbol{C}^+$ of $x \mapsto x^\alpha$ on $\boldsymbol{R}^*_+$.

CONVENTION. From now on any statement involving "$\cdots z \to \infty$" will mean that a fixed $\xi_0 > 0$ is chosen and that "$\cdots y \to \infty$" with $z = \xi_0 + iy$.

4.4.2. *Notation. For any $f : \boldsymbol{C}^+ \to \boldsymbol{C}$ bounded by an exponential when $z \to \infty$ we define, for any $(\sigma, t) \in \boldsymbol{R}^*_+ \times \boldsymbol{R}$,*

$$\hat{f}(\sigma, t) = \int_{\boldsymbol{R}} f(\xi_0 + iy) \exp(ity - \sigma(y - \sigma^{-1/2})^2\, dy.$$

4.4.3. DEFINITIONS. *Given a strictly increasing sequence $\{t_n\}_{n \in \boldsymbol{R}} \subset \boldsymbol{R}$ and maps $f, f_n : \boldsymbol{C}^+ \to \boldsymbol{C}$, one will write*

$$f(z) =_F \sum_{n=0}^{\infty} \exp(-zt_n) f_n(z)$$

if:

(i) $\forall t \notin \{t_n\} : \hat{f}(\sigma, t) = O(\sigma^{-1/2})$,

(ii) $\forall n : \hat{f}(\sigma, t_n) = \hat{f}_n(\sigma, 0) \exp(-\xi_0 t_n) + O(\sigma^{-1/2})$.

Also, for α a real number:

$$f(z) \underset{F_\alpha}{=} \sum_{n=0}^{\infty} \exp(-zt_n) f_n(z)$$

if

$$z^\alpha f(z) \underset{F}{=} \sum_{n=0}^{\infty} \exp(-zt_n)\, z^\alpha f_n(z).$$

A function g is said to be of type T_α if it can be written

$$\begin{aligned} g(z) &= \sum_{j=1}^{N} a_j z^{\alpha_j} && \textit{for } \mathrm{Im}(z) \geqq 0, \\ &= \overline{g(\bar{z})} && \textit{for } \mathrm{Im}(z) < 0 \end{aligned}$$

with $-\alpha < \alpha_1 < \cdots < \alpha_N$ and $a_j \in \boldsymbol{C}$.

4.4.4. PROPOSITION. *If all the f_n are of type T_α in $f(z) =_{F_\alpha} \sum_{n=0}^{\infty} \exp(-zt_n) f_n(z)$, then they are uniquely determined by the function f.*

This is easy, but it would have been false if we had taken $\exp(ity - \sigma y^2)$ instead of $\exp(ity - \sigma(y - \sigma^{-1/2})^2)$ in the definition of $\hat{f}$. We now state easy properties of $\hat{\ }$ which will be useful later on:

4.4.5. PROPOSITION. (i) *For any f which is bounded when $z \to \infty$: $\hat{f}(\sigma, t) = O(\sigma^{-1/2})$. Let*

$$P : z \mapsto \sum_{p=a}^{b} a_p z^{p/2}, \qquad a, b, p \in \boldsymbol{Z},\ a_p \in \boldsymbol{C}.$$

Then:

(ii) *if* $b \leqq 0$, $|\hat{P}(\sigma, t)| \leqq (\pi/\sigma)^{1/2} \sum_{b}^{p=a} |a_p|\, \xi_0^{p/2}$;

(iii) *if* $a \geqq 0$ *and* $t > 0$, *then* $\hat{P}(\sigma, t) = O(\sigma^{-1/2})$, *and this is true uniformly when* $|t| \geqq t_0 > 0$.

4.5. THE MAIN THEOREM.

4.5.1. THEOREM (**[13]** AND **[14]**). *Let* (M, g) *be a r.m. which is* $G^{\cdot\cdot}$ *and* $\mathscr{L} = \{L_p\}_{p \in N}$. *Then*:

(i) *the series* $Z(z) = \sum_i e^{-\lambda_i/z}$ *converges for every* $z \in \boldsymbol{C}^+$;

(ii) $Z(z) =_F \operatorname{Volume}(M, g)\, (z/4\pi)^{d/2} + \sum_{p=1}^{\infty} a_p \exp(ij_{L_p}\pi/2) z^{1/2} \exp(-zL_p^2/4)$ *with* $a_p > 0$ *for every* p.

From 4.2.2 and 4.4.4 we deduce in particular the:

4.5.2. COROLLARY. *For a* $G^{\cdot\cdot}$ *r.m., the spectrum determines* $\mathscr{L}$.

We will now devote most of our time to a sketch of the proof of 4.5.5. For generalizations, comments and applications, see § 5. We make only one comment here: (ii) in 4.5.1 is the companion of the asymptotic expansion 3.4.3, where $t \to 0+$ is replaced by $z \to 0$ in the following sense: z goes to zero on the circle $\{(\xi_0 + iy)^{-1} : y \in \boldsymbol{R}\}$:

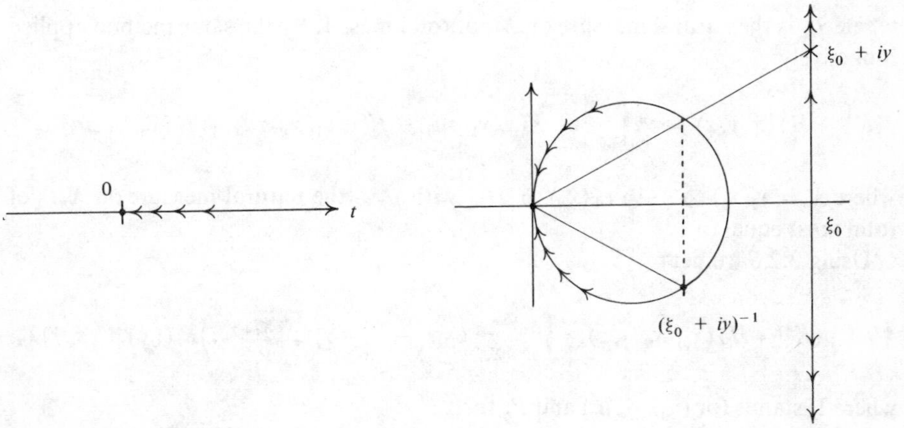

4.6. THE FSCHE.

4.6.1. DEFINITION. *A map* $F : M \times M \times \boldsymbol{C}^+ \to \boldsymbol{C}$ *is a fundamental solution of the complex heat equation (FSCHE) if*:

(i) F *is* C^0 *in the three variables,* C^2 *on the two first and holomorphic in the third, and then* $\Delta_2 F + \partial F/\partial z = 0$;

(ii) *for every continuous* $f : M \to \boldsymbol{C}$ *and every* α *with* $|\alpha| < \pi/2$, *then*

$$\lim_{z \to 0, |\operatorname{Arg}(z)| \leqq \alpha} \int_M F(m, n, z) f(n)\, v(n) = f(m) \quad \textit{for every } m.$$

By extending to C^+ the whole proof of 3.1.2, and K_k, H_k being the extensions to C^+ of the K_k, H_k of 3.2.5, we get

4.6.2. THEOREM. *The FSCHE exists and is unique. For any $k > d/2$, it is given by*

$$F = \sum_{n=0}^{\infty} (-1)^n K^{*n} * H_k.$$

In particular the series $Z(z) = \sum_i e^{-\lambda_i/z}$ converges for every $z \in C^+$ and is also given by

$$Z(z) = \sum_{n=0}^{\infty} (-1)^n \int_M (K_k^{*n} * H_k)(m, m, 1/z)\, v(m).$$

From now on we will pick $k = 2d + 1$ and drop the k in the K_k, H_k; the precise value $2d + 1$ is only used when proving $\sum_{n \geqq n_0} |I_n^+| = O(\sigma^{-1/2})$ in (iv) of 4.9.1.

4.7. COMPUTING $K^{*n} * H$. From 3.3.1 and 4.3 we have

$$\begin{aligned}(A * B)(m, n, z) &= \int_0^z \int_M A(m, q, \theta)\, B(q, n, z - \theta)\, v(q)\, d\theta \\ &= z \int_0^1 \int_M A(m, q, \theta z)\, B(q, n, z(1-\theta))\, v(q)\, d\theta \\ &= z \int_{M \times X_1} A(m, q, u_0 z)\, B(q, n, u_1 z)\, v(q) \mu_1,\end{aligned}$$

where μ_1 is the natural measure on X_1, of total mass 1. By the same method applied inductively:

$$(K^{*n} * H)(x, y, z) = z^n \int_{M^n \times X_{n+1}} K(x, x_1, zu_0) \cdots K(x_{n-1}, x_n, zu_{n-1})\, H(x_n, y, zu_n)\, \lambda_n$$

where $\lambda_n = v_1(x_1) \otimes \cdots \otimes v_n(x_n) \otimes \mu_{n+1}$ with μ_{n+1} the natural measure on X_{n+1} of total mass equal to 1.

Using 3.2.6 we get

$$\text{4.7.1.} \quad (K^{*n} * H)(x_0, x_{n+1}, z) = \int_{M^n \times X_{n+1}} \exp\left(-\frac{1}{4z} \sum_{i=0}^{n} \frac{\overline{x_i x_{i+1}}^2}{u_i}\right) P_n((x_i), U, z^{-1}) \lambda_n$$

where U stands for $(u_0, \cdots, u_n)$ and P_n for

$$\text{4.7.2.} \quad \begin{aligned} P_n((x_i), U, z) = z^{-n} \prod_{i=0}^{n-1} &\left[\left(\frac{z}{4\pi u_i}\right)^{d/2} \sum_{j=-1}^{k} V_j(x_i, x_{i+1}) \left(\frac{u_i}{z}\right)^j\right] \\ &\cdot \left[\left(\frac{z}{4\pi u_n}\right)^{d/2} \sum_{j=0}^{k} U_j(x_n, x_{n+1}) \left(\frac{u_n}{z}\right)^j\right].\end{aligned}$$

Note that, as a polynomial in z (we will always mean by *polynominal in z* a polynomial with positive and negative powers of $z^{1/2}$), P_n is of the highest possible degree, equal to $\frac{1}{2}(n + 1)d$ with corresponding coefficient

4.7.3. $$\left[\prod_{i=0}^{n-1} \frac{V_{-1}(x_i, x_{i+1})}{u_i(4\pi\, u_i)^{d/2}}\right] \frac{U_0(x_n, x_{n+1})}{(4\pi u_n)^{d/2}}.$$

Another way of writing the P_n's is:

4.7.4. $$P_n((x_i), U, z) = \sum_{j_0,\cdots,j_n} V_{j_0}(x_0, x_1) \cdots V_{j_{n-1}}(x_{n-1}, x_n) \cdot U_{j_n}(x_n, x_0)\, Q_{j_0\cdots j_n}(U)\, z^{n(d/2-1)-\Sigma_{i=0}^{n} j_i + d/2}$$

Now if in $(x_i)_{i=0,\cdots,n+1}$ we put $x_{n+1} = x_0$, then the exponential in 4.7.1 works on

$$\sum_{i=0}^{n} \frac{\overline{x_i\, x_{i+1}}^{\,2}}{u_i} = E_U((x_i)) \quad \text{from 4.3.1.}$$

This, combined with 4.6.2, yields finally

4.7.5. $$Z(z) = \sum_{n=0}^{\infty}(-1)^n \int_{M_0^{n+1}\times X_{n+1}} \exp\left(-\tfrac{1}{4}z\, E_U((x_i))\right) P_n((x_i), U, z)\, \lambda_{n+1},$$

where the integral is in fact only over M_0^{n+1} instead of $M \times M^n$ in view of the property of the support of the U_i and V_i (cf. 3.2.6).

4.7.6. *The formula* 4.7.5 *is the key to the proof; for it displays precisely the energy function and on the other hand the stationary phase technique* (*see* 4.8 *for precise statements*) *shows that integrals like those entering into* 4.7.5 *have a behavior for* $z \to \infty$ *related to the critical points of* E_U. The proof itself is complicated by technical difficulties: the polynomial P_n's have negative powers in z and also in the u_i's, moreover X_{n+1} is not compact.

4.7.7. We write $P_n = \sum_s p_{n,s}\, z^s$ and separate P into $P_n = P_n^+ + P_n^-$:

$$P_n^+ = \sum_{s\geq 0} p_{n,s}\, z^s, \qquad P_n^- = \sum_{s<0} p_{n,s}\, z^s.$$

We have to study $\hat{Z}$ (see 4.4.3 and 4.5), so we introduce the obvious expression:

$$\hat{Z}(\sigma, t) = \sum_{n=0}^{\infty}(-1)^n (I_n^+(\sigma, t) + I_n^-(\sigma, t)).$$

Now, the negative powers are easy to handle, since elementary estimates based on 4.4.5 (ii) give

4.7.8. For every t,

$$\sum_{n=0}^{\infty} |I_n^-| = O(\sigma^{-1/2}).$$

For the part I_n^+ we have first (see 4.4.3 and 4.6.1) to prove

$$\sum_{n=0}^{\infty} |I_n^+| = O(\sigma^{-1/2})$$

whenever $t \notin \mathscr{L}^2/4$; and afterwards, to see what happens when $t = L^2/4$ for some $L \in \mathscr{L}$.

4.8. THE STATIONARY PHASE TECHNIQUE. Setting

4.8.1. $$D_n = \int_{M_0^{n+1} \times X_{n+1}} \exp(-z\,\tilde{E}/4)\, P_n^+ \lambda_{n+1},$$

we have to study the behavior, up to $O(\sigma^{-1/2})$, of $I_n^+ = \hat{D}_n$ when $z \to \infty$. The basic tool is the:

4.8.2. THEOREM (THE STATIONARY PHASE [14]). *Let S be any C^∞ manifold of dimension N with some measure dx, $f : S \to \boldsymbol{R}$ a C^∞ function on S, $g : S \to \boldsymbol{C}^+$ a function on S with compact support. Then:*

(i) *if f_S has no critical point in the support of g, we have $\int_S z^b \exp(-zf(x))\, g(x) dx$ bounded when $z \to \infty$ for every real number b.*

(ii) *if the critical points of f in* supp (g) *form a nondegenerate critical submanifold W, of dimension w and index j_W, then*

$$\int_S \exp(-zf(x))\, g(x) dx = (2\pi/z)^{(N-W)/2} \exp(ij_W\, \pi/2 - zf(W))(a + r(z)),$$

with $a > 0$ and $r(z) \to 0$ when $z \to \infty$.

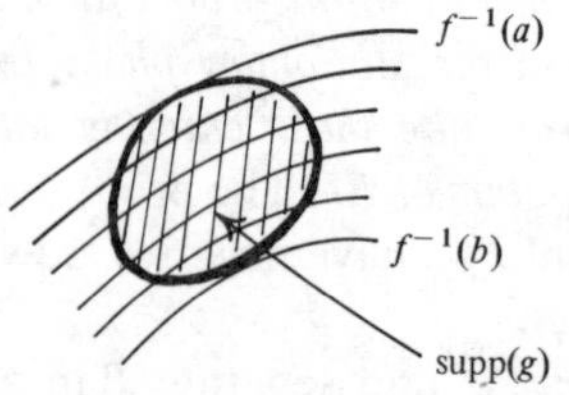

To prove (i), since there are no critical points of f in supp (g), we can make the change of variables $f(x) = t$ and write, using Fubini's theorem, the integral we are interested in as:

$$\int_a^b \left(\int_{f(x)=t} g(f^{-1}(t)) \right) z^b \exp(-zt) dt = \int_a^b z^b u(t) \exp(-zt)\, dt,$$

with u a C^∞ function; then (i) follows by integration by parts, because $|\exp(-zt)| = \exp(-\xi_0 t)$ (see 4.4.1). When f has critical points, one uses partitions of unity and Morse-type coordinate charts to reduce the problem to study the integral

$$\int_{R^N} \exp(z(x_1^2 + \cdots + x_{j_W}^2 - x_{j_W+1}^2 - \cdots - x_{N-W}^2)) h(x_1, \cdots, x_N)\, dx_1 \cdots dx_N,$$

and again everything is reduced to the Gauss integral.

4.8.3. The idea of the proof of 4.5 now is as follows: to study $I_n^+ = \hat{D}_n$ we write it as

$$I_n^+ = \int_{M_0^{n+1} \times X_{n+1}} \exp(-\xi_0 \tilde{E}/4) \exp(iy(t - \tilde{E}/4) - \sigma(y - \sigma^{-1/2})^2)\, P_n^+ \lambda_{n+1} \otimes dy,$$

where the important part is the term $iy(t - \tilde{E}/4)$, because 4.4.5 (iii) shows that if we are bounded away from t for $\tilde{E}/4$, then we are $O(\sigma^{-1/2})$; and, on the other hand, close to t we have compact support and we can apply 4.8.2 to 4.8.1. The remaining technical problems are now: to break the thing into two parts, one with compact support and the other away from the given t; to take care of the fact we have not one I_n^+ but a series, to deal with the noncompactness of X_{n+1} coupled with the negative powers for the u_i's. These technical problems are overcome with the two Lemmas 4.9.1 and 4.9.3. To be more clear we break the proof into two cases: $t \notin \mathscr{L}^2/4$ and $t \in \mathscr{L}^2/4$ (see 4.4.3).

4.9. The case $t \notin \mathscr{L}^2/4$. We will cut $\sum_{n=0}^{\infty} |I_n^+|$ into two parts: $\sum_{n\leqq n_0} + \sum_{n>n_0}$ for a suitable n_0 and will show that each term of $\sum_{n\leqq n_0}$ is $O(\sigma^{-1/2})$ while the series $\sum_{n>n_0}$ also is $O(\sigma^{-1/2})$. To that effect (and many others) we have the

4.9.1. Lemma. *A $t \notin \mathscr{L}^2/4$ being given, there exist two real numbers ρ_0, ρ_1 with $0 < \rho_0 < \rho_1 < \rho$, a function $\varphi : \boldsymbol{R} \to \boldsymbol{R}$ with support in $[t-\beta, t+\beta]$ and equal to 1 near t, and an integer n_0 such that, if we take the function η in 3.2.4 with that choice of ρ_0, ρ_1, then we have:*

(i) $\forall\, n > n_0$: supp $P_n^+ \subset \boldsymbol{C}^+ \times \{((x_i), U) : E_U((x_i)) \geqq 4(t+1)\}$;

(ii) $[t-\beta, t+\beta] \cap \mathscr{L}^2/4 = \varnothing$;

(iii) $\forall\, n \in \{0, \cdots, n_0\} : [n\rho_0, n\rho_1] \cap [2(t-\beta)^{1/2}, 2(t+\beta)^{1/2}] = \varnothing$;

(iv) $\hat{Z}(\sigma, t) = \sum_{n=0}^{n_0} \hat{J}_n + O(\sigma^{-1/2})$, *where*

$$J_n(z) = (-1)^n \int_{M_0^{n+1} \times X_{n+1}} \exp(-z\,\tilde{E}/4)\,(\varphi \circ \tilde{E}/4)\, P_n^+ \lambda_{n+1}.$$

The choices are made in the following order: $\rho_0, n_0, \varphi, \rho_1$. The key one is the choice of n_0. For ρ_0 pick any $\rho_0 < \rho$ with $2\,t^{1/2} \notin \rho_0$. To find n_0, for $(x_i) \in M_0^{n+1}$ define

$$N((x_i)) = \#\{i \in \{0, \cdots, n-1\} : \overline{x_i\, x_{i+1}} < \rho_0\}.$$

By 4.3.1, $E_U((x_i)) \geqq (N((x_i))\rho_0)^2$; and, because we are only interested in P_n^+ and, with 4.7.4, $\sum_{i=0}^{n} j_i \leqq n(d/2 - 1) + d/2$; moreover 3.2.6 and the definition of $N((x_i))$ imply

$$\sum_{i=0}^{n} j_i \geqq (n - N((x_i)))\,(2d+1) - N((x_i))$$

(remember $k = 2d+1$, 4.6). Hence, using the three inequalities above, it is trivial to find n_0 which implies (i) of 4.9.1. Now it is easy to pick β with $[t-\beta, t+\beta] \cap (\boldsymbol{N}\rho_0)^2/4 = \varnothing$ and $[t-\beta, t+\beta] \cap \mathscr{L}^2/4 = \varnothing$ since $\mathscr{L}$ is closed. Then one builds a function φ as usual; and, finally, choose ρ_1 with (iii).

Now we have to prove (iv); first $\sum_{n\geqq n_0} |I_n^+| = O(\sigma^{-1/2})$ follows from 4.9.1 (i) and estimates based on 4.4.5 (ii). Again from 4.9.1 (i), 4.4.5 (iii) and now also 4.9.1 (ii), it follows that, for every $n \in \{0, \cdots, n_0\} : \hat{A}_n(\sigma, t) = O(\sigma^{-1/2})$, when we set

$$A_n = \int_{M_0^{n+1} \times X_{n+1}} \exp(-z\tilde{E}/4)\,(1 - \varphi \circ \tilde{E}/4)\, P_n^+ \lambda_{n+1};$$

this ends the proof of (iv) and so the proof of Lemma 4.9.1 is now complete.

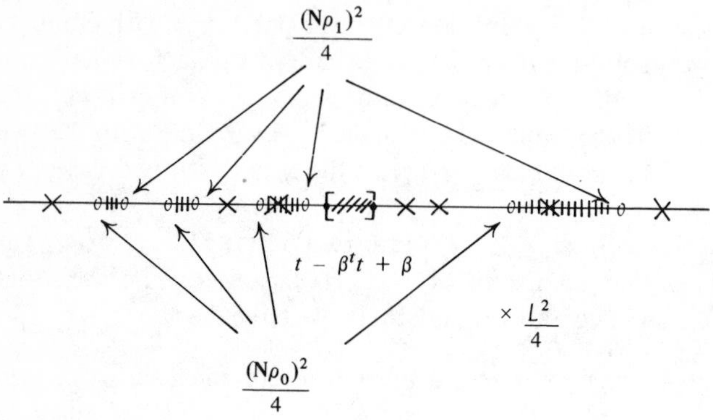

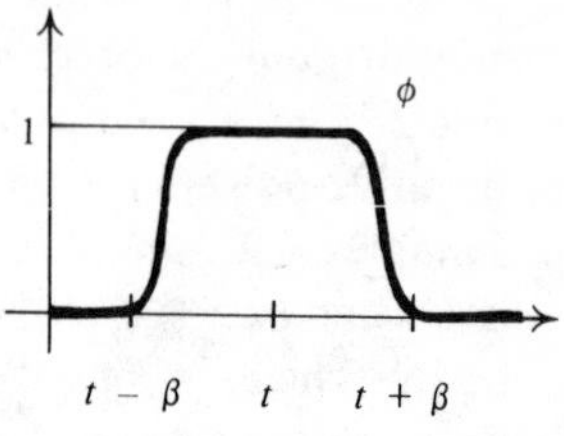

4.9.2. To finish the case $t \in \mathscr{L}^2/4$, the result 4.9.1 (iv) shows that it remains only to prove: for every $n \in \{0, \cdots, n_0\}$ one has $\hat{J}_n(\sigma, t) = O(\sigma^{-1/2})$. This would be achieved by 4.8.2(i) with $S = M_0^{n+1} \times X_{n+1}$, $f = \tilde{E}/4$ and $g = (\varphi \circ \tilde{E}/4)P_n^+$, since by 4.3.2 and 4.9 (ii) we know has no critical point in supp(g). But g does not have a compact support in the noncompact manifold S. So again we will have to cut J_n into two pieces; that is achieved by the:

4.9.3. LEMMA. *For every $n \in \{0, \cdots, n_0\}$ there exists $\psi : \boldsymbol{R} \to \boldsymbol{R}$ with compact support, equal to* 1 *near* 0, *such that $(\varphi \circ \tilde{E}/4)(1 - \psi(\overline{x_n x_0}))P_n^+$ has compact support in $S = M_0^{n+1} \times X_{n+1}$ and*

$$A_n = \int_S \exp(-z\tilde{E}/4)(\varphi \circ \tilde{E}/4)\psi(\overline{x_n x_0})\, P_n^+ \lambda_{n+1}$$

is bounded in z.

See **[14]** for the proof, which uses basically (iii) of 4.9.1. Then we are done, since now if we cut J_n $(n \in \{0, \cdots, n_0\})$ into $J_n = (-1)^n(A_n + B_n)$ with

$$B_n = \int_S \exp(-z\tilde{E}/4)(\varphi \circ \tilde{E}/4)(1 - \psi)\, P_n^+ \lambda_{n+1},$$

then $\hat{A}_n$ is $O(\sigma^{-1/2})$ by 4.4.5 (i) and 4.9.3 and $\hat{B}_n$ is $O(\sigma^{-1/2})$ by 4.4.5 (i), 4.8.2 (ii) and 4.3.2.

4.10. THE CASE $t \in \mathscr{L}^2/4$. We now only indicate quickly the modifications which have to take place when $t = L^2/4$ with $L \in \mathscr{L}$. Basically there is only one modification: (ii) of 4.9.1 is replaced by

4.10.1. $$\mathscr{L}^2/4 \cap [t - \beta, t + \beta] = \{L^2/4\}, \qquad L \in \mathscr{L}.$$

Then everything goes smoothly as in the preceding proof, except at the end in the behavior of the B_n's and the $\hat{B}_n$'s $(n = 0, \cdots, n_0)$: now E has critical points in supp$(\varphi \circ \tilde{E}/4)$. We still cut J_n into two pieces A_n and B_n, but first we can manage in Lemma 4.9.3 to have $\psi = 0$ at critical points of E. So we have only to study (see 4.7.7):

$$4.10.2. \qquad B_n = \sum_{s=0}^{(n+1)d/2} \int_S z^s \exp(-z\,\tilde{E}/4)(\varphi \circ \tilde{E}/4)(1-\psi)p_{n,s}\,\lambda_{n+1}.$$

By 4.3.2 and 4.10.1 the critical points of $\tilde{E}$ in S form two nondegenerate critical submanifolds $h^{-1}(W_+)$ and $h^{-1}(W_-)$(see 4.2.1) both of dimension $w = n+1$ and index j_L (see 4.2.2). Since the dimension of S is $N = \dim S = (n+1)d + n$, we have for (ii) of 4.8.2: $(N - w/2) = ((n+1)d - 1)/2$; and since 4.8.2 implies we have only to consider the term with $s = (n+1)d/2$ in 4.10.2, we see that

$$4.10.3. \qquad B_n = z^{(n+1)d/2}\left(\frac{2\pi}{z}\right)^{((n+1)d-1)/2} \exp(ij_L\pi/2 - zL^2/4)(a_n^+ + a_n^- + r(z))$$

with $r(z) \to 0$ when $z \to \infty$ and the obvious $\pm$ notations.

Now to get 4.5.1 it remains only to sum up 4.10.3 from $n = 0$ to $n = n_0$ and check $a_n^+ > 0$, $a_n^- > 0$ for every $n = 0, \cdots, n_0$. But, by 4.7.2, the term $P_{n,s}$ with $s = (n+1)d/2$ is necessarily

$$(1/4\pi)^{(n+1)d/2}\, V_{-1}(x_0, x_1) \cdots V_{-1}(x_{n-1}, x_n)\, U_0(x_n, x_0),$$

whose sign is $(-1)^n$ when $\rho_0 < \overline{x_i\, x_{i+1}} < \rho_1$ and $\overline{x_n\, x_0} < \rho_1$ by 3.2.6.

5. Other results of [14].

5.1. A STATEMENT FOR $G^{\cdot}$ RIEMANNIAN MANIFOLDS. The Theorem 4.5.1 will not apply to very nice r.m. like spheres, or symmetric spaces for example. But note those r.m. are $G^{\cdot}$ in the 4.2.1 sense. In fact we stated and proved 4.5.1 only for $G^{\cdot\cdot}$ r.m. for the sake of simplicity of computation. But since the stationary phase (4.8.2) applies to any nondegenerate critical submanifold, it is clear that the whole proof of 4.5.1 carries over for a $G^{\cdot}$ r.m.

Moreover, if one notes that the first term in (ii) of 4.5.1 is the first term of the asymptotic expansion 3.4.3, one is tempted to ask if there exists a generalization of 4.5.1 where in every term of the right-hand sum, the part $a_p \exp(ij_{L_p}\pi/2)z^{1/2}$ would be replaced by a polynomial in $z^{1/2}$ of order as big as desired. That in fact is indeed the case, and this was the reason for introducing the notation "$=_{F_\alpha}$" in 4.4.3; the proof is analogous to the proof of 4.5.1 the general statement being the following:

5.1.1. THEOREM. *Let (M, g) be a G. r.m. and $\mathcal{L} = \{L_p\}$, $p \in N$. Then, for every $\alpha \geqq 0$, we have*

$$Z(z) =_{F_\alpha} \sum_{p=0}^{\infty} f_p^\alpha(z) \exp(-z\,L_p^2/4),$$

where the f_p^α's are of the following type: if, for a given L_p, the critical points of E consist (see 4.2) of $\bigcup_{\lambda=1}^{A} W_\lambda$ where every W_λ is a nondegenerate critical sub-

manifold of nullity n_λ *and index* j_λ, *then* $f_p^\alpha = \sum_{\lambda=1}^{\Lambda} P_\lambda^\alpha$ *where*

$$P_\lambda^\alpha(z) = \exp(ij_\lambda \pi/2)(a_0 z^{(n_\lambda+1)/2} + a_1 z^{(n_\lambda-1)/2} + \cdots + a_n z^{-[2\alpha-1]/2}) \quad \text{if } \mathrm{Im}(z) \geqq 0$$

$$= \overline{P_\lambda^\alpha(\bar{z})} \quad \text{if } \mathrm{Im}(z) < 0,$$

and $a_0 > 0$.

5.1.2. *Notes.* First the f_p^α are uniquely determined by Spec(M, g); see 4.4.4. Second f_0^α, corresponding to the trivial periodic geodesics of length 0 (see 4.2), yields the coefficients of the asymptotic expansion 3.4.3. Third: Be careful to note that *in general one cannot obtain* $\mathscr{L}$ *from the spectrum* by using only 5.1.1; suppose in fact there are for some $L \in \mathscr{L}$ exactly two periodic geodesics (both of nullity 0) of length L but with indices differing by two; then the two corresponding terms in $\exp(-zL^2/4)$ will cancel and L will never show up in 5.1.1. But there might be some day another method which will recover $\mathscr{L}$ from the spectrum.

5.2. EXAMPLES.

5.2.1. For the standard sphere (S^3, g_0), the formula of 5.1.1 agrees with the formula of [**4**, p. 26].

5.2.2. FLAT TORI. They are $G^\cdot$; every periodic geodesic corresponds to an element of the lattice Λ on one hand, and on the other hand such a periodic geodesic yields exactly one nondegenerate critical submanifold of nullity always $d-1$ and of index always 0; whence

$$\text{for every } \alpha: \quad P_\lambda^\alpha(z) = a_\alpha (z/4\pi)^{d/2} \text{ Volume }(M, g),$$

for some $a_\alpha \in \boldsymbol{R}$ and the sum in 5.1.1 looks like the Poisson summation formula; see [**8**, p. 157] for example.

5.2.3. When (M, g) is of strictly negative curvature, then every periodic geodesic is of index and nullity 0. In particular (M, g) is $G^\cdot$, so 5.1.1 will read:

$$Z(z) =_{F_\alpha} \mathrm{Volume}(M, g)(z/4\pi)^{d/2} + \sum_\gamma a_\gamma z^{1/2} \exp(-z L_\gamma^2/4),$$

where the summation is over the set of nontrivial periodic geodesics γ, and a_γ always positive. Note that, in the present case, one can read $\mathscr{L}$ from the spectrum. When $d = 2$ that formula is a weak version of an asymptotic expansion that one can find in [**12**].

5.2.4. If not only $d = 2$ but moreover (M, g) is of constant curvature, then the spectrum still gives $\mathscr{L}$; it is in this special case that the first historical statement of that type occurred, by H. Huber in [**20**]; see also [**26**].

5.2.5. We wish here to warn the hopeful reader: even in good cases where one can recover $\mathscr{L}$ from the spectrum, that is not enough to get the local injectivity of the Spec map, cf. 2.4. For, recall the existence on S^2 of riemannian structures different from the standard one g_0 and having however all their periodic geodesics of same length 2π (and their multiples) with always same nullity, and same indexes: those are the Zoll's surfaces (see [**5**, p. 145]) and it is not too hard to find nontrivial C^∞-families of such structures.

5.2.6. Let us again try to apply 5.1.1 to the problem of isospectral deformations (see 2.4.2) in the very special case $d = 2$ and (M, g) of strictly negative curvature. Restricting us to conformal deformations $f(t) \cdot g$ with $f: M \times \boldsymbol{R} \to \boldsymbol{R}^*$, the condition

$$\mathrm{Spec}(M, f(t) \cdot g) = \mathrm{Spec}(M, f(0) \cdot g)$$

for every t will yield, thru 5.1.1, that every periodic geodesic is of constant length, from which we get easily that $\int_\gamma f'(0) d\gamma = 0$ for every such periodic geodesic γ of (M, g). But it does not look, at the present time, as if this would imply $f'(0) = 0$.

5.3. Other Statements. The reader will find in **[14]** the following extensions of 5.1.1.

5.3.1. An equivariant version relative to the case where we are given on (M, g) a fixed isometry T; then for the $G^{\cdot}$ case, if we define $Z_T(z) = \sum_\lambda t_\lambda \exp(-\lambda/z)$ with t_λ being the trace of the action of T on $\{f: \Delta f = \lambda f\}$, there exists a formula like in 5.1.1 for Z_T, where the P_λ^α's are replaced by suitable ones involving T.

5.3.2. A formula giving $E(x, y, 1/z)$ for any $x, y \in M$ and involving the set of the lengths of all geodesics joining x to y.

5.3.3. An extension of 5.1.1 to any laplacian of the Combet's type **[15**, p. 247**]** for vector bundle on (M, g) with $G^{\cdot}$. Combet's laplacians include the laplacian for exterior forms. A corollary of that extension is: for a $G^{\cdot\cdot}$ r.m. the knowledge of the spectra $\mathrm{Spec}^p(M, g)$ (see 3.5.4) for $p = 0, \cdots, d$, besides $\mathscr{L}$, determines the parallel transport along all periodic geodesics.

Bibliography

1. R. Abraham, *Bumpy metrics*, Proc. Sympos. Pure Math., vol. 14, Amer. Math. Soc., Providence, R. I., 1970, pp. 1–3. MR **42** #6875.

2. M. Atiyah, R. Bott and V. K. Patodi, *On the heat equation and the index theorem*, Invent. Math. **19** (1973), 279–330.

3. R. Balian and C. Bloch, *Distributions of eigenfrequencies for the wave in a finite domain.* III, Ann. Physics **69** (1972), 76–160. MR **44** #7147.

4. A. Benabdallah, *Noyau de diffusion sur les espaces homogènes compacts*, Bull. Soc. Math. France **101** (1973), 265–283.

5. M. Berger, *Lecture notes on geodesics in Riemannian geometry*, Tata Institute of Fundamental Research Lectures on Math., no. 33, Tata Institute of Fundamental Research, Bombay, 1965. MR **35** #6100.

6. ———, *Eigenvalues of the Laplacian*, Proc. Sympos. Pure Math., vol. 16, Amer. Math. Soc., Providence, R. I., 1970, pp. 121–125. MR **41** #9141.

7. ———, *Sur les premiéres valeurs propres des variétés riemanniennes*, Compositio Math. **26** (1973), 129–149.

8. M. Berger, P. Gauduchon and E. Mazet, *Le spectre d'une variété riemannienne*, Lecture Notes in Math., vol. 194, Springer-Verlag, Berlin and New York, 1971. MR **43** #8025.

9. R. S. Cahn, *The asymptotic expansion of the zeta-function of a compact semi-simple Lie group* (to appear).

10. J. Chazarain, *Spectre d'un operateur elliptique et bicaractéristiques périodiques*, C. R. Acad. Sci. Paris, Juillet 1973.

11. ———, *Formule de Poisson pour les variétés riemanniennes*, Invent. Math. **24** (1974), 65–82.

12. Y. Colin de Verdière, *Spectre du laplacien et longueurs des géodésiques périodiques*, I. Compositio Math. **27** (1973), 83–106.

13. Y. Colin de Verdière, *Spectre du laplacien et longueur des géodésiques périodiques*, C. R. Acad. Sci. Paris Sér. A–B **276** (1973), A1517–A1519.

14. ———, *Spectre du laplacien et longueurs des géodésiques périodiques*. II, Compositio Math. **27** (1973), 159–184.

15. E. Combet, *Paramétrix et invariants sur les variétés compactes*, Ann. Sci. École Norm. Sup. (4) **3** (1970), 247–271. MR **44** #7589.

16. H. Duistermaat, *The spectrum and periodic geodesics*, these PROCEEDINGS.

17. P. Flaschel and W. Klingenberg, *Riemannsche Hilbertmannifaltigkeiten, periodätische Geodätische*, Lecture Notes in Math., vol. 282, Springer-Verlag, Berlin and New York, 1972.

18. P. Gilkey, *Spectral geometry and the Kaehler condition for complex manifolds, for complex* Math. **26** (1974), 231–258.

19. V. Guillemin, *On a spectral property of elliptic operators with periodic bicharacteristic flow manifolds*, Inventions (preprint).

20. H. Huber, *Zur analytischen Theorie hyperbolischer Raumformen und Bewegungsgruppen*, Math. Ann. **138** (1959), 1–26. MR **22** #99.

21. V. F. Lazutkin, *Asymptotics of the eigenvalues of the Laplacian . . .*, Math. USSR Izvestija **7** (1973), 439–466.

22. M. Kac, *Can one hear the shape of a drum*?, Amer. Math. Monthly **73** (1966), no. 4, part II, 1–23, MR **34** #1121.

23. ———, *On applying mathematics*: *Reflexions and examples*, Quart. Appl. Math. **30** (1972), 17–29.

24. Y. Kitaoka, *On the relation between the positive definite quadratic forms with the same representation numbers*, Proc. Japan Acad. **47** (1971), 439–441.

25. G. Louchard, *Mouvement Brownien et valeurs propers du Laplacien*, Ann. Inst. H. Poincaré Sect. B. **4** (1968), 331–342. MR **41** #6308.

26. H. P. McKean, *Selberg's trace formula as applied to a compact Riemann surface*, Comm. Pure Appl. Math. **25** (1972), 225–246.

27. H. P. McKean, Jr. and I. M. Singer, *Curvature and the eigenvalues of the Laplacian*, J. Differential Geometry **1** (1967), no. 1, 43–69. MR **36** #828.

28. J. Milnor, *Morse theory*, Ann. of Math. Studies, no. 51, Princeton Univ. Press, Princeton, N.J., 1968. MR **29** #634.

29. ———, *Eigenvalues of the Laplace operator on certain manifolds*, Proc. Nat. Acad. Sci. U.S.A. **51** (1964), 542. MR **28** #5403.

30. M. A. Naĭmark, *Linear differential operators*, GITTL, Moscow, 1954; English transl., Ungar, New York, 1968. MR **16**, 702; **41** #7485.

31. V. K. Patodi, *Curvature and the fundamental solution of the heat operator*, J. Indian Math. Soc. **34** (1970), 269–285.

32. L. E. Payne, *Isoperimetric inequalities and their applications*, SIAM Rev. **9** (1967), 453–488. MR **36** #2058.

33. T. Sakai, *On eigen-values of Laplacian and curvature of Riemannian manifold*, Tôhoku Math. J. (2) **23** (1971), 589–603. MR **46** #2603.

34. ———, (to appear).

35. I. Singer, *Geometry of the spectra*. II, these PROCEEDINGS.

36. S. Tanno, *Eigenvalues of the Laplacian of Riemannian manifolds* (to appear).

37. R. T. Waechter, *On hearing the shape of a drum*: *An extension to higher dimensions*, Proc. Cambridge Philos. Soc. **72** (1972), 439–447. MR **46** #4019.

38. F. W. Warner, *Foundations of differentiable manifolds and Lie groups*, Scott, Foresman, Glenview, Ill., 1971.

39. A. Weinstein, *Application des opérateurs intégraux de Fourier aux spectres des variétés riemanniennes*, C. R. Acad. Sci. Paris Sér. A **279** (1974), A229–A230.

UNIVERSITÉ PARIS VII

Proceedings of Symposia in Pure Mathematics
Volume 27, 1975

CONSTANT SCALAR CURVATURE METRICS FOR COMPLEX MANIFOLDS

MELVYN S. BERGER*

Introduction. This article is concerned with the construction of a Hermitian metric of constant scalar curvature on a given compact complex manifold M. The Hermitian metric is required to be compatible with the complex structure of M. Thus, our topic is closely related to the natural problem of finding the Hermitian metric with the "simplest" curvature properties on a given complex manifold.

The topic in complex dimension one can be traced back to Monge and Liouville, who formulated and solved a local version of the problem. The actual global problem reached paramount importance in the late nineteenth century with the work of Poincaré and Klein on the uniformization of algebraic curves. The notion of universal covering surface (introduced by H.A. Schwarz) and the associated Riemann mapping theorem for simply connected Riemann surfaces proved sufficient to solve the uniformization problem. When coupled with the Gauss-Bonnet theorem, these results also yielded complete results on the construction of constant Gauss curvature metrics on any Riemann surface.

Until recently, analogous results on higher dimensional complex manifolds have been noticeably lacking. Perhaps the principal reasons for this situation were (a) there is no known sharp analogue for higher dimensional algebraic varieties as yet of the uniformization theorem for curves; (b) the generalized Gauss-Bonnet theorem for higher dimensional complex manifolds M does not seem useful in restricting the sign of a possible constant curvature metric by the topology of M, alone, as in the Riemann surface case.

In this article we shall discuss various approaches to the following general

AMS (MOS) subject classifications (1970). Primary 53C55, 58E15.

*Research partially supported by AFOSR grant 73–2437, and a National Science Foundation grant.

problem:

(π_1) Given a compact complex manifold M, construct a metric $\tilde{g}$ of constant scalar curvature $\tilde{R}$ (compatible with the complex structure of M) and relate the sign of $\tilde{R}$ to the given complex structure defined on M.

Because of its general applicability we shall concentrate on describing a direct approach to (π_1) via partial differential equations. This approach is entirely consistent with the early approach of Monge-Liouville and was in fact mentioned by Poincaré in [1] well before the prerequisite technique of studying partial differential equations on manifolds had been developed. Our discussion of the problem (π_1) (contained in §§ II and III) will be based on this approach via partial differential equations together with the following "complex" analogue of the Gauss-Bonnet formula for a compact Hermitian manifold (M, g) with scalar curvature $R(x)$ and $\dim_c M = N$

$$(0.1) \quad \int_M R(x) = b_N \int_M c_1 \wedge \omega^{N-1} \text{ where } \begin{cases} c_1 = \text{first Chern class of } M, \\ \omega = \text{fundamental form associated with } (M, g), \\ b_N = \text{absolute positive constant,} \end{cases}$$

and recent extensions of this formula by S. T. Yau [2]. The formula (0.1) is valid, provided one uses the so-called Hermitian connection on M ([3], [4]). The use of this connection is crucial in our work and is "natural" for (π_1) since the Hermitian connection is the unique connection compatible with a given complex structure on M. The Hermitian connection coincides with the Levi-Civita connection, for Kähler manifolds but on Hermitian (non-Kähler) manifolds, the results based on either of the two connections differ substantially. This distinction is described in §IV. (Further motivation for passing outside the class of Kähler manifolds and the class of Levi-Civita connections can be found in § II.)

In §V we apply our results to algebraic surfaces by utilizing the Enriques-Kōdaira classification theory of algebraic surfaces. For "most" regular algebraic surfaces M it is possible to prove, as in the one-dimensional case, that (i) a given Hermitian g on M can be deformed to a metric $\tilde{g}$ of constant scalar curvature $\tilde{R}$, and (ii) sgn $\tilde{R}$ depends only on the complex structure of M and not on the particular Hermitian metric g defined on M.

To date, the approach to (π_1) via partial differential equations has not been successful for cases in which $\tilde{R} > 0$. It is not clear whether this difficulty lies in the nature of the deformation problem itself or in the present state of the art of partial differential equations. Moreover our work can be considered as a necessary first step in the more difficult problem of deforming a given Hermitian metric g on a complex manifold M to an Einstein-Hermitian metric $\tilde{g}$. Despite a clever but unfortunately erroneous attempt by Aubin [5], this problem remains an important unsolved issue.

I. The problem in complex dimension one.

I.A. *Monge's problem.* Perhaps the earliest study of the construction of metrics of constant curvature is due to Monge, who posed the following "local" problem:

Under what circumstances does the metric $ds^2 = \lambda(x, y)\,[dx^2 + dy^2]$ have constant total (i.e. Gauss) curvature K?

This problem was solved by Liouville and his solution can be found in a Note in a late edition of Monge's classic *Application de analyse à geometrie*. Setting $u = \log \lambda$, Monge observed that $u(x, y)$ must satisfy the partial differential equation

$$\text{(I.1)} \qquad \Delta u = -2Ke^u \quad \text{where } \Delta = \frac{\partial^2}{\partial x^2} + \frac{\partial^2}{\partial y^2}.$$

Liouville was able to integrate this equation completely. In fact he found the general solution to Monge's problem could be written in the form

$$\text{(I.2)} \qquad ds^2 = 4a^2 \exp(2U)\,\{1 \pm \exp 2U\}^{-2}\,\{dU^2 + dV^2\}; \qquad (K = \pm 1/a^2)$$

where U and V are the real and imaginary parts of an *arbitrary* complex-valued function $\zeta(z)$ with $z = x + iy$, so that

$$\text{(I.3)} \qquad dU^2 + dV^2 = \zeta(z)\bar{\zeta}(z)\,\{dx^2 + dy^2\}.$$

The fact that the function $u(x, y)$ defined by (I.2) — (I.3) satisfies (I.1) is proven in §VI. The fact that it represents the general solution of (I.1) is essentially read off from the fact that $u(x, y)$ contains the two arbitrary functions U and V. Unfortunately Liouville's solution is not suitable for a problem more global than Monge's. Indeed, for example, it is easily shown that, for $K < 0$, no smooth solution of (I.1) can be defined on $\boldsymbol{R}^2$.

I. B. *Relation to the uniformization theorem.* Nonetheless, a few decades later the problem of determining globally defined metrics of constant Gauss curvature on a Riemann surface arose in the work of Klein and Poincaré on the uniformization of algebraic curves. This uniformization problem consists of finding a parametrization $x = x(t)$ for the curve defined by polynomial equation $F(w, z) = 0$, where w and z are complex variables and the parameter t is so chosen as to vary over a simply connected domain of the complex plane. It was found that the uniformization problem could be solved by finding on every compact orientable Riemannian 2-manifold (M, g) a metric $\tilde{g}$ of constant Gauss curvature leaving the complex structure of M unchanged. Indeed, assuming this metric g is known, then the argument proceeds as follows: On an appropriate compact Riemann surface S the relation $F(w, z) = 0$ can be written $w = f(z)$. Now if we can represent S as the quotient of a domain D in the complex plane by a discontinuous group Γ (acting without fixed points), then the canonical surjection mapping $\sigma : D \to D/\Gamma = S$ is easily shown to be analytic and single-valued. Thus $z = \sigma(t)$ and $w = f(\sigma(t))$, for $t \in D$, determines the desired uniformization. Now this representation of S, equipped with the metric g by D/Γ is precisely the content of the Clifford-Klein space problem for two-dimensional manifolds of constant Gauss curvature which was solved at the end of the nineteenth century. Thus, the problem is reduced to finding the metric $\tilde{g}$. Unfortunately this approach to the uniformization problem was not completed by Poincaré since the necessary analytic tools had not been sufficiently developed in his time. In fact, a complete reversal of the two problems occurred; by utilizing

H. A. Schwarz's notion of covering space the topological difficulties associated with the nonsimple-connectedness of the Riemann surface associated with $f(x, y) = 0$ were surmounted by Koebe and Poincaré in 1907. The resulting Riemann mapping theorem for simply-connected Riemann surfaces was then shown to yield both the proof of the uniformization of algebraic curves and the existence of the constant curvature metric $\tilde{g}$ mentioned above.

We illustrate this approach via complex function theory in the simplest case in which (M, g) is a simply connected compact orientable Riemannian 2-manifold, and $S(M)$ is the corresponding Riemann surface. One seeks a single-valued (nonconstant) meromorphic function f defined on $S(M)$ that takes the value $z = \infty$ exactly once. Then since a single-valued (nonconstant) meromorphic function takes on all values the same number of times f can be considered as a complex-analytic homeomorphism of $S(M)$ onto the complex sphere. Consequently f determines a conformal diffeomorphism between (M, g) and the sphere (S^2, g_1) with the canonical metric of constant Gauss curvature 1. In order to find the meromorphic function f defined on $S(M)$ it is of great interest to use the Dirichlet principle (in a form due to Hilbert [6]). We choose a point p on M with local parameter z; then we construct a harmonic function U which is smooth everywhere except at some neighborhood $N(P)$ where it behaves like Re $(1/z)$. This harmonic function U is found by minimizing the Dirichlet integral over all C^1 functions in $M - N(p)$ and which behave like U in $N(p)$. (Moreover the differential $dU + i * dV$ is meromorphic on $S(M)$ with the singularity $d(1/z)$ at p.) Thus we can state

(I.4) THEOREM. *A compact orientable 2-manifold (M, g) can be given a conformally equivalent metric $\tilde{g}$ of constant Gauss curvature unity if and only if M is simply connected.*

The necessity of the simple-connectedness of M for the validity of this theorem is an immediate consequence of the Gauss-Bonnet formula, since in the case of complex dimension one the Euler characteristic $\chi(M)$ of M is positive if and only if M is simply connected.

I.C. *The partial differential approach.* For 2-manifolds M that are nonsimply connected the approach to (π_1) via complex function theory requires the introduction of the universal covering surface $\tilde{M}$ of M. Thus it is useful to give (even in complex dimension one) an alternate approach to the construction of metrics of constant Gauss curvature on nonsimply connected manifolds that (a) does not depend on the covering surface $\tilde{M}$ and (b) can be extended to the higher dimensional case where the complex function theory approach breaks down.

This approach, presented in my 1969 paper [7], makes use of a nonlinear Dirichlet principle for the solution of a semilinear elliptic partial differential equation defined on the compact manifold. Interestingly enough the equation is analogous to the Monge-Liouville equation (I.1) (but, of course, its solvability could never be achieved by Liouville's method of "quadrature," but requires a more sophisticated existential argument).

To derive this equation we begin by saying that two C^∞ metrics g and $\tilde{g}$ are (pointwise) conformally equivalent if there is a C^∞ function σ defined on M such that $\tilde{g} = c^{2\sigma}g$. Then in terms isothermal parameters (u, v) locally defined on M, with $g = \lambda(u, v)[du^2 + dv^2]$ the Gauss curvature $K(u, v)$ of M can be written

$$K = -(1/2\lambda)\{(\log \lambda)_{uu} + (\log \lambda)_{vv}\}. \tag{I.5}$$

Consequently the Gauss curvature $\tilde{K}$ of $\tilde{g}$ can be written in terms of the Laplace-Beltrami operator Δ on (M, g) as

$$\tilde{K} = e^{-2\sigma}\{K - \Delta\sigma\}.$$

Thus to find a (pointwise) conformally equivalent metric $\tilde{g}$ with Gauss curvature $\tilde{K} = \text{const}$ one must find a function $\sigma \in C^\infty(M)$ that satisfies the partial differential equation

$$\Delta\sigma - K(x) + \tilde{K}e^{2\sigma} = 0; \qquad \tilde{K} = \text{const}. \tag{I.6}$$

Once again the Gauss-Bonnet formula implies that

$$\operatorname{sgn} \tilde{K} = \operatorname{sgn} \chi(M). \tag{I.6$'$}$$

In order to complete Theorem (I.4) we now state the following result concerning the equation (I.6).

(I.7) THEOREM. *The equation* (I.6) *is solvable with*

(i) $\tilde{K}$ *a negative constant if and only if* $\int_M K(x) < 0$. *Moreover, in this case, the situation is unique.*

(ii) $K = 0$ *if and only if* $\int_M K(x) = 0$. *Moreover we conclude from* (I.7)

(I.8) THEOREM. *The manifold* (M, g) *admits a (pointwise) conformally equivalent metric of*

(i) *constant negative Gauss curvature if and only if the Euler characteristic of* M, $\chi(M) < 0$, *and*

(ii) *zero Gauss curvature if and only if* $\chi(M) = 0$.

PROOF OF (I.8) FROM (I.7). The result follows immediately from (I.7) and the Gauss-Bonnet formula

$$\int_M K(x) = 2\pi\chi(M). \tag{I.9}$$

REMARK ON THE PROOF OF (I.7). We shall postpone the complete proof of (I.7) until §VI, since there will be no essential simplification in treating all the cases $N \geqq 1$ simultaneously. However it is useful to point out at this stage that we solve (I.6) by utilizing the nonlinear Dirichlet principle of minimizing the functional

$$I(u) = \int_M \{\tfrac{1}{2}|\nabla|^2 + K(x)u - \tfrac{1}{2}\tilde{K}\exp 2u\}\, dV \tag{I.10}$$

(where ∇u denotes the gradient of u with respect to (M, g)) over a sufficiently large admissible class of functions C so chosen that

(i) in case $\chi(M) \leqq 0$, $\inf_C I(u)$ is attained by an element $\tilde{u} \in C$,

(ii) the minimizer $\tilde{u}$ of $I(u)$ can be proven sufficiently smooth to satisfy (I.6) (after a possible redefinition on a set of measure zero), since any smooth critical point of $I(u)$ necessarily satisfies (I.10).

In case $\chi(M) < 0$, the appropriate class C will turn out to be the Sobolev space $W_{1,2}(M, g)$ consisting of all functions $u(x)$ such that $u(x)$ and $|\nabla u|$ are in $L_2(M, g)$. In case $\chi(M) = 0$, the class C may be chosen to be the closed subspace of $W_{1,2}(M, g)$ consisting of those functions of mean value zero over M.

A gap in this approach consists in finding a proof of (I.4) by partial differential equations. This difficulty can be traced to the fact that if we attempt to solve (I.6) *with* $\tilde{K} > 0$, *any critical point of* $I(u)$ *will necessarily be a saddle point of the functional* $I(u)$. To demonstrate this, suppose σ is a relative or absolute minimum of $I(u)$ defined by (I.10) with $\tilde{K} = 1$. Then we shall find a contradiction to this by showing that for any (nonzero) constant c,

$$I(\sigma + \varepsilon c) < I(\sigma) \quad \text{for } \varepsilon \text{ sufficiently small.}$$

Indeed

$$\delta^2 I(\sigma, c) = \frac{d^2}{d\varepsilon^2} I(\sigma + \varepsilon c)\Big|_{\varepsilon=0} = -2c^2 \int_M \exp 2\sigma < 0$$

so that

$$I(\sigma + \varepsilon c) - I(\sigma) = \tfrac{1}{2}\varepsilon^2 \delta^2 I(\sigma, c) + O(\varepsilon^3) < 0.$$

Consequently σ cannot be an absolute or relative minimum for $I(u)$.

REMARK. In an addendum to our paper [7], we attempted to overcome the difficulty of studying the saddle points of $I(u)$ for $\tilde{K} > 0$ by formulating the following natural isoperimetric problem associated with (I.6): Minimize the functional $J(u) = \int_M \{\frac{1}{2} |\nabla u|^2 - K(x) u\}$ subject to the constraint $\int_M e^{2u} = 4\pi$. If the compact manifold M is simply connected, any solution of this isoperimetric problem will satisfy (I.6) with $\tilde{K} = 1$. Unfortunately the estimates needed to solve this isoperimetric problem are very difficult to obtain (cf. Moser [**20**]) and to date this approach has succeeded only in the analogous problem of prescribing Gauss curvature function on the projection space (P^2, g_1) with its canonical metric.

II. Discussion of the general problem.

II. A. *Statement of the problem.* We now turn to the formulation of a problem, analogous to the construction of metrics of constant Gauss curvature of §I, but valid for *all* compact complex manifolds. However, in the higher dimensional case, in addition to the earlier mentioned difficulties (the lack of the existence of a Riemann mapping for simply-connected complex manifolds and the inadequacy of generalized Gauss-Bonnet theorem for the problem (π_1)), we add the facts that, unlike the one dimensional case, there are many complex manifolds that do not admit a Kähler structure, while there are many analogues of the Gauss curvature function in the higher dimensional case. In this article we shall concentrate on the *simplest* analogue of the Gauss curvature function, namely the scalar curvature function $R(x)$ associated with a Hermitian metric g defined on a given complex manifold M. Moreover we shall seek a deformation $\tilde{g}$ of g that possesses the following properties

(i) $\tilde{g}$ leaves the complex structure of (M, g) invariant,

(ii) for complex dimension one, the deformation reduces to the conformal deformation of §I,

(iii) the deformation reduces to scalar curvature function $R(x)$ relative to (M, g) to a constant $\tilde{R}$,

(iv) the sign of $\tilde{R}$ should "generally" depend only on the complex structure of M and not on the particular metric g.

We shall satisfy these requirements by considering the following specialization of (π_1):

$(\tilde{\pi}_1)$ Find a metric $\tilde{g}$ (pointwise) conformally equivalent to a given Hermitian metric g on a compact complex manifold M and such that the scalar curvature $\tilde{R}$ of the deformed metric $\tilde{g}$ (computed relative to the Hermitian connection of $(M, \tilde{g})$ is a constant.

II.B. *Motivation.* In order to motivate this specialization of (π_1) we note the following facts concerning possible deformations $\tilde{g}$ of a metric g on a complex manifold M to constant scalar curvature:

(1) *The results of Lichnerowicz-Matsushima* (see **[8]**), who prove that if $(M, \tilde{g})$ is a Kähler manifold and the deformation $\tilde{g}$ takes place within the class of Kähler metrics, then the resulting Kähler manifold $(M, \tilde{g})$ must be such as to satisfy stringent relations between the Lie algebra of holomorphic vector fields and the Lie algebra of infinitesimal isometries.

(2) *A result of S.T. Yau* **[2]**. Using the result just mentioned it can be shown that large classes of rational and ruled algebraic surfaces do not admit constant scalar curvature Kähler metrics.

(3) *A result of N. Hitchin* **[10]**. Hitchin showed that if M is a $K3$ surface (i.e, an algebraic surface whose first Betti number is zero and whose first Chern c is a torsion class) admits a *Riemannian* metric g of constant zero scalar curvature, then g must be a *Ricci-flat* Kähler metric.

(4) *S.T. Yau's Gauss-Bonnet theorem for total scalar curvature of compact Hermitian manifold* (M, g) **[2]**. Yau's result can be stated as follows: given a compact Hermitian manifold (M, g) with scalar curvature $R(x)$ computed relative to the Hermitian connection on (M, g), and $\beta = \int_M R(x)\, dV$, then

(i) $\beta > 0$ implies all the plurigenera[1] of the complex manifold M vanish.

(ii) $\beta = 0$ implies that either all the plurigenera[1] of M vanish, or the first Chern class of M is a torsion class and no plurigenera of M can exceed 1.

(iii) $\beta < 0$ implies that the first Chern class of M is not a torsion class.

The results (1) and (2) above indicate the necessity of passing outside the class of Kähler manifold in studying the problem $(\tilde{\pi}_1)$. On the other hand, the results

[1] See §II. D, p. 161.

(3) and (4) indicate advantages of using the Hermitian connection of the deformed metric $\tilde{g}$ in computing the scalar curvature of $(M, \tilde{g})$.

II.C. *Formulation of* $(\tilde{\pi}_1)$ *in terms of partial differential equations.* Now we observe that the problem $(\tilde{\pi}_1)$ can be formulated entirely in terms of partial differential equations as in the case of complex dimension one described in I.C.

Let M be a C^∞ complex compact manifold of complex dimension N with a Hermitian metric g defined (in local coordinates) by setting $ds^2 = \sum_{\alpha,\beta} 2g_{\alpha\bar{\beta}}\, dz^\alpha d\bar{z}^\beta$. Then if σ is a real C^∞ function defined on M, we consider the Hermitian metric $\tilde{g}$ defined by setting $d\tilde{s}^2 = e^{2\sigma}\, ds^2$. Then the Hermitian scalar curvatures R and $\tilde{R}$ of (M, g) and $(M, \tilde{g})$ respectively are related by the formula

$$\text{(II.1)} \qquad \tilde{R} = e^{-2\sigma}\{R - 2N\,\Box\sigma\}$$

where $\Box$ denotes the complex Laplace operator defined on (M, g). This formula is derived as follows. By using the Hermitian connection, the components of the Ricci tensor $\tilde{R}\alpha\tilde{\beta}$ relative to (M, g) are given by the expression [**4**, p. 18]

$$\text{(II.2)} \qquad \tilde{R}_{\alpha\beta} = -\partial^2 \log \tilde{G} \,/\, \partial z^\alpha\, \partial\bar{z}^\beta \quad \text{with } \tilde{G} = \det|\tilde{g}_{\alpha\beta}|.$$

Since $\tilde{G} = e^{2N\sigma}\, G$, we find that in terms of the components of the Ricci tensor $R_{\alpha\beta}$ for (M, g)

$$\text{(II.3)} \qquad \tilde{R}_{\alpha\beta} = R_{\alpha\beta} - 2N\partial^2\sigma \,/\, \partial z^\alpha\, \partial z^\beta.$$

Since the desired scalar curvatures are the traces of their respective Ricci tensors, we find

$$\text{(II.4)} \qquad \tilde{R}e^{2\sigma} = R - 2N\,\Box\sigma,$$

where $\Box u = \sum g^{\alpha\bar{\beta}}\, \partial^2 u \,/\, \partial z^\alpha \partial\bar{z}^\beta$. Clearly (II.4) yields (II.1) and if $\tilde{R} = \text{const}$, (II.4) can be regarded as a partial differential equation for σ, which is completely analogous to (I.6).

Before ending this section it is useful to make a few remarks regarding (II.4). First, if (M, g) is a Kähler manifold, it is well known that $\Box = \frac{1}{2}\Delta$, where Δ is the standard real Laplace-Beltrami operator relative to (M, g) regarded as a Riemannian manifold. Consequently if $N = 1$, (II.4) reduces to the equation (I.6) derived in §I. Secondly, by integrating (II.4) over M we find that if $\tilde{R} = \text{const}$, then (II.4) implies

$$\text{(II.5)} \qquad \operatorname{sgn} \tilde{R} = \operatorname{sgn} \int_M R(x) = \operatorname{sgn} \int_M c_1 \wedge w^{N-1}.$$

Thus sgn $\tilde{R}$ is a *conformal invariant* of (M, g) since it depends only on the undeformed Hermitian manifold (M, g). This result is clearly an analogue of (I.6′). Actually we see in the next section that sgn $\tilde{R}$ is often independent of the particular Hermitian metric g defined on the complex manifold M. In fact, for Kähler metrics on non-singular projective algebraic varieties $\int_M c_1 \wedge w^{N-1}$ can be interpreted as the degree of the canonical divisor of M (apart from an absolute negative constant). Moreover the sign of this degree often has interesting invariance properties (see for example [**9**, p. 127].

II.D. *Summary of invariants of the complex structure of a compact manifold M used in the sequel.* In complex dimension one, the only numerical invariant of a compact complex manifold M is its genus p and this integer happens also to be a topological invariant of M. For such complex manifolds of higher dimension there are many more such invariants of the complex structure of M. These invariants are generally not topological invariants of M, but they are crucial to the understanding of our results on the problem $(\tilde{\pi}_1)$. The major invariants that we use in the sequel are:

(1) *The first Chern class* c_1 of a complex manifold M, which we regard as a (1-1) form defined on M if K is the canonical line bundle of M, K has the cohomology class c_1.

(2) *The plurigenera* P^i *of* M defined by the formula $P^i = \dim H_0(M, K^i)$, where K^i denotes the i-fold tensor product of the line bundle K. If $i = 1$, and $\dim_c M = N$, P is equal to the number of independent holomorphic N-forms defined on M. Thus if $N = 1$, P is equal to the genus p of a Riemann surface.

Moreover it will be of importance to investigate exactly when $\operatorname{sgn} \int_M c_1 \wedge w^{N-1}$ (mentioned in the last subsection) will be an invariant of the complex structure of M.

For our discussion of complex surfaces M (of complex dimension 2), the following invariants will play an important role:

(i) the irregularity of M = first Betti number of M.

(ii) c_1^2, the self-intersection of c_1 with itself.

Moreover in investigating the dependence of $\operatorname{sgn} \tilde{R}$ of the Hermitian structures on M we shall make use of the Serre duality theorem and the Riemann-Roch theorems for algebraic surfaces (see **[11]**).

III. Results on the general problem. In order to utilize the discussion of II.C to resolve $(\tilde{\pi}_1)$, we shall distinguish three mutually exclusive cases for a given compact Hermitian manifold (M, g): namely the total scalar curvature of (M,g), $\int_M R(x)dV$, is zero, negative or positive.

III.A. *Zero total scalar curvature of* (M, g). This case is analogous to the case of the Riemann surfaces of genus one, and, in fact, based on our discussion of II.C we can prove the following result.

(III.1) THEOREM. (a) *The compact Hermitian manifold* (M, g) *admits a (pointwise) conformally equivalent metric of constant scalar curvature zero if and only if the total scalar curvature of* (M, g) *is zero.* (b) *If some plurigenus of* M *is not zero and the cohomology class of the first Chern class* c_1 *of* M *is zero, such a deformation exists and is independent of the particular Hermitian metric* g *defined on* M.

PROOF OF (a). By virtue of (II.4) in II.C, the result is equivalent to determining necessary and sufficient conditions for the solvability of the inhomogeneous Poisson equation

$$\text{(III.2)} \qquad N\Box\sigma = R(\alpha) \quad \text{where} \begin{cases} \Box \text{ is the complex Laplacian on } (M, g), \\ N = \dim_c M. \end{cases}$$

Clearly a necessary condition for the solvability of this equation is that $\int_M R(x)\,dV = 0$ in accord with (II.5). Actually this condition is also sufficient. This can be demonstrated as an immediate consequence of the ellipticity and selfadjointness of the operator $\Box$ (in an appropriate inner product space) as in the text [**12**, pp. 102], or in VI.C below, by regarding the functions $u \in C^\infty(M)$ as smooth (0, 0) forms relative to (M, g). Indeed since the mean value of $R(x)$ over (M, g) is zero, $R(x)$ is orthogonal in the L_2 sense to the constant functions on (M, g). But the constant functions on (M, g) are precisely the harmonic functions on (M, g) relative to $\Box$. Consequently, by the Hodge-Kōdaira decomposition $R(x)$ is in the image of $\Box$.

PROOF OF (b). By virtue of Yau's complex analogue Gauss-Bonnet theorem mentioned in II.B, we observe that if $[c_1] = 0$ the total scalar curvature of M relative to any Hermitian metric cannot be negative. On the other hand, Yau's results shows that if some plurigenera of M do not vanish, the total scalar curvature of (M, g) cannot be positive.

III.B. *Negative total scalar curvature of* (M, g). As in the case of Riemann surfaces of genus greater than one we first state

(III.3) LEMMA. *The equation* (II.4) *is solvable with $\tilde{R}$ a negative constant if and only if* $\int_M R(x)\,dV < 0$.

The necessity of the condition $\int_M R(x)\,dV < 0$ follows from (II. 5). Our sufficiency proof of this lemma will be given in full detail in §VI; however its basic idea is easily sketched here. We attempt to write down a functional $I(u)$ whose critical points (if they exist) coincide with the solutions of (II.4). Moreover since $\int_M R(x)\,dV < 0$, it suffices to prove that inf $I(u)$ (over a sufficiently large admissible class C of real-valued functions σ defined on M) is attained. In order to write down the functional $I(u)$, it is essential to find an inner product space in which the complex Laplace operator $\Box$ is selfadjoint. This can easily be done, as in [**13**], in case (M, g) is a Kähler manifold. However in the general Hermitian case, we proceed using the notation of [**12**], by noting that the complex Laplacian $\Box$ (acting on functions) can be written

$$\Box = -\{\bar{\partial}\bar{\partial}^T + \bar{\partial}^T\bar{\partial}\} = -\bar{\partial}^T\bar{\partial},$$

where (in terms of local complex coordinates on a coordinate patch)

$$\bar{\partial}u = \sum_\beta \frac{\partial u}{\partial \bar{z}^\beta}\, d\bar{z}^\beta; \qquad \frac{\partial}{\partial \bar{z}^\beta} = \frac{1}{2}\left\{\frac{\partial}{\partial x^\beta} + i\frac{\partial}{\partial y^\beta}\right\}$$

and $\bar{\partial}^T$ is the formal adjoint of $\bar{\partial}$ relative to the L_2 inner product of two (0,1) forms w and η defined by setting $(w, \eta) = \int_M \omega \wedge * \bar{\eta}$ so that, for a real function u, $(\bar{\partial}u, \eta) = (u, \bar{\partial}^T\eta)$. Consequently we set

$$I(u) = N(\bar{\partial}u, \bar{\partial}u) + \int_M \{R(x)\,u - \tfrac{1}{2}\tilde{R}e^{2u}\}\,dV. \tag{III.4}$$

Then, if u is a smooth critical point of $I(u)$, for any $v \in C^\infty(M, g)$ we find

$$0 = \frac{d}{d\varepsilon} I(u + \varepsilon v)\Big|_{\varepsilon=0} = 2N(\bar{\partial}^T\bar{\partial}u, v) + \int_M [R(x) - \tilde{R}e^{2u}]v.$$

Consequently u satisfies

(III.5) $$2N\Box u - R(x) + \tilde{R}e^{2u} = 0,$$

so that u satisfies the appropriate differential equation, and $I(u)$ defined by (III.4) is an appropriate functional for the minimization problem.

Lemma (III.3) is critical in proving

(III.6) THEOREM. (a) *The compact Hermitian manifold (M, g) admits a (pointwise) conformally equivalent metric of negative constant scalar curvature if and only if the total scalar curvature of (M, g) is also negative.* (b) *If some plurigenus of M is not zero and the cohomology class of M is not zero, such a deformation exists and is independent of the particular Hermitian metric g defined on M. The same result is true if some plurigenus of M is at least two.*

PROOF OF (a). This result is an immediate consequence of Lemma (III.3) and the discussion of §II.C.

PROOF OF (b). Again we make use of Yau's results in **[2]** described in II.2. If some plurigenus of M is not zero, then the total scalar curvature of M relative to any Hermitian metric must be nonpositive, but if either $[c_1] \neq 0$, or some plurigenus of M is at least two, the total scalar curvature of (M, g) must be strictly negative.

III.C. *Positive total scalar curvature of (M, g).* This case is analogous to the class of Riemann surfaces of genus zero (i.e., simply connected Riemann surfaces). As was mentioned in II.C, this case cannot at present be treated by the partial differential equation approach. On the other hand, for higher dimensional complex manifolds, neither an approach via analytic function theory, nor other method yet known seems to yield a general method for the construction of a Hermitian metric of constant positive scalar curvature on a Hermitian manifold of total positive scalar curvature. Thus in this case we have only the following result:

(III.7) THEOREM. (a) *Suppose a compact Hermitian manifold (M, g) of positive total scalar curvature can be deformed to a conformally equivalent metric of constant scalar curvature $\tilde{R}$; then $\tilde{R} > 0$ and all the plurigenera of M vanish.* (b) *Moreover such a deformation to constant positive scalar curvature will be independent of the particular Hermitian metric g defined on the complex manifold M if $[c_1] \neq 0$.*

PROOF OF (a). This is an immediate consequence of (II.5) and Yau **[2]**.

PROOF OF (b). If a conformal deformation $\tilde{g}$ with $\tilde{R} > 0$ exists, all the plurigenera of M vanish, and if in addition $[c_1] \neq 0$, the total scalar curvature of any Hermitian g on M cannot be nonpositive by Yau's result **[2]**.

In connection with (III.7) we state the following

CONJECTURE. There is a Kähler metric g on a rational algebraic surface M, whose total scalar curvature is positive but such that g cannot be conformally deformed to a Hermitian metric of constant positive scalar curvature (cf. §V, especially (V.1)).

IV. Comparison with Riemannian geometry.

IV.A. *Basis for comparison.* As mentioned in the Introduction, the transformation of the scalar curvature function $R(x)$ of a Hermitian manifold (M, g) under a conformal deformation using the *Levi-Civita* (or Riemannian) *connection* is quite distinct from the formula (II.1) if $\dim_c M > 1$. Indeed if $\tilde{g} = e^{2\sigma} g$, then relative to the Levi-Civita connection of (M, g) (regarded as a Riemannian manifold of dimension d) the deformed scalar curvature function $R(x)$ is given by the well-known formula (cf. [**14**, p. 90])

(IV.I) $$\tilde{R} = e^{-2\sigma}\{R - (d-1)\Delta\sigma - (d/2-1)(d-1)\,|\nabla\sigma|^2\}$$

where Δ is the Laplace-Beltrami operator relative to the Riemannian manifold (M, g).

An inspection of the right-hand side of (IV.1) shows that, in sharp contrast to the equation (II.4), this formula exhibits a *sharply defined qualitative change when d is increased above* 2. Actually for $d > 2$ by computing the scalar curvature of $(M, \tilde{g})$ relative to the Levi-Civita connection the complex structure of M, the complex structure of M has been changed. Thus our approach of §II must be regarded as the appropriate one for complex manifolds.

This fact is clarified by studying the possibility of using the Levi-Civita connection in the problem (π_1) and $(\tilde{\pi}_1)$. Of course, for $d = 2$, there is no distinction and our discussion of §I applies equally well to real and complex manifolds. The problems (π_1), $(\tilde{\pi}_1)$ for Riemannian geometry and $d > 2$ were first studied by Yamabe [**15**], and later by Trudinger [**16**], Eliasson [**17**], and Kazdan-Warner [**18**]. In these papers the following results were shown:

(IV.2) Let $\lambda_1(g)$ be the smallest eigenvalue of the perturbed Laplace-Beltrami operator $L_g = c_d\,\Delta - R(x)$ defined relative to (M, g) where $c_d = 4(d-1)(d-2)^{-1}$ $(d > 2)$ (i.e. $\lambda_1(g)$ is the smallest real number such that $[L_g + \lambda_1(g)]\,u = 0$ has a nontrivial solution).

Then (i) *any compact Riemannian manifold* (M, g) can be deformed to a Riemannian metric $\tilde{g}$ of constant negative scalar curvature.

(ii) A compact Riemannian manifold (M, g) can be (pointwise) conformally deformed to a metric of constant negative (zero) curvature if and only if $\lambda_1(g) < 0$ $[\lambda_1(g) = 0]$.

(iii) sgn $\lambda_1(g)$ is invariant under (pointwise) conformal deformations of $\tilde{g}$.

(The proof that a Riemannian manifold (M, g) $(d > 2)$ with $\lambda_1(g) > 0$ admits a conformal deformation to a metric of constant positive curvature is still unresolved.)

From (IV.2) we draw the following comparisons. (a) The result (IV.2(i)) demonstrates the lack of rigidity of the scalar curvature function (relative to Levi-Civita connection). The fact (II.B (4)) of Yau [**2**] demonstrates that such a result is false for the Hermitian connection of a complex manifold.

(b) Generally speaking, the number $\lambda_1(g)$ is not independent of the Riemannian metric g on the manifold M so the conformal invariant sgn $\lambda_1(g)$ generally has no deep connection with the intrinsic geometry of the manifold M. Indeed it is relatively straightforward to construct two Riemannian metrics g_1 and g_2 on S^3 such that $\lambda_1(g_i)$ $(i = 1, 2)$ have different signs. On the other hand the complex analogue of $\lambda_1(g)$ for the Hermitian connection sgn $\int_M c_1 \wedge w^{N-1}$ = sgn $\tilde{R}$ has the interesting invariance properties described in §III.

IV.B. *Remarks on the proof of* IV.2. The result (ii) is most easily established by a nonlinear Dirichlet principle analogous to our proof of (III.3). The result (i) then follows from (ii) making an initial deformation of the metric g to a metric g^* such that $\lambda_1(g^*) < 0$. This can always be achieved by example by choosing g^* such that the total scalar curvature (relative to the Levi-Civita connection of g^*) $\int_M R^*(x) < 0$ (see for example Avez **[19]**). The argument needed to establish (iii) is based on setting $u = \exp(d/2 - 1)\,\sigma$ in (IV.1) to obtain

(IV.3) $$\tilde{R} = u^{-\beta}\{R(x) - u^{-1} c_d \Delta u\}, \qquad \beta = 4(d-2)^{-1}.$$

Thus $\tilde{R} = 0$ if and only if $\lambda_1(g) \equiv 0$, since the positivity of u implies u must be the eigenfunction associated with the smallest eigenvalue of L_g. More generally, suppose $\lambda_1(g) \neq 0$, and that u satisfies (IV.3) with $\tilde{R}=$ const. Then $v = u$ is a positive nontrivial solution of the linear equation

$$c_d \Delta v - R(x)v + \tilde{R}u^{\beta} v = 0.$$

Thus $\tilde{R}$ can be considered as the smallest eigenvalue of L_g relative to the weighted L_2 norm $\int_M u^\beta v^2$. The Rayleigh variational principle for this smallest eigenvalue then shows that sgn $\lambda_1(g)$ = sgn $\tilde{R}$.

V. **Application to algebraic surfaces.** Here we consider the application of our results to nonsingular projective algebraic varieties of complex dimension 2 (i.e. algebraic surfaces). An algebraic surface is called regular if its first Betti number is zero. Such *regular* algebraic surfaces can be divided into four classes:

(1) rational surfaces (all of whose plurigenera vanish),

(2) elliptic surfaces (which possess some nonvanishing plurigenus but for which $[c_1] \neq 0$ and such that no plurigenera are greater than one),

(3) $K3$ surfaces (with $[c_1] = 0$),

(4) surfaces of general type (with the properties that $[c_1] \neq 0$ and such that some plurigenus is larger than one).

Based on the results of §II we can prove

(V.1) THEOREM. (i) *An algebraic surface of general type (or any elliptic surface) M always admits a Hermitian metric of constant negative scalar curvature. Moreover any Hermitian metric g on M can be (pointwise) conformally deformed to a Hermitian metric of constant negative scalar curvature.*

(ii) *A K3 surface always admits a Hermitian metric of constant zero scalar curvature. Moreover any Hermitian metric g on M can be (pointwise) conformally deformed to a Hermitian metric of constant zero scalar curvature.*

(iii) *A rational surface M always admits a Hermitian metric g of positive total curvature. Moreover any Kähler metric g on M will have total positive scalar curvature provided* $c_1^2 \geqq 0$.

PROOF. (i) follows immediately from the results of II.B, our definitions of elliptic surface and surface of general type, and the result (2). (ii) follows from the results of II.A (III.1). The first part of (iii) was proved by Yau in [**2**]. The second part of (iii) follows by noting that if K denotes the canonical line bundle of M by the Serre duality theorem $H^2(M, \theta(-K)) = H^0(M, \theta(2K)) = 0$ (since all the plurigenera of M vanish). Consequently, the Riemann-Roch theorem implies that (for rational surfaces M)

$$\chi(-K) = H^0(M, \theta(-K)) - H^1(M, \theta(-K)) = c_1^2 + \text{Todd genus of } M$$
$$= c_1^2 + 1 > 0.$$

Hence $H^0(M, \theta(-K)) \neq 0$ and so $\int_M c_1 \wedge \omega > 0$ for Kähler manifolds with fundamental class ω.

More generally, we have the following result concerning the invariance properties of the sign of the possible Hermitian metrics of constant scalar curvature R defined on M.

(V.2) *Suppose $c_1^2 > 0$, and a Kähler metric g defined on an algebraic surface can be conformally deformed to a Hermitian metric of constant scalar curvature $\tilde{R}$; then sgn $\tilde{R}$ is independent of the Kähler metric g.*

PROOF. The result follows from (II.5) and a fact from [**9**, p. 127] which implies that sgn $\int_M c^1 \wedge w$ is independent of the Kähler form w defined on M.

VI. **Some proofs.** In this section we complete the proofs that were omitted in the earlier sections.

VI.A. *Liouville's quadrature of Monge's problem.* We show that the equation (I.1) is satisfied by

$$\text{(VI.1)} \qquad \lambda = 4a^2\, \zeta(u)\, \psi'(v) \exp\{\zeta(u) + \psi(v)\} [1 \pm \exp(\zeta(u) + \psi(v))]^{-2}$$

where

$$\text{(VI.2)} \qquad u = x + iy, \quad v = x - iy, \quad \log \lambda = u, \quad K = \pm a^{-2}.$$

and the functions ζ, ψ, are arbitrary complex conjugates. Indeed, performing the transformations (VI.1) on (VI.2), reduces this equation to the form

$$\text{(VI.3)} \qquad \partial^2 \log \lambda \,/\, \partial u\, \partial v \pm \lambda \,/\, 2a^2 = 0.$$

Now by differentiating (VI.1) with respect to u and v, we find, after some computation, that λ satisfies (VI.3). Thus if

$$\zeta(u) = U + iV \quad \text{and} \quad \psi(u) = U - iV$$
$$\zeta(u)\, \psi'(v) \{dx^2 + dy^2\} = dU^2 + dV^2$$

so that the desired metric has the form (I.2).

VI.B. *The "Poincaré shift" for the study of equation* (III.5). In the study of the equation (III.5) it is sometimes useful to study a transformed equation with the inhomogeneous term $R(x)$ replaced by a constant $\bar{R}$, namely its mean value. This device was mentioned by Poincaré in [**1**]. To describe this device we write a solution w of the equation $2N\square u - R(x) + \tilde{R}e^{2u} = 0$ in the form $u = v + w$ where v and w satisfy

(VI.4) $$2N\square v - R(x) + \bar{R} = 0 \quad \text{where } \bar{R} = \frac{1}{\mathrm{vol}(M, g)} \int_M R(x),$$

(VI.5) $$2N\square w - \bar{R} + \tilde{R}e^{2v} e^{2w} = 0.$$

Clearly the linear inhomogeneous equation (VI.4) is solvable by our discussion of (III.2) in §III. Moreover the solution of this equation is unique *apart from an additive constant.*

Now we call the equation (VI.5) the "Poincaré shift" associated with (III.5) since the inhomogeneous term $R(x)$ has been replaced by the constant $\bar{R}$. The advantage of this shift is that various bounds for (III.5) are often more easily deduced by fixing v appropriately from (VI.4) and studying (VI.5). (See Step (β) of the next subsection.)

VI.C. PROOF OF LEMMA III.3. We prove that the equation (III.5) is solvable with $\tilde{R}$ a negative constant provided $\int_M R(x)\, dV < 0$. To this end we observe that by scaling it suffices to demonstrate the solvability of (III.5) with $\tilde{R} = -1$. Then, utilizing our preliminary discussion of the problem in III.B, we shall establish (III. 3) by means of the following three steps:

STEP (α). The infimum of the functional

$$I(u) = N(\bar{\partial} u, \bar{\partial} u) + \int_M \{R(x)u + \tfrac{1}{2} e^{2u}\}\, dV$$

over the Hilbert space functions $\tilde{W} = \{u \mid \| u \|^2 = (\bar{\partial} u, \bar{\partial} u) + \int_M u^2\, dV\}$ is attained by an element $\bar{u} \in \tilde{W}$.

STEP (β). The function $\bar{u}$ described in Step (α) can be chosen to be essentially bounded.

STEP (γ). The function $\bar{u}$ can be chosen to be of class C^∞ and so satisfies (III.5) with $\tilde{R} = -1$.

PROOF OF STEP (α). We show that (i) $I(u)$ is bounded below for $u \in \tilde{W}$, (ii) if $b = \inf_{\tilde{W}} I(u)$, then the set $\Sigma = I^{-1}[b, b+1]$ is bounded and weakly sequentially closed in the Hilbert space $\tilde{W}$, and (iii) $I(u)$ is lower semicontinuous on Σ with respect to weak convergence in $\tilde{W}$. Once these facts are established, any minimizing sequence $\{u_n\}$ for $I(u)$ over $\tilde{W}$ has a weakly convergent subsequence (which we again label $\{u_n\}$) with weak limit $\bar{u} \in \Sigma$. Indeed the Hilbert space structure of $\tilde{W}$, (i) and (ii) show that for sufficiently large n, $\{u_n\} \subset \Sigma$, $\{\|u_n\|\}$ is uniformly bounded. Then $I(\bar{u}) = b$, since, by (iii), $I(u) \leqq \lim I(u_n)$, so $\bar{u}$ is the desired minimizer of $I(u)$ over $\tilde{W}$.

Now we establish the facts (i) — (iii). To verify (i) we write an arbitrary element u of $\tilde{W}$ in the form $u = u_0 + m$ where m is the mean value of u over (M, g) and u_0 has mean value zero. Then

$$I(u) = N(\bar{\partial} u_0, \bar{\partial} u_0) + \int_M R(x)u_0 + m \int_M R(x) + \tfrac{1}{2} e^{2m} \int_M e^{2m_0}.$$

Combining this fact with the "Poincaré inequality" for $\square$ $(\bar{\partial} u, \bar{\partial} u) \geqq \rho \|u_0\|_{L_2(M)}$ for ρ an absolute positive constant and the fact that $e^{2u_0} \geqq 1 + 2u_0$, for any $\varepsilon > 0$

we find

$$I(u) \geqq N(\bar{\partial} u_0, \bar{\partial} u_0) - \varepsilon^{-1} \| R(x) \|_{L_2} - \varepsilon \| u_0 \|_{L_2} \tag{VI.6}$$
$$+ m \int R(x) + \tfrac{1}{2} \operatorname{vol}(M, g) e^{2m}$$

$$\geqq (N\rho - \varepsilon) \| u_0 \|^2_{L_2} + g(m) \tag{VI.7}$$

where $g(m) \to \infty$ as $| m | \to \infty$. Consequently choosing $\varepsilon = N$, we find $I(u)$ is bounded below. We can establish (ii) in the same way by showing that $I(u) \to \infty$ as $\|u\|_{\tilde{w}} \to \infty$. To this end we observe that an equivalent norm on $\tilde{W}$ can be chosen to be $\|u\|^2 = (\bar{\partial} u, \bar{\partial} u) + m^2$.

Now if $\|u\| \to \infty$ either $(\bar{\partial} u, \bar{\partial} u) \to \infty$ with $|m|$ and $\| u \|_{L_2}$ bounded (in which case $I(u) \to \infty$ by (VI.6)), or $\|u\|_{L_2} + m^2 \to \infty$ in which case $I(u) \to \infty$ by (VI.7) with $N\rho > \varepsilon$.

Now to verify the weak lower semicontinuity of $I(u)$ we observe that if $u_n \to u$ weakly in $\tilde{W}$, $u_n \to u$ strongly in $L_2(M, g)$ by the complex analogue of Rellich's lemma, so that $\int_M R(x)\, u_n \to \int_M R(x)\, u$. Moreover Fatou's theorem implies each of the terms $(\bar{\partial} u, \bar{\partial} u)$ and $\int_M e^{2w}$ are weakly lower semicontinuous since the integrands in both cases are positive.

REMARK. We note that the fact (α) just established is also valid for the "Poincaré shift" of (III.5) with $\tilde{R}$ a negative constant. Indeed the proof is completely similar to the one just given.

PROOF OF STEP (β). We first assume $\sup_M R(x) < 0$, and establish the existence of a finite real number $k > 0$ such that for any element $u \in \Sigma$, the truncated function u_k defined by setting

$$\begin{aligned} u_k &= \inf(u, k), & u > 0, \\ &= \sup(u, -k), & u \leqq 0, \end{aligned}$$

is such that $I(u_k) \leqq I(u)$. For then, $u_k \in \Sigma$ and the minimizing sequence $\{u_n\}$ (and so $\bar{u}$) of Step (α) can be chosen to consist of functions with ess $\sup_M | u | \leqq k$. To find such a positive number k we find that, for *any* positive number k, $u \in \tilde{W}$, $u_k \in \tilde{W}$ and in fact $(\bar{\partial} u_k, \bar{\partial} u_k) \leqq (\bar{\partial} u, \bar{\partial} u)$. (This fact is easily established by considered local coordinates patches.) Consequently, it suffices to consider the effect of the transformation $u \to u_k$ for the functional

$$J(u) = \int_M \{R(x)\, u + \tfrac{1}{2}\, \varepsilon^{2u}\}\, dV.$$

Since M is compact and $\sup_M R(x) < 0$ (by assumption) there are two positive numbers p and P such that $-p \leqq R(x) \leqq -P$ and consequently as $| u | \to \infty$, the integrand of $J(u)\, f(u) = R(x)u + \frac{1}{2} e^{2u} \to \infty$. Thus there is a positive number k_1 such that $f(k_1) \leqq f(u)$ for $u \geqq k_1$, and a positive number k_2 such that $f(-k_2) \geqq f(u)$ for $u \leqq -k_2$. Consequently the desired number k can be chosen to be $\sup(k_1, k_2)$.

Now if $\sup_M R(x) > 0$, we apply the argument just given to the "Poincaré shift" equation (VI.5) (discussed above in VI.B)

$$2N\Box u - \tilde{R} - e^{2v} e^{2u} = 0 \text{ with } \bar{R}\, [\operatorname{vol}(M, g)]^{-1} \int_M R(x)\, dV < 0$$

and the associated functional

$$\tilde{I}(u) = N(\bar{\partial}u, \bar{\partial}u) + \int_M \{\bar{R}u + e^{2v}e^{2u}\}\, dV,$$

and so we conclude the minimizer $\tilde{u}$ of $\tilde{I}(u)$ can be chosen to be essentially bounded. Consequently since a minimizer of $I(u)$ can be written $u = v + \tilde{u}$ (where v is a solution of the Poisson equation (VI.4) for $\square$) we conclude that the function u is also uniformly bounded.

PROOF OF STEP (γ). Let $v \in C^\infty(M, g)$. Then (since $\bar{u}$ is essentially bounded so that the associated integrands are finite) after a short computation we find (as in (III.B))

$$0 = \frac{d}{d\varepsilon} I(u + \varepsilon v)\Big|_{\varepsilon=0} = 2N(\bar{\partial}u, \bar{\partial}v) + \int_M \{R(x)v + e^{2\bar{u}}\, v\}\, dV.$$

Consequently $\bar{u}$ can be considered as a weak solution of the following *linear* nonhomogeneous elliptic equation

$$2N\square u = R(x) + e^{2\bar{u}}.$$

Since $\bar{u} \in L_\infty(M, g)$, the L_p regularity theory for linear elliptic equations implies that u has generalized second derivatives in $L_p(M, g)$ for every $p > 1$. Consequently by the Sobolev imbedding theorem (after a possible redefinition on a set of measure zero) $\bar{u} \in C_1, \alpha\, (M, g)$ $(0 < \alpha < 1)$ (the space of functions with Hölder continuous first derivatives of exponent α). Thus by the Schauder regularity theory $\bar{u} \in C_2, \alpha(M, g)$ and by iteration $u \in C^\infty(M, g)$. Since v is an arbitrary C^∞ function, $\bar{u}$ satisfies (III.5) with $\tilde{R} = -1$.

BIBLIOGRAPHY

1. H. Poincaré, *Les fonctions fuchsiennes et l'equation $\Delta u = e^u$*, Oeurves. Vol. I, pp. 512–591.

2. S.T. Yau, *On the curvature of compact Hermitian manifolds*, Invent. Math. **25** (1974), 213–240.

3. S. S. Chern, *Complex manifolds without potential theory*, Van Nostrand Math. Studies, no. 15, Van Nostrand, Princeton, N.J., 1967. MR **37** #940.

4. S. Kobayashi, *Hyperbolic manifolds and holomorphic mappings*, Pure and Appl. Math., no. 2, Dekker, New York, 1970. MR **43** #3503.

5. T. Aubin, *Metriques riemanniennes et courbure*, J. Differential Geometry **4** (1970), 383–424. MR **43** #5452.

6. D. Hilbert, *Zur Theorie der konformen Abbildungen*, Werke III.

7. M. S. Berger, *On the conformal equivalence of compact 2-dimensional manifold*, J. Math. Mech. **19** (1969/70), 13–18. MR **39** #6219.

8. S. Kobayashi, *Transformation groups in differential geometry*, Springer-Verlag, New York, 1972.

9. D. Mumford, *Lectures on curves on an algebraic surface*, Ann. of Math. Studies, no. 59, Princeton Univ. Press, Princeton, N.J., 1967. MR **35** #187.

10. N. Hitchin, *On compact 4-dimensional Einstein manifolds*, J. Differential Geomety **9** (1974), 435–442.

11. F. Hirzebruch, *Topological methods in algebraic geometry*, 3rd ed., Die Grundlehren der math. Wissenschaften, Band 131, Springer-Verlag, New York, 1966. MR **34** #2573.

12. J. Morrow and K. Kodaira, *Complex manifolds*, Holt, Rinehart and Winston, New York, 1971. MR **46** #2080.

13. M. S. Berger, *Two minimax problems in the calculus of variations*, Proc. Sympos. Pure Math., vol. 23, Amer. Math. Soc., Providence, R.I., 1973, pp. 261–267.

14. L. Eisenhart, *Riemannian geometry*, Princeton Univ. Press, Princeton, N.J., 1926.

15. H. Yamabe, *On a deformation of Riemannian structures on compact manifolds*, Osaka Math. J. **12** (1960), 21–37. MR **23** #A2847.

16. N. Trudinger, *Remarks concerning the conformal deformation of Riemannian structures on compact manifolds*, Ann. Scuola Norm. Sup. Pisa (3) **22** (1968), 265–274. MR **39** #2093.

17. H. Eliasson, *On variations of metrics*, Math. Scand. **29** (1971), 317–327.

18. J. Kazdan and F. Warner, *Scalar curvature and conformal curvature of Riemannian structure* (preprint).

19. A. Avez, *Valeur moyenne du scalaire de courbure sur une variété compacte*, C. R. Acad. Sci. Paris **256** (1963), 5271–5273. MR **27** #1909.

20. J. Moser, *On a nonlinear problem in differential geometry* (to appear).

BELFER GRADUATE SCHOOL, YESHIVA UNIVERSITY

Proceedings of Symposia in Pure Mathematics
Volume 27, 1975

SOBOLEV INEQUALITIES FOR RIEMANNIAN BUNDLES

M. CANTOR*

1. Introduction. Sobolev inequalities play a major role in the study of differential operators and nonlinear functional analysis. The inequalities are the primary tools in the study of the properties of spaces of functions with Sobolev topologies; for example, the Schauder ring theorem. There are theorems involving continuity and closure of composition in such spaces [3, Chapter 2]. The latter theorems involve application of the inequalities to vector fields. It is this case and its generalization which this paper studies, where R^n is replaced by an arbitrary Riemannian manifold satisfying certain geometric conditions.

While the Sobolev inequalities over R^n have been known for some time, the usual proofs use transform methods and are therefore hard to generalize to manifolds. In 1959 Nirenberg [5] presented particularly elegant proofs, due to himself and other authors, which were more geometric. These techniques are the basis for the proofs of this paper.

Throughout, M denotes a complete Riemannian n-dimensional manifold without boundary. The canonical volume form on M is denoted dV. A Riemannian bundle is a map $\pi: E \to M$, a vector bundle with a specified smooth metric $(,)$. ∇ is a connection on E satisfying $d(V, W)X_m = (\nabla_{X_m}V, W) + (V, \nabla_{X_m}W)$, where V and W are sections of E and $X_m \in T_mM$. ∇^n is the iterated covariant derivative. In most applications E is a tensor bundle over M. $C_0^\infty(E)$ is the space of C^∞ sections of E with compact support. For $x \in M$, $V \in C_0^\infty(E)$, the quantity $|\nabla^i V(x)|$ is the norm of $\nabla^i V(x)$ in the canonical norm of $L^k(TM: E)$.

AMS (MOS) subject classifications (1970). Primary 58C99, 58D99, 58B20.

**Current address*: Department of Mathematics, Duke University, Durham, N. C.

The author wishes to thank Jerrold Marsden for his help and encouragement and E. Calabi for his assistance.

2. Statement of results.

DEFINITION 1. Let $V \in C_0^\infty(E)$. Then define

(i) (C^k norm) $\|V\|_k = \sum_0^k \sup |\nabla^i V(x)|$;

(ii) (L_k^p, Sobolev norm) $|V|_{p,k} = \sum_0^k (\int_M |\nabla^i V(x)|^p \, dV)^{1/p}$;

(iii) (Hölder norm) For $0 < \theta < 1$,

$$[V]_{\theta,k} = \sum_0^k \sup_{x,y \in M} \sup_{c \in G(x,y)} \frac{|\tau(c)\nabla^i V(x) - \nabla^i V(y)|}{d(x,y)^\theta},$$

where $G(x, y) = \{$length-minimizing geodesics joining x and $y\}$, $\tau(c)$ is parallel translation along c from $\pi^{-1}(y)$ to $\pi^{-1}(x)$, and $d(x, y)$ is the distance from x to y.

Each of the functions defined in Definition 1 is a norm on the vector space $C_0(E)$.

Also, denote $\| \ \|_0 = \| \ \|$, $| \ |_{p,0} = | \ |_p$, and $[\ ]_{\theta,0} = [\ ]_\theta$.

DEFINITION 2. $C^k(E)$ (resp. $L_k^p(E)$, $C^{k+\theta}(E)$) is the completion of $C_0^\infty(E)$ with respect to $\| \ \|_k$ (resp. $| \ |_{p,k}$, $[\ ]_{\theta,k}$).

We note that if $0 < \theta < 1$, then $C^{k+1}(E) \subset C^{k+\theta}(E) \subset C^k(E)$ and the inclusion is continuous.

We state the following hypotheses:

C1. The injective radius R of M is bounded away from zero.

C2. The sectional curvature is C^∞ bounded on TM.

THEOREM 1. *Let M satisfy* C1 *and* C2. *Then if $p > 1$ and $s > n/p + k$, there is a constant C such that for all $f \in L_s^p(E)$, we have $\|f\|_k \leqq C|f|_{p,s}$, where C is independent of f.*

THEOREM 2. *Let M satisfy* C1 *and* C2 *and $r > n$. Then there is a constant C such that for each $f \in L_s^r(E)$, we have $[f]_{1-n/r,s} \leqq C|f|_{r,s+1}$, where C is independent of f.*

REMARK. For Theorems 1 and 2, C2 can be weakened to require that only $K_x(V, W) < \delta$.

THEOREM 3. *Let M satisfy* C1 *and* C2, $0 \leqq j \leqq m$, *and $q, r \geqq 1$. Then if $m/p = j/r + (m-j)/q$, there is a constant C such that for each $f \in L_m^r(E) \cap L^q(E)$, we have $|\nabla^j f|_p \leqq C|f|_{r,m}^{j/m} |f|_q^{1-j/m}$, where C is independent of f.*

In the next theorem we adopt the following notation, due to Nirenberg [5].

NOTATION. For $p > 0$, the definition of $| \ |_p$ remains as in Definition 1. For $p < 0$ and $f \in C_0^\infty(E)$, let $h = [-n/p]$, $-\alpha = h + n/p$ and

$$|f|_p = \|\nabla^h f\|_\alpha, \quad \alpha = 0,$$
$$|f|_p = [\nabla^h f]_\alpha, \quad \alpha > 0.$$

THEOREM 4. *Let M satisfy* C1 *and* C2, *and let $0 \leqq j \leqq m$. Then if $m - j - n/r$ is not a nonnegative integer and $1/p = 1/r + (j-m)/n$, there is a constant C such* *that for $f \in L^r_m(E)$, it follows that $|\nabla^j f|_p \leqq C|\nabla^j f|_{r,m-j}$, where C is independent of f.*

Under some circumstances, one can interpolate between these inequalities. We do have the standard interpolation lemma:

LEMMA. *If $-\infty < \lambda \leqq \mu \leqq \nu < \infty$ and $\lambda \geqq 0$ or $\nu \leqq 0$, then for all $u \in L^{1/\lambda}(E) \cap L^{1/\nu}(E)$,*

$$u \in L^{1/\mu}(E) \quad \text{and} \quad |u|_{1/\mu} \leqq C\,|u|_{1/\lambda}^{(\nu-\mu)/(\nu-\lambda)}\,|u|_{(1/\nu)}^{(\mu-\lambda)/(\nu-\lambda)}.$$

In the classical case the assumption that λ, μ, and ν all have the same sign is unnecessary [**5**, p. 126]. Under what conditions it is necessary is not known to the author.

There are several applications of this lemma. For example,

THEOREM 5. *Let M satisfy* C1 *and* C2 *and let $1 \leqq q,\ r \leqq \infty$ and $0 \leqq j \leqq m$. If $j/n + 1/r - m/n \geqq 0$ and for $j/m \leqq a \leqq 1$,*

$$1/p = j/n + a(1/r - m/n) + (1-a)/q,$$

then for $u \in L^r_m(E) \cap L^q(E)$, it follows that $|\nabla^j u|_p \leqq C\,|u|^a_{r,m}\,|u|_q^{1-a}$, where C is independent of u.

3. Outline of the proofs. The fundamental idea behind all of the theorems is the reduction of the argument to a local one. Using conditions C1 and C2 and standard comparison results (Lemma 2, below), one gets the necessary uniform bounds on the exp maps. For Theorems 3 and 4 we use a triangulation method of Cheeger and Calabi presented by Professor Calabi at this conference.

The local arguments employ the relationship between parallel translation in a bundle and covariant differentiation. In particular, we have this lemma:

LEMMA 1. *Let $\pi\colon E \to M$ be a bundle with connection ∇, and $c\colon A \to M$ a smooth curve. Then, for $V \in C^1(E)$,*

$$\frac{d^n}{ds^n}\left(\tau^s_t V(c(s))\right)\Big|_{s=s'} = \tau^{s'}_t \nabla^n V(c'(t+s'),\cdots,c'(t+s')).$$

This is a simple generalization of the standard formula for the covariant derivative [**4**, p. 114].

The necessary estimates on the $\exp_x$ maps follow from this technical lemma:

LEMMA 2. *Let $m \in M$, $v \in T_m M$, $|v| = 1$, and $c(t) = \exp_m tv$. Suppose, for $0 \leqq t \leqq t_0$, there is no conjugate point on c. Let $J \exp_m$ be the Jacobian determinant of $\exp_m$.*

(1) *If, for each $0 \leqq t \leqq t_0$, the Ricci curvature* $\operatorname{Tr} R_{c'(t)} \geqq (n-1)a^2$, *then*

$$J \exp_m tv \leqq ((\sin at)/at)^{n-1} \quad \textit{for } t \in [0, t_0]$$

and

$$\sup_{|w|=1} \left|D \exp(tv)(w)\right| \leqq (\sin at)/at.$$

(2) *If, for each* $0 \leqq t \leqq t_0$, *and y normal to c at* $c(t)$, *we have the sectional curvature* (0.4.16), $k_{c(t)}(c'(t), y) \leqq b^2$,

$$J \exp_m tv \geqq ((\sin bt)/bt)^{n-1}$$

and

$$\inf_{|w|=1} |D \exp_m (tv)(w)| \geqq \frac{\sin bt}{bt} \quad \text{for } t \in [0, t_0].$$

If a^2 *or* b^2 *is negative, use the complex extension of* sin; *if* a *or* $b = 0$, *replace* $(\sin at)/at$ *or* $(\sin bt)/bt$ *by* 1.

For a proof (and the statement of a more general theorem) see [**1**, p. 253]. The analytic content is a generalization of the Rauch comparison theorem.

COROLLARY 1. *Let M satisfy* C2. *Also, let* $S > 0$ *be the conjugate radius of M. Then, if* $0 < \alpha < 1$, *there are constants* A_1 *and* A_2 *such that for each* $m \in M$ *and* $v \in T_m M$, *with* $|v| < \alpha S$, $0 < A_1 < J \exp_m (v) \leqq A_2$.

This is immediate since C2 implies the Ricci curvatures are bounded.

Another application of Lemma 2 is the following:

LEMMA 3. *Let* $\pi: E \to M$ *satisfy* C1 *and* C2. *Let U be a normal neighborhood of* $x \in M$ *of radius S less than the conjugate radius of M. Also, let* $(e^1, \cdots, e^n)$ *be an orthonormal basis of* $T_x M$. *Define the norms* $\| \ \|_i$ *on* $L(TM, E)|_U$ *by* $\|f(y)\|_1 = \sum |f(y_i)|$, *where* $y_i = D \exp_x(\exp^{-1}(y)(e^i))$, $\|f(y)\|_2 = \sup_{|v|=1} |f(y)(v)|$, *and* $\|f(y)\|_3 = \langle f(y), f(y) \rangle^{1/2}$, *where* $\langle\, ,\rangle$ *is the canonical metric on* $L(TM, E)$. *Then there are constants* $C_1(s)$, $C_2(s)$, $D_1(s)$, *and* $D_2(s)$ *depending only on s such that for all* $y \in U$, $C_1(s)\|f(y)\|_1 \leqq \|f(y)\|_2 \leqq C_2(s)\|f(y)\|_1$ *and* $D_1(s)\|f(y)\|_1 \leqq \|f(y)\|_3 \leqq D_2(s)\|f(y)\|_1$.

This lemma follows from the application of the inequalities of Lemma 2 involving $D \exp_x$.

We shall proceed with the proofs of theorems.

PROOF OF THEOREM 1. We assume $k = 0$ and $U \in C_0^\infty(E)$. The general case follows inductively.

Let $R > 0$ be the constant specified in C1. Let m be any member of M. Put spherical normal coordinates about m on the ball $B_m(R)$. Let (s, θ) be spherical coordinates on $T_m M$. In particular, $s \to \exp_m(s, \theta)$ is a geodesic through m. Let τ_0^s be the appropriate parallel translation map in E.

Now, let $g: \boldsymbol{R} \to \boldsymbol{R}$ be a C^∞ function satisfying $g(t) = 1$ for $t < R/2$, $g(t) = 0$, $t \geqq 3R/4$, and the ith derivative of g, $g^{(i)}$, satisfies $|g^{(i)}(t)| < A$ for all $i \leqq s$.

Fix $\theta_0 \in S_0(1) \subset T_m M$. Now

$$U(m) = U(\exp_m(0, \theta_0)) = -\int_0^R \frac{\partial}{\partial \rho}(\tau_0^\rho g(\rho)\, U(\exp_m(\rho, \theta_0)))\, d\rho.$$

Note $\tau_0^\rho g(\rho)\, U(\exp(\rho, \theta_0))$ maps $\boldsymbol{R} \to T_m M \simeq \boldsymbol{R}^n$, so the above equality is simply elementary calculus. Integrate by parts to get

$$U(m) = -\frac{\rho\partial\tau_0^\rho g(\rho)\, U(\exp(\rho,\theta))}{\partial\rho}\Big|_0^R + \int_0^R \frac{\rho\partial^2\tau_0^\rho g(\rho)\, U(\exp(\rho,\theta))}{\partial\rho^2}\, d\rho.$$

But

$$\begin{aligned}\frac{\partial\tau_0^\rho g(\rho)\, U(\exp_m(\rho,\theta))}{\partial\rho}\Big|_{\rho=R} &= g'(R)\,\tau_0^R\, U(\exp_m(R,\theta)) + g(R)\frac{\partial\tau_0^\rho\, U(\exp_m(\rho,\theta))}{\partial\rho}\Big|_{\rho=R}\\ &= 0+0=0.\end{aligned}$$

So,

$$U(m) = -\int_0^R \frac{\rho\partial^2\tau_0^\rho g(\rho)\, U(\exp_m(\rho,\theta))}{\partial\rho^2}\, d\rho.$$

In fact, inductively,

$$U(m) = \frac{(-1)^{s-1}}{(s-1)!}\int_0^R \rho^{s-1}\frac{\partial^s}{\partial\rho^s}(\tau_0^\rho g(\rho)\, U(\exp_m(\rho,\theta)))\, d\rho.$$

Applying Lemma 1, and the fact that τ is an isometry,

$$|U(m)| \leqq \frac{(-1)^{s-1}}{(s-1)!}\int_0^R \rho^{s-1}|\nabla^s g(\rho)\, U(\exp(\rho,\theta))|\, d\rho.$$

Integrate both sides with respect to dS_θ on the unit sphere in $T_m M$ to get

$$\left[\int_{S_m(1)} dS_\theta\right]|U(m)| \leqq \frac{1}{(s-1)!}\times\int_{S_m(1)}\int_0^R \rho^{s-n}\,|\nabla^s g(\rho)\, U(\exp(\rho,\theta))|\rho^{n-1}\, d\rho\, dS_\theta.$$

Now, $\int_{S_m(1)} dS_\theta = V_n$, volume of unit sphere in $\boldsymbol{R}^n$. So, using this and Hölder's inequality with respect to the measure $\rho^{n-1}\, d\rho\, dS_\theta$,

$$\begin{aligned}|U(m)| \leqq \frac{1}{V_n(s-1)!}&\left[\int_{S_m(1)}\int_0^R \rho^{(s-n)p/(p-1)}\,\rho^{n-1}\, d\rho\, dS_\theta\right]^{p-1/p}\\ &\times\left[\int_0^R\int_{S_m(1)} |\nabla^s g(\rho)\, U(\rho,\theta)|^p\,\rho^{n-1}\, d\rho\, dS_\theta\right]^{1/p}.\end{aligned}$$

Now

$$\left[\int_{S_m(1)}\int_0^R \rho^{(s-n)p/(p-1)}\,\rho^{n-1}\, d\rho\, dS_\theta\right]^{1-1/p} = \left[\frac{R^{(s-1)/(p-1)p+1}}{(s-1)/(p-1)p+1}V_n\right]^{1-1/p},$$

since $s > n/p$ implies $(s-n)p/(p-1)+n-1 > -1$. This is a constant which is independent of U and m. So,

$$|U(m)| \leqq D\left[\int_0^R\int_{S_m(1)} |\nabla^s g(\rho)\, U(\exp(\rho,\theta))|^p\,\rho^{n-1}\, d\rho\, dS_\theta\right]^{1/p}.$$

Now $dV = J\exp_m(\rho,\theta)\rho^{n-1}\, d\rho\, dS_\theta$. So,

$$\frac{dV}{J\exp_m(\rho,\theta)} = \rho^{n-1}\,d\rho\,dS_\theta.$$

Now, applying Corollary 1 of Lemma 2, there is a constant E such that $1/(J\exp_m(\rho,\theta)) < E$. So,

$$|U(m)| \leqq F\left[\int_{B_m(R)} |\nabla^s g(\rho)\, U(\exp(\rho,\theta))|^p\, dV\right]^{1/p},$$

where F is independent of U and m. By the product rule for covariant differentiation,

$$\begin{aligned}|U(m)| &\leqq F\left[\sum_{i=0}^{s}\binom{s}{i}\int_{B_m(R)} |g^{(s-i)}(\rho)||\nabla^i U(\exp(\rho,\theta))|dV\right]^{1/p}\\ &\leqq C\left[\sum_{i=0}^{s}\int_{B_m(1)} |\nabla^i U(\exp(\rho,\theta))|^p\, dV\right]^{1/p}\\ &\leqq C\left[\sum_{i=0}^{s}\int_M |\nabla^i U(y)|^p\, dV\right]^{1/p} \leqq C|U|_{p,s}.\end{aligned}$$

Since this holds for each $m \in M$, we have $\sup_{m\in M}|U(m)| = \|U\|_0 \leqq C|U|_{p,s}$. Q.E.D.

PROOF OF THEOREM 2. We let $s = 0$ and $f \in C_0^\infty(E)$. Let R be the constant specified in C1, and $x, y \in M$. There are two cases:

Case 1: $d(x,y) < R$. Because of C1 there is a normal spherical neighborhood about x containing y. Using spherical coordinates, $y = \exp_x(\rho_0, \theta_0)$. The unique length-minimizing geodesic joining x and y is $c\colon \rho \to \exp_x(\rho, \theta_0)$. Let τ^ρ be parallel translation along this curve. Then

$$f(x) - \tau^{\rho_0} f(y) = \int_0^{\rho_0} \frac{\partial}{\partial\rho}(\tau^\rho f(\exp_x(\rho,\theta_0)))d\rho;$$

using Lemma 1, we conclude

$$|f(x) - \tau^{\rho_0} f(y)| \leqq \int_0^{\rho_0} |\nabla_{c'(\rho)} f(\exp_x(\rho,\theta_0))|\, d\rho \leqq \int_0^{\rho_0} |\nabla f(\exp_x(\rho,\theta_0))|\, d\rho.$$

Integrate both sides of this on the unit sphere S in T_xM. So,

$$S_n|f(x) - \tau^{\rho_0} f(y)| \leqq \int_S\int_0^{\rho_0} |\nabla f(\exp(\rho,\theta))|\, d\rho\, dS_\theta,$$

where S_n is the volume of the standard unit sphere in $\boldsymbol{R}^n$. Now,

$$S_n|f(x) - \tau^\rho f(y)| \leqq \int_S\int_0^{\rho_0} |\nabla f \exp(\rho,\theta)| \frac{1}{\rho^{n-1}}\rho^{n-1}\, d\rho\, dS_\theta,$$

and by Hölder's inequality,

$$\begin{aligned}S_n|f(x) - \tau^\rho f(y)| \leqq &\left[\int_S\int_0^{\rho_0} |\nabla f(\exp_x(\rho,\theta))|^r \rho^{n-1}\, d\rho\, dS_\theta\right]^{1/r}\\ &\times\left[\int_S\int_0^{\rho_0} \rho^{(1-n)(r/r-1)}\,\rho^{n-1}\, d\rho\, dS_\theta\right]^{1-1/r}\end{aligned}$$

Now, since $r > n$, we have $(1-n)r/(r-1)+n-1 > -1$ and

$$\left[\int_S\int_0^{\rho_0} \rho^{(1-n)r/(r-1)+n-1}\, d\rho\, dS_\theta\right]^{1-1/r} = \rho_0^{(r-n)/r}\left[S_n\frac{r-1}{r-n}\right]^{1-1/r}.$$

Now, $\rho_0 - d(x, y)$ and $(r-n)/r = 1 - n/r$ and we have constant, F, such that

$$\frac{|f(x) - \tau^{\rho_0} f(y)|}{d(x, y)^{1-n/r}} \leqq F\left[\int_S\int_0^{\rho_0} |\nabla f(\exp(\rho, G))|^r \rho^{n-1}\, d\rho\, dS_\theta\right]^{1/r}.$$

Now $\rho^{n-1}\, d\rho\, dS_\theta = 1/(J\exp_x(\rho, \theta))$, and by Lemma 3, there is a constant E such that $1/(J\exp(\rho, \theta)) < E$. So, using covariant notation,

$$\begin{aligned}\frac{|f(x) - \tau^{\rho_0} f(y)|}{d(x, y)^{1-n/r}} &\leqq FE^{1/r}\left[\int_S\int_0^{\rho_0} |\nabla f(m)|^r\, dV\right]^{1/r}\\ &\leqq C|\nabla f|_r \leqq C|f|_{r,1}.\end{aligned}$$

Case 2: $d(x, y) \geqq R$. Let τ be parallel translation along any geodesic joining x and y. Then

$$|f(x) - \tau f(y)| \leqq |f(x)| + |\tau f(y)| \leqq |f(x)| + |f(y)|.$$

Now, since $r > n$, $1 > n/r$ by Theorem 1, there is a constant C_2, such that for any $m \in M$, $|f(m)| \leqq C_2|f|_{r,1}$. Hence $|f(x) - \tau f(y)| \leqq 2C_2|f|_{r,1}$ and, since $d(x, y) \geqq R$,

$$\frac{1}{d(x,y)^{1-n/r}} \leqq \left(\frac{1}{R}\right)^{1-n/r} \quad\text{and}\quad \frac{|f(x) - \tau f(y)|}{d(x, y)^{1-n/r}} \leqq \left(\frac{1}{R}\right)^{1-n/r} 2C_2|f|_{r,1}.$$

Hence

$$\sup_{x,y\in M}\ \sup_{\tau\in G(x,y)} \frac{|f(x) - \tau f(y)|}{d(x, y)^{1-n/r}} = [f]_{1-n/r,0} \leqq C|f|_{r,1},$$

and the general case ($f \in L_1^r(E)$) follows by completion. For $s > 0$, just apply the above result to each $\nabla^i f$ for $i \leqq s$. Q.E.D.

Before proceeding with the proofs of Theorems 3 and 4, we state some results needed in the proofs of both theorems.

DEFINITION 3. Let $c = [-1, 1]^n$. A *paving* of an n-dimensional manifold M is a collection of smooth maps $\{\varphi_i : c \to M\}_{i=1}^{\infty}$ such that

(1) $\operatorname{int}\varphi_i(c) \cap \operatorname{int}\varphi_j(c) = \emptyset$ if $i \neq j$,

(2) $\bigcup_{i=1}^{\infty} \varphi_i(c) = M$.

LEMMA 4. *Let M satisfy* C1 *and* C2. *Take the cut radius R to be less than half the conjugate radius. Let $c = [-1, 1]^n$ and $B = \partial c$. Then there is a paving $\{\varphi_i : c \to M\}_1^{\infty}$ of M and an open cover $\{U_i\}_1^{\infty}$ of M such that if $x_i = \varphi_i(0)$, U_i is a neighborhood of x_i contained in a sphere of radius R satisfying:*

(a) *If $\varphi_i(B) = \{(\theta, h_i(\theta)) | \theta \in S_1(x_i)\}$ in spherical normal coordinates, there is an $\varepsilon > 0$ such that for each i, $\{(\theta, h_i(\theta) + \varepsilon) | \theta \in S_1(x_i)\} \subset U_i$.*

(b) $\{\exp_{x_i}^{-1} \circ \varphi_i : c \to \boldsymbol{R}^n\}_1^\infty$ *is uniformly* C^∞ *bounded from above and uniformly* C^0 *bounded from below, i.e., there is a constant* $F > 0$ *such that for each* i *and* $\theta \in S_1(x_i)$, $|\exp_{x_i}^{-1}(\theta, h_i(\theta))| \geqq F$.

This result follows from a triangulation method of Cheeger and Calabi [2].

NOTATION. Let $\pi : E \to M$, and U an open set in M. Then $\pi^{-1}(U)$ is a vector bundle over U. $C_0^k(\pi^{-1}(U))$ is $\{f \in C^k(E) : f \text{ has compact support in } U\}$. Also,

$$|f|_{p,k,U} = \left[\sum_{i=0}^{k} \int_U |\nabla^i f|^p \, dV\right]^{1/p}.$$

So, $|f|_{p,k,M} = |f|_{p,k}$.

LEMMA 5. *Let* M *satisfy* C1 *and* C2, *and let* $\{\varphi_i\}$ *be the paving and* $\{U_i\}$ *the cover specified in Lemma* 4. *Then there is a constant* A *such that for all* i, *and* $f \in C^\infty(\pi^{-1}(U_i))$, *there is an* $F_i \in C_0^\infty(\pi^{-1}(U_i))$ *satisfying*

(1) $\nabla^j F_i|_{\varphi_i(c)} = \nabla^j f|_{\varphi_i(c)}$;

(2) $|F_i|_{p,k} \leqq A|f|_{p,k,U_i}$.

The proof of this lemma involves constructing bump functions based on the triangulation of Lemma 4.

LEMMA 6. *Let* M *satisfy* C1 *and* C2 *and let* $\{\varphi_i\}$ *and* $\{U_i\}$ *be the paving and cover specified by Lemma* 4. *Then there is a natural number* N *such that for each* $x \in M$, x *belongs to at most* N *of* U_i. *In particular, if* $g : M \to R$ *is integrable,* $\sum_{i=1}^\infty \int_{U_i} g \, dV \leqq N \int_M g \, dV$.

PROOF OF THEOREM 3. We can assume $u \in C_0^\infty(E)$. Now, let $\{\varphi_i : c \to M\}$ be the paving of M specified by Lemma 4, $c_i = \varphi_i(c)$, $x_i = \varphi_i(0)$, and $\{U_i\}$ the neighborhoods of the c_i given in the lemma. We shall break the proof into several steps.

Step 1. Let $U_i \in \{U_i\}$, and $2/p = 1/r + 1/q$, $r, q \geqq 1$. Then there is a constant C_1, not depending on i, such that for each $f \in C_0^\infty(\pi^{-1}(U_i))$,

$$\int_{U_i} |\nabla f|^p \, dV \leqq C_1 \left[\int_{U_i} |\nabla^2 f|^r \, dV\right]^{p/2r} \left[\int_{U_i} |f|^q \, dV\right]^{p/2q}$$

PROOF. We put orthonormal coordinates on $T_{x_i}M$. Using the coordinates on $T_{x_i}M$, we set up normal coordinates $\{y^1, \cdots, y^n\}$ on U_i, where $\{(\partial/\partial y^j)(\exp_{x_i}(0)) = e_j\}$ are orthonormal. Also, let τ be the appropriate parallel translation map.

Let $y \in \operatorname{span}(e_1, \cdots, \hat{e}_j, e_{j+1}, \cdots, e_n) = H_j$. Let $x = \exp_{x_i}(y)$, and temporarily put orthonormal coordinates on $\pi^{-1}(x)$ and let $\bar{p}_m : \pi^{-1}(x) \to \boldsymbol{R}$ be the projection onto the mth coordinate.

Now, if $w_{jm}(y, t) = \bar{p}_m \tau_0^t f(\exp_{x_i}(te^j + y))$, we have, since $\bar{p}_m$ is linear, following Lemma 1,

$$\frac{d^k}{dt^k} w_{jm}(y, t) = \bar{p}_m \tau_0^t \nabla^k f(\exp_{x_i}(te^j + y) \overbrace{(y_j'(y, t) \cdots)}^{k}),$$

where $y_j'(y, t) = (d/dt) \exp_{x_i}(t(\partial/\partial y_j)(x_i) + y)$.

Since f has compact support in U_i, we may assume $w_{jm}(y, t)$ is defined for all $t \in \boldsymbol{R}$.

We apply the statement of Theorem 3 to functions from $\boldsymbol{R}$ to $\boldsymbol{R}$ [5, p. 129ff.] to conclude

$$\int_{-\infty}^{\infty} \left|\frac{d}{dt} w_{jm}(y,t)\right|^p dt \leqq D\left[\int_{-\infty}^{\infty} \left|\frac{d^2}{dt^2} w_{jm}(y,t)\right|^r dt\right]^{p/2r} \times \left[\int_{-\infty}^{\infty} |w_{jm}(y,t)|^q \, dt\right]^{p/2q}.$$

So, we have, for each m,

$$\int_{-\infty}^{\infty} |\bar{p}_m \tau_0^t \nabla_{y_j'} f(\exp_{x_i} te^j + y)|^p \, dt \leqq D\left[\int_{-\infty}^{\infty} |\bar{p}_m \tau_0^t \nabla^2 f(\exp_{x_i} te^j + y)(y_j'(t), y_j'(t))|^r \, dt\right]^{p/2r} \times \left[\int_{-\infty}^{\infty} |\bar{p}_m \tau_0^t f(\exp_{x_i} te^j + y)|^q \, dt\right]^{p/2q}.$$

Using Hölder's inequality with $1 = p/2r + p/2q$, we get

$$\int_{-\infty}^{\infty} \sum_m |\bar{p}_m \tau_0^t \nabla_{y'} f(\exp_{x_i} te^j + y)|^p \, dt \leqq D\left[\int_{-\infty}^{\infty} \sum_m |\bar{p}_m \tau_0^t \nabla^2 f(\exp_{x_i} te^j + y)(y_j'(t), y_j'(t))|^r \, dt\right]^{p/2r} \times \left[\int_{-\infty}^{\infty} \sum_m |\bar{p}_m \tau_0^t f(\exp_{x_i} te^j + y)|^q \, dt\right]^{p/2q}.$$

Since the $\bar{p}_m$ are projections with respect to orthonormal coordinates, we have

$$\int_{-\infty}^{\infty} |\nabla'_{y_j(y,t)} f(\exp_{x_i}(te^j + y))|^p \, dt \leqq E\left[\int_{-\infty}^{\infty} |\nabla^2 f(\exp_{x_i}(te^j + y))|^r \, dt\right]^{p/2r} \times \left[\int_{-\infty}^{\infty} |f(\exp_{x_i}(te^j + y))|^q \, dt\right]^{p/2q}.$$

Recall that τ_0^t is length preserving.

Integrate both sides on H_j. Again, since f has compact support in U_i, we may integrate on all of H_j. Let $d\bar{V}$ be the volume element on $\boldsymbol{R}^n$. Hence,

$$\int_{\exp_{x_i}^{-1}(U_i)} |\nabla'_{y_i} f(\exp_{x_i}(v))|^p \, d\bar{V} \leqq E\left[\int_{\exp_{x_i}^{-1}(U_i)} |\nabla^2 f(\exp_{x_i}(v))| \, d\bar{V}\right]^{p/2r} \times \left[\int_{\exp_{x_i}^{-1}(U_i)} |f(\exp_{x_i}(v))| \, d\bar{V}\right]^{p/2q}.$$

Again, we have used Hölder's inequality. Recall $dV = J \exp_x d\bar{V}$.

Now apply Corollary 1 of Lemma 2 to set $A_1 \leqq J \exp_{x_i} \leqq A_2$, where A_1 and

A_2 are independent of i. So,

$$\frac{1}{A_2}\int_{U_i} |\nabla_{y_j} f(v)|^p \, dV$$
$$\leqq E\left[\int_{U_i} |\nabla^2 f(v)|^r \frac{dV}{J\exp_x}\right]^{p/2r}\left[\int_{U_i} |f(v)|^q \frac{dV}{J\exp_x}\right]^{p/2q}$$
$$\leqq \frac{E}{A_1}\left[\int_{U_i} |\nabla^2 f(v)|^r \, dV\right]^{p/2r}\left[\int_{U_i} |f(v)|^q \, dV\right]^{p/2q}.$$

Since this holds for all j, and $y_j(y, t) = D\exp(\exp_x(te^j + y))(e^j)$, we sum over j and apply Lemma 3 to obtain

$$\int_{U_i} |\nabla f|^p \, dV \leqq \frac{A_2 DEn}{A_1}\left[\int_{U_i} |\nabla^2 f(v)|^r \, dV\right]^{p/2r}\left[\int_{U_i} |f(v)|^q \, dV\right]^{p/2q}.$$

Step 2. Let $0 \leqq j \leqq m$, $q, r \geqq 1$, and $m/p = j/r + (m-j)/q$. Then there is a constant, C_2, such that for each $U_i \in \{U_i\}$ and $f \in C_0^\infty(\pi^{-1}(U_i))$,

$$|\nabla^j f|_p \leqq C_2 |\nabla^m f|_r^{j/m} |f|_q^{1-j/m}.$$

PROOF. This follows from Step 1 by induction on j and m.

Step 3. Let $f \in C_0^\infty(E)$ and $m/p = j/r + (m-j)/q$. Then

$$|\nabla^j f|_p \leqq C|f|_{r,m}^{j/m} |f|_q^{1-j/m}.$$

PROOF. For $f \in C_0^\infty(E)$ and for each i, let F_i be a section of $\pi^{-1}(U_i)$ specified by Lemma 5. Since $F_i \in C_0^\infty(\pi^{-1}(U_i))$, by Step 2, there is a constant C_1 independent of i and F_i such that

$$|\nabla^j F_i|_p \leqq C_1 |\nabla^m F_i|_r^{j/m} |f|_q^{1-j/m}.$$

Now

$$|\nabla^j f|_p^p = \sum_i |\nabla^j f|_{p,c_i}^p = \sum_i |\nabla^j F_i|_{p,c_i}^p \leqq \sum_i |\nabla^j F_i|_p^p$$
$$\leqq \sum_i C|\nabla^m F_i|^{pj/m} |f|_q^{p(1-j/m)}$$
$$\leqq \sum_i C_1\left[\int_{U_i} |\nabla^m F_i|^r \, dV\right]^{pj/mr}\left[\int_{U_i} |F_i|^q \, dV\right]^{p(1-j/m)/q}.$$

Since $pj/mr + p(1-j/m)/q = 1$, we apply Hölder's inequality to obtain

$$|\nabla^j f|_p^p \leqq C_1\left[\sum_i \int_{U_i} |\nabla^m F_i|^r \, dV\right]^{pj/mr}\left[\sum_i \int_{U_i} |F_i|^q \, dV\right]^{p(1-j/m)/q}$$
$$\leqq C_1\left[\sum_i \int_{U_i} \sum_{a=0}^{m} |\nabla^a F_i|^r \, dV\right]^{pj/mr}\left[\sum_i \int_{U_i} |F_i|^q \, dV\right]^{p(1-j/m)/q}.$$

Now, by Lemma 5, we have

$$|\nabla^j f|_p^p \leqq AC_2\left[\sum_i \int_{U_i} \sum_{a=0}^{m} |\nabla^a f|^r \, dv\right]^{pj/mr}\left[\sum_i \int_{U_i} |f|^q \, dv\right]^{p(1-j/m)/q}.$$

Now, apply Lemma 6 to each of the summations, and obtain

$$|\nabla^j f|_p^p \leqq AC_2 N\left[\int_M \sum_{a=0}^m |\nabla^a f|^r \, dV\right]^{pj/mr}\left[\int_M |f|^q \, dV\right]^{p(1-j/m)/q}$$

Take the pth root of both sides and conclude

$$|\nabla^j f|_p \leqq C|f|_{r,m}^{j/m}|f|_q^{1-j/m}. \quad \text{Q.E.D.}$$

PROOF OF THEOREM 4. It is sufficient to consider $f \in C_0^\infty(E)$. Now, let $\{\varphi_i : c \to M\}$ be the paving of M specified by Lemma 4, $c_i = \varphi_i(c)$, $x_i = \varphi_i(0)$, and $\{U_i\}_1^\infty$ the neighborhoods of the c_i given in the lemma. We shall break the proof into several steps.

Step 1. Let $U_i \in \{U_i\}_1^\infty$ and $f \in C_0^\beta(\pi^{-1}(U_i))$. Let $n > 1$ and $1/p = 1 - 1/n$ ($r = 1$, $j = 0$, $m = 1$). Then $|f|_p \leqq C_1 |\nabla f|_1$, where C_1 is independent of U_i and f.

PROOF. Put orthonormal coordinates on $T_{x_i} M$ and let $a = \exp_{x_i}(a^1, \cdots, a^n) \in U_i$. Let $c^j(s) = \exp_{x_i}(a^1, \cdots, a^j + s, \cdots, a^n)$. Note that the c^j's are the coordinate curves on U with respect to a normal coordinate system. So, if $c^j(s) \in U_i$, $(c^j)'(s) = e^j(c(s))$, where $(e^1(c(s)), \cdots, e^n(c(s)))$ is the coordinate frame field.

Let τ_0^s be parallel translation in E along c^j, so $s \to \tau_0^s f(c^j(s))$ is a curve in $\pi^{-1}(a)$, and hence

$$f(a) = f(c^j(0)) = \int_{-\infty}^0 \frac{d}{dt} \tau_0^t f(c^j(t)) \, dt.$$

Since f has compact support in U_i, we can assume the integral is defined on all of $\boldsymbol{R}$ and $\lim_{s \to \pm\infty} f(c^j(s)) = 0$. Now, by Lemma 1,

$$\frac{d}{dt} \tau_0^t f(c^j(t)) = \tau_0^t \nabla_{e^j(c^j(t))} f(c^j(t)),$$

and so, since τ is norm preserving

$$|f(a)| \leqq \int_{-\infty}^0 |\nabla_{e^j(c^j(s))} f(c^j(s))| \, ds.$$

By an identical argument,

$$|f(a)| \leqq \int_0^\infty |\nabla_{e^j(c^j(s))} f(c^j(s))| \, ds.$$

Hence

$$|2f(a)| \leqq \int_{-\infty}^\infty |\nabla_{e^j(c^j(s))} f(c^j(s))| \, ds.$$

This holds for each j, so

$$|2f(a)|^{n/(n-1)} \leqq \prod_{j=1}^n \left[\int_{-\infty}^\infty |\nabla_{e^j(c^j(s))} f(c^j(s))| \, ds^j\right]^{1/n-1},$$

where ds^j is the length element along the jth coordinate axis of $T_{x_i}M$.

Since $c^j(s) = \exp_{x_i}(a^1,\cdots, a^j + s,\cdots, a^n)$, both sides of the inequality are functions of a. Also, the jth factor of the product on the right side is constant on $c^j(a, s) = \exp_{x_i}(a^1, \cdots, a^j + s,\cdots, a^n)$ for each a.

Again, since f has compact support in U_i, we can integrate both sides of the inequality on $T_{x_i}M$ and get

$$\int_{T_xM} |2f(\exp_{x_i}(s^1,\cdots,s^n))|^{n/(n-1)}\, d\bar{V}$$
$$\leqq \int_{-\infty}^{\infty}\cdots\int_{-\infty}^{\infty}\left[\prod_{j=1}^{n}\int_{-\infty}^{\infty} |\nabla_{e^j} f(\exp_{x_i}(s^1,\cdots,s^n))|\, ds^j\right]^{1/n-1} ds^1\cdots ds^n,$$

where $d\bar{V}$ is the volume element on $\boldsymbol{R}^n$. Claim

$$\int_{-\infty}^{\infty}\cdots\int_{-\infty}^{\infty}\left[\prod_{j=1}^{n}\int_{-\infty}^{\infty} |\nabla_{e^j} f(\exp_{x_i}(s^1,\cdots, s^n))|\, ds^j\right]^{1/n-1} ds^1\cdots ds^n$$
$$\leqq \prod_{j=1}^{n}\left[\int_{-\infty}^{\infty}\cdots\int_{-\infty}^{\infty} |\nabla_{e^j} f(\exp_{x_i}(s^1,\cdots,s^n))| ds^1\cdots ds^n\right]^{1/n-1}.$$

This is shown exactly as in the classical case [**5**, p. 128].

Thus

$$(*)\qquad \left[\int_{T_xM} |2f(\exp_{x_i}(s^1,\cdots,s^n))|^{n/(n-1)}\, d\bar{V}\right]^{(n-1)/n} \leqq \sum_{i=1}^{n}\left[\int_{T_xM} |\nabla_{e^j} f(\exp_{x_i}(s^1,\cdots,s^n))|\, d\bar{V}\right]^{1/n}.$$

Now $p = n/(n-1)$, and the geometric mean is less than or equal to the arithmetic mean. Hence

$$\left[\int_{T_xM} |f(\exp_{x_i}(s^1,\cdots,s^n))|^p\, dV\right]^{1/p} \leqq \frac{1}{2n}\int_{T_xM}\sum_{j=1}^{n} |\nabla_{e^j} f(\exp_{x_i}(s^1,\cdots,s^n))| dV,$$

since $e^j(a) = D\exp_{x_i}(\exp_{x^{-1}}(a))(e^{-j})$, where $\{e^{-j}\}$ is the orthonormal basis of $T_{x_i}M$, we may apply Lemma 3 to conclude there is an A independent of f, U_i, and $a \in U_i$, such that

$$\sum_{j=1}^{n} |\nabla_{e^j(a)} f(a)| \leqq A|\nabla f(a)|.$$

Also, $d\bar{V} = dV(\exp_{x_i} V)/J\exp_{x_i}(V)$. So, by Corollary 1 of Lemma 2, on U_i there are B_1 and B_2 such that $B_1 \leqq 1/(J\exp_{x_i}) \leqq B_2$. Hence

$$\left[\int_M |f|^p\, dV\right]^{1/p} = \left[\int_{U_i} |f|^p\, dV\right]^{1/p} \leqq \left[\int_{\exp^{-1}_{x_i}(U_i)} |f\circ\exp_{x_i}|^p J\exp_x dV\right]^{1/p}$$
$$\leqq \frac{1}{B_1}\left[\int_{\exp^{-1}_{x_i}(U_i)} |f\circ\exp_{x_i}|^p\, d\bar{V}\right]^{1/p}$$
$$\leqq \frac{A}{2nB_1}\int_{\exp^{-1}(U_i)} |\nabla f\circ(\exp_{x_i})|\, d\bar{V}$$
$$\leqq \frac{AB_2}{2nB_1}\int_{U_i} |\nabla f|\, dV = C\int_M |\nabla f|\, dV.$$

Step 2. Let $1/p = 1/r - 1/n$, $r \geqq 1$, $r < n$; then, for $f \in C_0^\infty(\pi^{-1}(U_i))$, $|f|_p \leqq C_1 |\nabla f|_r$, where C_1 is independent of i and f.

PROOF. We use this simple lemma:

LEMMA. *Let V be a section of a Riemannian vector bundle over M. Let C be a curve in M. Then*

$$\frac{d}{dt}|V(c(t))|^n\Big|_{t=s} \leqq n|V(c(s))|^{n-1}|\nabla_{c'(s)} V(c(s))|.$$

Thus if $\{e^i\}$ is the coordinate on U_i based on normal coordinates, we have

$$e^i(y)(|f|^{(n-1)r/n-r}(y)) \leqq \frac{(n-1)r}{n-r}|f(y)|^{n(r-1)/(n-r)}|\nabla_{e^i(y)} f(y)|. \tag{1}$$

Now, applying inequality (*) from Step 1 to the real-valued function $|f|^{(n-1)r/(n-r)}$,

$$\left[\int_M |f|^{(n-1)r/n-r(n/n-1)}\, d\bar{V}\right]^{1-1/n} \leqq D_1 \prod_{i=1}^n \left[\int |e_i|\,|f|^{(n-1)r/(n-r)} d\bar{V}\right]^{1/n}.$$

So, using (1), we get

$$\left[\int |f|^{nr/(n-r)}\, d\bar{V}\right]^{1-1/n} \leqq D_2 \frac{(n-1)r}{n-r} \prod_{i=1}^n \left[\int |f|^{n(r-1)/(n-r)}\, |\nabla_{e_i} f|\, d\bar{V}\right]^{1/n}.$$

Using $1/r + (r-1)/r = 1$, and Hölder's inequality,

$$\left[\int |f|^{nr/(n-r)}\, d\bar{V}\right]^{1-1/n} \leqq D_3 \prod_{i=1}^n \left[\int |f|^{nr/(n-r)}\, d\bar{V}\right]^{(r-1)/rn} \left[\int |\nabla_{e^i} f|^r\, d\bar{V}\right]^{1/rn}.$$

Since $(n-1)/n - (r-1)/r = (n-r)/nr = 1/p$, we have

$$\left[\int |f|^p\, d\bar{V}\right]^{1/p} \leqq D_3 \prod_{i=1}^n \left[\int |\nabla_{e^i} f|^r\, d\bar{V}\right]^{1/rn}.$$

This is inequality (*) of Step 1 with $r > 1$. The remainder of the proof of this step proceeds exactly as in Step 1.

Step 3. Let $r \geqq 1$; then, for $1/p = 1/r - 1/n$, where $1 - n/r \neq 0$, there is a constant C_2 such that, for all $f \in C_0^\infty(E)$, $|f|_p \leqq C_2 |f|_{r,1}$.

PROOF. Since $1/p < 1/r < 1$, if $p > 0$, then $p > 1$. Hence there are only two cases to consider: $p < 0$ or $p > 1$. However, the case $p < 0$ $(r > n)$ follows immediately from Theorem 1. Thus we need only consider $p > 1$ $(r < n)$.

Now, for each i, let F_i be a section of $\pi^{-1}(U_i)$ specified by Lemma 5.

Since each $F_i \in C_0^\infty(\pi^{-1}(U_i))$, by Step 2, there is a constant C_1 independent of i and F_i such that $|F_i|_p \leqq C_1 |\nabla F_i|_r$. Now

$$|f|_p^p = \sum_i |f|_{p,c_i}^p = \sum_i |F_i|_{p,c_i}^p$$
$$\leqq C_1 \sum_i |\nabla F_i|_{r,1,c_i}^p \leqq C_1 \sum_i (|F_i|_{r,1,U_i}^r)^{p/r}.$$

Now, $p/r = n/(n-r) > 1$, so

$$|f|_p^p \leqq C_1(\sum_i |F_i|_{r,1,U_i}^r)^{p/r} \leqq C_1 A(\sum_i |f|_{r,1,U_i}^r)^{p/r}.$$

Now, applying Lemma 6, we have $|f|_p^p \leqq C_2^p(|f|_{r,1}^r)^{p/r} = C_2^p(|f|_{r,1})^p$. Take the pth root of both sides.

Step 4. Let $r \geqq 1$, $m \geqq 1$, and $1/p = 1/r - m/n$, where $m - n/r$ is not a nonnegative integer. Then there is a constant C such that, for $f \in C_0^\infty(E)$, $|f|_p \leqq C|f|_{r,m}$.

PROOF. This follows by induction on m. Q.E.D.

BIBLIOGRAPHY

1. R. Bishop and R. Crittenden, *Geometry of manifolds*, Pure and Appl. Math., vol. 15, Academic Press, New York, 1964. MR **29** #6401.
2. E. Calabi, private communication.
3. M. Cantor, Thesis, University of California, Berkeley, Calif., 1973.
4. S. Kobayashi and K. Nomizu, *Foundations of differential geometry*. Vol. I, Interscience, New York, 1963. MR **27** #2945.
5. L. Nirenberg, *On elliptic partial differential equations*, Ann. Scuola Norm. Sup. Pisa (3) **13** (1959), 115–162. MR **22** #823.

UNIVERSITY OF CALIFORNIA, BERKELEY

Proceedings of Symposia in Pure Mathematics
Volume 27, 1975

EIGENFUNCTIONS AND EIGENVALUES OF LAPLACIAN

SHIU-YUEN CHENG

1. Introduction. Let (M, g) be a compact C^∞ Riemannian manifold with or without boundary. The Laplacian operator Δ acting on functions is locally given by

$$-\frac{1}{g^{1/2}}\frac{\partial g^{1/2} g^{ij}}{\partial x^i}\frac{\partial}{\partial x^j}, \quad \text{where } g = \det(g_{ij}).$$

We shall be interested in studying the following two eigenvalue problems:

Fixed membrane problem ($\partial M \neq \emptyset$)

$$\Delta\varphi = \mu\varphi, \qquad \varphi = 0 \quad \text{on } \partial M. \tag{1}$$

Free membrane problem ($\partial M = \emptyset$).

$$\Delta\psi = \lambda\psi. \tag{2}$$

It is well known that both (1) and (2) have discrete spectra. We list the eigenvalues of problem (1) as $0 < \mu_1 < \mu_2 \leqq \mu_3 \leqq \cdots$ and eigenvalues of (2) as $0 = \lambda_0 < \lambda_1 \leqq \lambda_2 \leqq \lambda_3 \leqq \cdots$. Also, we have a complete orthonormal basis of $L^2(M)$ (the space of square integrable functions) consisting of eigenfunctions. This gives us the eigenfunction expansion, at least in the L^2 sense, and an immediate consequence is the so-called minimum principle:

MINIMUM PRINCIPLE.

$$\mu_n = \inf_{\substack{\varphi=0 \text{ on } \partial M;\\ \varphi\perp\varphi_i; 1\leqq i\leqq n-1}} \left(\int (d\varphi, d\varphi) \Big/ \int \varphi^2\right),$$

where $\Delta\varphi_i = \mu_i\varphi_i$ and φ attains the minimum iff φ is smooth and satisfies $\Delta\varphi = \mu_n\varphi$.

AMS (MOS) subject classifications. Primary 35P15, 58G99.

$$\lambda_n = \inf_{\psi \perp \psi_i; 0 \leqq i \leqq n-1} \left(\int (d\psi, d\psi) \Big/ \int \psi^2 \right),$$

where $\Delta\psi_i = \lambda_i \psi_i$ and ψ attains the minimum iff ψ is smooth and satisfies $\Delta\psi = \lambda_n \psi$. The expression $\int(df, df)/\int f^2$ is often called the Rayleigh-Ritz quotient.

From the minimum principle we can easily derive Courant's nodal domain theorem. To state it, we need some definitions.

DEFINITION. Let $\Delta f = \lambda f$; then $f^{-1}(0)$ is called the *nodal set* of f and each component of $M - f^{-1}(0)$ is called a *nodal domain* of f. When dim $M = 2$ we also call $f^{-1}(0)$ the *nodal lines* of f.

COURANT'S NODAL DOMAIN THEOREM. *Let φ_n be the nth eigenfunction of problem* (1); *then* $\sharp$(*nodal domains of* φ_n) $\leqq n$.

Let ψ_n be the nth eigenfunction of problem (2); *then* $\sharp$(*nodal domains of* ψ_n) $\leqq n + 1$.

REMARKS. (i) For dim $M = 2$, it is proved by A. Pleijel [**11**] that the above inequalities are attained only finitely many times.

(ii) One can easily see that $\sharp$(nodal domains of φ_2) $= 2$ and $\sharp$(nodal domains of ψ_1) $= 2$.

(iii) We can also observe that each eigenfunction is the 1st eigenfunction of problem (1) when restricting it to each of its nodal domains. Therefore, it is interesting to know the behaviour of the nodal set, and this is the subject of § 2.

2. Nodal lines. We shall study the nodal set of an eigenfunction when dim $M = 2$. The theorems and corollaries are stated in terms of simplicity and clarity but not for maximum generality, and some of them can be generalized to dim $M > 2$ (for details see [**7**]).

Our first question is about the regularity of the nodal lines. This amounts to the study of the local behaviour of the zero set of a solution to an elliptic equation (in our case we have $(\Delta - \lambda)f = 0$). For this purpose, we use the theorem of Lipman Bers [**2**]. His theorem applies for domains in E^m; to apply it we use an exponential map to pull our Laplacian and eigenfunction back to a small open set about the origin. Then we have the following theorem:

THEOREM 2.1. (i) *The critical points are isolated on the nodal lines of an eigenfunction.*

(ii) *The nodal lines consist of a number of C^2-immersed circles.*

(iii) *When the nodal lines meet they form an equi-angular system.*

(iv) *When the nodal lines meet, the geodesic curvatures are zero there.*

(v) *Let M be a compact surface of genus g, and ψ be the* 1*st eigenfunction; then* $\sharp$(*immersed circles of* ψ) $\leqq 2g + 1$.

COROLLARY 2.2. *Let g be any Riemannian metric on S^2. Then the nodal line of a* 1*st eigenfunction is a smooth simple closed curve.*

COROLLARY 2.3. *Let the assumption be the same as in Corollary* 2.2. *Then the multiplicity of the* 1*st eigenvalue is at most* 3.

COROLLARY 2.4. *Let g be any Riemannian metric on the torus. Then the multiplicity of the* 1*st eigenvalue is at most* 10.

COROLLARY 2.5. *Let g be any Riemannian metric on* S^2 *such that* $S^1 \times \{-1, 1\}$ *acts on* (S^2, g) *as isometries. There exists a nodal line of a* 1*st eigenfunction which is a geodesic.*

REMARK. The theorems tell us that there are topological restrictions to multiplicities of eigenvalues. Moreover, if we know that every minimal immersion of S^2 into S^3 is by 1st eigenfunctions then we obtain the famous Almgren-Calabi theorem that every minimal immersion of S^2 into S^3 must lie on a great circle.

3. Estimates of eigenvalues. We can use Theorem 2.1 and an isoperimetric inequality of Burago and Zalgaller [**4**] to obtain the following theorem:

THEOREM 3.1. *Let* (M, g) *be a compact surface of nonnegative curvature; then* $1/4(\text{diam } M)^2 \leqq \lambda_1$.

PROOF. We can assume M is orientable, otherwise we pass to its double cover and observe that the 1st eigenvalue of the double cover is not greater than the original one. If M is flat, then all its eigenvalues are known (see [**1**, p. 147]) and the theorem is immediately seen to be true. If the curvature of M is positive somewhere, then $M \approx S^2$. By Corollary 2.2, the nodal line of a 1st eigenfunction cuts M into two parts each homeomorphic to a disc. The inequality of Burago and Zalgaller says that : On a compact simply connected region D on a Riemann surface we have $\text{area}(D) \leqq \text{length}(D) \times R + \frac{1}{2}(\omega^+ - 2\pi)/R^2$ where R is the radius of the largest disc inscribed in D and ω^+ is the integral of the positive part of the curvature over D.

Therefore, we take the part with total curvature $\leqq 2\pi$, and the above isoperimetric inequality implies the number h [**6**] defined by Cheeger satisfies $h \geqq 1/\text{diam}(M)$; then the well-known theorem of Cheeger [**6**] tells us that

$$\lambda_1 \geqq h^2/4 \geqq 1/4(\text{diam } M)^2. \quad \text{Q.E.D.}$$

REMARK. If $\dim M > 2$, we do not know whether one can still give a lower bound to λ_1 in terms of $\text{diam}(M)$ under the assumption that sectional curvature is always nonnegative.

For problem (1) on plane domains, there are numerous results (e.g. see [**12**]). Things work there because the Euclidean space is highly symmetric and the coordinate functions are harmonic. For a minimal hypersurface in Euclidean space, the coordinate functions are also harmonic. Hence, we can generalize the method of Pólya-Payne-Weinberger [**10**] and obtain the following theorem:

THEOREM 3.2. *Let* M *be a compact domain on a minimal hypersurface in* E^{m+1}; *then*

$$\mu_{n+1} \leqq \mu_n + 4\sum_{i=1}^{n} \mu_i \Big/ mn \quad \textit{or} \quad (\mu_{n+1})/\mu_n \leqq 1 + 4/m.$$

PROOF. Let $x^1,\cdots,x^{m+1}$ denote the coordinates in E^{m+1} w.r.t. an orthonormal coordinate frame.

Set

$$a_{\alpha ik} = \int x^\alpha \varphi_i \varphi_k \quad \text{where } \Delta\varphi_i = \mu_i \varphi_i,\ 1 \leqq i \leqq n \text{ and } \int \varphi_i \varphi_k = \delta_{ik}. \tag{3}$$

Then the functions

$$U_{\alpha i} = x^\alpha \varphi_i - \sum_{k=1}^{n} a_{\alpha ik} \varphi_k \tag{4}$$

are all orthogonal to $\varphi_1,\cdots,\varphi_n$ and vanish at the boundary. Then

$$\Delta U_{\alpha i} = \mu_i x^\alpha \varphi_i - 2(dx^\alpha, d\varphi_i) - \sum_{k=1}^{n} a_{\alpha ik} \mu_k \varphi_k.$$

Hence by (3), (4),

$$\begin{aligned} \int U_{\alpha i} \Delta U_{\alpha i} &= \mu_i \int x^\alpha \varphi_i U_{\alpha i} - 2\int U_{\alpha i}(dx^\alpha, d\varphi_i) \\ &= \mu_i \int (U_{\alpha i})^2 - 2\int U_{\alpha i}(dx^\alpha, d\varphi_i), \end{aligned}$$

so that

$$\mu_{n+1} \leqq \frac{\int U_{\alpha i} \Delta U_{\alpha i}}{\int U_{\alpha i}^2} = \mu_i + \frac{-2\int U_{\alpha i}(dx^\alpha, d\varphi_i)}{\int (U_{\alpha i})^2}$$

and

$$\mu_{n+1} - \mu_n \leqq \frac{-2\int U_{\alpha i}(dx^\alpha, d\varphi_i)}{\int (U_{\alpha i})^2},$$

and then

$$\mu_{n+1} - \mu_n \leqq \frac{\sum_\alpha \sum_i (-2) \int U_{\alpha i}(dx^\alpha, d\varphi_i)}{\sum_\alpha \sum_i \int (U_{\alpha i})^2}. \tag{5}$$

We have

$$\sum_\alpha \sum_i (-2) \int U_{\alpha i}(dx^\alpha, d\varphi_i) = \sum_\alpha \sum_i (-2) \int \Big(x^\alpha \varphi_i - \sum_k a_{\alpha ik} \varphi_k \Big)(dx^\alpha, d\varphi_i)$$

and

$$\begin{aligned} \int \varphi_k (dx^\alpha, d\varphi_i) &= \int (dx^\alpha, d\varphi_i \varphi_k) - \int \varphi_i (dx^\alpha, d\varphi_k) \\ &= \int (\Delta x^\alpha) \varphi_i \varphi_k - \int \varphi_i (dx^\alpha, d\varphi_k) \\ &= -\int \varphi_i (dx^\alpha, d\varphi_k). \end{aligned}$$

Since $a_{\alpha ik}$ is symmetric w.r.t. i, k we have $\sum_i \sum_k a_{\alpha ik} \int \varphi_k(dx^\alpha, d\varphi_i) = 0$ so that

$$
\begin{aligned}
\sum_\alpha \sum_i -2 \int U_{\alpha i}(dx^\alpha, d\varphi_i) &= \sum_\alpha \sum_i (-2) \int x^\alpha \varphi_i (dx^\alpha, d\varphi_i) \\
&= \sum_\alpha \sum_i (-\tfrac{1}{2}) \int (d(x^\alpha)^2, d\varphi_i^2) \\
&= \sum_i (-\tfrac{1}{2}) \int \Big(\Delta \sum_\alpha (x^\alpha)\Big) \varphi_i^2 \\
&= \sum_i (-\tfrac{1}{2}) \int (-2m)\, \varphi_i^2 \\
&= mn.
\end{aligned} \tag{6}
$$

From this, we use the Schwarz inequality to derive

$$
\begin{aligned}
m^2 n^2 &= 4\Big(\sum_\alpha \sum_i \int U_{\alpha i}(dx^\alpha, d\varphi_i)\Big)^2 \\
&\leqq 4\Big(\sum_\alpha \sum_i \int (U_{\alpha i})^2\Big)\Big(\sum_\alpha \sum_i (dx^\alpha, d\varphi_i)\Big).
\end{aligned}
$$

We see that $\sum_\alpha \int (dx^\alpha, d\varphi_i)^2$ is invariant if we perform an orthonormal change of coordinates, so we derive that $\sum_\alpha \int (dx^\alpha, d\varphi_i)^2 = \int (d\varphi_i, d\varphi_i) = \mu_i$. Whence

$$
\sum_\alpha \sum_i \int (U_{\alpha i})^2 \geqq m^2 n^2 \Big/ 4 \sum_i \mu_i. \tag{7}
$$

Substituting (6), (7) into (5), we complete the proof of the theorem

$$
\mu_{n+1} - \mu_n \leqq mn \Big/ \Big(m^2 n^2 \Big/ 4 \sum_i \mu_i\Big) = (4/mn) \sum_i \mu_i \quad \text{Q.E.D.}
$$

For the vibration of an inhomogenous membrane, we also have harmonic coordinate functions, so the same method applies and we obtain

COROLLARY 3.3. *Let ρ be a positive function on a bounded plane domain D. Then for the eigenvalue problem of the inhomogenous membrane $\Delta\varphi = \mu\rho\varphi$, $\varphi = 0$ on ∂D we have*

$$
\mu_{n+1} \leqq \mu_n + \left(\frac{\rho_{\max}}{\rho_{\min}}\right)^2 \frac{2}{n} \sum_i \mu_i \quad \text{or} \quad \mu_{n+1} \leqq \left(1 + 2\left(\frac{\rho_{\max}}{\rho_{\min}}\right)^2\right)\mu_n.
$$

COROLLARY 3.4. *Let M be a compact domain on the standard sphere S^2 with the induced metric; then*

$$
\mu_{n+1} \leqq \mu_n + \left(\frac{2}{1 - \cos c}\right)^4 \frac{2}{n} \sum_i \mu_i \quad \text{or} \quad \frac{\mu_{n+1}}{\mu_n} \leqq 1 + 2\left(\frac{2}{1 - \cos c}\right)^4
$$

where c is the radius of the largest inscribed geodesic circle contained in $S^2 \backslash M$.

We now prove a comparison theorem for the 1st eigenvalue using a beautiful observation of Barta. The proof of Barta's theorem is so simple that we also include it here.

THEOREM 3.5 (BARTA). *Let (M, g) be a Riemannian manifold with boundary and $f \in C^\infty(M)$ such that f is positive inside M and equal to zero on the boundary; then*

$$\inf (\Delta f / f) \leqq \mu_1 \leqq \sup(\Delta f / f). \tag{8}$$

PROOF. Let φ_1 be the 1st eigenfunction, i.e. $\Delta\varphi_1 = \mu_1\varphi_1$ and $\varphi_1 = 0$ an ∂M. We can assume that φ_1 is positive inside M. Then we write $\varphi_1 = f + h$, and hence $h = 0$ on ∂M also. Then

$$\begin{aligned} \mu_1 &= \frac{\Delta\varphi_1}{\varphi_1} = \frac{\Delta(f+h)}{f+h} = \frac{\Delta f}{f} + \frac{\Delta(f+h)}{f+h} - \frac{\Delta f}{f} \\ &= \frac{\Delta f}{f} + \frac{f\Delta h - h\Delta f}{f(f+h)}. \end{aligned} \tag{9}$$

Observe that the second term on the right of (9) changes sign or vanishes identically in M because $f(f+h)$ is positive inside M and $\int_M f\Delta h - h\Delta f = 0$ by Stoke's theorem. Since the left-hand side of (9) is a constant, (8) is immediately seen to be true. To state our next theorem, we define $V_n(k, r)$ to denote a geodesic ball with radius r in the n-dimensional simply connected space form with sectional curvature k.

THEOREM 3.6. *Suppose $B(x_0; r)$ is a geodesic ball within the cut-locus of x_0 in an n-dimensional Riemannian manifold. Then we have the following*:

(i) *If Ricci curvature $\geqq (n-1)k_m$, then μ_1 of $B(x_0; r) \leqq \mu_1$ of $V_n(k_m, r)$.*

(ii) *If sectional curvature $\leqq k_M$, and $V_n(k_M, r)$ lies with the cut-locus, then μ_1 of $V_n(k_M, r) \leqq \mu_1$ of $B(x_0; r)$.*

PROOF. Let φ_1 be the nonnegative 1st eigenfunction of $V_n(k_m, r)$. Since all simply connected space forms are two-point homogenous, φ_1 is a radial function. Barta's theorem tells us that

$$\mu_1 \text{ of } B(x_0; r) \leqq \sup(\Delta\varphi_1/\varphi_1).$$

φ_1 being radial, $\Delta\varphi_1$ takes a very simple form (see [**1**, p. 134])

$$\Delta\phi_1 = -\frac{d^2\varphi_1}{dr^2} - \left(\frac{\theta'}{\theta} + \frac{n-1}{r}\right)\frac{d\varphi_1}{dr}$$

where $\theta = (\det(g_{ij}))^{1/2}$ and (g_{ij}) is calculated with respect to the normal coordinates. However,

$$-\frac{d_2\varphi_1}{dr^2} - \left(\frac{\theta'(k_m)}{\theta(k_m)} + \frac{n-1}{r}\right)\frac{d\varphi_1}{dr} = \mu_1 x \qquad [\mu_1 \text{ of } V(k_m, r)]$$

and obviously $\theta(k_m)$ is calculated in the space form with curvature k_m. Assume that $d\varphi_1/dr \leqq 0$; then $\sup(\Delta\varphi_1/\varphi_1) \leqq \mu_1$ of $V_n(k_m, r)$ if and only if

$$\theta'/\theta \leqq \theta'(k_m)/\theta(k_m). \tag{10}$$

However, (10) is true by the theorem on p.253 of [**3**]. Therefore, μ_1 of $B(x_0; r) \leqq \mu_1$ of $V_n(k_m, r)$ is true and the proof of μ_1 of $V_n(k_M, r) \leqq \mu_1$ of $B(x_0; r)$ goes exactly the same. To complete the proof of our theorem, it remains to prove the following lemma:

LEMMA 3.7. *Let $V_n(k, r)$ be within the cut-locus and φ_1 is the nonnegative 1st eigenfunction of $V_n(k, r)$; then $d\varphi_1/dr \leqq 0$.*

PROOF. φ_1 satisfies

$$\frac{d^2\varphi_1}{dr^2} + \left(\frac{\theta'(k)}{\theta(k)} + \frac{n-1}{r}\right)\frac{d\varphi_1}{dr} + \mu_1\varphi_1 = 0 \tag{11}$$

and $\varphi_1(r) = 0 = \varphi_1'(0)$.

Considering φ_1 as a smooth function defined on $[-r, r]$, φ_1 takes a local maximum at 0. Suppose $(d\psi_1/dr)(t) = 0$ with $t \in (0, r]$, then from (11) and that $\varphi_1 > 0$ on$(-r, r)$ we know that φ_1 must have a local maximum at t. Then on $[0, t]$, either φ_1 is a constant function, or there will be a local minimum. However, both possibilities are impossible and so $d\varphi_1/dt \neq 0$ on $(0, r]$. But φ_1 is positive on $[0, r)$ and $\varphi_1(r) = 0$, so $d\varphi_1/dt \geqq 0$ is impossible. Hence we see that $d\varphi_1/dt \leqq 0$ on $[0, r]$.

REMARKS. (i) Barta's theorem enables us to give a lower bound of $B(x_0; r)$ without using symmetrization. However, when the domain is not a geodesic ball, symmetrization seems to be the only technique (see **[8]**).

(ii) I. Chavel and E. Feldman **[5]** give an upper bound of μ_1 by comparing a geodesic ball with nonnegative Ricci curvature with a geodesic ball in the Euclidean space. They use the Reid comparison theorem for this purpose. However, one can give a proof of the Reid comparison theorem by using Barta's theorem; see **[8]**.

Now, as a consequence of Theorem 3.6, we can prove the following:

THEOREM 3.8. *Let M be a complete Riemannian manifold of dimension n such that the exponential map at some point is a diffeomorphism.*

(i) *If sectional curvative $\leqq -k$, $k > 0$ then*

$$\inf_{f\in C_0^\infty(M)} \int (df, df) \Big/ \int f^2 \geqq (n-1)^2 k/4.$$

(ii) *If Ricci curvature $\geqq -(n-1)l$, $l > 0$ then*

$$\inf_{f\in C_0^\infty(M)} \int (df, df) \Big/ \int f^2 \leqq (n-1)^2 l/4.$$

PROOF. Suppose $x_0 \in M$ such that the exponential map at x_0 is a diffeomorphism. Observe that, if $r > r'$ then

$$\inf_{f\in C_0^\infty(B(x_0;r))} \int (df, df) \Big/ \int f^2 \leqq \inf_{f\in C_0^\infty(B(x_0;r))} \int (df, df) \Big/ \int f^2.$$

Therefore,

$$\inf_{f\in C_0(M)} \frac{\int (df, df)}{\int f^2} = \lim_{r\to\infty}\left(\inf_{f\in C_0^\infty(B(x_0;r))} \frac{\int (df, df)}{\int f^2}\right)$$
$$= \lim_{r\to\infty}(\mu_1 \text{ of } B(x_0; r)).$$

Suppose $-(n-1)l \leqq$ Ricci curvature $\leqq -(n-1)k$ with $k, l > 0$. Theorem 3.6 then implies that

$$\inf_{f \in C_0^\infty(H_n(l))} \frac{\int (df, df)}{\int f^2} \geqq \inf_{f \in C_0^\infty(M)} \frac{\int (df, df)}{\int f^2} \geqq \inf_{f \in C_0^\infty(H_n(k))} \frac{\int (df, df)}{\int f^2}$$

where $H_n(a)$, with $a > 0$, denotes the n-dimensional simply connected space form with sectional curvature identically equal to $-a$. However, it is easy to see that

$$\inf_{f \in C_0^\infty(H_n(a))} \frac{\int (df, df)}{\int f^2} = \frac{(n-1)^2 a}{4}$$

(see e.g. **[9]**) and then our theorem follows immediately.

REMARK. The result (i) of Theorem 3.8 was obtained by H. P. McKean **[9]**.

We can also estimate the nth eigenvalue of a compact manifold M without boundary by using Theorem 3.6.

Let $\psi_1, \cdots, \psi_{n-1}$ be the first n eigenfunctions such that

$$\Delta \psi_i = \lambda_i \psi_i, \qquad \int \psi_i \psi_k = \delta_{ik}, \qquad 0 \leqq i \leqq n-1.$$

Let $i(M)$ denote the injectivity radius of M. We can easily construct $n+1$ geodesic balls $B(x_i; c_n)$ with center at x_i and radius equal to $c_n = \min\ [i(M), \operatorname{diam}(M)/(2n)]$ such that they are pairwise disjoint. On each $B(x_i; c_n)$, let φ_i denote the first eigenfunctions of the fixed membrane problem on $B(x_i; c_n)$. Extend φ_i to be zero outside $B(x_i; c_n)$. Elementary linear algebra tells us that $\exists a_1, \cdots, a_{n+1}$ not all zero such that

$$\int \psi_j \left(\sum_{i=1}^{n+1} a_i \varphi_i \right) = 0, \qquad \forall j = 0, \cdots, n-1.$$

Since $B(x_i; c_n)$ are pairwise disjoint $\sum_{i=1}^{n+1} a_i \varphi_i \not\equiv 0$. Plugging this function in the Rayleigh-Ritz quotient we have

$$\lambda_n \leqq \sum a_i^2 \int (d\varphi_i, d\varphi_i) \Big/ \sum a_i^2 \int \varphi_i^2.$$

We see that, if $\int (d\varphi_i, d\varphi_i) / \int \varphi_i^2 \leqq c\ \forall i$, then $\lambda_n \leqq c$, too.

Hence, the problem of estimating λ_n is reduced to estimating the upper bound of the 1st eigenvalue of the fixed membrane problem on any geodesic ball with radius c_n.

THEOREM 3.9. *Suppose the Ricci curvature* $\geqq (n-1)k$; *then* $\lambda_n \leqq \mu_1$ *of* $V_m(c_n, k)$, $m = \dim M$.

REMARK. (i) We can obtain an explicit formula of the μ_1 of $V_m(c_n, k)$ by consulting a book on Legendre functions. Here we only need the qualitative result. For example, letting $n \to \infty$ then eventually $i(M) \geqq \operatorname{diam}(M)/(2n)$, then we have the following conclusion:

$$\lambda_n = O(n^2) \quad \text{as } n \to \infty.$$

Of course, this is far from being sharp; actually H. Weyl [**13**] proved that

$$\lambda_n = O(n^{2/m}), \quad m = \dim M.$$

(ii) We can also obtain a proof of the following theorem of Toponogov: Suppose sectional curvature of $M^m \geqq c > 0$ and diam $(M^m) = \pi/c^{1/2}$, then M^m is isometric to the standard sphere with curvature c. From the assumption we see that Ricci curvature $\geqq (m - 1)cg$, where g is the fundamental tensor of M^m. Then, Lichnerowicz's formula (see [**1**, p. 179]) shows that $\lambda_1 \geqq mc$. Obata showed that if equality holds then the manifold is isometric to a sphere (see [**1**, p. 179]). Now it remains to find a good upper bound for λ_1. Using comparison theorems we can show that if p and q realize the diameter then every geodesic which emerges from p passes through q. So

$$\lambda_1 \leqq \mu_1 \text{ of } B(p; \operatorname{diam}(M)/2) \leqq \mu_1 \text{ of } V_m(\pi/2\, c^{1/2}, c) = mc.$$

So $\lambda_1 = mc$ and then M must be isometric to a sphere by Obata's result.

Acknowledgment. The author wishes to thank his advisor Professor S. S. Chern for his continuous help and encouragement.

References

1. M.Berger, P. Gauduchon and Edmond Mazet, *Le spectre d'une variété riemannienne*, Lecture Notes in Math., vol. 194, Springer-Verlag, Berlin and New York, 1971. MR **43** #8025.

2. L. Bers, *Local behavior of solutions of general linear elliptic equations*, Comm. Pure Appl. Math. **8** (1955), 473–496. MR **17**, 743.

3. R. Bishop and R. Crittenden, *Geometry of manifolds*, Pure and Appl. Math., vol. 15, Academic Press, New York, 1964. MR **29** #6401.

4. Ju. D. Burago and V. A. Zalgaller, *Isoperimetric problems for regions on a surface having restricted width*, Trudy Mat. Inst. Steklov. **76** (1965), 81–87=Proc. Steklov Inst. Math. **76** (1965), 100–108. MR **34** #1972.

5. I. Chavel and E. Feldman, *The first eigenvalue of the laplacian on manifolds of nonnegative curvature*, these PROCEEDINGS.

6. J. Cheeger, *A lower bound for the smallest eigenvalue of the laplacian*, Problem in Analysis, Princeton Univ. Press, Princeton, N.J., 1970.

7. S. Y. Cheng, *Eigenfunctions and nodal sets* (to appear).

8. ———, Thesis, Berkeley, 1974.

9. H. P. McKean, *An upper bound to the spectrum of Δ on a manifold of negative curvature*, J. Differential Geometry **4** (1970), 359–366. MR **42** #1009.

10. L. E. Payne, G. Pólya and H. F. Weinberger, *On the ratio of consecutive eigenvalues*, J. Mathematical Phys. **35** (1956), 289–298. MR **18**, 905.

11. A. Pleijel, *Remarks on Courant's nodal line theorem*, Comm. Pure Appl. Math. **9** (1956), 543–550. MR **18**, 315.

12. G. Pólya and G. Szegö, *Isoperimetric inequalities in mathematical physics*, Ann. of Math. Studies, no. 27, Princeton Univ. Press, Princeton, N. J., 1951. MR **13**, 270.

13. I. M. Singer and H. P. McKean, Jr., *Curvature and the eigenvalues of the Laplacian*, J. Differential Geometry **1** (1967), 43–69. MR **36** #828.

UNIVERSITY OF CALIFORNIA, BERKELEY

Proceedings of Symposia in Pure Mathematics
Volume 27, 1975

MINAKSHISUNDARAM'S COEFFICIENTS ON KAEHLER MANIFOLDS

HAROLD DONNELLY

Suppose we are given a differential operator P on a manifold M. Certain natural conditions on P guarantee the existence of the fundamental solution to the associated heat equation. It is well known that the coefficients in the asymptotic expansion of this solution give links between the spectrum of P and the geometry of M. Several authors ([**1**], [**4**], [**6**]) have calculated these coefficients for the real Laplacian Δ on a compact Riemannian manifold M.

In this paper we consider the complex Laplacian $\Box$ acting on forms of type (p, q) for a complex manifold M. We calculate the coefficients for M Kaehler. § 1 is devoted to preliminaries. In § 2 we recall some required facts concerning invariants on Kaehler manifolds. The actual calculation takes place in § 3 and geometric applications are given in the following section. There we obtain the result that a Kaehler manifold with the same spectrum as CP^n is isometric to CP^n. This has also been demonstrated independently by Peter Gilkey and John Sacks [**3**] using somewhat different methods. The last section is devoted to counterexamples. In particular we give two isospectral Kaehler manifolds which are not holomorphically equivalent.

We would like to thank V. K. Patodi for suggesting the problem and for several helpful conversations. Also we would like to thank Professor Chern for his advice and encouragement. Our gratitude is expressed to Peter Gilkey for his comments concerning the original manuscript. In particular he informed us of his theorem (2.1) appearing in § 2 of this paper which saved us much unnecessary labor.

AMS (MOS) subject classifications (1970). Primary 53C99.

1. Fundamental solution of the heat equation. Let M^d be a compact Hermitian manifold of complex dimension d and Hermitian metric h. If $\bar\partial : \Lambda^{p,q} \to \Lambda^{p,q+1}$, $0 \leq p, q \leq d$, is the usual operator then set $\delta = * \# \bar\partial * \#$ where $*$ is the Hodge star operator and $\#$ is complex conjugation. The complex Laplacian is given by $\Box = -(\bar\partial\delta + \delta\bar\partial)$.

$\Box$ is a negative selfadjoint operator and is well known to have eigenvalues $0 \geqq \lambda_0^{p,q} \geqq \lambda_1^{p,q} \geqq \cdots$ where each eigenvalue is repeated up to multiplicity. If $\phi_i^{p,q}$ are the corresponding eigenfunctions then

$$\sum \exp\{t\lambda_i^{p,q}\}\, \phi_i^{p,q}(z') \otimes \phi_i^{p,q}(z) \tag{1.1}$$

converges uniformly on compact sets in $(0, \infty) \times M \times M$ to the fundamental solution of the heat equation

$$(\partial/\partial t - \Box_z)\phi = 0.$$

It is well known that there exists an asymptotic expansion for (1.1) given by

$$(2\pi t)^{-d} \sum t^i\, U^{i,p,q}(z', z) + O(t).$$

Set $a_{i,p,q} = \operatorname{Tr} U^{i,p,q}(z', z')$. Then

$$\sum \exp\{\lambda_n^{p,q} t\}\, \langle \phi_n^{p,q}(z'), \phi_n^{p,q}(z') \rangle = (2\pi t)^{-d} \sum t^i a_{i,p,q}(z') + O(t). \tag{1.2}$$

It is well known that $a_{0,p,q}(z') = \binom{d}{p}\binom{d}{q}$ and this may be verified by considering product manifolds. $a_{1,p,q}$ and $a_{2,p,q}$ will be computed in § 3 and then applied to deduce geometric properties of the eigenvalues of $\Box$.

2. Invariance theory on Kaehler manifolds. A complete description of the invariants of order one and two on Kaehler manifolds has been given by Peter Gilkey [3].

THEOREM 2.1 (GILKEY). *Let M^d be a Kaehler manifold of complex dimension d. Then*

(1) *$\tau = \sum R_{ijij}$ is a basis for the invariants of order one.*

(2) *If $d=1$, then τ^2, $\Delta\tau = \sum R_{ijij/kk}$ are a basis for the invariants of order two.*

(3) *If $d \geqq 2$, then τ^2, $\Delta\tau$, $* \Omega^{d-2} c_1^2$, $* \Omega^{d-2} c_2$ are a basis for the invariants of order two. Here c_1, c_2 are the Chern classes of M, Ω is the Kaehler form of M, and $*$ is the Hodge star operator.*

We wish to use a different basis from (3) in the theorem. Consider the invariants $\|R\|^2 = \sum R_{ijkl}^2$, $\|\rho\|^2 = \sum_{j,l} (\sum R_{ijil})^2$. Then claim τ^2, $\Delta\tau$, $\|\rho\|^2$, $\|R\|^2$ are a basis for the invariants of order two on M^d for $d \geqq 2$. Because of the theorem we need only show that these invariants are independent.

First remark that on a product $M = M_1 \times M_2$ we have the following identities:

$$\begin{aligned} \tau &= \tau_1 + \tau_2, & \|\rho\|^2 &= \|\rho_1\|^2 + \|\rho_2\|^2, \\ \Delta\tau &= (\Delta\tau)_1 + (\Delta\tau)_2, & \|R\|^2 &= \|R_1\|^2 + \|R_2\|^2. \end{aligned} \tag{2.2}$$

Also on a one manifold the invariants degenerate, and we have the relationships

(2.3) $$\|R\|^2 = 2\|\rho\|^2 = \tau^2.$$

Now suppose there is a dependence relation

$$\alpha\|R\|^2 + \beta\,\|\rho\|^2 + \gamma\tau^2 + \varepsilon\,\Delta\tau = 0$$

for all M^d Kaehler for some fixed $d \geqq 2$.

Taking M to be a product of one manifolds and using (2.2), (2.3) yields $\varepsilon=\gamma=0$, $2\alpha+\beta=0$. Now on CP^n complex projective space with its standard metric of constant holomorphic sectional curvature four $\|\rho\|^2 = 8n\,(n+1)^2$, $\|R\|^2 = 32n \cdot (n+1)$. Substitution yields $\alpha = \beta = 0$. Thus $\|R\|^2$, $\|\rho\|^2$, $\Delta\tau$, τ^2 form a basis for the invariants of order two on a Kaehler manifold M^d when $d \geqq 2$.

3. Minakshisundaram's coefficients. By the invariance theory of the preceding section we have

$$\begin{aligned} a_{1,p,q} &= K(d,p,q)\tau, \\ a_{2,p,q} &= K_1(d,p,q)\tau^2 + K_2(d,p,q)\,\|\rho\|^2 \\ &\quad + K_3(d,p,q)\,\|R\|^2 + K_4(d,p,q)\,\Delta\,\tau. \end{aligned}$$

In this section we will compute the $K(d, p, q)$ and the $K_i(d, p, q)$ explicitly.

Note that for a product $M^d = M_1^{d_1} \times M_2^{d_2}$ of manifolds we have the relations (2.2) along with

$$e^{p,q}(t,x,y) = \sum_{p_1+p_2=p;\; q_1+q_2=q} e_1^{p_1,q_1}(t,x,y) \wedge e_2^{p_2,q_2}(t,x,y)$$

where $e^{p,q}(t, x, y)$ is the asymptotic expansion of Minakshisundaram obtained in § 1. This gives

$$a_{s,p,q} = \sum_{p_1+p_2=p;\; q_1+q_2=q}\ \sum_{i+j=s} a^1_{i,p_1,q_1} a^2_{j,p_2,q_2}$$

and in particular

(3.1) $$a_{1,p,q} = \sum_{p_1+p_2=p;\; q_1+q_2=q} \left[\binom{d_1}{p_1}\binom{d_1}{q_1} a^2_{1,p_2,q_2} + \binom{d_2}{p_2}\binom{d_2}{q_2} a^1_{1,p_1,q_1}\right].$$

(3.2) $$a_{2,p,q} = \sum_{p_1+p_2=p;\; q_1+q_2=q} \left[\binom{d_1}{p_1}\binom{d_1}{q_1} a^2_{2,p_2,q_2} + a^1_{1,p_1,q_1} a^2_{1,p_2,q_2} + \binom{d_2}{p_2}\binom{d_2}{q_2} a^1_{2,p_2,q_2}\right].$$

Expansion of (3.1) with M_2 flat gives

(3.3) $$K(d,p,q) = \sum_{p_1+p_2=p;\; q_1+q_2=q} \binom{d-d_1}{p_2}\binom{d-d_1}{q_2} K(d_1,p_1,q_1).$$

Now in the Kaehler case the complex Laplacian is just half the real Laplacian. Thus the formulas of **[6]** give immediately

(3.4) $$\begin{aligned} &K(d,0,0) = 1/12, \\ K_1(d,0,0) = 1/288, &\qquad K_2(d,0,0) = -1/720, \\ K_3(d,0,0) = 1/720, &\qquad K_4(d,0,0) = 1/720. \end{aligned}$$

The fantastic cancellation of [5] gives for one manifolds

$$a_{1,0,0} - a_{1,0,1} = 2\pi\,(*\tfrac{1}{2}c_1)$$

where c_1 is the first Chern class. Further it may be shown by computation that $*c_1 = \frac{1}{4}\pi\,\tau$ so

$$K(1,0,0) - K(1,0,1) = 1/4, \qquad K(1,0,1) = -1/6. \tag{3.5}$$

On a Kaehler manifold complex conjugation gives an isomorphism $\Lambda^{q,p} \sim \Lambda^{p,q}$ which commutes with the action of the Laplacian. Applying this to a one manifold yields

$$K(1,1,0) = -1/6, \qquad K(1,1,1) = 1/12. \tag{3.6}$$

Taking $d_1 = 1$ in (3.3) and using (3.4), (3.5), and (3.6) yields

$$\begin{aligned} K(d,p,q) = &\frac{1}{12}\binom{d-1}{p}\binom{d-1}{q} - \frac{1}{6}\binom{d-1}{p}\binom{d-1}{q-1} \\ &- \frac{1}{6}\binom{d-1}{p}\binom{d-1}{q} + \frac{1}{12}\binom{d-1}{p-1}\binom{d-1}{q-1}. \end{aligned} \tag{3.7}$$

This completes the computation of $a_{1,p,q}$.

Now consider $K_i(d,p,q)$ on a one manifold $d = 1$. We have the relations (2.3) and thus $K_i(1,p,q)$ are not well determined. Set

$$a_{2,0,1} = A\,|\rho|^2 + B\,\Delta\,\tau.$$

Using the isomorphism $\Lambda^{0,1} \sim \Lambda^{1,0}$ the calculations of Patodi in the real case [6] yield

$$A = 1/120, \qquad B = 1/80. \tag{3.8}$$

By duality

$$\begin{aligned} a_{2,1,0} &= \frac{1}{120}\,|\rho|^2 - \frac{1}{80}\,\Delta\tau, \\ a_{2,1,1} &= \frac{1}{120}\,|\rho|^2 + \frac{1}{120}\,\Delta\tau \end{aligned} \tag{3.9}$$

on one manifolds. This determines our invariants on one manifolds.

Expanding (3.2) we get

$$\begin{aligned} K_1(d,p,q)\,&(\tau_1^2 + \tau_2^2 + 2\tau_1\tau_2) + K_2(d,p,q)(|\rho_1|^2 + |\rho_2|^2) \\ &+ K_3(d,p,q)\,(|R_1|^2 + |R_2|^2) + K_4(d,p,q)\,(\Delta\tau_1 + \Delta\tau_2) \\ = &\sum_{p_1+p_2=p;\ q_1+q_2=q} \binom{d_1}{p_1}\binom{d_1}{q_1}[K_1(d_2,p_2,q_2)\,(\tau_2)^2 + K_2(d_2,p_2,q_2)\,|\rho_2|^2 \\ &\qquad + K_3(d_2,p_2,q_2)\,|R_2|^2 + K_4(d_2,p_2,q_2)\,\Delta\tau_2] \\ &+ K(d_1,p_1,q_1)\,K(d_2,p_2,q_2)\,\tau_1\tau_2 \\ &+ \binom{d_2}{p_2}\binom{d_2}{q_2}[K_1(d_1,p_1,q_1)\,(\tau_1)^2 + K_2(d_1,p_1,q_1)\,|p_1|^2 \\ &\qquad + K_3(d_1,p_1,q_1)\,|R_1|^2 + K_4(d_1,p_1,q_1)\,\Delta\tau_1]. \end{aligned} \tag{3.10}$$

Taking M_1 to be a one manifold in (3.10) and using (3.4), (3.7), (3.8), (3.9) gives

$$
\begin{aligned}
K_1(d,p,q) &= \frac{1}{2}\sum_{p_1+p_2=p;\ q_1+q_2=q} K(d_1,p_1,q_1)\,K(d_2,p_2,q_2)\\
&= \frac{1}{288}\binom{d-2}{p}\binom{d-2}{q} - \frac{1}{72}\binom{d-2}{p}\binom{d-2}{q-1} + \frac{1}{72}\binom{d-2}{p}\binom{d-2}{q-2}\\
&\quad - \frac{1}{72}\binom{d-2}{q}\binom{d-2}{p-1} + \frac{1}{72}\binom{d-2}{q}\binom{d-2}{p-2} + \frac{5}{144}\binom{d-2}{p-1}\binom{d-2}{q-1}\\
&\quad - \frac{1}{72}\binom{d-2}{p-1}\binom{d-2}{q-2} - \frac{1}{72}\binom{d-2}{p-2}\binom{d-2}{q-1} + \frac{1}{288}\binom{d-2}{p-2}\binom{d-2}{q-2}.
\end{aligned}
\tag{3.11}
$$

$$
\begin{aligned}
K_4(d,p,q) &= \sum_{p_1+p_2=p;\ q_1+q_2=q} K_4(d_1,p_1,q_1)\binom{d_2}{p_2}\binom{d_2}{q_2}\\
&= \frac{1}{120}\binom{d-1}{p}\binom{d-1}{q} - \frac{1}{80}\binom{d-1}{p}\binom{d-1}{q-1}\\
&\quad - \frac{1}{80}\binom{d-1}{p-1}\binom{d-1}{q} + \frac{1}{120}\binom{d-1}{p-1}\binom{d-1}{q-1}.
\end{aligned}
\tag{3.12}
$$

Now consider manifolds of complex dimension two. Using the isomorphism $\Lambda^{0,1} \sim \Lambda^{1,0}$ we deduce easily from the calculations of Patodi [6] that:

$$
\begin{aligned}
K_1(2,0,1) &= -1/72, & K_2(2,0,1) &= 43/720,\\
K_3(2,0,1) &= -11/1440, & K_4(2,0,1) &= -1/240.
\end{aligned}
\tag{3.13}
$$

By the fantastic cancellation of [5] we have

$$a_{2,0,0} - a_{2,0,1} + a_{2,0,2} = 4\pi^2\,(1/12)*(c_1{}^2 + c_2) \tag{3.14}$$

where c_1, c_2 are the Chern classes of the tangent bundle.

It may be shown by calculation that

$$*\, c_1^2 = (1/16\pi^2)\,(\tau^2 - 2\,|\rho|^2), \qquad *\, c_2 = (1/32\pi^2)\,(|R|^2 - 4|\rho|^2 + \tau^2).$$

Substitution in (3.14) gives

$$a_{2,0,0} - a_{2,0,1} + a_{2,0,2} = (1/96\pi^2)\,(|R|^2 - 8\,|\rho|^2 + 3\tau^2).$$

This then yields

$$
\begin{aligned}
K_1(2,0,0) - K_1(2,0,1) + K_1(2,0,2) &= 1/32,\\
K_2(2,0,0) - K_2(2,0,1) + K_2(2,0,2) &= -1/12,\\
K_3(2,0,0) - K_3(2,0,1) + K_3(2,0,2) &= 1/96,\\
K_4(2,0,0) - K_4(2,0,1) + K_4(2,0,2) &= 0.
\end{aligned}
$$

Therefore using (3.4) and (3.13),

$$
\begin{aligned}
K_1(2,0,2) &= 1/72, & K_2(2,0,2) &= -1/45,\\
K_3(2,0,2) &= 1/720, & K_4(2,0,2) &= -1/80.
\end{aligned}
$$

By duality we obtain the $K_i(2,1,0)$, $K_i(2,2,1)$, $K_i(2,2,0)$, and $K_i(2,1,2)$. Then since $\Lambda^2 = \Lambda^{0,2} \oplus \Lambda^{1,1} \oplus \Lambda^{2,0}$ we deduce the $K_i(2,1,1)$ from the calculations of Patodi [6]:

$$K_1(2,1,1) = 5/144, \qquad K_3(2,1,1) = 4/45,$$
$$K_2(2,1,1) = -77/360, \qquad K_4(2,1,1) = -1/120.$$

From (3.10) with M_1 a two manifold

$$
\begin{aligned}
K_2(d,p,q) &= \sum_{p_1+p_2=p;\ q_1+q_2=q} K_2(d_1,p_1,q_1)\binom{d_2}{p_2}\binom{d_2}{q_2}\\
&= \frac{-1}{720}\binom{d-2}{p}\binom{d-2}{q} + \frac{43}{720}\binom{d-2}{p}\binom{d-2}{q-1} - \frac{1}{45}\binom{d-2}{p}\binom{d-2}{q-2}\\
&\quad + \frac{43}{720}\binom{d-2}{q-1}\binom{d-2}{p} - \frac{1}{45}\binom{d-2}{p-2}\binom{d-2}{q} - \frac{1}{720}\binom{d-2}{p-2}\binom{d-2}{q-2}\\
&\quad + \frac{43}{720}\binom{d-2}{p-2}\binom{d-2}{q-1} + \frac{43}{720}\binom{d-2}{p-1}\binom{d-2}{q-2} - \frac{77}{360}\binom{d-2}{p-1}\binom{d-2}{q-1}.
\end{aligned}
\tag{3.15}
$$

$$
\begin{aligned}
K_3(d,p,q) &= \sum_{p_1+p_2=p;\ q_1+q_2=q} K_3(d_1,p_1,q_1)\binom{d_2}{p_2}\binom{d_2}{q_2}\\
&= \frac{1}{720}\binom{d-1}{p}\binom{d-1}{q} - \frac{11}{1440}\binom{d-2}{p}\binom{d-2}{q-2} + \frac{1}{720}\binom{d-2}{p}\binom{d-2}{q-2}\\
&\quad - \frac{11}{1440}\binom{d-2}{p-1}\binom{d-2}{q} + \frac{1}{720}\binom{d-2}{p-2}\binom{d-2}{q} + \frac{1}{720}\binom{d-2}{p-2}\binom{d-2}{q-2}\\
&\quad - \frac{11}{1440}\binom{d-2}{p-1}\binom{d-2}{q} - \frac{11}{1440}\binom{d-2}{p}\binom{d-2}{q-1} + \frac{4}{45}\binom{d-2}{p-1}\binom{d-2}{q-1}.
\end{aligned}
\tag{3.16}
$$

THEOREM 3.1. *The coefficients $a_{1,p,q}$ and $a_{2,p,q}$ of Minakshisundaram are*

$$
\begin{aligned}
a_{1,p,q} &= K(d,p,q)\,\tau,\\
a_{2,p,q} &= K_1(d,p,q)\,\tau^2 + K_2(d,p,q)\,|\rho|^2\\
&\quad + K_3(d,p,q)\,|R|^2 + K_4(d,p,q)\,\Delta\tau
\end{aligned}
$$

where $K(d,p,q)$, $K_i(d,p,q)$ are given by (3.7), (3.11), (3.12), (3.15), (3.16) *above.*

4. Applications to geometry.

PROPOSITION 4.1. *For $d > 2$,*

$$\sum (-1)^p a_{1,0,p} = 0, \qquad \sum (-1)^p a_{2,0,p} = 0,$$
$$\sum (-1)^{p+q} a_{1,p,q} = 0, \qquad \sum (-1)^{p+q} a_{2,p,q} = 0,$$

special cases of the fantastic cancellation in [**4**] *and* [**5**].

THEOREM 4.2. *Let M and M' be Kaehler manifolds whose complex Laplacians have the same spectrum acting on forms of type (p,q). Then M and M' have the same dimension and volume. Further*

$$\int_M \tau = \int_{M'} \tau', \qquad \int_M |R|^2 = \int_{M'} |R'|^2,$$
$$\int_M |\rho|^2 = \int_{M'} |\rho'|^2, \qquad \int_M \tau^2 = \int_{M'} |\tau'|^2.$$

PROOF. It is standard to conclude from the spectrum on functions that M and M' have the same dimension and volume. Also by isospectrality:

(i) $$\int a_{1,p,q} = \int a'_{1,p,q},$$

(ii) $$\int a_{2,p,q} = \int a'_{2,p,q}.$$

Theorem 3.1 gives $\int \tau = \int \tau'$ immediately from (i) with $(p, q) = (0, 0)$. From (ii) with $(p, q) = (0, 0), (0, 1), (0, 2)$ we get the system of equations

$$\frac{1}{288}\int \tau^2 - \frac{1}{720}\int |\rho|^2 + \frac{1}{720}\int |R|^2 = \frac{1}{288}\int |\tau'|^2 - \frac{1}{720}\int |\rho'|^2 + \frac{1}{720}\int |R'|^2,$$

$$-\frac{1}{72}\int \tau^2 + \frac{43}{720}\int |\rho|^2 - \frac{11}{1440}\int |R|^2 = -\frac{1}{72}\int |\tau'|^2 + \frac{43}{720}\int |\rho'|^2 - \frac{11}{1440}\int |R'|^2,$$

$$\frac{1}{72}\int \tau^2 - \frac{1}{45}\int |\rho|^2 + \frac{1}{720}\int |R|^2 = \frac{1}{72}\int |\tau'|^2 - \frac{1}{45}\int |\rho'|^2 + \frac{1}{720}\int |R'|^2.$$

Algebra yields the required result.

COROLLARY 4.3. *If M and M' are isospectral compact Kaehler manifolds then*

(i) *M has constant scalar curvature c if and only if M' has constant scalar curvature c.*

(ii) *M is Einstein with scalar curvature c if and only if M' is Einstein with scalar curvature c.*

(iii) *M has constant holomorphic sectional curvature c if and only if M' has constant holomorphic sectional curvature c.*

PROOF. (i) By Schwarz' inequality $(\int \tau)^2 \leqq \int \tau^2 \operatorname{vol}(M)$ with equality only if τ is constant. The result then follows from Theorem 4.2.

(ii) For manifolds of dimension one (ii) is a consequence of (i). So assume dimension greater than one. Then if $d = \dim M$, $\int |\rho|^2 \geqq (1/2d) \int \tau^2$ with equality if and only if M is Einstein. The result then follows from Theorem 4.2 and (i).

(iii) Let $\{e_i, e_i^*\}$ be any canonical orthonormal basis at a point in M' and R'_{ijkl} the components of the curvature tensor with respect to this basis. $R'_{ijkl} = R_{ijkl} + S_{ijkl}$ where

$$R_{ijkl} = (c/4)\,[(g_{ik}\, g_{jl} - g_{il}\, g_{jk}) + (g_{ik^*}\, g_{jl^*} - g_{il^*}\, g_{jk^*}) + 2\, g_{ij^*}\, g_{kl^*}]$$

are the components of the curvature tensor with respect to any canonical orthonormal basis $\{f_j, f_j^*\}$ at any point in a space of constant holomorphic sectional curvature c. Since $\int_M \tau = \int_{M'} \tau'$ we have $\int_{M'} \sum S_{ijij} = 0$. Furthermore

$$\begin{aligned} \|R'\|^2 &= \sum (R'_{ijkl})^2 \\ &\geqq \sum (R'_{ijij})^2 + \sum (R'_{ijji})^2 + \sum_{j\neq i^*} (R'_{iji^*j^*})^2 + \sum_{j\neq i^*} (R'_{ijj^*i^*} + \sum_{j\neq i, i^*} (R'_{ii^*jj^*})^2 \\ &\geqq \|R\|^2 + 2 \sum_{j\neq i^*} S^2_{ijij} + 2 \sum S^2_{ijij} + \sum_{j\neq i, i^*} S^2_{ii^*jj^*} + 4c \sum S_{ijij} \\ &\geqq \|R\|^2 + 4c \sum S_{ijij}. \end{aligned}$$

Since $\int_{M'} \|R'\|^2 = \int_M \|R\|^2 = \int_{M'} \|R\|^2 + 4c \sum S_{ijij}$ all the above inequalities must be equalities. This shows $R'_{ijkl} = R_{ijkl}$. M' has constant holomorphic sectional curvature c.

REMARK. Peter Gilkey recalls to us the well-known consequence of the index theorem that CP^n cannot cover any space. Thus a manifold isospectral with CP^n must be isometric with CP^n.

5. Counterexamples. A Kaehler metric in M is said to be of restricted type if its Kaehler form represents an integral class.

In [2] we find conditions that a complex torus admit a Kaehler metric of restricted type. Let $\theta = C^m/\Gamma$ be a complex torus with generators $\pi_\lambda = (\pi_\lambda^1, \cdots, \pi_\lambda^m)$, $\lambda = 1, \cdots, 2m$. Then θ admits a Kaehler metric of restricted type if and only if there exists a skew symmetric $2m \times 2m$ matrix G with integral elements such that

$$(5.1) \qquad \begin{aligned} &(-1)^{1/2t}\, \pi\, G^{-1}\, \pi \quad \text{is positive definite,} \\ &{}^t\pi\, G^{-1}\, \pi = 0 \end{aligned}$$

where $\pi = \pi_\lambda^k$ is a $2m \times m$ matrix formed with generators as rows. We use them to demonstrate the following:

PROPOSITION 5.1. (i) *There exist isospectral Kaehler manifolds such that one has a Kaehler metric of restricted type and the other does not.*

(ii) *There exist isospectral Kaehler manifolds which are not holomorphically equivalent.*

PROOF. Consider the complex tori with generating matrices

$$\pi(T_1) = \begin{pmatrix} 1 & 0 \\ 0 & 1 \\ \alpha + n\sqrt{-1} & \beta + p\sqrt{-1} \\ \beta + p\sqrt{-1} & \gamma + m\sqrt{-1} \end{pmatrix},$$

$$\pi(T_2) = \begin{pmatrix} 1 & 0 \\ 0 & 1 \\ \alpha + p\sqrt{-1} & \beta + n\sqrt{-1} \\ \beta + m\sqrt{-1} & \gamma + p\sqrt{-1} \end{pmatrix}.$$

Require $m, n > 0$, $p > 0$ to be integers such that $(n - m)p$ and $m(m + n) + 2p^2$ are nonzero and relatively prime to each other. Also specify that $(m + n)(mn + p^2) > 2|p|(m^2 + n^2 + p^2)$. Let $\alpha, \gamma, \alpha\gamma, 1$ be linearly independent over the integers and set $\beta = -p(\alpha + \gamma)/(n + m)$.

Then T_1 with metric $dz_1 \otimes d\bar{z}_1 + dz_2 \otimes d\bar{z}_2$ is a Kaehler manifold with metric of restricted type. T_2 with metric $dz_1 \otimes d\bar{z}_1 + dz_2 \otimes d\bar{z}_2$ is not of restricted type. In fact T_2 admits no Kaehler metric of restricted type as may be verified via a lengthy calculation using (5.1). Therefore T_1 is not holomorphically equivalent to T_2.

However it is easy to see that T_1 and T_2 are isometric as real manifolds. Therefore they are isospectral on p forms. Since T_1 and T_2 are flat tori their spectra on (p, q) forms are also the same.

BIBLIOGRAPHY

1. Marcel Berger, Paul Gauduchon and Edmond Mazet, *Le spectre d'une variété riemannienne*, Lecture Notes in Math., vol. 194, Springer-Verlag, Berlin and New York, 1971. MR **43** #8025.

2. S. S. Chern, *Complex manifolds without potential theory*, Van Nostrand Math. Studies, no. 15, Van Nostrand, Princeton, N. J., 1967. MR **37** #940.

3. Peter Gilkey, Private conversation.

4. H. P. McKean, Jr. and I. M. Singer, *Curvature and the eigenvalues of the Laplacian*, J. Differential Geometry **1** (1967), 43–69. MR **36** #828.

5. V. K. Patodi, *An analytic proof of Riemann-Roch-Hirzebruch theorem for Kaehler manifolds*, J. Differential Geometry **5** (1971), 251–283. MR **44** #7502.

6. ———, *Curvature and the fundamental solution of the heat operator*, J. Indian Math. Soc. **34** (1970), 269–285.

UNIVERSITY OF CALIFORNIA, BERKELEY

Proceedings of Symposia in Pure Mathematics
Volume 27, 1975

THE SPECTRUM OF POSITIVE ELLIPTIC OPERATORS AND PERIODIC GEODESICS

J. J. DUISTERMAAT* AND V. W. GUILLEMIN

Let X be a compact C^∞ manifold of dimension n without boundary, and $P = P(x, i^{-1}\partial/\partial x)$ an elliptic pseudodifferential operator of order $m > 0$ on X which is positive in the sense that

$$(Pu, u) \geqq c(u, u) \qquad \text{for } u \in C^\infty(X) \tag{1}$$

and some constant $c > 0$. We assume that the total symbol of P on local coordinates is equal to an asymptotic sum of smooth functions $P_{m-j}(x, \xi)$ on $R^n \times R^n/\{0\}$ which are homogeneous of degree $m - j$ in ξ, $j = 0, 1, 2, \cdots$. The principal part P_m then defines in an invariant way a homogenous C^∞ function $p = p(x, \xi)$ of degree m on $T^*X/0$; here T^*X = cotangent bundle of X, 0 = zero section in T^*X. Note that (1) implies that $p(x, \xi) > 0$. P will be regarded as an operator acting on $\frac{1}{2}$-densities in X, and as such it has an invariantly defined subprincipal symbol given by

$$P_{m-1} - (2i)^{-1} \sum_{j=1}^{n} \partial^2 P_m / \partial x_j \partial \xi_j \tag{2}$$

on local coordinates (see Duistermaat and Hörmander [4, § 5.2]). For simplicity we will assume here that the subprincipal symbol of P vanishes, as is always the case for a real selfadjoint operator P.

Let $\lambda_1 \leqq \lambda_2 \leqq \cdots$ denote the eigenvalues of P. Using the description of the operator

AMS (MOS) subject classifications (1970). Primary 35P20; Secondary 58G15.

*Partially supported by NSF grants GP-20095 and GP-3475-X.

(3) $$U(t) = \exp(-itP^{1/m})$$

for small $|t|$ as a Fourier integral operator, Hörmander [6] proved that

(4) $$\# \{j; \lambda_j \leqq \lambda\} = (2\pi)^{-n} \cdot \operatorname{vol}(B^*X) \cdot \lambda^{n/m} + O(\lambda^{(n-1)/m})$$

for $\lambda \to \infty$. Here

(5) $$B^*X = \{(x, \xi) \in T^*X; p(x, \xi) \leqq 1\}$$

and we use the canonical volume in T^*X. (Actually Hörmander proved in [6] the corresponding asymptotic result for the whole spectral function of P on a paracompact manifold X.) We want to show here some applications of the following extension to all $t \in R$ of the description of $U(t)$ as a Fourier integral operator allowed by the global theory of Fourier integral operators as given in Hörmander [7], Duistermaat and Hörmander [4].

THEOREM 1. *U, regarded as an operator: $C^\infty(X) \to C^\infty(\boldsymbol{R} \times X)$, is a Fourier integral operator of order $-\frac{1}{4}$, defined by the canonical relation*

(6) $$C = \{((t, \tau), (x, \xi), (y, \eta)); \tau + q(x, \xi) = 0, (x, \xi) = \Phi^t(y, \eta)\}.$$

Here Φ^t denotes the flow of the Hamilton field H_q of the function $q(x, \xi) = p(x, \xi)^{1/m}$,

(7) $$H_q = \sum_{j=1}^{n} \partial_q/\partial \xi_j \cdot \partial/\partial x_j - \partial_q/\partial x_j \cdot \partial/\partial \xi_j$$

on local coordinates.

An immediate consequence of Theorem 1 is that the function

(8) $$\hat{\sigma}(t) = \sum_{j=1}^{\infty} \exp(-it\mu_j), \qquad \mu_j = \lambda_j^{1/m},$$

can only have singularities at the periods of the periodic H_q-solutions. The analysis of the (big) singularity at $t = 0$ leads to (4). In order to analyze the other singularities, we multiply $\hat{\sigma}$ by a function $\hat{\rho} \in C_0^\infty(\boldsymbol{R})$ and investigate the growth order of the inverse Fourier transform

(9) $$(2\pi)^{-1} \int e^{i\mu t} \cdot \hat{\rho}(t) \cdot \hat{\sigma}(t)\, dt = (\rho * \sigma)(\mu) = \sum_{j=1}^{\infty} \rho(\mu - \mu_j).$$

Let γ be a periodic H_q-solution with period T in

(10) $$S^*X = \{(x, \xi) \in T^*X;\ q(x, \xi) = 1\}.$$

Let S be a hypersurface in S^*X through $\gamma(0)$ which is transversal to $H_q(\gamma(0))$, obtained as the inverse image of a hypersurface in X transversal to $d_\xi q(\gamma(0))$. Then the map assigning to each $(x, \xi) \in S$ the point of return to S of a nearby H_p-solution starting at (x, ξ) is a local diffeomorphism: $S \to S$ with $\gamma(0)$ as a fixed point, called the Poincaré map of the H_q-flow along the periodic solution γ. Denote the differential of this map at $\gamma(0)$ by $\mathscr{P}_\gamma$.

THEOREM 2. *Suppose that $I-\mathscr{P}_\gamma$ is invertible for every periodic H_q-solution curve γ with period $T_\gamma \in \operatorname{supp} \hat{\rho}$. Then*

$$(\rho*\sigma)(\mu) \sim \sum_{k=0}^{\infty} \sum_{\gamma} (2\pi)^{-1} e^{\mu T_\gamma} \cdot T_\gamma \rho(T_\gamma) \cdot i^{-\sigma_\gamma} \cdot a_{\gamma k} \cdot \mu^{-k} \tag{11}$$

for $\mu \to \infty$, with

$$a_{\gamma 0} = \left| \det (I - \mathscr{P}_\gamma) \right|^{-1/2} \tag{12}$$

and σ_γ is an integer, in the Riemannian case equal to the Morse index of γ on the null space of $d_\xi q(\gamma(0))$. Here the summation in (11) is extended over the finitely many periodic H_q-solutions γ in S^*X, two such solutions being identified with each other if they have the same orbit and are considered on a time interval of the same length equal to the period T_γ.

If (X, g) is a Riemannian manifold and $P = -\Delta^\delta + c$ then $H_q = \frac{1}{2}H_p$ on S^*X and this is just the geodesic spray in S^*X. So if $I - \mathscr{P}_\gamma$ is invertible for every periodic geodesic and the periods of different periodic geodesics are distinct, then $\hat{\sigma}(t)$ has its singularities exactly at the lengths of the periodic geodesics and we recover the theorem of Colin de Verdière **[2]** that in this case the lengths of the periodic geodesics are determined by the spectrum of the Laplacian. (Here antipodal H_q-solution curves are identified with each other which does no harm because for these the numbers T_γ, σ_γ, $a_{\gamma k}$ are the same.) However (11) also gives $|\det (I - \mathscr{P}_\gamma^l)|$ for all $l \in \mathbf{Z}$ for each Poincaré map $\mathscr{P}_\gamma$ of a periodic H_p-solution. Now from these numbers one can for instance extract the number

$$\Pi \left| \alpha_j \right| \tag{13}$$

where α_j are the eigenvalues of $\mathscr{P}_\gamma$ and the product is taken over the j for which $|\alpha_j| \geqq 1$. So the spectrum determines not only the lengths of the periodic geodesics but also whether the geodesic flow along them is of elliptic type ($|\alpha_j| = 1$ for all j) or hyperbolic type ($|\alpha_j| > 1$ for some j), as was suggested by Cotsaftis **[3]**. We also note here that independently Chazarain **[1]** obtained a formula like (11) in the more general case that the periodic H_q-solutions form submanifolds of S^*X in a certain nondegenerate way, however without a geometric interpretation of the coefficients in the top order terms.

Another application of Theorem 1 deals with the question of whether Riemannian manifolds all of whose geodesics are closed (see Weinstein **[8]** for more information about such manifolds) can be characterized by properties of the spectrum of their Laplacian.

THEOREM 3. *Suppose that all H_q-solution curves are periodic with period $T > 0$. Write*

$$\nu_k = 2\pi T^{-1}(k + \alpha/4), \qquad k = 1, 2, \cdots. \tag{14}$$

Here α = integral of the cohomology class of Keller-Maslov-Arnol'd (in the *terminology of Hörmander [7]) over an arbitrary periodic H_q-solution curve γ of period T. Then there exist positive constants K, $\tilde{\mu}$ such that*

$$(15)\quad \begin{aligned} &\#\{j; |\mu_j - \mu| \leqq K, |\mu_j - \nu_k| \leqq K \cdot \varepsilon^{-1/2} k^{-1} \text{ for some } k \in \boldsymbol{Z}_+\} \\ &\qquad \geqq (1-\varepsilon) \cdot \#\{j; |\mu_j - \mu| \leqq K\} \quad \text{for all } \varepsilon > 0,\ \mu \geqq \tilde{\mu}. \end{aligned}$$

Suppose conversely that there exists a constant K such that for each $\varepsilon > 0$ we have a number $\tilde{\mu}$ such that

$$(16)\quad \begin{aligned} &\#\{j; |\mu_j - \mu| \leqq K, |\mu_j - \nu_k'| \leqq \varepsilon \text{ for some } k \in \boldsymbol{Z}_+\} \\ &\qquad \geqq (1-\varepsilon) \cdot \#\{j; |\mu_j - \mu| \leqq K\} \quad \text{for all } \mu \geqq \tilde{\mu}. \end{aligned}$$

Here $\nu_k' = 2\pi T^{-1}(k + \alpha'/4)$, α' a fixed real number. Then all H_q-solutions are periodic with period T, $\alpha' = \alpha \pmod 4$ and (16) can be replaced by the stronger estimate (15).

REMARK. If $d_\xi^2 q$ is positive definite on the null space of $d_\xi q$ (as in the case of the Laplacian on a Riemannian manifold) then α = number of conjugate points along γ = Morse index of γ. (For a modern proof of the latter identity see for instance the Appendix of Karcher in [5].)

Excluding some pathological cases these estimates can be improved as follows.

PROPOSITION 4. *Suppose that the H_q-flow is periodic with period $T > 0$ and that the union of the subperiodic orbits (= orbits of solutions with a period T/l for some $l \in \boldsymbol{Z}_+$, $l \geqq 2$) is equal to zero. Then there exists a constant K such that for every $\varepsilon > 0$ there is a $\tilde{k} \in \boldsymbol{Z}_+$ with the properties that*

$$(17)\quad |\#\{j; |\mu_j - \nu_k| \leqq K \cdot \varepsilon^{-1/2} \cdot k^{-1}\} - 2\pi T^{-1} c_0 \nu_k^{n-1}| \leqq \varepsilon \nu_k^{n-1}$$

and

$$(18)\quad \#\{j; K \cdot \varepsilon^{-1/2} \cdot k^{-1} \leqq |\mu_j - \nu_k| \leqq \pi T^{-1}\} \leqq \varepsilon \cdot \nu_k^{n-1}$$

for all $k \geqq \tilde{k}$. Here ν_k is defined as in (14) and

$$(19)\quad c_0 = (2\pi)^{-n} \operatorname{vol}(B^*X)/n.$$

PROPOSITION 5. *Assume that (9) is of order $o(\mu^{n-1})$ as $\mu \to \infty$ for every $\hat{\rho} \in C_0^\infty(\boldsymbol{R})$ such that $0 \notin \operatorname{supp} \hat{\rho}$. Then there exists for every $\varepsilon > 0$, $\delta > 0$ a constant $\tilde{\mu}$ such that*

$$(20)\quad |\#\{j; |\mu_j - \mu| \leqq \tfrac{1}{2}\varepsilon\} - \varepsilon c_0 \mu^{n-1}| \leqq \delta \mu^{n-1}$$

for all $\mu \geqq \tilde{\mu}$.

If the H_q-flow is not periodic then the condition in Proposition 5 will be satisfied unless there is a subset A of S^*X of positive measure on which Φ^t, for some $t > 0$ depending on the point in A, has a contact of infinite order with the identity. This highly exceptional, rather pathological situation cannot occur if for instance X and $q(x, \xi)$ are real analytic. So with exception of some pathological situations, we can say that if the H_q-flow is periodic then the spectrum clusters around the points

ν_k in a rather strong asymptotic sense, and if it is not, then the spectrum is fairly evenly distributed. We also note that summing (20) one obtains

$$\# \{j;\, \lambda_j \leqq \lambda\} = (2\pi)^{-n} \cdot \mathrm{vol}\, (B^*X) \cdot \lambda^{n/m} + o(\lambda^{(n-1)/m}), \tag{21}$$

improving (4) just a tiny little bit. The infinite tail of ρ (due to the assumption that $\hat{\rho}$ has compact support) is an obstacle for getting any asymptotic information about the spectrum in terms of finite λ_j-intervals which is better than modulo a term of order $o(\lambda^{(n-1)/m})$.

The authors finally want to thank Alan Weinstein for many stimulating discussions on this subject.

References

1. J.Chazarain, *Formule de Poisson pour les variétés Riemanniennes*, preprint, Nice, August 1973. See also: *Spectre d'un opérateur elliptiques et bicaractéristiques périodiques*, C.R. Acad. Sci. Paris, July 1973.

2. Y.Colin de Verdière, *Spectre du laplacien et longueurs des géodésiques périodiques*. I, II (preprints). See also C. R. Acad. Sci. Paris **275** (1972), 805–808; **276** (1973), 1517–1519.

3. M.Cotsaftis, *Une propriété des orbites périodiques des systèmes hamiltoniens non linéaires*, C. R. Acad. Sci. Paris **275** (1972), 911–914.

4. J.J. Duistermaat and L.Hörmander, *Fourier integral operators*. II, Acta Math. **128** (1972), 183–269.

5. P.Flaschel und W.Klingenberg, *Riemannsche Hilbertmannigfaltigkeiten. Periodische Geodätische*, Lecture Notes in Math., vol. 282, Springer-Verlag, Berlin and New York, 1972.

6. L.Hörmander, *The spectral function of an elliptic operator*, Acta Math. **121** (1968), 193–218.

7. ———, *Fourier integral operators*. I, Acta Math. **127** (1971), 79–183.

8. A. Weinstein, *On the volume of manifolds all of whose geodesics are closed*, these Proceedings.

Katholieke Universiteit

Massachusetts Institute of Technology

Proceedings of Symposia in Pure Mathematics
Volume 27, 1975

RANDOM WALK ON THE FUNDAMENTAL GROUP

JAMES EELLS*

In this lecture my purpose is to show how Brownian motion on a Riemannian manifold induces random walk on its fundamental group, and on its holonomy group—and to note some elementary geometric consequences.

There are several different sorts of interactions and applications, e.g.,

(1) with Kesten's program ([17]-[19]) examining the influence of random walks on the amenability, growth, and entropy of countable (and Lie) groups; see also [1], [10],

(2) with Milnor-Švarc's results ([21], [24]) relating those concepts to Riemannian geometric restrictions,

(3) between the space of ends of the fundamental group and the Martin (or other) boundary of the random walk ([19], [4]),

(4) to relationships between the potential theories of the manifold and its fundamental group—e.g., between the spectral properties of the Laplacian and the energy levels of geodesics. These are in the course of preparation, and will be published elsewhere.

1. Riemannian Wiener measure. (A) For simplicity of exposition we restrict our attention to a compact connected oriented Riemannian manifold M. Let Δ denote its Laplace-Beltrami operator, and $L = \frac{1}{2}\Delta - \partial/\partial t$ the associated heat operator. Then L has a unique fundamental solution $h: M \times M \times \boldsymbol{R}(>0) \to \boldsymbol{R}(>0)$

AMS (MOS) subject classifications (1970). Primary 53C20; Secondary 60J15.

*Written while the author enjoyed the hospitality of the Institut des Hautes Études Scientifiques. He has benefited by conversations with P. de la Harpe, H. Kesten, J. Milnor, and H. Widom—and now records his thanks.

with the following properties:

(1) h is a smooth function, with $h_t(x, y) = h(x, y; t) = h(y, x; t)$ for all $x, y \in M$ and $t > 0$.

(2) $L_x h = 0 = L_y h$; here L_x is L operating on the first variable of h.

(3) $\int_M h_t(x, y)\, dy = 1$ for all $x \in M$ and $t > 0$.

(4) Chapman-Kolmogoroff identity: $h_{t+s}(x, y) = \int_M h_t(x, z) h_s(z, y)\, dz$. We call h the *heat density* of M.

(B) Now fix points $a, b \in M$ and let $\mathscr{C}_{ab}(M)$ denote the space of continuous maps $I \to M$ carrying $(0, 1) \to (a, b)$, topologized with the topology of uniform convergence. Then $\mathscr{C}_{ab}(M)$ is a complete separable metrizable space. It is elementary that its Borel σ-algebra (the σ-algebra generated by the open sets) is also generated by the fibred sets $\rho_{\boldsymbol{t}}^{-1}(B) \subset \mathscr{C}_{ab}(M)$, where $\boldsymbol{t} = (0 < t_1 < \cdots < t_n < 1)$ and $\rho_{\boldsymbol{t}}$: $\mathscr{C}_{ab}(M) \to M^{\boldsymbol{t}}$ $(= M \times \cdots \times M$, n copies) is the evaluation map $\rho_{\boldsymbol{t}}(x) = (x(t_1), \cdots, x(t_n))$; and B is a Borel subset of $M^{\boldsymbol{t}}$.

We define [7] the function w_{ab} on Borel $(\mathscr{C}_{ab}(M))$ as follows : For any such fibred set $\rho_{\boldsymbol{t}}^{-1}(B)$ we let

$$w_{ab}[\rho_{\boldsymbol{t}}^{-1}(B)] = \int_B h_{t_1}(a, m_1) h_{t_2 - t_1}(m_1, m_2) \cdots \\ h_{t_n - t_{n-1}}(m_{n-1}, m_n)\, h_{1-t_n}(m_n, b)\, dm_1 \cdots dm_n.$$

Now if $\boldsymbol{s} \subset \boldsymbol{t}$, then for B a Borel subset of $M^{\boldsymbol{s}}$, the Chapman-Kolmogoroff identity insures that the two possible definitions agree:

$$w_{ab}[\rho_{\boldsymbol{s}}^{-1}(B)] = w_{ab}[\rho_{\boldsymbol{t}}^{-1}(\pi^{-1}(B))],$$

where π denotes the indicated projection:

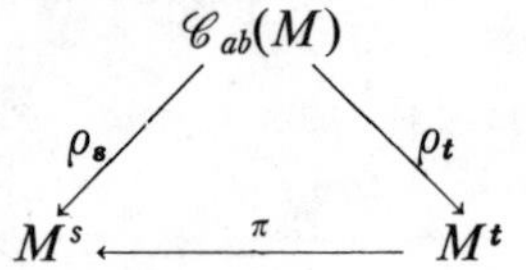

A variation in the classical construction of Wiener [25] shows that w_{ab} *determines a (countably additive) regular Borel measure on* $\mathscr{C}_{ab}(M)$; *and*

$$w_{ab}[\mathscr{C}_{ab}(M)] = h_1(a, b). \tag{1}$$

Furthermore, every nonvoid open subset of $\mathscr{C}_{ab}(M)$ *has strictly positive* w_{ab}*-measure.*

It is also true that the subset $\mathscr{C}_{ab}^{\alpha}(M)$ of α-Hölder continuous paths has full w_{ab}-measure in $\mathscr{C}_{ab}(M)$ for $\alpha < \frac{1}{2}$; and that the subset of paths having a derivative at least one point has w_{ab}-measure 0.

(C) There is a simple relationship between w_{ab} and the Wiener measure w_a on $\mathscr{C}_a(M)$, the space of continuous paths starting at a and ending anywhere on M: The locally trivial fibration $\rho_1 : \mathscr{C}_a(M) \to M$ has $\mathscr{C}_{ab}(M)$ as fibre over b. *For any Borel subset* $C \subset \mathscr{C}_a(M)$ *we have*

$$w_a[C] = \int_M w_{ab}[C \cap \mathscr{C}_{ab}(M)]\, db;$$

and $\rho_1(w_a) = h_1(a, m)\, dm$, *where* dm *is the Riemannian volume of* M. See [7].

(D) The universal covering space $\sigma : \tilde{M} \to M$ has the representation as the quotient of $\mathscr{C}_a(M)$ under the equivalence relation of fixed endpoint homotopy of paths:

$$\begin{array}{ccc} \mathscr{C}_a(M) & & \\ \downarrow \rho_1 & \searrow & \\ & & \tilde{M} \\ & \swarrow \sigma & \\ M & & \end{array}$$

The homotopy class of the constant path $\tilde{a} : I \to a$ serves as base point in $\tilde{M}$; and the uniqueness of path lifting identifies the fibre $\sigma^{-1}(a) \subset \tilde{M}$ with the fundamental group $\pi_1(M, a) = \pi_0(\Omega_a(M))$, the discrete (finitely presented) group of the components of the loop space $\Omega_a(M) = \mathscr{C}_{aa}(M)$. With its induced Riemannian structure, $\tilde{M}$ is a complete manifold on which $\pi_1(M, a)$ acts principally as a group of covering isometries.

If $\tilde{h}$ denotes the heat density of $\tilde{M}$, then $\tilde{h}_t(\tilde{x}\gamma, \tilde{y}\gamma) = \tilde{h}_t(\tilde{x}, \tilde{y})$ for all $\tilde{x}, \tilde{y} \in \tilde{M}$ and $\gamma \in \pi_1(M, a)$. Furthermore, $\sum_\gamma \tilde{h}_t(\tilde{x}, \tilde{y}\gamma) : \tilde{M} \times \tilde{M} \to \boldsymbol{R}(> 0)$ factors through σ:

$$\begin{array}{ccc} \tilde{M} \times \tilde{M} & \longrightarrow & \boldsymbol{R}(>0) \\ \downarrow \sigma & \nearrow h_t & \\ M \times M & & \end{array}$$

to give the heat density h on M. From the canonical identification $\mathscr{C}_a(M) = \mathscr{C}_{\tilde{a}}(\tilde{M})$ we obtain the lifted Wiener measure $\tilde{W}_{\tilde{a}}$ on $\mathscr{C}_{\tilde{a}}(\tilde{M})$, and $\sigma\tilde{w}_{\tilde{a}} = w_a$. Moreover,

$$\tilde{w}_{\tilde{a}, \tilde{a}\gamma}[\mathscr{C}_{\tilde{a}, \tilde{a}\gamma}(\tilde{M})] = \tilde{h}_1(\tilde{a}, \tilde{a}\gamma). \tag{2}$$

2. The random walk. (A) For any $\gamma \in \pi_1(M, a)$ we let $\Omega_a^\gamma(M)$ denote the corresponding component. Then define $p: \pi_1(M, a) \to \boldsymbol{R}(> 0)$ by $p(\gamma) = w_{aa}[\Omega_a^\gamma(M)]/w_{aa}[\Omega_a(M)]$. From that definition (or from formula (3) below) we see that *p is a symmetric probability measure on the countable group $\pi_1(M, a)$*: $p(\gamma^{-1}) = p(\gamma)$ for all $\gamma \in \pi_1(M, a)$, and $\sum_\gamma p(\gamma) = 1$.

Since $w_{aa}[\Omega_a^\gamma(M)] = \tilde{w}_{\tilde{a}, \tilde{a}\gamma}[\mathscr{C}_{\tilde{a}, \tilde{a}\gamma}(\tilde{M})]$, (1) and (2) provide the alternative definition

$$p(\gamma) = \tilde{h}_1(\tilde{a}, \tilde{a}\gamma)/h_1(a, a). \tag{3}$$

Although the notation does not display it, the probability p depends in a geometrically significant way on (a) the Riemannian structure of M; (b) the initial point a:

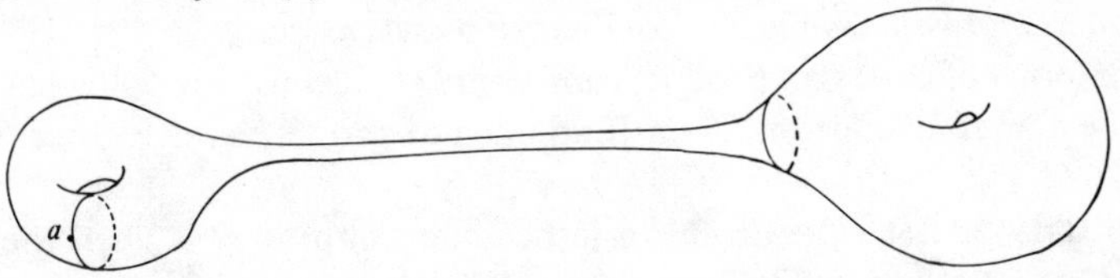

(c) on time, if we replace 1 in the right member of (3) by $t > 0$. In analogy with Feynman integration, it might be interesting to see what happens to p if we complexify t, and then let $t \to i$ in $\mathrm{Re}(t) > 0$.

(B) Setting $\Gamma = \pi_1(M, a)$ and $P = (P_{\alpha\beta})$, where

$$P_{\alpha\beta} = p(\alpha^{-1}\beta) = P_{\beta\alpha}, \tag{4}$$

we obtain a stochastic matrix, $\sum_{\beta \in \Gamma} P_{\alpha\beta} = 1$, which is Γ-invariant: $P_{\gamma\alpha,\gamma\beta} = P_{\alpha\beta}$ for all $\alpha, \beta, \gamma \in \Gamma$. Thus we obtain the traditional form of random walk on a countable group ([**18**], [**6**]). Furthermore, it is *irreducible*, in the sense that any element of Γ can be reached from any other element in a finite number of steps (one, in our case). It follows [**16**] that (Γ, p) is either

TRANSIENT. $\sum_{n \geq 1} P^{(n)}_{\alpha\alpha} < \infty$ for all $\alpha \in \Gamma$;

or

RECURRENT. $\sum_{n \geq 1} P^{(n)}_{\alpha\alpha} = \infty$ for all $\alpha \in \Gamma$, where $P^{(n)}$ denotes the nth power of P.

(C) The following properties are common to all symmetric probabilities $p > 0$ on the designated countable group Γ.

(1) The matrix P defines a bounded linear operator on every $l^r(\Gamma, \boldsymbol{C})$ for $1 \leqq r \leqq \infty$, by

$$(Pf)(\alpha) = (p * f)(\alpha) = \sum_{\gamma \in \Gamma} P_{\alpha\gamma} f(\gamma).$$

Otherwise said, P acts as the Gauss-Weierstrass transform [**11**]. Taking $r = 2$, the spectrum of P lies in $[-1, +1]$. A theorem of Kesten [**18**] asserts that $+1$ *lies in the spectrum of P if and only if Γ is amenable*. (I.e., there is a bounded linear form $m : l^\infty(\Gamma) \to \boldsymbol{R}$ such that $f \geqq 0$ implies $m(f) \geqq 0$, $\|m\| = 1$, and $m(L_\alpha f R_\gamma) = m(f)$ for all $f \in l^\infty(\Gamma)$ where L_α (resp., R_γ) denotes left translation by α (resp. right translation by γ).) Furthermore, Kesten has remarked to me that *if Γ is infinite and amenable, then $+1$ is not an isolated point in* Spec P.

Perron-Frobenius theory [**9**] yields: *If $\|P\| \in$ Spec P, then it is an eigenvalue of multiplicity* 1, *and the associated eigenspace is spanned by an $f \in l^2(\Gamma, \boldsymbol{R})$ which is strictly positive.*

(2) *Any random walk on a finite or cyclic group is recurrent* [**23**].

(3) *If Γ is abelian, then it admits a recurrent random walk iff* rank $\Gamma \leqq 2$ (Theorem of Dudley, see [**23**]). *Also p is transient iff $\int_{\hat{\Gamma}} \mathrm{Re}(1 - \hat{p}(\lambda))^{-1}\, d\lambda < \infty$* [**20**].

(4) *If a finitely generated group Γ admits a recurrent random walk, then Γ is amenable* ([**19**], [**5**]).

(5) *If Γ is a free group on $k \geqq 2$ generators, then p is transient. Its spectral radius is $= ((2k - 1)/k^2)^{1/2}$. If a countable group Γ contains a subgroup free on two generators, then p is transient. If Γ can be expressed nontrivially as a free product of two subgroups, not both isomorphic to $\boldsymbol{Z}_2$, then p is transient.*

Thus, in terms of cartesian products and connected sums, a topological manifold M must be rather irreducible if its fundamental group has a recurrent random walk.

(D) EXAMPLE 1. Let Γ be a rank n lattice subgroup of $\boldsymbol{R}^n$. Then the quotient $T^n = \boldsymbol{R}^n/\Gamma$ is a flat torus. The heat density of its universal covering space $\boldsymbol{R}^n$ is

$$\tilde{h}_t(\tilde{x}, \tilde{y}) = (2\pi t)^{-n/2} \exp(-|\tilde{x} - \tilde{y}|^2/2t); \tag{5}$$

and that of T^n itself is

(6) $$h_t(x, y) = (2\pi t)^{-n/2} \sum_{\gamma \in \Gamma} \exp(-|x - y - \gamma|^2/2t),$$

which is a constant multiple of $\theta(x - y, e^{-t})$, where

$$\theta(x, q) = \sum_{\gamma} \exp(i\langle\gamma, x\rangle)q^{|\gamma|^2}$$

is the third theta function of Jacobi. Our random walk on Γ is given by

$$p(\gamma) = \exp(-|\gamma|^2/2) \Big/ \sum_{\alpha \in \Gamma} \exp(-|\alpha|^2/2).$$

The spectrum of P is a closed subinterval of $(-1, +1)$ containing $+1$.

EXAMPLE 2. The uniformization theorem gives a classification of Riemann surfaces M into the following types: *Elliptic*, if M is conformally isomorphic to the projective line; *parabolic*, if its universal cover is conformally isomorphic to $\boldsymbol{C}$; and *hyperbolic*, if its universal cover is conformally isomorphic to the upper half plane $U = \operatorname{Im}(z > 0)$.

Consider Brownian motion on such a surface M, relative to any Hermitian metric of the given conformal class—first, on its universal cover $\tilde{M}$. Putting aside the elliptic case, we have the parabolic case, with Laplacian $\Delta = \partial^2/\partial x^2 + \partial^2/\partial y^2$, and with heat density (5), taking $n = 2$. And the hyperbolic case, with Laplacian $\Delta = y^2(\partial^2/\partial x^2 + \partial^2/\partial y^2)$, and heat density

(7) $$\tilde{h}_t(x, y) = 2^{1/2} e^{-t/8}/(2\pi t)^{-3/2} \int_a^\infty \frac{b \exp(-b^2/2t)\, db}{(\cosh b - \cosh a)^{1/2}},$$

where $a =$ hyperbolic distance (x, y). In any case, if we write $M = \tilde{M}/\Gamma$ for a Fuchsian group Γ, then we have its heat density $h_t(x, y) = \sum_{\gamma \in \Gamma} \tilde{h}_t(x, y\gamma)$. Kakutani ([**13**], [**14**], [**15**]) has shown that the associated *Brownian motion on $\tilde{M}$ is recurrent iff M is parabolic*. We note that *our random walk on $\pi_1(M, a)$ is recurrent iff M is parabolic.*

(E) A random walk on the integral cohomology group $H^1(M)$ can be constructed as follows, based on a recent theorem of P. Baxendale. Consider the Sobolev space $\mathscr{L}^2_r(M, \boldsymbol{C})$ of functions, with inner product given by the following sum of $\mathscr{L}^2_0$-inner products of iterated covariant differentials:

$$\langle\varphi, \psi\rangle_r = \sum_{i=0}^{r} \langle\nabla^i\varphi, \nabla^i\psi\rangle_0.$$

Baxendale [**3**] has shown that for $r > \dim M/2$ *there is a unique reproducing kernel associated with $\mathscr{L}^2_r(M, \boldsymbol{C})$ which is the covariance of a canonically defined Gaussian measure γ_r on $\mathscr{C}(M, \boldsymbol{C})$.*

Now the restriction of γ_r to the open subset $\mathscr{C}(M, \boldsymbol{C} - 0)$ determines (after normalization) an irreducible random walk on the group $\pi_0(\mathscr{C}(M, \boldsymbol{C} - 0))$ of components. But that group is canonically identified with $H^1(M)$. It would be interesting to know whether the random walk so constructed on $H^1(M)$ is significantly related to that on $H_1(M)$ obtained as the image of p under the epimorphism

$\pi_1(M, a) \twoheadrightarrow H_1(M)$ defined by abelianizing $\pi_1(M, a)$.

3. Random walk on the holonomy group. (A) The notion of stochastic parallel transport on M was introduced by K. Itô **[12]** via the solutions of a certain system of stochastic differential equations. In particular by endpoint conditioning we can define horizontal lifts into the principal bundle of M (more generally **[8]**, into any bundle over M with Riemannian connection) of almost all paths in $\Omega_a(M)$.

(B) It can be shown that *such parallel transport induces a measurable map* (*defined almost everywhere*) $\eta : \Omega_a(M) \twoheadrightarrow \Phi_a(M)$ *onto the holonomy group of* M *at* a (defined as usual by C^k-parallel transport ($k \geqq 1$) around C^k-loops at a). If we normalize w_{ab} on $\Omega_a(M)$ by setting

$$v_a = w_{aa}/w_{aa}[\Omega_a(M)],$$

then *its image* $P = \eta v_a$ *defines a random walk on the Lie group* $\Phi_a(M)$, in the sense of (**[2]**, **[10]**).

Finally, let $\Phi_a^0(M)$ denote the restricted holonomy group (= the subgroup of $\Phi_a(M)$ obtained by parallel transport around the null homotopic loops); then $\Phi_a^0(M)$ is the identity component of $\Phi_a(M)$, and we have the commutative diagram of surjections.

$$\begin{array}{ccc} \Omega_a(M) & \xrightarrow{\ \eta\ } & \Phi_a(M) \\ \big\downarrow \xi & & \big\downarrow \theta \\ \pi_1(M, a) & \xrightarrow[\ \zeta\ ]{} & \Phi_a(M)/\Phi_a^0(M) \end{array}$$

Therefore, *the induced random walks on the component group* $\Phi_a(M)/\Phi_a^0(M)$ *coincide*:

$$\zeta(p) = \theta(P). \tag{8}$$

I do not know an example of a complete manifold with irreducible holonomy having random walk (8) transient on its component group.

REFERENCES

1. A. Avez, *Entropie des groupes de type fini*, C. R. Acad. Sci. Paris Sér. A-B **275** (1972), A1363–A1366.

2. R. Azencott and P. Cartier, *Martin boundaries of random walks on locally compact groups*, Proc. Sixth Berkeley Sympos. Math. Statist. and Probability, Univ. of California Press, Berkeley, 1970/71, pp. 87–129.

3. P. Baxendale, *Gaussian measures on function spaces*, Thesis, Warwick University, 1973.

4. P. Cartier, *Fonctions harmoniques sur un arbre*, Sympos. Math. IX, Bologna, 1972, pp. 203–270.

5. M. M. Day, *Convolutions, means, and spectra*, Illinois J. Math. **8** (1964), 100–111. MR **28** #2447.

6. E. B. Dynkin and M. B. Malyutov, *Random walk on groups with a finite number of generators*, Dokl. Akad. Nauk SSSR **137** (1961), 1042–1045 = Soviet Math. Dokl. **2** (1961), 399–402. MR **24** #A1751.

7. J. Eells and K.D. Elworthy, *Wiener integration on certain manifolds*, Some Problems in Non-Linear Analysis, C.I.M.E. IV, 1970, pp. 67–94.

8. J. Eells and P. Malliavin, *Diffusion processes in Riemannian bundles* (in preparation).

9. W. G. Faris, *Invariant cones and uniqueness of the ground states for Fermion systems*, J. Mathematical Phys. **13** (1972), 1285–1290.

10. H. Furstenberg, *Random walks and discrete subgroups of Lie groups*, Advances in Probability **1** (1971), 3–63.

11. I. I. Hirschman and D. V. Widder, *The convolution transform*, Princeton Univ. Press, Princeton, N. J., 1955. MR **17**, 479.

12. K. Itô, *The Brownian motion and tensor fields on Riemannian manifold*, Proc. Internat. Congress Math. (Stockholm, 1962), Inst. Mittag-Leffler, Djursholm, 1963, pp. 536–539. MR **31** #772.

13. K. Itô and H. P. McKean, Jr., *Diffusion processes and their sample paths*, Die Grundlehren der math. Wissenschaften, Band 125, Academic Press, New York; Springer-Verlag, Berlin, 1965. MR **33** #8031.

14. S. Kakutani, *Two-dimensional Brownian motion and the type problem of Riemann surfaces*, Proc. Japan Acad. **21** (1945), 138–140 (1949). MR **11**, 257.

15. ———, *Random walk and the type problem of Riemann surfaces*, Contributions to the Theory of Riemann Surfaces, Ann. of Math. Studies, no. 30, Princeton Univ. Press, Princeton, N. J., 1953, pp. 95–101. MR **15**, 25.

16. J. G. Kemeny, J. L. Snell and A. W. Knapp, *Denumerable Markov chains*, Van Nostrand, Princeton, N.J., 1966.

17. H. Kesten, *Symmetric random walks on groups*, Trans. Amer. Math. Soc. **92** (1959), 336–354. MR **22** #253.

18. ———, *Full Banach mean values on countable groups*, Math. Scand. **7** (1959), 146–156. MR **22** #2911.

19. ———, *The Martin boundary of recurrent random walks on countable groups*, Proc. Fifth Berkeley Sympos. Math. Statist. and Probability (Berkeley, Calif., 1965/66), vol. II: Contributions to Probability Theory, part 2, Univ. of California Press, Berkeley, Calif., 1967, pp. 51–74. MR **35** #4988.

20. H. Kesten and F. Spitzer, *Random walk on countably infinite abelian groups*, Acta Math. **114** (1965), 237–265.

21. J. Milnor, *A note on curvature and fundamental group*, J. Differential Geometry **1** (1968), 1–7.

22. L. Robin, *Fonctions sphériques de Legendre et fonctions sphéroidales*. Tome I, Gauthier-Villars, Paris, 1957. MR **19**, 954.

23. M. Rosenblatt, *Markov processes. Structure and asymptotic behavior*, Die Grundlehren der math. Wissenschaften, Band 184, Springer-Verlag, Berlin and New York, 1971.

24. A. S. Švarc, *A volume invariant of coverings*, Dokl. Akad. Nauk SSSR **105** (1955), 32–34. (Russian) MR **17**, 781.

25. N. Wiener, *Generalized harmonic analysis*, Acta Math. **55** (1930), 117–258.

UNIVERSITY OF WARWICK

Proceedings of Symposia in Pure Mathematics
Volume 27, 1975

LINEARIZATION STABILITY OF NONLINEAR PARTIAL DIFFERENTIAL EQUATIONS

ARTHUR E. FISCHER[1] AND JERROLD E. MARSDEN[1]

0. Introduction. In this article we study solutions to systems of nonlinear partial differential equations that arise in riemannian geometry and in general relativity. The systems we shall be considering are the scalar curvature equations $R(g) = \rho$ and the Einstein equations Ric $({}^{(4)}g) = 0$ for an empty spacetime. Here g is a riemannian metric and $R(g)$ is the scalar curvature of g, ρ is a given function, ${}^{(4)}g$ is a Lorentz metric on a 4-manifold and Ric $({}^{(4)}g)$ denotes the Ricci curvature tensor of ${}^{(4)}g$.

To study the nature of a solution to a given system of partial differential equations, it is common to linearize the equations about the given solution, solve the linearized equations, and assert that the solution to these linearized equations can be used to approximate solutions to the nonlinear equations in the sense that there exists a curve of solutions to the full equations which is tangent to the linearized solution. This assertion, however, is not always valid. In our study of the above equations we give precise conditions on solutions guaranteeing that such an assertion is valid—at these solutions, the equations are called *linearization stable.* We also give examples of solutions which are not linearization stable. Although such solutions are exceptional, they still point up the need to exercise caution when such sweeping assumptions are made.

The term "stable" has the general meaning that a stated property is not destroyed when certain perturbations are made, as in structural stability or dynam-

---

AMS (MOS) subject classifications (1970). Primary 53C25, 53C50, 58B20, 83C05; Secondary 35B20, 35J60, 35J70, 58F10, 83C10, 83C15.

[1]Partially supported by NSF grants GP-39060 and GP-15735.

ical (Liapunov) stability. For linearization stability the persistent property is "finding solutions in a given direction". If the equations are linearization stable, this property is not destroyed when we pass from the linearized equations to the nonlinear ones by adding on the "higher order terms".

The general set-up is as follows: let X and Y be Banach spaces or Banach manifolds of maps over a compact manifold M, and let $\Phi: X \to Y$ be a given map, e.g., a nonlinear differential operator between X and Y; we assume Φ itself is a differentiable map. Thus for $y_0 \in Y$

$$\Phi(x) = y_0 \tag{1}$$

is a system of partial differential equations.

Let $T_x X$ denote the tangent space to X at $x \in X$ and let

$$D\Phi(x) = T\Phi(x)\colon T_x X \to T_{\Phi(x)} Y$$

be the derivative of Φ at x. Thus to each solution x_0 of (1),

$$D\Phi(x_0)\cdot h = 0, \qquad h \in T_{x_0} X \tag{2}$$

is the associated system of *linearized equations* about x, and a solution $h \in T_{x_0}X$ to (2) is an *infinitesimal deformation* (or *first order deformation*) of the solution x_0 to (1).

Working in charts, the sum $x_0 + h$ is then to first order also a solution to (1), since

$$\begin{aligned}\Phi(x_0 + h) &= \Phi(x_0) + D\Phi(x_0)\cdot h + \text{higher order terms}\\ &= y + \text{higher order terms.}\end{aligned}$$

We now ask if there exist a $\delta > 0$ and a curve $x(\lambda)$, $|\lambda| < \delta$, of exact solutions of (1), $\Phi(x(\lambda)) = y_0$ which is tangent to h at x_0; i.e., such that $x(0) = x_0$ and $x'(0) = h$. If there exists such a curve for each solution h to (2) we say that the equations (1) are *linearization stable* at x_0; the curve $x(\lambda)$ is a *finite deformation* tangent to h.

We apply our general set-up and procedures to study the linearization stability and instability of the scalar curvature equation in riemannian geometry, and the Einstein empty space field equations of general relativity. We also study the possibility of isolated unstable solutions to these equations.

In the course of proving our results, we shall prove that several subsets of the space of riemannian metrics $\mathcal{M}$ and its cotangent bundle $T^*\mathcal{M}$ are actually submanifolds; these submanifolds are of interest to geometers and relativitists.

We summarize our main results as follows:

I. Riemannian geometry. *Let M be a a compact C^∞ manifold,* dim $M \geqq 2$, *and* $\rho\colon M \to \boldsymbol{R}$ *a C^∞ map.*

I.A. LINEARIZATION STABILITY. *The equation* $R(g) = \rho$ *is linearization stable about any solution g_0 if either*

(a) dim $M = 2$;

(b) dim $M \geqq 3$ *and* ρ *is not a constant* $\geqq 0$; *or*

(c) dim $M \geqq 3$, $\rho = 0$ *and* Ric $(g_0) \neq 0$ (*here* Ric (g_0) *is the Ricci tensor of* g_0).

If dim $M \geqq 3$, *and either*

(d) $\rho =$ *constant* > 0, *and* (M, g_0) *is a standard n-sphere in* $\mathbf{R}^{n+1}$; *or*

(e) $\rho = 0$ *and* g_0 *is flat,*

then $R(g) = \rho$ *is not linearization stable around* g_0.

These results occur in suitable H^s, $W^{s,p}$, *or* C^∞ *function spaces.*

I.B. SUBMANIFOLDS OF $\mathscr{M}$. *Let* $\mathscr{M}_\rho = \{g \in \mathscr{M}: R(g) = \rho\}$ *denote the set of riemannian metrics with prescribed scalar curvature. Assume that one of the following conditions is satisfied:*

(a) dim $M = 2$;

(b) dim $M = 3$, ρ *is not a positive constant;*

(c) dim $M \geqq 4$, ρ *is not a positive constant, and if* $\rho = 0$, *then there exists a flat riemannian metric on M.*

Then $\mathscr{M}_\rho$ *is a* C^∞ *closed submanifold of* $\mathscr{M}$.

Also, the space $\mathscr{F}$ *of flat riemannian metrics on M is a* C^∞ *closed submanifold of* $\mathscr{M}$, *and has the structure of a homogeneous fibre bundle.*

I.C. ISOLATION. *If* $g_F \in \mathscr{F}$, *there exists a neighborhood* $U_{g_F} \subset \mathscr{M}$ *such that if* $g \in U_{g_F}$ *and* $R(g) \geqq 0$, *then g is also flat. Consequently,* $\mathscr{M}_0$ *is the disjoint union of the closed submanifolds* $\mathscr{M}_0 - \mathscr{F}$ *and* $\mathscr{F}$, *under the hypothesis that* $\mathscr{F} \neq \varnothing$ (*not needed if* dim $M = 2$ *or* 3).

II. General relativity.

II.A. LINEARIZATION STABILITY OF THE EINSTEIN EQUATIONS. *Let V be a spacetime,* dim $V = 4$, *and* ${}^{(4)}g$ *a Lorentz metric satisfying Einstein's empty space equations* Ric $({}^{(4)}g) = 0$. *Let M denote a compact spacelike hypersurface with metric g and second fundamental form k. Assume that* (g, k) *satisfies conditions*

$C_{\mathscr{H}}$: *If* $k = 0$, *then g is not flat;*

C_δ: *If* $L_X g = 0$ *and* $L_X k = 0$, *then* $X = 0$ (L_X *denotes the Lie derivative by the vector field X*);

C_{tr}: tr k (= *the trace of k*) *is constant on M.*

If M is noncompact, one imposes asymptotic conditions, and we assume

$\mathscr{A}$: *g is complete and asymptotically flat, k is asymptotically zero;*

$\tilde{C}_\delta$: *If* $L_X g = 0$ *and* $L_X k = 0$, *and X is asymptotically zero, then* $X = 0$.

Then Ric $({}^{(4)}g)$ *is linearization stable on the appropriate maximal Cauchy development of M.*

II.B. SUBMANIFOLDS OF $T^*\mathscr{M}$ AND LINEARIZATION STABILITY OF THE CONSTRAINT EQUATIONS. *Let* $\mathscr{C}_{\mathscr{H}} = \{(g, \pi) \in T^*\mathscr{M}: \mathscr{H}(g, \pi) = 0\}$ *and* $\mathscr{C}_\delta = \{(g, \pi) \in T^*\mathscr{M}: \delta_g \pi = 0\}$ *denote the set of solutions to the Hamiltonian and divergence constraint on a compact spacelike hypersurface M. Replace k with* π *in conditions* $C_{\mathscr{H}}$, C_δ, *and* C_{tr}.

II.B.1. *The set* $\mathscr{C}_{\mathscr{H}} = (\mathscr{C}_{\mathscr{H}} - \mathscr{F} \times \{0\}) \cup (\mathscr{F} \times \{0\})$ *is the disjoint union of submanifolds of* $T^*\mathscr{M}$, *and* $\mathscr{F} \times \{0\}$ *is closed.* $\mathscr{C}_{\mathscr{H}}$ *itself, however, need not be a submanifold as* $\mathscr{C}_{\mathscr{H}} - \mathscr{F} \times \{0\}$ *is not closed.*

The Hamiltonian constraint equation $\mathscr{H}(g, \pi) = 0$ *is linearization stable about a solution* (g, π) *if and only if* (g, π) *satisfies condition* $C_{\mathscr{H}} \Leftrightarrow (g, \pi) \notin \mathscr{F} \times \{0\}$.

II.B.2 *Let* $(g, \pi) \in \mathscr{C}_\delta$ *satisfy condition* C_δ. *Then* $\mathscr{C}_\delta$ *is a* C^∞ *submanifold of* $T^*\mathscr{M}$ *in a neighborhood of* (g, π). *The divergence constraint* $\delta_g \pi = 0$ *is linearization stable about a* $(g, \pi) \in \mathscr{C}_\delta$ *if and only if condition* C_δ *is satisfied.*

II.B.3. *The constraint set* $\mathscr{C} = \mathscr{C}_{\mathscr{H}} \cap \mathscr{C}_\delta$ *is a* C^∞ *submanifold in a neighborhood of those* $(g, \pi) \in \mathscr{C}$ *that satisfy conditions* $C_{\mathscr{H}}$, C_δ, *and* C_{tr}.

II.C. ISOLATION. *There are no isolated solutions tot he empty space constraint equations of general relativity. However, the solutions* $\mathscr{F} \times \{0\} \subset T^*\mathscr{M}$ *are isolated among those solutions which also satisfy* $\mathrm{tr}\, \pi = 0$.

As a consequence of I.C, in a neighborhood U_{g_F} of a flat metric, (i) there are no metrics with $R(g) \geqq 0$ and $R(g) \neq 0$, and (ii) $R(g) = 0$ implies g is flat, so the flat solutions of $R(g) = 0$ are isolated solutions. This result extends to a full neighborhood of the flat metrics the "second order version" of these results obtained by Kazdan-Warner **[21]** and Brill **[5]**. Our isolation result I.C. was inspired by the work of Brill.

In **[6]**, Brill-Deser obtain to second order similar results for the linearization stability and the isolation problem of the constraint equations, and in **[10]**, **[11]**, Choquet-Bruhat and Deser prove that Minkowski space (which satisfies conditions $\mathscr{A}$ and $\tilde{C}_\delta$) is linearization stable.

The following simple but useful test for proving linearization stability will be our basic technique.

THEOREM. *Let* X, Y *be Banach manifolds and* $\Phi\colon X \to Y$ *be* C^1. *Let* $x_0 \in X$ *and* $\Phi(x_0) = y_0$. *Suppose that* $T\Phi(x_0)$ *is surjective with splitting kernel. Then the equation* $\Phi(x) = y_0$ *is linearization stable about* x_0.

PROOF. From the implicit function theorem, the set $\Phi^{-1}(y_0)$ is a C^1 submanifold near x_0 with tangent space the kernel of $T\Phi(x_0)$. Thus $h \in T_{x_0} X$ is a first order deformation iff $h \in \ker T\Phi(x_0)$ iff $h \in T_{x_0}(\Phi^{-1}(y_0))$, and since $\Phi^{-1}(y_0)$ is a submanifold, there exists a curve $x(\lambda) \in \Phi^{-1}(y_0)$ which is actually tangent to h. □

REMARKS. 1. If $T\Phi(x_0)$ is surjective, then Φ maps a neighborhood of x_0 onto a neighborhood of y_0.

2. In this paper we work with Hilbert manifolds so that the splitting condition is automatic. However, the Banach space context is important as the spaces $W^{s,p}$ are often needed; see I.2.1, Remark 4.

In the cases of immediate concern, the hard part of the linearization stability problem will be to establish surjectivity of the appropriate map under as minimal assumptions as possible. This is done pretty much on an *ad hoc* basis by using various elliptic operator methods.

If $T\Phi(x_0)$ is not surjective, the equation $\Phi(x) = y_0$ may still be linearization stable about x_0. This actually happens for the equation $R(g) = \rho$, $\rho =$ constant $\geqq 0$, on a 2-dimensional manifold M; see I.2.3, Remark 1.

On the other hand, even if $\Phi^{-1}(y_0)$ is a submanifold of X, the equation $\Phi(x) = y_0$ need not be linearization stable around a solution x_0.

EXAMPLE 1. Let $\Phi\colon \boldsymbol{R}^2 \to \boldsymbol{R}$, $(x, y) \mapsto x(x^2 + y^2)$. Then $\Phi^{-1}(0) = \{(0, y) : y \in \boldsymbol{R}\}$

$= \{y\text{-axis}\}$ is a submanifold of $\boldsymbol{R}^2$, and $d\Phi(0, 0) = 0$, so that any $h = (h_1, h_2)$ is a solution to the linearized equations $d\Phi(0, 0) \cdot h = 0$. However, if $h_1 \neq 0$, h cannot be tangent to any curve $(x(\lambda), y(\lambda)) \in \Phi^{-1}(0)$ of exact solutions of $\Phi(x, y) = 0$.

REMARK. The difficulty in this example is that the tangent space $T_{(0,0)}(\Phi^{-1}(0)) = \boldsymbol{R}$ is smaller than the formal tangent space $\ker(d\Phi(0, 0)) = \boldsymbol{R}^2$; the formal tangent space therefore contains nonintegrable directions. A similiar phenomenon occurs for the equation $R(g) = 0$ ($\dim M \geqq 3$) which is linearization unstable about a flat solution but whose solution set $\mathscr{M}_0$ is a C^∞ submanifold; see I.5.2, Remark 4.

In the applications we shall see that if equation (1) is not linearization stable at a solution x_0, and if h is a solution to the linearized equations (2) at x_0, then it is possible to get an extra condition on h in order that it be tangent to a curve of exact solutions of (1). For suppose $x(\lambda)$ is a curve of solutions to (1) with $(x(0), x'(0)) = (x_0, h)$ and h a solution to (2). Differentiating $\Phi(x(\lambda)) = y_0$ twice and evaluating at $\lambda = 0$ gives

$$\text{(2)} \qquad \frac{d\Phi}{d\lambda}(x(\lambda))\bigg|_{\lambda=0} = D\Phi(x(\lambda)) \cdot x'(\lambda)\bigg|_{\lambda=0} = D\Phi(x_0) \cdot h = 0$$

and

$$\text{(3)} \qquad \begin{aligned} \frac{d^2\Phi}{d\lambda^2}(x(\lambda)) &= D^2\Phi(x(\lambda)) \cdot (x'(\lambda), x'(\lambda))\bigg|_{\lambda=0} + D\Phi(x(\lambda)) \cdot x''(\lambda)\bigg|_{\lambda=0} \\ &= D^2\Phi(x_0) \cdot (h, h) + D\Phi(x_0) \cdot x''(0) = 0, \end{aligned}$$

so that (2) and (3) are necessary conditions on the derivatives $x'(0) = h$ and $x''(0)$ of a curve $x(\lambda) \in \Phi^{-1}(y_0)$. Condition (2), the linearized equations, is a condition on h alone; (3) is then a condition on $x''(0)$ in terms of a solution h. If, however, the second term involving $x''(0)$ could be made to drop out, (3) might provide a "second order" condition on h. For example, if $\Phi : X \to \boldsymbol{R}$ and x_0 is a critical point of Φ, $d\Phi(x_0) = 0$, but $d^2\Phi(x_0) \neq 0$, then (3) becomes

$$\text{(4)} \qquad d^2\Phi(x_0) \cdot (h, h) = 0,$$

a second order condition on h, which may not be implied by the first order equations (2). In that case solutions h to (2) which do *not* also satisfy (4) *cannot* be tangent to any curve $x(\lambda) \in \Phi^{-1}(y_0)$. When this occurs, we say that (4) is an "extra (second order) condition" on h.

If $d^2\Phi(x_0) = 0$, then of course (4) does not generate an extra condition on h. However, if (1) is unstable, we can get an extra condition by going to higher order deformations. For example, by considering third order derivatives, we have

$$\text{(5)} \qquad \begin{aligned} \frac{d^3\Phi(x(\lambda))}{d\lambda^3}\bigg|_{\lambda=0} &= d^3\Phi(x_0) \cdot (h, h, h) \\ &\quad + 3d^2\Phi(x_0) \cdot (h, x''(0)) + d\Phi(x_0) \cdot x'''(0) = 0. \end{aligned}$$

Thus if $d\Phi(x_0) = 0$, $d^2\Phi(x_0) = 0$, but $d^3\Phi(x_0) \neq 0$, then there is an extra third order condition

$$\text{(6)} \qquad d^3\Phi(x_0) \cdot (h, h, h) = 0$$

on first order deformations h.

EXAMPLE 2. Let $\Phi: \boldsymbol{R}^2 \to \boldsymbol{R}, (x, y) \mapsto x^2 - y^2$. Then $\Phi^{-1}(0) = \{(x, y) : y = \pm x\}$, $d\Phi(0, 0) = 0$, and the extra second order condition on $h = (h_1, h_2)$ is $d^2\Phi(0,0) \cdot (h, h) = 2(h_1^2 - h_2^2) = 0 \Leftrightarrow h_2 = \pm h_1$.

EXAMPLE 3. Let $\Phi(x, y) = x^3 + xy^2$ as in Example 1. Then $d\Phi(0, 0) = 0$ and $d^2\Phi(0, 0) = 0$. But we now have a third order condition on $h = (h_1, h_2)$,

$$d^3\Phi(0, 0) \cdot (h, h, h) = 6h_1(h_1^2 + h_2^2) = 0 \Leftrightarrow h = (0, h_2).$$

REMARKS. 1. If in these examples the extra condition on h is satisfied, then h is actually the tangent to a curve $(x(\lambda), y(\lambda)) \in \Phi^{-1}(0)$. In our later applications, we will not always know if satisfaction of the extra condition will be sufficient to find a curve $x(\lambda) \in \Phi^{-1}(y_0)$ tangent to h; see the end of I.4.

2. In Example 2, when $d^2\Phi(x_0) \neq 0$, the third order equations (5) do not provide an extra condition on first order deformations.

In the main applications we have in mind, the following example will be more generic:

EXAMPLE 4. Let $X = H^s(T^p_q(M))$ and $Y = H^{s-k}(T^{p'}_{q'}(M))$ be the linear spaces of Sobolev sections of tensor bundles over a compact manifold M with volume element $d\mu$ (see I.1), and let

$$\Phi: H^s(T^p_q(M)) \to H^{s-k}(T^{p'}_{q'}(M))$$

be a nonlinear differential operator of order k. If for $y_0 \in Y$, $\Phi(x) = y_0$ is not linearization stable about a solution $x_0 \in X$, then $\gamma_{x_0} = D\Phi(x_0)$ is not surjective, and this condition can be expressed via the Fredholm alternative by the fact that

$$\ker \gamma^*_{x_0} \neq \{0\},$$

where $\gamma^*_{x_0} : H^s(T^{p'}_{q'}(M)) \to H^{s-k}(T^p_q(M))$ is the L_2-adjoint of γ_{x_0}, defined by

$$\int_M (\gamma^*_{x_0} f) \cdot h \, d\mu = \int_M f \cdot (\gamma_{x_0} h) \, d\mu.$$

Thus if $f \in \ker \gamma^*_{x_0}$, $f \neq 0$, transvecting (3) with f and integrating over M gives the extra condition on h,

$$\int f D^2\Phi(x_0) \cdot (h, h) \, d\mu = 0, \tag{7}$$

since the term involving $x''(0)$,

$$\int f \cdot (D\Phi(x_0) \cdot x''(0)) \, d\mu = \int (\gamma^*_{x_0} f) \cdot x''(0) \, d\mu = 0,$$

drops out for $x''(0)$ arbitrary.

REMARKS. 1. The extra condition (7) is now an integrated condition, corresponding to the fact that $D\Phi(x_0)^*$ is an L_2-adjoint; i.e. we do not get an extra condition on h until we integrate (3) with $f \in \ker (D\Phi(x_0))^*$ so that the term involving $x''(0)$ drops out. Equation (3) itself is just a pointwise condition on the acceleration $x''(0)$ in terms of h and does not lead to an extra condition on h.

2. In this example and in our later applications there will be an "independent" extra condition for each dimension in $\ker (D\Phi(x_0))^*$, as is evident from (7) (providing (7) really is an extra condition; see Remark 3). Note that in Examples 2 and 3, $\ker (d\Phi(0, 0))^* = \boldsymbol{R}$, and there is one extra condition on h as expected.

3. It may happen that solutions to the linearized equations automatically satisfy (7) for all $f \in \ker \gamma_{x_0}^*$; e.g., this occurs when $D\Phi(x_0)$ is not surjective but $\Phi(x) = y_0$ is linearization stable. This actually happens for the equation $R(g) = \rho =$ constant $\geqq 0$ when $\dim M = 2$. Of course, in these cases we can never find an extra condition on first order deformations h by considering higher order variations of $\Phi(x) = y_0$.

4. In certain circumstances, the extra integrated condition implied by linearization instability can be converted to a pointwise condition on first order deformations. This situation signals that an even stronger type of instability is occurring; viz., that $x_0 \in S \subset \Phi^{-1}(y_0)$, and that S is an isolated subset of solutions; i.e., there exist no solutions near S which are not in S. This situation actually occurs in a neighborhood of the flat solutions of $R(g) = 0$; see I.5.

In Part II of this paper we will consider the problem of linearization stability of the Einstein equations, a system of nonlinear evolution equations. The problem of linearization stability for a nonlinear evolution equation is interesting when there are some nonlinear constraints on the initial data of the form $\Phi(x) = y_0$. Then linearization stability of Φ implies that the corresponding evolution equation, say

$$\dot{x} = F(x), \tag{8}$$

is also linearization stable, if it satisfies suitable uniqueness and existence theorems. Indeed, we can argue formally as follows: let $\Phi(x_0) = y_0$ and let $x(t)$ satisfy $\dot{x} = F(x)$, $x(0) = x_0$. Let $h(t)$ satisfy the *linearized evolution equations*

$$\dot{h}(t) = DF(x(t)) \cdot h(t) \tag{9}$$

with initial condition $h(0)$ satisfying the linearized constraint equation

$$D\Phi(x_0) \cdot h(0) = 0.$$

Let $x_0(\lambda)$ be a curve through x_0 with $(d/d\lambda)\, x_0(\lambda)|_{\lambda=0} = h(0)$. Now solve (8) with initial data $x_0(\lambda)$ to get a one-parameter family of solutions $x(t, \lambda)$,

$$\dot{x}(t, \lambda) = F(x(t, \lambda)), \tag{10}$$

with $x(t, 0) = x(t)$ and $x(0, \lambda) = x_0(\lambda)$. Then $x(t, \lambda)$ is a curve of solutions of (8) which is tangent to the linearized solution $h(t)$; i.e., $(x(t, \lambda), dx(t, \lambda)/d\lambda|_{\lambda=0} = (x(t), h(t))$. For by differentiating (10),

$$\frac{d}{d\lambda} \dot{x}(t, \lambda)\Big|_{\lambda=0} = \frac{d}{dt} \cdot \frac{d}{d\lambda} x(t, \lambda)\Big|_{\lambda=0}$$

$$= DF(x(t, \lambda)) \cdot \frac{dx}{d\lambda}(t, \lambda)\Big|_{\lambda=0} = DF(x(t)). \frac{dx}{d\lambda}(t, \lambda)\Big|_{\lambda=0}$$

and

$$\frac{dx}{d\lambda}(0, \lambda)\Big|_{\lambda=0} = \frac{dx_0}{d\lambda}(\lambda)\Big|_{\lambda=0} = h(0);$$

thus

$$\frac{dx}{d\lambda}(t, \lambda)\Big|_{\lambda=0} = h(t)$$

since both sides satisfy (9) with the same initial conditions.

For these reasons, linearization stability of the Einstein equations reduces to the linearization stability of the constraint equations on the initial Cauchy data (g, π) (although in the above argument, one needs uniqueness of solutions for the equations (8) and (9); for the Einstein equations one has uniqueness only up to coordinate-transformation so that a further argument is needed; see II.5). Since these constraint equations involve the scalar curvature rather than the Ricci curvature of a riemannian metric g, the linearization stability of the Einstein equations is similar to the scalar curvature equation in the riemannian case. Thus because of the dynamical aspect of Lorentz manifolds (see e.g. **[1]**, **[14]**), Ricci-flat Lorentz manifolds are in some ways more manageable than Ricci-flat riemannian manifolds. For example, we have not been able to establish whether $\mathrm{Ric}(g) = 0$ is linearization stable around a nonflat Ricci-flat solution; indeed the existence of complete riemannian nonflat Ricci-flat metrics is unknown.

During the course of this work, many people were consulted. We would especially like to thank Professors J. P. Bourguignon, D. Ebin, J. Guckenheimer, R. Palais, A. Weinstein and J. A. Wolf. The final form of Theorem I.2.1 was obtained jointly with J. P. Bourguignon and I.3.1 with A. Weinstein.

For further details of the topics and results presented in this paper, see **[16]**, **[18]**.

I. DEFORMATIONS OF RIEMANNIAN STRUCTURES

I.1. Some preliminaries. Throughout, M will denote a C^∞ compact connected oriented n-manifold without boundary, $n \geqq 2$, $T^p_q(M)$ the bundle of p-contravariant and q-covariant tensors on M, and $C^k(T^p_q(M))$ the C^k sections, $0 \leqq k \leqq \infty$. Let g_b be a fixed "background" C^∞ riemannian metric on M with associated covariant derivative $\nabla\colon C^k(T^p_q(M)) \to C^{k-1}(T^p_{q+1}(M))$ and volume element $d\mu_b$. Then for sections $t_1, t_2 \in C^\infty(T^p_q(M))$ and for integer $s \geqq 0$, we let

$$(t_1, t_2)_{H^s} = \sum_{0 \leqq l \leqq s} \int_M \langle \nabla^l t_1(x), \nabla^l t_2(x)\rangle_x \, d\mu_b(x),$$

the H^s inner product on sections $C^\infty(T^p_q(M))$. In the above $\nabla^l = \nabla\circ\nabla\circ\cdots\circ\nabla$ is the covariant derivative iterated l times, and $\langle\ ,\ \rangle_x$ is the pointwise inner product on tensors at x. We let $\|\ \|_s$ denote the associated H^s norm, and let $H^s(T^p_q(M))$ denote the completion of $C^\infty(T^p_q(M))$ in this norm. Thus $H^s(T^p_q(M))$ is the Sobolev space of H^s sections. While the norm depends on g_b, the space and the topology do not.

We will make use of the following two basic Sobolev theorems that are standard in functional analysis (see **[27]** for more information and references):

(i) For $s > n/2 + k$, $H^s(T^p_q(M)) \subset C^k(T^p_q(M))$, and $\|t\|_{C^k} \leqq (\text{constant})\ \|t\|_{H^s}$ for $t \in H^s(T^p_q(M))$, where

$$\|t\|_{C^k} = \sup_{x\in M} \{ \|t(x)\|_x, \|\nabla t(x)\|_x, \cdots, \|\nabla^k t(x)\|_x\}.$$

(ii) For $s > n/2$, H^s is a ring under pointwise multiplication; i.e., if B: $T^p_q(M) \times T^{p'}_{q'}(M) \to T^{p''}_{q''}(M)$ is some pointwise bilinear map, then B induces a continuous bilinear map of the corresponding Sobolev spaces, and

$$\|B(t_1, t_2)\|_{H^s} \leqq \text{const}\ \|t_1\|_{H^s}\ \|t_2\|_{H^s}.$$

We shall refer to this property as the Schauder ring property. More generally, if $s > n/2$, multiplication from $H^s \times H^k$ to H^k is continuous, $0 \leqq k \leqq s$.

For $s > n/2$, we let

$H^s = H^s(M; \boldsymbol{R})$, the H^s functions on M;
$\mathscr{X}^s = H^s(T^1_0(M))$, the H^s vector fields;
$S^s_2 = H^s(T^0_{2,\text{sym}}(M))$, the H^s 2-covariant symmetric tensor fields;
$S^2_s = H^s(T^2_{0,\text{sym}}(M))$, the H^s 2-contravariant symmetric fields;
$\mathscr{D}^{s+1} = \{$the group of H^{s+1} diffeomorphisms on $M\}$;
$\mathscr{M}^s = \mathscr{M}^0 \cap S^s_2$, the space of H^s riemannian metrics on M, where $\mathscr{M}^0$ is the space of continuous riemannian metrics.

$\mathscr{M}^s \subset S^s_2$ is an open convex cone, so that for $g \in \mathscr{M}^s$, $T_g\, \mathscr{M}^s = S^s_2$ and $T\mathscr{M}^s = \mathscr{M}^s \times S^s_2$, the tangent bundle of $\mathscr{M}^s$.

For $g \in \mathscr{M}^s$, we let $d\mu_g$ denote the associated volume element.

As in Ebin [13], when the "s" is omitted on any of the above spaces, "∞" will be understood; e.g., $\mathscr{M} = \bigcap_{s>n/2} \mathscr{M}^s$ is the space of all C^∞ riemannian metrics on M.

For $g \in \mathscr{M}^{s+1}$, $X \in \mathscr{X}^{s+1}$, and $h \in S^s_2$, we let $L_X g \in S^s_2$ be the Lie derivative of g with respect to X, and $\delta_g h \in \mathscr{X}^s$ the divergence of h. In local coordinates, $L_X g = X_{i|j} + X_{j|i}$ and $\delta_g h = -\, g^{ik}\, h^j_{k|j}$ (a vertical bar denotes covariant derivative with respect to the metric g).

Let $\alpha_g : \mathscr{X}^{s+1} \to S^s_2$ denote the map $X \mapsto L_X g$. Using g to define the L_2 inner products

$$(X_1, X_2) = \int_M \langle X_1(x), X_2(x)\rangle_x\, d\mu_g(x),$$

$$(h_1,h_2) = \int_M \langle h_1(x), h_2(x)\rangle_x\, d\mu_g(x);$$

we can see from Stokes' theorem that α_g and $2\delta_g$ are adjoints:

$$(\alpha_g X, h) = 2(X, \delta_g h), \qquad X \in \mathscr{X}^{s+1},\ h \in S^{s+1}_2.$$

Also α_g has injective symbol, so from Berger-Ebin [4] (modified to use an H^{s+1} metric), S^s_2 splits as a L_2 orthogonal direct sum

(1) $$S_2^s = \mathring{S}_2^s \oplus \alpha_g(\mathscr{X}^{s+1})$$

where $\mathring{S}_2^s = \ker \delta_g = \{h \in S_2^s : \delta_g h = 0\}$, the divergence free symmetric tensors. Thus each $h \in S_2^s$ can be decomposed as

(2) $$h = \mathring{h} + L_X g$$

where $\delta_g \mathring{h} = 0$ and where the pieces $\mathring{h}$ and $L_X g$ are unique and orthogonal in the L_2 metric. We will refer to this decomposition as the *canonical decomposition* of $h \in S_2^s$.

Note that the splitting in (1) is not valid if g is only of class H^s, since then $\alpha_g(X) = L_X g$ need only be of class H^{s-1}. If however g is a flat riemannian H^s metric, then the splitting (1) is still valid (see I.3).

In the decomposition (2), the vector field X satisfies

(3) $$\delta h = \delta L_X g = \varDelta X + (d\delta X)^\# - 2\text{Ric}(g)\cdot X$$

where $(d\delta X)^\#$ denotes the vector corresponding to the 1-form $d\delta X$; in general we will let $t^\#$ denote the totally contravariant form of any tensor t, and $t^\flat$ denote the totally covariant form; e.g., if t is of type $\binom{1}{2}$, $(t^\#)^{ijk} = g^{jb}\, g^{kc}\, t^i{}_{jk}$ and $(t^\flat)_{ijk} = g_{ia}\, t^a{}_{jk}$, $\varDelta X = ((d\delta + \delta d)X^\flat)^\#$ is the Laplace-de Rham operator acting on vector fields, $\delta X = -\operatorname{div} X = -X^i{}_{|i}$ is the divergence of X, $\text{Ric}(g) = R_{ij}$ is the Ricci tensor of g and $\text{Ric}(g)\cdot X = R^i{}_j X^j$. The above computation uses the contracted Ricci commutation formula $X^j{}_{|i|j} - X^j{}_{|j|i} = R_{ij} X^j$, and the Weitzenböck formula for the Laplace-de Rham operator (see e.g. Nelson **[29]**), $\varDelta X = \bar{\varDelta} X + \text{Ric}(g)\cdot X$ where $\bar{\varDelta} X = -g^{jk} X^i{}_{|j|k}$ is the "rough Laplacian" on vector fields. Throughout we use the conventions of Nelson **[29]** and Lichnerowicz **[25]**.

If $X = \operatorname{grad} \phi$ is a gradient vector field, then $L_X g = 2 \text{ Hess } \phi = 2\nabla\nabla\phi = 2\phi_{|i|j}$, the Hessian or double covariant derivative of ϕ, and $(\varDelta X)^\flat + d\delta X = \varDelta d\phi + d\varDelta\phi = 2\, d\varDelta\phi$. Thus we have the identity

(4) $$\delta \text{ Hess } \phi = (d\varDelta\phi)^\# - \text{Ric}(g)\cdot d\phi$$

which we shall need in the proof of I.2.1. By contracting the Bianchi identities one also has

(5) $$\delta \text{ Ric}(g) = -\tfrac{1}{2}(d(R(g))^\#$$

where $R(g) = R^i{}_i$ is the scalar curvature of g.

Finally, as in **[13]**, we let $A : \mathscr{D}^{s+1} \times \mathscr{M}^s \to \mathscr{M}^s, (\eta, g) \mapsto \eta^* g$ denote the right group action of $\mathscr{D}^{s+1}$ on $\mathscr{M}^s$ by pullback of metrics (locally this is just coodinate transformation of g by η). Fixing $g \in \mathscr{M}^s$, let $\mathscr{O}_g = \{\eta^* g : \eta \in \mathscr{D}^{s+1}\}$ be the orbit of g. For $g \in \mathscr{M}^{s+1}$, $\mathscr{O}_g$ is a C^1 submanifold and $T_g \mathscr{O}_g = \alpha_g(\mathscr{X}^{s+1})$. Thus in the decomposition of $h \in T_g\mathscr{M}^s$, $h = \mathring{h} + L_X g$, the $L_X g$ piece is tangent to the orbit which is the direction of isometric changes of g, while the orthogonally directed $\mathring{h}$ piece is in the direction of "true" geometric deformations.

For $g \in \mathscr{M}^s$, the orbit map $\psi_g : \mathscr{D}^{s+1} \to \mathscr{M}^s$, $\varphi \mapsto \varphi^* g$ is only continuous and need not be C^1. Thus $\mathcal{O}_g$ need only be a C^0 submanifold. This is another reason why the splitting (1) is not valid for $g \in \mathcal{M}^s$ (unless g is a flat H^s metric); however for $g \in \mathcal{M}^{s+1}$, the orbit map ψ_g is C^1.

I.2. Deformations of the scalar curvature equation. Let $s > n/2 + 1$, $g \in \mathcal{M}^s$ and let $R(g)$ denote the scalar curvature of g. Consider $R(\cdot)$ as a map

$$R(\cdot) : \mathcal{M}^s \to H^{s-2}, \qquad g \mapsto R(g).$$

Since $R(g)$ is a rational combination of g and its first two derivatives, we see from the Schauder ring property that if $s > n/2 + 2$, R is a C^∞ map. This is also true for $s > n/2 + 1$ if we use the fact that the second derivatives of g occur linearly.

Thus we consider $R(\cdot)$ to be a nonlinear second order differential operator.

For $g \in \mathcal{M}^s$, $s > n/2 + 1$, we let

$$\gamma_g = DR(g) : T_g \mathcal{M}^s \approx S_2^s \to T_{R(g)} H^{s-2} \approx H^{s-2}$$

denote the derivative of $R(\cdot)$. A classical computation, given for example in **[25]**, shows that for $h \in S_2^s$

$$\gamma_g(h) = \Delta \mathrm{tr}\, h + \delta\delta h - h \cdot \mathrm{Ric}(g).$$

Here $\mathrm{tr}\, h = g^{ij} h_{ij}$ is the trace of h, $\Delta \mathrm{tr}\, h = -g^{ij}(\mathrm{tr}\, h)_{|i|j}$ is the Laplacian, $\delta\delta h = h^{ij}{}_{|i|j}$ is the double covariant divergence, $\mathrm{Ric}(g) \in S_2^{s-2}$ is the Ricci tensor of g, and $h \cdot \mathrm{Ric}(g) = \langle h, \mathrm{Ric}(g)\rangle = g^{ij} g^{ab} h_{ia} R_{jb}$ is the pointwise contraction.

We let $\gamma_g^* : H^s \to S_2^{s-2}$ be the L_2-adjoint of γ_g, defined by

$$\int_M \langle \gamma_g^* f, h\rangle \, d\mu_g = \int_M \langle f, \gamma_g h\rangle \, d\mu_g.$$

Using Stokes' theorem, $\gamma_g^* f$ is easily computed to be

$$\gamma_g^* f = g\Delta f + \mathrm{Hess}\, f - f\, \mathrm{Ric}(g).$$

A somewhat remarkable property of the scalar curvature map is that locally it is almost always a surjection.

I.2.1. THEOREM. *Let $g \in \mathcal{M}^s$, $s > n/2 + 1$, and suppose that*
(1) *$R(g)$ is not a positive constant, and*
(2) *if $R(g) = 0$, then $\mathrm{Ric}(g) \not\equiv 0$.*
Then $R(\cdot) : \mathcal{M}^s \to H^{s-2}$ maps any neighborhood of g onto a neighborhood of $R(g)$.

PROOF. First assume $s < \infty$. It then suffices by the standard implicit function theorem to show that $\gamma_g : S_2^s \to H^{s-2}$ is surjective. From elliptic theory (see e.g. **[4]**), this follows if $\gamma_g^* : H^s \to S_2^{s-2}$ is injective and has injective symbol. For $\xi_x \in T_x^* M$ (the cotangent space at x), the symbol $\sigma_{\xi_x}(\gamma_g^*) : \boldsymbol{R} \to T_x^* M \otimes_{\mathrm{sym}} T_x^* M$ ($\otimes_{\mathrm{sym}}$ means the symmetric tensor product) is given by $s \mapsto (-g\, \|\xi_x\|^2 + \xi_x \otimes \xi_x)s$, $s \in \boldsymbol{R}$, which is clearly injective for $\xi_x \neq 0$ and $n \geqq 2$.

To show γ_g^* injective, assume $f \in \ker \gamma_g^*$, so that

(a) $$\gamma_g^* f = g\Delta f + \text{Hess}\, f - f\,\text{Ric}(g) = 0.$$

Taking the trace yields

(b) $$(n-1)\,\Delta f = R(g) f.$$

First we consider case (2). Thus if $R(g) = 0$, $\Delta f = 0$, so f is a constant. Thus from (a), $f\,\text{Ric}(g) = 0$, and since $\text{Ric}(g) \not\equiv 0, f = 0$. Thus in case (2), γ_g^* is injective.

Now consider case (1). We shall show that if $f \not\equiv 0$, $R(g) = \text{constant} > 0$. Taking the divergence of (a) yields

(c) $$-d\Delta f + \delta\,\text{Hess} f + df\cdot\text{Ric}(g) - f\delta(\text{Ric}(g)) = 0$$

(where $df\cdot\text{Ric}(g) = f_{|j} R^j{}_i$) and using the identities $\delta\,\text{Hess} f - d\Delta f + df\cdot\text{Ric}(g) = 0$, and $\delta\,\text{Ric}(g) = -(1/2)d(R(g))$, (c) reduces to $-(1/2)fd(R(g)) = 0$. If f is never zero, then $d(R(g)) = 0$ so $R(g) = \text{constant}$, and in fact from (b) an eigenvalue of the Laplacian, and thus $\geqq 0$. If $R(g) = 0$, then $f = \text{constant} \neq 0 \Rightarrow \text{Ric}(g) = 0$, contradicting (2), and $R(g) = \text{constant} > 0$ contradicts (1). Thus find $x_0 \in M$ such that $f(x_0) = 0$. We must have $df(x_0) \neq 0$; indeed, if $df(x_0) = 0$, let $\gamma(t)$ be a geodesic starting at x, and let $h(t) = f(\gamma(t))$. Combining (a) and (b) we have

(d) $$\text{Hess}\, f = (\text{Ric}(g) - (1/(n-1))\, gR(g))f.$$

Hence $h(t)$ satisfies the linear second order differential equation

$$\begin{aligned} h''(t) &= (\text{Hess} f)_{\gamma(t)} \cdot (\gamma'(t), \gamma'(t)) \\ &= \{(\text{Ric}(g) - (n-1)^{-1} g\, R(g))_{\gamma(t)} \cdot (\gamma'(t), \gamma'(t))\}\, h(t) \end{aligned}$$

with $h(0) = 0$ and $h'(0) = df(\gamma(0))\cdot\gamma'(0) = 0$. Thus f is zero along $\gamma(t)$ and so by the Hopf-Rinow theorem on all of M. Thus df cannot vanish on $f^{-1}(0)$ so 0 is a regular value of f and so $f^{-1}(0)$ is an $n-1$ dimensional submanifold of M. Hence $d(R(g)) = 0$ on an open dense set and hence everywhere. Thus as before $R(g) = \text{constant} \geqq 0$, contradicting (1) or (2). Thus if (1) and (2) hold, $f \equiv 0$, γ_g^* is injective, γ_g is surjective, and $R(\cdot)$ is locally surjective.

The case $s = \infty$ requires some additional arguments. One needs to show that the image neighborhood of $R(g)$ can be chosen independent of s. This is possible because one can construct local right inverses for R by maps independent of s; they depend only on the right inverse for the derivative and the geometry of the space. The idea is similar to one occurring in Ebin [**13**] and works quite generally when we have L_2 orthogonal splittings for elliptic operators. □

Note. We thank J. P. Bourguignon for pointing out the substantial improvement that $\gamma_g^* f = 0$, $f \not\equiv 0$ implies $R(g) = \text{constant}$. Previously we had condition (1) replaced with the condition $R(g) \leqq 0$. Bourguignon's argument appears in [**37**].

REMARKS. (1) If $\text{Ric}(g) = 0$, $\ker \gamma_g^* = \{\text{constant functions on } M\}$, so that γ_g^* is not surjective. It is not known if there exist any nonflat Ricci-flat complete riemannian metrics. If there are none, then condition (2) can be replaced by (2)′: If $R(g) = 0$, then g is not flat.

If dim $M = 3$, then $\mathrm{Ric}(g) = 0$ implies that g is flat so that (2) can be replaced by (2)′.

(2) If $M = S^n$ has the metric g_0 of a standard sphere of radius r_0 in $\boldsymbol{R}^{n+1}$, then

$$\mathrm{Ric}(g_0) = \left(\frac{n-1}{r_0^2}\right) g_0 \quad \text{and} \quad R(g_0) = \frac{n(n-1)}{r_0^2}$$

so that $f \in \ker \gamma_{g_0}^*$ if

$$\mathrm{Hess}\, f = \left(\mathrm{Ric}(g_0) - \frac{1}{n-1} g_0 R(g_0)\right) f = -\frac{f}{r_0^2} g_0.$$

But the eigenfunctions of the Laplacian with first nonzero eigenvalue n/r_0^2 also satisfy $\mathrm{Hess}\, f = -(f/r_0^2) g_0$. Hence again γ_{g_0} is not surjective, and $\ker \gamma_{g_0}^* = \{f \in H^s\colon \Delta f = (n/r_0^2) f\}$.

Conversely, a theorem of Obata [**31**], [**32**] states that if a riemannian manifold admits a solution $f \not\equiv 0$ of $\mathrm{Hess}\, f = -c^2 f g$, then the manifold is isometric to a standard sphere in $\boldsymbol{R}^{n+1}$ of radius $1/c$.

If in Theorem I.2.1., $R(g) = \text{constant} > 0$ and γ_g is not surjective, then there exists a solution $f \not\equiv 0$ of

(e) $$\mathrm{Hess}\, f = (\mathrm{Ric}(g) - (n-1)^{-1} g\, R(g)) f.$$

This equation is similiar to Obata's equation, and it is reasonable to conjecture that a solution $f \not\equiv 0$ of (e) implies that the space is a sphere. For example if g is an Einstein space, $\mathrm{Ric}(g) = \lambda g$ with $\lambda > 0$, then the space is a standard sphere. In fact, if $\mathrm{Ric}(g)$ is parallel (e.g., g is a product of Einstein spaces) then Obata's proof goes through and proves that (e) has solutions only on the standard sphere.

Thus it is reasonable to conjecture that *γ_g is surjective unless (M, g) is flat, or unless (M, g) is a standard sphere*. This would be quite a nice result.

(3) Also note that among the spaces with positive constant scalar curvature, if $R(g)/(n-1)$ is *not* an eigenvalue of the Laplacian, $\Delta f = (R(g)/(n-1)) f$, then γ_g is surjective.

(4) If $R(g)$ is not a constant $\geqq 0$, then from surjectivity of $DR(g)$, it follows that $R(\cdot)$ is locally surjective in an H^s neighborhood of g. Thus if ρ is sufficiently near $R(g)$ in the H^s topology, ρ is the scalar curvature of some metric. An analogous theorem holds using $W^{s,p}$ spaces. In fact, Kazdan and Warner [**22**] have pointed out that local surjectivity of $R(\cdot)$, together with an approximation lemma, can be used to prove their results concerning what functions can be realized as scalar curvatures for dim $M \geqq 2$. A variant of this technique yields, for example: *If $n \geqq 3$ and there is a $g \in \mathcal{M}^s$ with $R(g) = 0$, $\mathrm{Ric}\,(g) \neq 0$, then $R(\cdot)\colon \mathcal{M}^s \to H^{s-2}$ is surjective; i.e., every function can be realized as a scalar curvature of some metric.* This follows from local surjectivity of $R(\cdot)$, together with [**22**, Theorem C].

(5) Using some recent sharp elliptic estimates of Nirenberg and Walker [**30**] one can extend much of the above work to noncompact manifolds satisfying suitable asymptotic conditions. We discuss this aspect in [**18**].

For $\rho : M \to \boldsymbol{R}$ a C^∞ map, $s > n/2 + 1$, we set

$$\mathscr{M}^s_\rho = \{g \in \mathscr{M}^s : R(g) = \rho\} \quad \text{and} \quad \mathscr{M}_\rho = \{g \in \mathscr{M} : R(g) = \rho\},$$

sets of metrics with prescribed scalar curvature ρ. Using I.2.1, we can now prove that under certain mild conditions on ρ, $\mathscr{M}^s_\rho$ is a C^∞ submanifold of $\mathscr{M}^s$.

I.2.2. THEOREM. *Let $\rho : M \to \boldsymbol{R}$ be a C^∞ map, and $s > n/2 + 1$. If either*
(a) dim $M = 2$, *or*
(b) dim $M \geqq 3$ *and ρ is not a constant $\geqq 0$,*
then $\mathscr{M}^s_\rho$ (respectively, $\mathscr{M}_\rho$) is a C^∞ closed submanifold of $\mathscr{M}^s$ (respectively, $\mathscr{M}$).

If dim $M \geqq 3$, $\rho = 0$, $g \in \mathscr{M}^s_0$ *(respectively, $g \in \mathscr{M}^0$) and* Ric$(g) \neq 0$ *(or if* dim $M = 3$, *g is not flat), then $\mathscr{M}^s_0$ (respectively, $\mathscr{M}_0$) is a C^∞ closed submanifold in a neighborhood of g.*

PROOF. The cases (b) and $\rho = 0$ are a direct consequence of the surjectivity of $DR(g)$ and the inverse function theorem.

If dim $M = 2$, we need only consider the case ρ = constant $\geqq 0$ since otherwise $DR(g)$ is surjective. If $\rho = 0$, then $\mathscr{M}^s_0 = \mathscr{F}^s = \{$the set of H^s flat riemannian metrics on $M\}$, which from I.3.3 is a C^∞ closed submanifold of $\mathscr{M}^s$; similiarly for $\mathscr{M}_0 = \mathscr{F}$.

If $g \in \mathscr{M}^s$ and $R(g) = \rho$ = constant > 0, then (M, g) is H^s isometric to (S^2, g_0), a standard 2-sphere in $\boldsymbol{R}^3$ with radius $r_0 = (2/\rho)^{1/2}$. Thus $\mathscr{M}^s_\rho = \mathscr{D}^{s+1}(g_0) = \{g \in \mathscr{M}^s : \varphi^* g = g_0$ for some $\varphi \in \mathscr{D}^{s+1}\} = \mathscr{O}_{g_0}$, the orbit through g_0, which by **[13]** is a C^∞ closed submanifold of $\mathscr{M}^s$ since g_0 is C^∞; similiarly $\mathscr{M}_\rho = \mathscr{D}(g_0)$ is a C^∞ closed submanifold of $\mathscr{M}$. □

REMARKS. 1. If $R(g) = \rho$ = constant $\geqq 0$, dim $M = 2$, $DR(g)$ is not surjective but $\mathscr{M}_\rho$ is still a submanifold, as remarked in the introduction; see also I.2.3, Remark 1.

2. We are not making any statements about whether $\mathscr{M}^s_\rho$ is empty or not. This is another question. For example, Lichnerowicz **[26]** has shown that for spin manifolds with $\hat{A}$ genus not zero, $\mathscr{M}_\rho = \varnothing$ if $\rho \geqq 0, \rho \not\equiv 0$. However Kazdan and Warner **[21]**, **[22]** have shown that if $n \geqq 3$ and ρ is negative somewhere, or if $n = 2$ and ρ satisfies a sign condition consistent with the Gauss-Bonnet formula, then $\mathscr{M}^s_\rho \neq \varnothing$.

3. If dim $M = 1$, then $M = S^1$ and $R(g) = 0$ for all metrics in $\mathscr{M}^s$. Thus, $\mathscr{M}^s_\rho = \varnothing$ if $\rho \neq 0$, $\mathscr{M}^s_0 = \mathscr{M}^s$, and $R(\cdot)$ is not locally surjective around any g.

As another application of I.2.1, we have the following result concerning the linearization stability of the equation $R(g) = \rho$.

I.2.3. THEOREM. *Let $g_0 \in \mathscr{M}^s$ (respectively, $g_0 \in \mathscr{M}$), and let $\rho = R(g_0)$. Assume that one of the following conditions is satisfied:*
(a) dim $M = 2$;
(b) dim $M \geqq 3$, *ρ is not a constant* $\geqq 0$;
(c) dim $M \geqq 3$, $\rho = 0$, *and* Ric$(g_0) \neq 0$ *(or if* dim $M = 3$, *g_0 is not flat).*
Then $R(g) = \rho$ is linearization stable about g_0; i.e. for any $h \in S^s_2$ (respectively, $h \in S_2$) satisfying the linearized equations

$$DR(g_0)\cdot h = \Delta \operatorname{tr} h + \delta\delta h - h\cdot \operatorname{Ric}(g_0) = 0$$

there exists a C^∞ *curve* $g(\lambda) \in \mathscr{M}_\rho^s$ *of exact solutions of* $R(g) = \rho$ *such that* $(g(0), g'(0)) = (g_0, h)$.

PROOF. Cases (b) and (c) are a consequence of I.2.1 and Theorem 1 in § 0, the case $s = \infty$, as before, requiring a special regularity argument.

Suppose dim $M = 2$; we will show directly that $R(g) = \rho$ is linearization stable about a solution $g_0 \in \mathscr{M}^s$ by showing that we can integrate any first order deformation h to a curve of exact solutions. We need only consider the case $\rho =$ constant ≥ 0; otherwise stability follows as in case (b). Thus (M, g_0) is H^s isometric to either a flat torus (if $\rho = 0$) or a standard 2-sphere S^2 with radius $r_0 = (2/\rho)^{1/2}$ (if $\rho > 0$). If $h \in S_2^s$, let $h = \mathring{h} + L_X g_0$ be the canonical decomposition of h. Such a decomposition always exists even if g_0 is only of class H^s; one decomposes the push-forward of h on the torus or sphere using their C^∞ riemannian structures and then pulls back this decomposition; see I.3.

Now suppose h is a solution to the linearized equations,

$$DR(g_0)\cdot h = DR(g_0)\cdot \mathring{h} = \Delta \operatorname{tr} \mathring{h} - \mathring{h}\cdot \operatorname{Ric}(g_0) = 0,$$

where we have used $DR(g_0)\cdot L_X g_0 = L_X(R(g_0)) = X\cdot d(R(g_0)) = 0$ since $R(g_0) = \rho = \text{constant} \geq 0$ (see also the proof of I.3.4). If $\rho = 0$, g_0 is flat; thus $\Delta \operatorname{tr} \mathring{h} = 0 \Rightarrow \operatorname{tr} \mathring{h} = \text{constant} \Rightarrow \nabla\mathring{h} = 0$, as can be seen by writing out $\delta\mathring{h} = 0$ and $d \operatorname{tr} \mathring{h} = 0$. But if $\mathring{h}$ is parallel, we can explicitly find a curve $g(\lambda)$ such that $g'(0) = \mathring{h} + L_X g_0$ by the method at the end of I.4.

If $R(g_0) = \rho = \text{constant} > 0$, then $\operatorname{Ric}(g_0) = \frac{1}{2}\rho g_0$, and $DR(g_0)\cdot\mathring{h} = \Delta \operatorname{tr} \mathring{h} - \frac{1}{2}\rho \operatorname{tr} \mathring{h} = 0 \Rightarrow \Delta \operatorname{tr} \mathring{h} = \frac{1}{2}\rho \operatorname{tr} \mathring{h}$. But the first nonzero eigenvalue on the standard sphere is $2/r_0^2 = \rho$. Thus $\operatorname{tr} \mathring{h} = 0$.

From the uniformization theorem, any two riemannian metrics on the manifold S^2 are conformally equivalent; i.e. if $\bar{g}, g \in \mathscr{M}$, then $\bar{g} = \varphi^*(pg)$ for some $\varphi \in \mathscr{D}$ and $p \in C^\infty(M; \mathbf{R})$, $p > 0$. Thus if $h \in T_g\mathscr{M}$, and $g(\lambda)$ is a curve tangent to h, $g(\lambda) = \varphi_\lambda^*(p(\lambda)\, g)$, $\varphi_0 = \mathrm{id}_M$, $p(0) = 1$. Thus any $h \in T_g \mathscr{M}$ is of the form

$$h = g'(0) = \varphi_\lambda^*(p'(\lambda)g + L_Y(p(\lambda)g))|_{\lambda=0} = fg + L_Y g, \qquad f \in C^\infty(M; \mathbf{R}),$$

which we write as $h = L_Y g + (\delta Y)\, g + \frac{1}{2}(\operatorname{tr} h)g$. Note that fg is not divergence free so that $h = fg + L_Y g$ is not the canonical decomposition.

Thus if $\operatorname{tr} h = 0$ and $\delta h = 0$, $\delta h = \delta(L_Y g + (\delta Y)g) = 0$. But since $\operatorname{tr} h = 0$,

$$0 = \int Y\cdot\delta h\, d\mu \frac{1}{2}\int L_Y g\cdot h\, d\mu = \frac{1}{2}\int (L_Y g + (\delta Y)g)\cdot h\, d\mu$$
$$= \frac{1}{2}\int h\cdot h\, d\mu \Rightarrow h = 0.$$

Thus if $\operatorname{tr} \mathring{h} = 0$, then the above argument implies that $\mathring{h} = 0$. Thus a first order deformation of $R(g) = \text{constant} > 0$ must be of the form $h = L_X g_0$ (using the canonical decomposition). Thus if φ_λ is the flow of X, $\varphi_0 = \mathrm{id}_M$, the curve $g(\lambda) = \varphi_\lambda^* g_0 \in \mathscr{M}_\rho$ since $R(\varphi_\lambda^* g_0) = \varphi_\lambda^*(R(g_0)) = (R(g_0))\circ\varphi_\lambda = \rho\circ\varphi_\lambda = \rho$, and $(g(0), g'(0))$

$= (g_0, L_X g_0)$. A similiar argument can be adapted for the H^s case. □

REMARKS. 1. If $\dim M = 2$, $\rho =$ constant $\geqq 0$, $R(g) = \rho$ is linearization stable about g_0 even though $DR(g)$ is not surjective. Note, however, that this is not implied by the submanifold structure of $\mathscr{M}_\rho$; cf. Example 1 of § 0.

2. If $\dim M \geqq 3$, and g_0 is flat, then $R(g) = 0$ is not linearization stable about g_0; similarly if (M, g_0) is a standard n-sphere S^n, $R(g) = \rho =$ constant > 0 is not linearization stable about g_0; see I.4.2 and I.4.2, Remark 5. Thus the only case that remains open is whether or not $R(g) = \rho =$ constant > 0 is linearization stable about a solution g_0, where (M, g_0) is not a standard n-sphere, $n \geqq 3$ or where g_0 is Ricci-flat but not flat.

3. If $\dim M \geqq 3$, $R(g_0) = \rho =$ constant > 0, but $\rho \neq (n-1)\lambda$, where λ is a nonzero eigenvalue of Δ, then $R(g) = \rho$ is linearization stable about g_0; this follows from the proof of I.2.1, since if $f \in \ker DR(g_0)^*$, $\Delta f = (\rho/(n-1))f$, so if $\rho/(n-1) \neq \lambda$, $f = 0$ and $DR(g_0)$ is surjective.

I.3. The space of flat Riemannian metrics. In the next section we shall study the set of metrics with zero scalar curvature, $\mathscr{M}_0^s = \{g \in \mathscr{M}^s : R(g) = 0\}$. This case is singular in the sense that the scalar curvature map $R(\cdot)$ fails to be a submersion at those g with $\mathrm{Ric}(g) = 0$. Thus $\mathscr{M}_0^s$ *may* not be a manifold at these points. This difficulty is investigated in I.5 and **[18]**.

For $\lambda \in \boldsymbol{R}$, $s > n/2 + 1$, let $\mathscr{E}_\lambda^s = \{g \in \mathscr{M}^s : \mathrm{Ric}(g) = \lambda g\}$ denote the set of Einstein metrics with "Einstein constant" λ. Thus $\mathscr{E}_0^s = \{g \in \mathscr{M}^s : \mathrm{Ric}(g) = 0\}$ is the set of Ricci-flat metrics and is part of the singular set for $R(\cdot)$. From Theorem I.2.2, $\mathscr{M}_0^s \backslash \mathscr{E}_0^s$ is a smooth submanifold of $\mathscr{M}^s$.

We let $\mathscr{F}^s$ denote the set of H^s flat riemannian metrics on M, and $\mathscr{H}^{s-1}$ the set of flat riemannian connections. For $\Gamma \in \mathscr{H}^{s-1}$ and $\varphi \in \mathscr{D}^{s+1}$ (= the group of H^{s+1} diffeomorphisms of M), we let $\varphi^*\Gamma \in \mathscr{H}^{s-1}$ denote the pullback of the connection Γ by φ.

Let ∇ be the covariant derivative associated with Γ and in local coordinates x^i on M, define the Christoffel symbols Γ^i_{jk} as usual by

$$\nabla_{\partial/\partial x^i} \frac{\partial}{\partial x_j} = \Gamma^k_{ij} \frac{\partial}{\partial x^k}.$$

Let $\bar\Gamma^i_{jk}$ be the Christoffel symbols of the pulled back connection $\bar\Gamma = \varphi^*\Gamma$ in the local coordinates $\bar x^j$ and let φ be locally represented by $\bar x^i(x^j)$. Then pullback of the connection just corresponds to the transformation rules of the Christoffel symbols,

$$\bar\Gamma^i_{jk} = \frac{\partial \bar x^i}{\partial x^c} \frac{\partial x^a}{\partial \bar x^j} \frac{\partial x^b}{\partial \bar x^k} \Gamma^c_{ab} - \frac{\partial x^a}{\partial \bar x^j} \frac{\partial x^b}{\partial \bar x^k} \frac{\partial^2 \bar x^i}{\partial x^a \partial x^b}.$$

The following is a sort of regularity theorem for flat metrics and connections:

I.3.1. THEOREM. *Let $\Gamma \in \mathscr{H}^{s-1}$, $s > n/2 + 1$. Then there exists a $\varphi \in \mathscr{D}^{s+1}$ such that $\varphi^*\Gamma \in \mathscr{H}$; i.e., $\varphi^*\Gamma$ is a C^∞ flat (riemannian) connection. Similarly, if $g_F \in \mathscr{M}^s$, then there exists a $\varphi \in \mathscr{D}^{s+1}$ such that $\varphi^* g_F \in \mathscr{F}$, the space of C^∞ flat riemannian metrics.*

We thank Alan Weinstein for pointing out the following proof. We first prove a local version:

LEMMA. *Let $\Gamma \in \mathscr{H}^{s-1}$. Then the coordinate change to normal coordinates is of class H^{s+1}. (One could use C^k spaces here as well.)*

PROOF. Let Γ^i_{jk} be the Christoffel symbols of Γ in a coordinate system x^j, and let $\bar{x}^i(x^j)$ be the coordinate change to normal coordinates so that $\bar{\Gamma}^i_{jk} = 0$. Thus from the transformation rules for the Christoffel symbols,

$$\frac{\partial^2 \bar{x}^i}{\partial x^j \, \partial x^k} = \frac{\partial \bar{x}^i}{\partial x^a} \Gamma^a_{jk}, \tag{1}$$

and the Christoffel symbols Γ^a_{jk} are of class H^{s-1}. Now we know by construction of normal coordinates from the exponential map that $\bar{x}^i(x^j)$ has the same differentiability as the Christoffel symbols (see e.g. Lang [23, p. 96]); thus $\bar{x}^i(x^j)$ is of class H^{s-1} and $\partial \bar{x}^i/\partial x^a$ is of class H^{s-2}. Thus from multiplication properties of Sobolev spaces (I.1), the right-hand side of (1) is also of class H^{s-2}. Thus $\bar{x}^i(x^k)$ is actually of class H^s, and from (1) again is actually of class H^{s+1}. □

Now we prove the first part of I.3.1. Using the exponential map we get a new differentiable structure on M in which the connection Γ is smooth. Call this manifold M_1. The identity is a map of class H^{s+1} (by the lemma) so can be H^{s+1} approximated by a C^∞ diffeomorphism $f: M \to M_1$. Pulling back Γ on M_1 by f gives a C^∞ flat connection on M which is H^s close and H^{s+1} diffeomorphic to the original connection.

The case of a flat metric can be proved the same way or can be deduced from the result for connections. □

This argument shows that if g is of class H^s and η^*g is of class H^s and η is class H^{s-1} (or just C^1) then η is automatically H^{s+1}. This result can be used to advantage in the uniqueness problem for the Einstein equations; see Fischer and Marsden **[15]**.

Although it is not known if there exist any complete nonflat Ricci-flat riemannian manifolds, it is known (Fischer and Wolf **[19]**, **[20]**) that a compact manifold cannot admit both flat and nonflat Ricci-flat riemannian metrics. This is established next in the H^s case; see **[19]**, **[20]** for other necessary conditions for a Ricci-flat metric to be flat.

I.3.2. THEOREM. *If M admits an H^s flat riemannian metric $g_F \in \mathscr{F}^s$, $s > n/2 + 1$, then every $g \in \mathscr{M}^s$ such that* $\mathrm{Ric}(g) = 0$ *is flat.*

PROOF. Let $\varphi \in \mathscr{D}^{s+1}$ be such that $\varphi^* g_F \in \mathscr{F}$. Then by one of the Bieberbach theorems [**35**, Theorem 3.3.1], there is a normal riemannian covering $\pi: T^n \to M$ of M by a flat n-dimensional torus T^n, and π is a C^∞ map.

Now suppose $g \in \mathscr{M}^s$, $\mathrm{Ric}(g) = 0$. Then the pulled back metric $\pi^* g$ on T^n is of class H^s and $\mathrm{Ric}(\pi^* g) = \pi^*(\mathrm{Ric}(g)) = 0$. But a Ricci-flat metric on T^n is flat **[4]**; indeed from Hodge's Theorem (using H^s metrics), there are n linearly independent harmonic vector fields and these are parallel since $\mathrm{Ric}(\pi^*g) = 0$ ($\bar{\Delta}X = \Delta X = 0 \Rightarrow \nabla X = 0$). Hence π^*g is flat. Since π is a local isometry, g is flat. □

Thus if $\mathscr{F}^s \neq \varnothing$, then $\mathscr{E}_0^s = \mathscr{F}^s$, so that $\mathscr{M}_0^s - \mathscr{E}_0^s = \mathscr{M}_0^s - \mathscr{F}^s$ is a smooth submanifold of $\mathscr{M}^s$.

We now show that $\mathscr{F}^s$ also is a smooth submanifold of $\mathscr{M}^s$; thus if $\mathscr{F}^s \neq \varnothing$, $\mathscr{M}_0^s = (\mathscr{M}_0^s - \mathscr{F}^s) \cup \mathscr{F}^s$ is the disjoint union of submanifolds. In I.5, we shall in fact show that $\mathscr{M}_0^s - \mathscr{F}^s$ is closed.

The space $\mathscr{F}^s$ (if not empty) has an interesting structure itself.

For $\Gamma \in \mathscr{H}^{s-1}$, let $I_\Gamma^{s+1} = \{\varphi \in \mathscr{D}^{s+1} : \varphi^*\Gamma = \Gamma\}$ denote the Lie group of affine transformations of Γ, and let $\mathscr{F}_\Gamma^s = \{g \in \mathscr{F}^s : \Gamma(g) = \Gamma\}$, the set of flat riemannian metrics whose Levi-Civita connection is Γ. Here we are letting $\Gamma(g)$ designate the Levi-Civita connection of the metric g. Thus if $g \in \mathscr{M}^s$, $\Gamma(g)$ is an H^{s-1} connection, and if $\varphi \in \mathscr{D}^{s+1}$, $\Gamma(\varphi^*g) = \varphi^*\Gamma(g)$. Thus if $\varphi \in I_\Gamma^{s+1}$ and $g \in \mathscr{F}_\Gamma^s$, $\Gamma(\varphi^*g) = \varphi^*\Gamma(g) = \varphi^*\Gamma = \Gamma$ so $\varphi^*g \in \mathscr{F}_\Gamma^s$. Thus I_Γ^{s+1} acts by pullback on $\mathscr{F}_\Gamma^s$, $A : I_\Gamma^{s+1} \times \mathscr{F}_\Gamma^s \to \mathscr{F}_\Gamma^s$ and this action is continuous.

For $g \in \mathscr{M}^s$, let $I_g^{s+1} = \{\varphi \in \mathscr{D}^{s+1} : \varphi^*g = g\}$ denote the isometry group of g, and let $\mathring{I}_g^{s+1}$ denote the connected component of the identity. Similarly, let $I_{\Gamma(g)}^{s+1}$ and $\mathring{I}_{\Gamma(g)}^{s+1}$ denote the affine group and connected component of the identity. Since M is compact, by a result of Yano **[34]** (adapted to H^s metrics), $\mathring{I}_g^{s+1} = \mathring{I}_{\Gamma(g)}^{s+1}$ so in the above action A, $\mathring{I}_\Gamma^{s+1}$ is a common normal isotropy group for all $g \in \mathscr{F}_\Gamma^s$. Thus A is not an effective action, so we let D be the discrete group $I_\Gamma^{s+1}/\mathring{I}_\Gamma^{s+1}$ and $\tilde{A} : D \times \mathscr{F}_\Gamma^s \to \mathscr{F}_\Gamma^s$ be the associated effective action by D.

Note that if $g \in \mathscr{F}_\Gamma^s$, $\varphi \in I_\Gamma^{s+1}$, $\varphi \notin I_g^{s+1}$, then $\varphi^*g \in \mathscr{F}_\Gamma^s$ as above but $\varphi^*g \neq g$. Thus φ^*g and g are distinct isometric metrics in $\mathscr{F}_\Gamma^s$. Thus $\mathscr{F}_\Gamma^s$ intersects the orbit $\mathscr{O}_g$ through g once for each coset in I_Γ^{s+1}/I_g^{s+1}.

For $\Gamma \in \mathscr{H}^{s-1}$, the homogeneous space $\mathscr{D}^{s+1}/I_\Gamma^{s+1}$ can be given the structure of a C^∞ manifold by using Theorem I.3.1 and methods of Ebin **[13]**.

I.3.3. THEOREM. *Let $\Gamma \in \mathscr{H}^{s-1}$, $s > n/2 + 1$. Then the space $\mathscr{H}^{s-1}$ of flat riemannian H^{s-1} connections is homeomorphic to the homogeneous space $\mathscr{D}^{s+1}/I_\Gamma^{s+1}$. Using the above action $\tilde{A}$, the associated homogeneous fiber bundle is*

$$\pi : \mathscr{F}^s \to \mathscr{H}^{s-1} \approx \mathscr{D}^{s+1}/I_\Gamma^{s+1}$$

where the projection $\pi(g) = \Gamma(g)$ is the Levi-Civita connection of g, and the fibers $\pi^{-1}(\Gamma) = \mathscr{F}_\Gamma^s$ are finite-dimensional manifolds. Thus $\mathscr{F}^s$ is the total space of a homogeneous fiber bundle, and, moreover, $\mathscr{F}^s$ is a smooth closed submanifold of $\mathscr{M}^s$.

PROOF. Since all C^∞ flat riemannian connections on a compact connected manifold are affinely equivalent **[35**, Theorem 3.3.1], $\mathscr{D}$ acts transitively on $\mathscr{H}$ so that $\mathscr{H} \approx \mathscr{D}/I_\Gamma$ for $\Gamma \in \mathscr{H}$. From Theorem I.3.1, if $\Gamma_1, \Gamma_2 \in \mathscr{H}^{s-1}$, there exist $\varphi_1, \varphi_2 \in \mathscr{D}^{s+1}$ such that $\varphi_1^*\Gamma_1, \varphi_2^*\Gamma_2 \in \mathscr{H}$. Thus from the transitivity of $\mathscr{D}$ on $\mathscr{H}$, there exists $\psi \in \mathscr{D}$ such that $\psi^*(\varphi_1^*\Gamma_1) = \varphi_2^*\Gamma_2$ so that $\Gamma_2 = (\varphi_1 \circ \psi \circ \varphi_2^{-1})^*\Gamma_1$, $\varphi_1 \circ \psi \circ \varphi_2^{-1} \in \mathscr{D}^{s+1}$. Thus $\mathscr{D}^{s+1}$ acts transitively on $\mathscr{H}^{s-1}$.

Let Ψ_x be the linear holonomy group of Γ at $x \in M$, a finite group [35, Corollary 3.4.7]. Then $\mathscr{F}^s_\Gamma$ is in one-to-one correspondence with the Ψ_x-invariant inner products on T_xM (this is proved as in [35, Theorem 3.4.5]). By exponentiation, the Ψ_x-invariant inner products on T_xM are diffeomorphic to the Ψ_x-invariant symmetric bilinear forms on T_xM, a finite-dimensional linear space so that $\mathscr{F}^s_\Gamma$ is a finite-dimensional manifold.

Let $\pi : \mathscr{F}^s \to \mathscr{H}^{s-1}$ map each g to its Levi-Civita connection $\Gamma(g) = \pi(g)$. Thus for $\Gamma \in \mathscr{H}^{s-1}$, $\pi^{-1}(\Gamma) = \mathscr{F}^s_\Gamma$. Using the action $\tilde{A}: D \times \mathscr{F}^s_\Gamma \to \mathscr{F}^s_\Gamma$ on each fiber $\mathscr{F}^s_\Gamma = \pi^{-1}(\Gamma)$, $\pi : \mathscr{F}^s \to \mathscr{H}^{s-1}$ can be given the structure of an associated homogeneous fiber bundle over the homogeneous space $\mathscr{H}^{s-1} \approx \mathscr{D}^{s+1}/I^{s+1}_\Gamma$, where the "twisting" of the bundle is given by $\tilde{A}$.

Using I.3.1 as before we get a C^∞ structure for $\mathscr{F}^s$. The map injecting $\mathscr{F}^s$ to $\mathscr{M}^s$ can be seen to be a smooth immersion with closed range so $\mathscr{F}^s$ is a closed C^∞ submanifold of $\mathscr{M}^s$. See [18] for details. □

REMARK. The set of isometry classes of flat riemannian metrics on M is in bijective correspondence with $\mathscr{F}^s/\mathscr{D}^{s+1} = \mathscr{F}^s_\Gamma/I^{s+1}_\Gamma = \mathscr{F}^s_\Gamma/D$. Although $\mathscr{F}^s_\Gamma$ is a finite-dimensional manifold, unfortunately the quotient space $\mathscr{F}^s_\Gamma/D$ is not a manifold, as $\tilde{A} : D \times \mathscr{F}^s_\Gamma \to \mathscr{F}^s_\Gamma$ does not act freely. In [36], Wolf describes the set $\mathscr{F}^s_\Gamma/D$ explicitly, as a double coset space. For example, if Γ is a flat riemannian connection on

$$M = T^n, \quad \mathscr{F}_\Gamma = O(n)\backslash \mathrm{GL}(n;\boldsymbol{R}), \quad I_\Gamma = T^n \cdot \mathrm{GL}(n, \boldsymbol{Z}),$$
$$D = I_\Gamma/\mathring{I}_\Gamma = \mathrm{GL}(n, \boldsymbol{Z}) \quad \text{and} \quad \mathscr{F}_\Gamma/D = O(n)\backslash \mathrm{GL}(n;\, R)/\mathrm{GL}(n;\, \boldsymbol{Z}).$$

Note that $\mathscr{F}_\Gamma = O(n)\backslash\mathrm{GL}(n;\boldsymbol{R})$ is isomorphic to $\mathrm{Pos}(n;\boldsymbol{R})$, the space of inner products on $\boldsymbol{R}^n$, since the linear holonomy group of T^n is the identity.

We can actually compute explicitly the tangent spaces of $\mathscr{F}^s$ and $\mathscr{F}^s_\Gamma$. First we remark that the splitting $S^s_2 = \mathring{S}^s_2 \oplus \alpha_g(\mathscr{X}^{s+1})$, $\mathring{S}^s_2 = \ker \delta_g$ for $g \in \mathscr{M}^{s+1}$ is valid for $g \in \mathscr{F}^s$ as was mentioned in I.1. and the proof of I.2.3. For by I.3.1, let $\varphi \in \mathscr{D}^{s+1}$ be such that $\bar{g} = \varphi^* g \in \mathscr{F}$. Let

$$S^s_2 = \ker \delta_{\bar{g}} \oplus \alpha_{\bar{g}}(\mathscr{X}^{s+1})$$

be the splitting of S^s_2 with respect to $\bar{g}$. This provides us with a splitting of S^s_2 with respect to g as follows. Suppose $h \in S^s_2$. Then $\bar{h} = \varphi^* h \in S^s_2 = \ker \delta_{\bar{g}} \oplus \alpha_{\bar{g}}(\mathscr{X}^{s+1})$; thus

$$\varphi^* h = \bar{h} = \bar{h}^\circ + L_X \bar{g} \quad \text{where } \delta_{\bar{g}}(\bar{h}^\circ) = 0.$$

Thus $(\varphi^{-1})^*\bar{h} = h = (\varphi^{-1})^*\bar{h}^\circ + (\varphi^{-1})^* L_X\bar{g} = \mathring{h} + L_{(\varphi)_* X}\, g$ where $\mathring{h} = (\varphi^{-1})^*(\bar{h})^\circ \in \mathring{S}^s_2$ since by pulling back $\delta_g(\mathring{h}) = 0$, $(\varphi^{-1})^*\, \delta_{\bar{g}}(\mathring{h}) = \delta_g(\varphi^{-1})^*(\bar{h}^\circ) = 0$. Note that even though the pushed forward vector field $(\varphi)_* X$ is only H^s (so $L_{\varphi_* X} g$ might be H^{s-1}), $L_{\varphi_* X}\, g = (\varphi^{-1})^* L_X\, \bar{g}$ is actually H^s.

Now we compute the tangent spaces of $\mathscr{F}^s$ and $\mathscr{F}^s_\Gamma$.

I.3.4. THEOREM. *For $g \in \mathscr{F}^s$, $s > (n/2) + 1$, let $\Gamma(g) = \Gamma$ denote the Levi-Civita connection of g. Then $T_g \mathscr{F}^s_\Gamma = \{h \in S^s_2 : \nabla h = 0\} = S^{s\|}_2$, the parallel symmetric 2-tensors, and $T_g \mathscr{F}^s = \{h : \nabla \mathring{h} = 0\} = S^{s\|}_2 \oplus \alpha_g(\mathscr{X}^{s+1})$, where $\mathring{h}$ is the divergence free piece of h.*

PROOF. To compute the tangent space of $\mathscr{F}^s_\Gamma$, we differentiate the condition $\Gamma(g(\lambda)) = \Gamma$ for curves $g(\lambda) \in \mathscr{F}^s_\Gamma$. This yields

$$\frac{d}{d\lambda}(\Gamma(g(\lambda)))\Big|_{\lambda=0} = D\Gamma(g)\cdot h = 0$$

where $h = dg(0)/d\lambda \in S^s_2$. In local coordinates, this condition is

$$\tfrac{1}{2}(h^i{}_{j|k} + h^i{}_{k|j} - h_{jk}{}^{|i}) = 0.$$

Taking the trace of this condition with respect to ik yields $d \operatorname{tr} h = 0$, and with respect to ij yields $\delta h + \frac{1}{2} d \operatorname{tr} h = 0$, and thus $\delta h = 0$.

Taking another covariant derivative, contracting with i, commuting covariant derivatives (since g is flat), and using $\delta h = 0$, gives

$$0 = h^i{}_{j|k|i} + h^i{}_{k|j|i} - h_{jk}{}^{|i}{}_{|i} = h^i{}_{j|i|k} + h^i{}_{k|i|j} - h_{jk}{}^{|i}{}_{|i} = - h_{jk}{}^{|i}{}_{|i} = \bar{\Delta} h$$

where $\bar{\Delta} h = - \nabla^i \nabla_i h$ is the rough Laplacian. However on a compact manifold $\bar{\Delta} h = 0$ implies that $\nabla h = 0$ (since $0 = \int h \cdot \bar{\Delta} h \, d\mu_g = \int (\nabla h)^2 \, d\mu_g \Rightarrow \nabla h = 0$).

To find the tangent space of $T_g \mathscr{F}^s$, let $g(\lambda) \in \mathscr{F}^s$ be a smooth curve through g with tangent h, and so in particular $\operatorname{Ric}(g(\lambda)) = 0$. Differentiating and using the variational equations for the Ricci tensor in **[25]** gives

$$\begin{aligned} 0 = \frac{d}{d\lambda} \operatorname{Ric}(g(\lambda))\Big|_{\lambda=0} &= D \operatorname{Ric}(g(\lambda))\cdot h(\lambda)\Big|_{\lambda=0} \\ &= D \operatorname{Ric}(g)\cdot h = \tfrac{1}{2}(\Delta_L h - 2\delta^*\delta h - \operatorname{Hess} \operatorname{tr} h) \end{aligned}$$

where $\Delta_L h = \bar{\Delta} h + R_{ik} h^k{}_j + R_{jk} h^k{}_i - 2R_i{}^a{}_j{}^b h_{ab}$ is the Lichnerowicz Laplacian **[25]** acting on symmetric 2-tensors, and δ^* is the adjoint of δ.

Let $h = \mathring{h} + L_X g$. Then

$$\begin{aligned} D \operatorname{Ric}(g)\cdot h = D \operatorname{Ric}(g)\cdot(\mathring{h} + L_X g) &= D\operatorname{Ric}(g)\cdot \mathring{h} + L_X \operatorname{Ric}(g) \\ &= D \operatorname{Ric}(g)\cdot \mathring{h} = \tfrac{1}{2}(\Delta_L \mathring{h} - \operatorname{Hess} \operatorname{tr} \mathring{h}) = 0 \end{aligned}$$

since $\operatorname{Ric}(g) = 0$ and hence $L_X \operatorname{Ric}(g) = 0$. To see that $D \operatorname{Ric}(g)\cdot L_X g = L_X \operatorname{Ric}(g)$, let ψ_t be the flow of X and differentiate the identity $\operatorname{Ric}(\psi_\lambda^* g) = \psi_\lambda^* \operatorname{Ric}(g)$ to get

$$\frac{d}{d\lambda} \operatorname{Ric}(\psi_\lambda^* g)\Big|_{\lambda=0} = D \operatorname{Ric}(\psi_\lambda^* g)\cdot \frac{d}{d\lambda}(\psi_\lambda^* g)\Big|_{\lambda=0} = D \operatorname{Ric}(g)\cdot L_X g$$

for the left-hand side and

$$\frac{d}{d\lambda} \psi_\lambda^* \operatorname{Ric}(g) = L_X \operatorname{Ric}(g)$$

for the right-hand side.

Taking the trace of $\Delta_L \mathring{h} -$ Hess tr $\mathring{h} = 0$ gives $\Delta \operatorname{tr} \mathring{h} = 0$ (since $\operatorname{tr}(\Delta_L \mathring{h}) = \Delta \operatorname{tr} \mathring{h}$) so that $\operatorname{tr} \mathring{h} =$ constant. Thus $\Delta_L \mathring{h} = 0$ and since g is flat, $\Delta_L \mathring{h} = \bar{\Delta} \mathring{h}$ and hence $\nabla \mathring{h} = 0$. □

It is interesting that a variation of the Ricci equations around a flat metric is already enough to imply that $\nabla \mathring{h} = 0$. This fact, discovered by Berger [2], is essentially the first order version of I.3.2; if $\mathscr{F}^s \neq \varnothing$, then $\mathscr{F}^s = \mathscr{E}_0^s$.

Note that $\mathscr{F}^s$ is an invariant subset under the action of $\mathscr{D}^{s+1}$ (i.e., $\varphi^* g_F \in \mathscr{F}^s$), whereas $\mathscr{F}_\Gamma^s$ is not, since if $g \in \mathscr{F}_\Gamma^s$, $\varphi^* g \in \mathscr{F}_{\varphi^*\Gamma}^s$. That $T_g \mathscr{F}^s$ contains $\alpha_g(\mathscr{X}^s)$ whereas $T_g \mathscr{F}_\Gamma^s$ does not is the infinitesimal version of this remark.

We also remark that since $T_g \mathscr{F}_\Gamma^s = S_2^{s+1} \subset \mathring{S}_2^{s+1}$, $\mathscr{F}_\Gamma^s$ is, in a neighborhood of g, orthogonal in the L_2 metric on $\mathscr{M}^s$ (see [13]) to the orbit $\mathscr{O}_g$ through g. Nonetheless, we have seen (after I.3.2) that $\mathscr{F}_\Gamma^s$ bends back to intersect $\mathscr{O}_g$ at metrics $\varphi^* g \in \mathscr{O}_g$, $\varphi \in I_\Gamma^{s+1}$, $\varphi \notin I_g^{s+1}$.

In I.5 we will need the following result regarding the volume elements of flat metrics:

I.3.5. THEOREM. *Let $g, \bar{g} \in \mathscr{F}_\Gamma^s$, $s > n/2 + 1$, and let $d\mu_g$ and $d\mu_{\bar{g}}$ be the volume elements of g and $\bar{g}$, respectively. Then there exists a constant $c > 0$ such that $d\mu_g = c\, d\mu_{\bar{g}}$.*

PROOF. Let (U, ψ_U) be a chart on M such that the Christoffel symbols of $\Gamma = \Gamma(g) = \Gamma(\bar{g})$ are $\Gamma^i_{jk} = 0$. Thus $\partial g_{ij}/\partial x^k = 0$ and $\partial \bar{g}_{ij}/\partial x^k = 0$ on U so that $(g \restriction U)_{ij} = c_{ij}$ and $(\bar{g} \restriction U)_{ij} = \bar{c}_{ij}$ are matrices of constants. In this coordinate chart $d\mu_g \restriction U = (\det \bar{c}_{ij})^{1/2}\, dx^1 \wedge \cdots \wedge dx^n$ and $d\mu_{\bar{g}} \restriction U = (\det \bar{c}_{ij})^{1/2}\, dx^1 \wedge \cdots \wedge dx^n$, so that $d\mu_g \restriction U = c_U (d\mu_{\bar{g}} \restriction U)$, = where $c_U = (\det \bar{c}_{ij})^{1/2}/(\det \bar{c}_{ij})^{1/2}$ is constant on U.

Now suppose (V, ψ_V) is another coordinate chart for which $U \cap V \neq \varnothing$. Then $d\mu_g \restriction V = c_V (d\mu_{\bar{g}} \restriction V)$ and on the intersection $U \cap V$, $d\mu_g \restriction U \cap V = c_U (d\mu_{\bar{g}} \restriction U \cap V) = c_V (d\mu_{\bar{g}} \restriction U \cap V)$, so that $c_U = c_V$. Thus there exists a global $c =$ constant > 0 such that $d\mu_g = c\, d\mu_{\bar{g}}$. □

I.4. Linearization instability around Ricci-flat metrics. If $g \in \mathscr{M}^s$ is Ricci-flat, then the map $R(\cdot) \to H^{s-2}$ is singular at g in the sense that $DR(g) : S_2^s \to H^{s-2}$ is not surjective. The failure of $DR(g)$ to be surjective suggests the equation $R(g) = 0$ is linearization unstable in a neighborhood of a Ricci-flat solution for $\dim M \geqq 3$; i.e. there will be first order deformations $h \in S_2^s$ that will not be tangent to any curve $g(\lambda)$ of exact solutions of $R(g) = 0$. In this section we analyze the structure of the map $R(\cdot)$ at Ricci-flat (and flat) metrics, and work out the extra condition that a first order deformation must satisfy in order that it be tangent to a curve of exact solutions of $R(g) = 0$.

Our main computation is contained in the following:

I.4.1. THEOREM. *Let $g \in \mathscr{M}^s$, $s > n/2 + 1$, $\mathrm{Ric}(g) = 0$. Then for $h \in T_g \mathscr{M}^s = S_2^s$,*

$$\int D^2 R(g) \cdot (h, h)\, d\mu_g = -\frac{1}{2} \int (h \cdot \Delta_L h)\, d\mu_g - \frac{1}{2} \int (d \operatorname{tr} h)^2\, d\mu_g + \int (\delta h)^2\, d\mu_g$$

where $\Delta_L h = \bar{\Delta} h - 2R_i{}^a{}_j{}^b h_{ab}$ is the Lichnerowicz Laplacian [25] for a Ricci-flat space.

If $g \in \mathscr{M}^{s+1}$ and $h = \mathring{h} + L_X g$, $\mathring{h} \in \mathring{S}_2^s$, $X \in \mathscr{X}^{s+1}$ is the canonical decomposition of h, then

$$\int D^2R(g)\cdot(h,h)\,d\mu_g = -\frac{1}{2}\int \mathring{h}\cdot\Delta_L\mathring{h}\,d\mu_g - \frac{1}{2}\int (d\operatorname{tr}\mathring{h})^2\,d\mu_g$$
$$+\,2\int (d\operatorname{tr}\mathring{h})\cdot(d\delta X)\,d\mu_g.$$

PROOF.

$$\begin{aligned}\int D^2R(g)\cdot(h,h)\,d\mu_g &= D\left(\int DR(g)\cdot h\,d\mu_g\right)\cdot h - \int (DR(g)\cdot h)(D(d\mu_g)\cdot h)\\ &= D\left(\int(\Delta\operatorname{tr}h + \delta\delta h - h\cdot\operatorname{Ric}(g))\,d\mu_g\right)\cdot h\\ &\quad - \int (DR(g)\cdot h)\,(D(d\mu_g)\cdot h)\\ &= -\,D\left(\int h\cdot\operatorname{Ric}(g)\,d\mu_g\right)\cdot h - \int(DR(g)\cdot h)(D(d\mu_g)\cdot h)\end{aligned}$$

since $\int(\Delta\operatorname{tr}h + \delta\delta h)\,d\mu_g \equiv 0$ for all (g, h) by Stokes' theorem. Since $\operatorname{Ric}(g) = 0$, all contributions due to the metric terms in the pointwise contraction $h\cdot\operatorname{Ric}(g) = g^{ab}g^{cd}h_{ac}R_{bd}$ are zero and so we have

$$\begin{aligned}\int D^2R(g)\cdot(h,h)\,d\mu_g &= -\int h\cdot(D\operatorname{Ric}(g)\cdot h)\,d\mu_g - \int(\Delta\operatorname{tr}h + \delta\delta h)\frac{1}{2}\operatorname{tr}h\,d\mu_g\\ &= -\frac{1}{2}\int h\cdot(\Delta_L h - 2\delta^*\delta h - \operatorname{Hess}\operatorname{tr}h)\,d\mu_g\\ &\quad -\frac{1}{2}\int(\Delta\operatorname{tr}h + \delta\delta h)\operatorname{tr}h\,d\mu_g\\ &= -\frac{1}{2}\int h\cdot\Delta_L h\,d\mu_g + \int(\delta h)^2\,d\mu_g + \frac{1}{2}\int\delta h\cdot d\operatorname{tr}h\,d\mu_g\\ &\quad -\frac{1}{2}\int(d\operatorname{tr}h)^2\,d\mu_g - \frac{1}{2}\int\delta h\cdot d\operatorname{tr}h\,d\mu_g\\ &= -\frac{1}{2}\int h\cdot\Delta_L h\,d\mu_g - \frac{1}{2}\int(d\operatorname{tr}h)^2\,d\mu_g + \int(\delta h)^2\,d\mu_g\end{aligned}$$

where we have used $D(d\mu_g)\cdot h = \frac{1}{2}(\operatorname{tr}h)\,d\mu_g$ and have integrated several times by parts.

Now suppose $g \in \mathscr{M}^{s+1}$, $\operatorname{Ric}(g) = 0$, and $h = \mathring{h} + L_Xg$. Then from the proof of I.3.4, $D\operatorname{Ric}(g)\cdot h = D\operatorname{Ric}(g)\cdot\mathring{h} = \frac{1}{2}(\Delta_L h - \operatorname{Hess}\operatorname{tr}\mathring{h})$, and similiarly $DR(g)\cdot h = DR(g)\cdot\mathring{h} = \Delta\operatorname{tr}\mathring{h}$. Thus from (1) above,

$$\begin{aligned}\int D^2R(g)\cdot(h,h)\,d\mu_g &= -\frac{1}{2}\int h\cdot(\Delta_L\mathring{h} - \operatorname{Hess}\operatorname{tr}\mathring{h})\,d\mu_g - \frac{1}{2}\int(\Delta\operatorname{tr}\mathring{h})(\operatorname{tr}h)\,d\mu_g\\ &= -\frac{1}{2}\int\mathring{h}\cdot\Delta_L\mathring{h}\,d\mu_g + \frac{1}{2}\int(\delta h - d\operatorname{tr}h)\cdot d\operatorname{tr}\mathring{h}\,d\mu_g\end{aligned}$$

where we have used the fact that for Einstein spaces $\delta\circ\Delta_L = \Delta\circ\delta$ (see [25]), so that $\Delta_L\mathring{h} \in \mathring{S}_2^{s-2}$, and so by orthogonality of $\mathring{S}_2^{s-2}$ and $\alpha_g(\mathscr{X}^{s+1})$, $\int h\cdot\Delta_L\mathring{h}\,d\mu_g = \int\mathring{h}\cdot\Delta_L\mathring{h}\,d\mu_g$.

Since $\mathring{h} = h + L_Xg$, $\delta h = \delta L_Xg = \Delta X + (d\delta X)^\# = (2d\delta X + \delta dX^\flat)^\#$ and $\operatorname{tr}h = \operatorname{tr}\mathring{h} - 2\delta X$. Thus $(\delta h)^\# - d\operatorname{tr}h = 4\,d\delta X + \delta dX^\flat - d\operatorname{tr}\mathring{h}$ and

$$\frac{1}{2}\int((\delta h)^{\#} - d\operatorname{tr} h)\cdot d\operatorname{tr}\mathring{h}\, d\mu_g = \frac{1}{2}\int(4d\delta X + \delta dX^{\flat} - d\operatorname{tr}\mathring{h})\cdot d\operatorname{tr}\mathring{h}\, d\mu_g$$
$$= -\frac{1}{2}\int(d\operatorname{tr}\mathring{h})^2\, d\mu_g + 2\int d\delta X\cdot d\operatorname{tr}\mathring{h}\, d\mu_g$$

since $\delta\, dX^{\flat}$ and $d\operatorname{tr} h$ are L_2 -orthogonal. □

The extra condition that a first order deformation of $R(g) = 0$ about a Ricci-flat solution must satisfy for it to be tangent to a curve of exact solutions is now easily computed.

I.4.2. THEOREM. *Let $g(\lambda) \in \mathcal{M}^s$, $\lambda \in (-\delta, \delta)$, $\delta > 0$, $s > n/2 + 1$ be a C^2 curve with $(g(0), g'(0)) = (g, h)$. Suppose $g \in \mathcal{M}^{s+1}$, and let $h = \mathring{h} + L_X g$ be the canonical decomposition of h. If $R(g(\lambda)) = 0$ and $\operatorname{Ric}(g) = 0$, then $\operatorname{tr}\mathring{h}$ = constant and*

$$\int \mathring{h}\cdot \Delta_L \mathring{h}\, d\mu_g = 0.$$

If $g \in \mathcal{F}^s$, then $\nabla\mathring{h} = 0$.

PROOF. Differentiating $R(g(\lambda)) = 0$ twice and evaluating at $\lambda = 0$ gives

(1) $$\frac{d}{d\lambda} R(g(\lambda))\Big|_{\lambda=0} = DR(g(\lambda))\cdot\frac{dg}{d\lambda}(\lambda)\Big|_{\lambda=0} = DR(g)\cdot h$$
$$= \Delta\operatorname{tr} h + \delta\delta h = \Delta\operatorname{tr}\mathring{h} = 0,$$

(2) $$\frac{d^2R}{d\lambda^2}(g(\lambda))\Big|_{\lambda=0} = D^2R(g(\lambda))\cdot\Big(\frac{dg}{d\lambda}(\lambda), \frac{dg}{d\lambda}(\lambda)\Big)\Big|_{\lambda=0} + DR(g(\lambda))\cdot\Big(\frac{d^2g}{d\lambda^2}\Big)(\lambda)\Big|_{\lambda=0}$$
$$= D^2R(g)\cdot(h, h) + DR(g)\cdot g''(0) = 0.$$

Here we are identifying $T\mathcal{M}^s$ with $\mathcal{M}^s \times S_2^s$, so $dg(\lambda)/d\lambda \in S_2^s$ and $d^2g(\lambda)/d\lambda^2 \in S_2^s$.

Integrating (2) over M (using the volume element $d\mu_g$) gives the extra condition

(3) $$\int D^2R(g)\cdot(h, h)\, d\mu_g = 0$$

since $\int DR(g)\cdot g''(0)\, d\mu_g = \int (\Delta \operatorname{tr}(g''(0)) + \delta\delta(g''(0)))\, d\mu_g = 0$ for all accelerations $g''(0)$.

From (1), $\operatorname{tr}\mathring{h}$ = constant, so that from I.4.1, (3) becomes

(4) $$-\frac{1}{2}\int \mathring{h}\cdot\Delta_L\mathring{h}\, d\mu_g - \frac{1}{2}\int(d\operatorname{tr}\mathring{h})^2\, d\mu_g + 2\int d\operatorname{tr}\mathring{h}\cdot d\delta X^{\flat}\, d\mu_g$$
$$= -\frac{1}{2}\int\mathring{h}\cdot\Delta_L\mathring{h}\, d\mu_g = 0.$$

If g is flat, $\Delta_L = \bar{\Delta}$, so that $0 = \int \mathring{h}\cdot\bar{\Delta}\mathring{h}\ d\mu_g = \int \mathring{h}\cdot\bar{\Delta}\mathring{h}\ d\mu_g = \int(\nabla\mathring{h})^2\ d\mu_g \Rightarrow \nabla\mathring{h} = 0$. □

REMARKS. 1. At a regular point $g \in \mathcal{M}_0^s$ where $DR(g)$ is surjective (so that $\operatorname{Ric}(g) \neq 0$), equation (2), when integrated over M, gives

$$\int D^2R(g)\cdot(h, h)\, d\mu_g - \int \operatorname{Ric}(g)\cdot g''(0)\, d\mu_g = 0,$$

an integrated condition on $g''(0)$ in terms of h, which at regular points g does not give an extra condition on h. It is only when $\mathrm{Ric}(g) = 0$ that the term involving $g''(0)$ drops out leaving an integrated extra condition on h.

2. That we get one extra condition on h when $\mathrm{Ric}(g) = 0$ corresponds to the fact that $\ker (DR(g))^* = \{\text{constant functions on } M\}$ is 1-dimensional, since as in § 0, Example 4, Remark 2, there is an extra condition for each dimension in $\ker(DR(g))^*$. In the case at hand, the equation $\int DR(g)\cdot g''(0)\, d\mu_g = 0$ (which leads to the extra condition on h) can be expressed as

$$\int (DR(g)^*1)\cdot g''(0)\, d\mu_g = 0 \text{ for all } g''(0) \text{ iff } 1 \in \ker (DR(g))^*.$$

That the extra condition on h is an integrated condition corresponds to the fact that $(DR(g))^*$ is an L_2-adjoint; i.e. we do not get an extra condition on h until we integrate (2) against an element of $\ker (DR(g))^*$; cf. Example 4, Remark 1.

3. If $g \in \mathscr{M}^s$, $\mathrm{Ric}(g) = 0$, but g is not of class H^{s+1}, h may not have a canonical decomposition. In this case, by using the first order condition $DR(g)\cdot h = \Delta \operatorname{tr} h + \delta\delta h = 0$ and equation (1) in the proof of I.4.1, the extra condition (3) can be expressed as

$$\begin{aligned}\int D^2 R(g)\cdot(h, h)\, d\mu_g &= -\int h\cdot(D\,\mathrm{Ric}(g)\cdot h)\, d\mu_g \\ &= -\frac{1}{2}\int h\cdot(\Delta_L h - 2\delta^*\delta h - \mathrm{Hess}\operatorname{tr} h)\, d\mu_g \\ &= -\frac{1}{2}\int h\cdot \Delta_L h\, d\mu_g + \frac{1}{2}\int (\delta h)^2\, d\mu_g \\ &\quad + \frac{1}{2}\int \delta h\cdot d\operatorname{tr} h\, d\mu_g = 0.\end{aligned}$$

4. Considering third and higher order derivatives of $R(g(\lambda)) = 0$, $\mathrm{Ric}\,(g(0)) = 0$, does not lead to any extra condition on the first order deformations. For example, differentiating

$$D^2R(g(\lambda))\cdot(h(\lambda), h(\lambda)) + DR(g(\lambda))\cdot g''(\lambda) = 0$$

(where $h(\lambda) = g'(\lambda)$) and evaluating at $\lambda = 0$ gives

$$D^3R(g)\cdot(h, h, h) + 3D^2R(g)\cdot(h, g''(0)) + DR(g)\cdot g'''(0) = 0. \tag{5}$$

Integrating over M, the last term again drops out, leaving

$$\int D^3 R(g)\cdot(h, h, h)\, d\mu_g + 3\int D^2R(g)\cdot(h, g''(0))\, d\mu_g = 0$$

as the extra integrated "third order" condition on $g''(0)$ (beyond that implied by the second order pointwise condition of equation (2)) that has to be satisfied for $g''(0)$ to be the acceleration of some curve of exact solutions of $R(g) = 0$. This of course is the analog of the second order phenomenon. This situation repeats, and in general there is an extra integrated condition on the nth order deformation that comes from the $(n + 1)$st order equations. However, these higher order equations do not provide any further conditions on the first order deformations in general.

In special cases, however, it is possible that third order variations can lead to extra conditions on first order deformations, as in § 0, Example 3.

5. We can also examine the linearization instability of the equation $R(g) = \rho = \text{constant} > 0$ around a solution (S^n, g_0), a standard sphere in $\boldsymbol{R}^{n+1}$ of radius $r_0 = (n(n-1)/\rho)^{1/2}$. In this case, from Remark 2 of I.2.1, $\ker(DR(g_0))^* = \{$eigenfunctions of $\Delta_{g_0}\}$ so that if $f \in \ker(DR(g_0))^*$,

$$\int f\, DR(g_0)\cdot g''(0)\, d\mu_g = \int (DR(g_0)^* f)\cdot g''(0)\, d\mu_g = 0$$

for all $g''(0)$. Thus multiplying (2) by f and integrating over M gives for each linearly independent eigenfunction f of Δ_{g_0} the extra condition

$$\int f D^2 R(g_0)\cdot(h, h)\, d\mu_{g_0} = 0$$

on a first order deformation h. That this really is extra is shown in **[18]**.

In the case that $g_F \in \mathscr{F}^s$, the integrated extra condition $\int \mathring{h}\cdot\bar{\Delta}\,\mathring{h}\, d\mu_{g_F} = 0$ can be converted to the very strong pointwise condition $\nabla\mathring{h} = 0$. This pointwise condition signals an even greater type of instability, viz., that the flat solutions of $R(g) = 0$ are isolated among all the solutions. This aspect of the map $R(\cdot)$ will be examined in the next section.

In the flat space case, if the extra condition $\nabla\mathring{h} = 0$ is satisfied, then we can explicitly integrate up any deformation $h = h_\| + L_X g_F$, $\nabla h_\| = 0$. Indeed, let $g_F^{-1}\cdot h_\|$ denote the 1-contravariant 1-covariant form of $h_\|$, let $\exp(g_F^{-1}\cdot h_\|)$ denote the pointwise exponentiation of $g_F^{-1}\cdot h_\|$, another tensor of type $\binom{1}{1}$, and let $g(\lambda) = g_F \exp(\lambda g_F^{-1}\cdot h_\|)$ denote the 2-contravariant form. In coordinates, $g_F \exp(\lambda g_F^{-1}\cdot h_\|) = (g_F)_{ik} \exp(\lambda\, g_F)^{kl}\,(h_\|)_{lj})$. Then $g(\lambda)$ is a C^∞ curve in $\mathscr{F}^s_{\Gamma(g_F)}$, defined for all $\lambda \in \boldsymbol{R}$, such that $(g(0), g'(0)) = (g_F, h_\|)$. If $\varphi_\lambda \in \mathscr{D}$, $\varphi_0 = \mathrm{id}_M$, is the flow of the vector field X, then $\bar{g}(\lambda) = \varphi_\lambda^*(g(\lambda)) \in \mathscr{F}^s$, $\bar{g}(0) = g_F$, and $\bar{g}'(0) = \varphi_\lambda^*(g'(\lambda) + L_X g(\lambda))|_{\lambda=0} = h_\| + L_X g_F$.

In the next section we shall see that $g(\lambda) \in \mathscr{F}^s$ is quite necessary if $g(\lambda) \in \mathscr{M}_0^s$ and $g(0)$ is flat.

If $\mathscr{F}^s = \varnothing$, and if there exist nonflat Ricci-flat metrics g_0 (so that $\dim M \geqq 4$), then we do not know if satisfaction of the extra condition $\int \mathring{h}\cdot\Delta_L \mathring{h}\, d\mu_{g_0} = 0$ is sufficient to find a curve $g(\lambda) \in \mathscr{M}_0^s$, $g(0) = g_0$, $g'(0) = h$. However, because third and higher order deformations of $R(g) = 0$ do not lead to any new conditions on h, we suspect that if the second order condition on h is satisfied, then there is a curve $g(\lambda) \in \mathscr{M}_0^s$ which is tangent to h.

One of the difficulties here is that the structure of the set $\mathscr{E}_0^s = \{g \in \mathscr{M}^s : \mathrm{Ric}(g) = 0\}$, if not empty, is unknown. In particular, we do not know if it is a manifold. The formal tangent space of $\mathscr{E}_0^s$ at $g \in \mathscr{M}^{s+1}$ is, from the proof of I.3.4, given by $\ker(D\ \mathrm{Ric}(g)) = \{h \in S_2^s : \Delta_L h = 0\} = \{$harmonic tensors of $\Delta_L\}$. Thus if $\mathscr{E}_0^s$ were a manifold, any h such that $\Delta_L h = 0$ is tangent to a curve in $\mathscr{E}_0^s$. This would partially answer the question of whether the extra condition of I.4.2 on a deformation h is sufficient to find a curve in $\mathscr{M}_0^s$ tangent to h.

We now give an example of the linearization instability of the equation $R(g) = 0$ on a flat 3-torus. Let Λ be the lattice generated by the standard basis $\{\hat{e}_i\}$ in $\boldsymbol{R}^3$, let $T^3 = \boldsymbol{R}^3/\Lambda$, and let g_F be the metric induced on T^3 from $\boldsymbol{R}^3$. Let S^1 be the circle with unit circumference, and let $f\colon S^1 \to \boldsymbol{R}$ be any smooth function, $f \neq$ constant. Set

$$h = h_{ij} = \begin{pmatrix} 0 & f(x_3) & 0 \\ f(x_3) & 0 & 0 \\ 0 & 0 & 0 \end{pmatrix}$$

Then $\delta h = 0$ and $\operatorname{tr} h = 0$ (taken with respect to g_F), but $\nabla h \neq 0$. Thus this h, although a solution to the linearized equations $DR(g_F)\cdot h = \Delta \operatorname{tr} h + \delta\delta h = 0$, is *not* tangent to any curve $g(\lambda) \in \mathscr{M}_0^s$, $g(0) = g_F$, since $\nabla h \neq 0$.

If $f(x_3) =$ constant, then the extra condition $\nabla h = 0$ is satisfied, and we can integrate h up to a curve

$$g(\lambda) = g_F \exp(\lambda g_F^{-1}\cdot h) = \begin{pmatrix} \cosh \lambda f & \sinh \lambda f & 0 \\ \sinh \lambda f & \cosh \lambda f & 0 \\ 0 & 0 & 1 \end{pmatrix}$$

of flat metrics on T^3 with $(g(0), g'(0)) = (g_F, h)$.

REMARKS. 1. The above expression for $g(\lambda)$ is also valid if $f \neq$ constant, but then $R(g(\lambda))$ need not be zero for $\lambda \neq 0$.

2. On a flat 2-torus T^2, $\delta h = 0$ and $\operatorname{tr} h =$constant imply $\nabla h = 0$ so that we cannot construct our example. This corresponds to the fact that $R(g) = 0$ is linearization stable if $\dim M = 2$.

I.5. Isolated solutions of $R(g) = 0$. In the flat space case, the emergence of a pointwise condition from the integrated extra condition signals that the flat solutions of $R(g) = 0$ may be isolated solutions. In this section we show that this expectation is correct. We then use this result to work out the structure of $\mathscr{M}_0$ in the case that $\mathscr{F} \neq \varnothing$.

That $R(g) = 0$ implies that g is flat if g is in a neighborhood of a flat metric is somewhat surprising in view of the fact that the scalar curvature is a relatively weak measure of the curvature.

Fix an H^s volume element $d\mu$ on M, and define

$$\Psi\colon \mathscr{M}^s \to \boldsymbol{R}, \qquad g \mapsto \int R(g)\, d\mu.$$

Note that Ψ is not the usual integrated scalar curvature (cf. II.2), since in general $d\mu \neq d\mu_g$.

I.5.1. THEOREM. *A metric $g \in \mathscr{M}^s$ is a critical point of Ψ if and only if* $\operatorname{Ric}(g) =$ 0 *and $d\mu = c\, d\mu_g$ for some constant $c > 0$. At a critical metric $g_c \in \mathscr{M}^s$, the Hessian $d^2\Psi(g_c)\colon S_2^s \times S_2^s \to \boldsymbol{R}$ of Ψ is given by*

$$d^2 \Psi(g_c) \cdot (h, h) = -\frac{1}{2} \int h \cdot \Delta_L h \, c \, d\mu_{g_c} - \frac{1}{2} \int (d \operatorname{tr} h)^2 c \, d\mu_{g_c} + \int (\delta h)^2 c \, d\mu_{g_c}.$$

PROOF. For $g \in \mathscr{M}^s$, let $p \in H^s(M; \boldsymbol{R})$, $p > 0$ be such that $d\mu = p \, d\mu_g$. Then

$$\begin{aligned} d\Psi(g) \cdot h &= \int DR(g) \cdot h \, d\mu = \int (\Delta \operatorname{tr} h + \delta\delta h - h \cdot \operatorname{Ric}(g)) p \, d\mu_g \\ &= \int (g\Delta p + \operatorname{Hess} p - p \operatorname{Ric}(g)) \cdot h \, d\mu_g = 0 \end{aligned}$$

for all $h \in S_2^s$ iff

(1) $$\gamma_g^*(p) = g\Delta p + \operatorname{Hess} p - p \operatorname{Ric}(g) = 0, \qquad p > 0.$$

From the proof of I.2.1, (1) $\Rightarrow R(g) = \text{constant} \geqq 0$. Contracting (1) gives

(2) $$(n-1) \Delta p = R(g) p, \qquad p > 0;$$

integrating (2) over M gives

$$0 = (n-1) \int \Delta p \, d\mu_g = R(g) \int p \, d\mu_g.$$

If $R(g) = \text{constant} > 0$, then $\int p \, d\mu_g = 0$, contradicting $p > 0$. Thus $R(g) = 0$, so from (2), $p = \text{constant} = c > 0$, so that $d\mu = c \, d\mu_g$. Since $p = \text{constant} > 0$, from (1), $\operatorname{Ric}(g) = 0$.

Now suppose g_c is a critical metric of Ψ, so $\operatorname{Ric}(g) = 0$, $d\mu = c \, d\mu_{g_c}$ Thus

$$d^2 \Psi(g_c) \cdot (h, h) = \int D^2R(g_c) \cdot (h, h) \, d\mu = c \int D^2R(g_c) \cdot (h, h) \, d\mu_{g_c}.$$

The theorem now follows from I.4.1. □

REMARKS. 1. In the above it is important that we hold the volume element fixed and then let $d\mu = c \, d\mu_g$ after we take the derivatives; see I.5.2, Remark 2.

2. That $d\mu = c \, d\mu_{g_c}$ at a critical point of Ψ allows us to compute $d^2 \Psi(g_c)$ from I.4.1. Otherwise, the computation of $d^2 \Psi(g) \cdot (h, h) = \int D^2R(g) \cdot (h, h) \, d\mu$ is considerably more complicated than the computation of $\int D^2R(g) \cdot (h, h) \, d\mu_g$ (unless $d\mu = c \, d\mu_g$), since $\int (\Delta \operatorname{tr} h + \delta\delta h) \, d\mu$ does not vanish for all (g, h) as in the proof of I.4.1.

3. Ψ need not have any critical points; e.g. if $d\mu$ is chosen so as not to be the volume element of any Ricci-flat metric.

To second order, Brill **[5]** shows that the flat solutions of $R(g) = 0$ are isolated, and Kazdan and Warner **[21**, §5] show that near a flat metric to second order there are no metrics with positive scalar curvature. The following extends these results to a full neighborhood of the flat metrics.

I.5.2. THEOREM. *Let $g_F \in \mathscr{F}^s$, $s > n/2 + 1$. Then there exists a neighborhood $U_{g_F} \subset \mathscr{M}^s$ of g_F such that if $g \in U_{g_F}$ and $R(g) \geqq 0$, then g is also in $\mathscr{F}^s$.*

PROOF. Let $d\mu_{g_F}$ denote the volume element of g_F, and let Γ denote its Levi-Civita connection. Let $\Psi : \mathscr{M}^s \to \boldsymbol{R}$, $\Psi(g) = \int R(g) \, d\mu_{g_F}$.

If $\bar{g}_F \in \mathscr{F}^s_\Gamma$, by I.3.5, $d\mu_{\bar{g}_F} = c \, d\mu_{g_F}$, $c = \text{constant} > 0$, so that $\mathscr{F}^s_\Gamma$ is a critical submanifold of Ψ (but not the entire critical submanifold), and at a critical point $\bar{g}_F \in \mathscr{F}^s_\Gamma$,

$$d^2 \Psi(\bar{g}_F) \cdot (h, h) = -\frac{1}{2}\int (\nabla h)^2 \, c\,d\mu_{\bar{g}_F} - \frac{1}{2}\int (d \operatorname{tr} h)^2 \, c\,d\mu_{\bar{g}_F} + \int (\delta h)^2 \, c\,d\mu_{\bar{g}_F}.$$

Let $S^s_{g_F}$ be a slice at g_F (see the remark preceding Theorem I.3.4) and let $\Psi_S = \Psi \upharpoonright S^s_{g_F}$. Since g_F is a critical point for Ψ, g_F is also a critical point for Ψ_S, and the Hessian of Ψ_S is given by restricting the Hessian of Ψ; thus from I.5.1., for $h \in T_{g_F} S^s_{g_F} = \{h \in S^s_2 : \delta_{g_F} h = 0\}$,

$$d^2\Psi_S(g_F) \cdot (h, h) = d^2 \Psi(g_F) \cdot (h, h) = -\frac{1}{2}\int (\nabla h)^2 \, d\mu_{g_F} - \frac{1}{2}\int (d \operatorname{tr} h)^2 \, d\mu_{g_F}.$$

Thus $d^2 \Psi_S(g_F)$ is negative-definite on a complement to $T_{g_F}\mathscr{F}^s_\Gamma = \{h \in S^s_2 : \nabla h = 0\}$ in $T_{g_F} S^s_{g_F}$. Thus since $\mathscr{F}^s_\Gamma$ is critical there exists a neighborhood $V \subset S^s_{g_F}$ of g_F such that $\Psi_S \leqq 0$ on V, and if $\Psi_S(g) = 0$, then $g \in \mathscr{F}^s_\Gamma$. Thus the critical submanifold $\mathscr{F}^s_\Gamma$ is isolated among the zeros of Ψ_S.

Let $U_{g_F} = \mathscr{D}^{s+1}(V) = \{\varphi^* g \in \mathscr{M}^s : \varphi \in \mathscr{D}^{s+1}, g \in V\}$ be the saturation of V. By the slice theorem [13], U_{g_F} fills out a neighborhood of g_F. Thus if $g \in U_{g_F}$ and $R(g) \geqq 0$, there exists a $\varphi \in \mathscr{D}^{s+1}$ such that $\varphi^* g \in V \subset S^s_{g_F}$ and thus $\Psi_S(\varphi^* g) \leqq 0$, since Ψ_S is negative on V. But $R(\varphi^* g) = R(g) \circ \varphi \geqq 0$ so that $\Psi_S(\varphi^* g) = \int R(\varphi^* g) \, d\mu_{g_F} = \int R(g) \circ \varphi \, d\mu_{g_F} \geqq 0$. Thus $\Psi_S(\varphi^* g) = 0$ so that $\varphi^* g \in \mathscr{F}^s_\Gamma$ and $g \in (\varphi^{-1})^* \mathscr{F}^s_\Gamma \subset \mathscr{F}^s$ is flat. □

REMARKS. 1. For dim $M = 2$, we need not restrict to a neighborhood U_{g_F}. Indeed, from the Gauss-Bonnet theorem, if $\mathscr{F}^s \neq \varnothing$, the Euler-Poincaré characteristic $\chi_M = 0$, so that for any metric g, $\int R(g) \, d\mu_g = 0$. Thus $R(g) \geqq 0 \Rightarrow R(g) = 0 \Rightarrow g$ is flat.

2. Usually one considers an integrated scalar curvature $\Phi(g) = \int R(g) \, d\mu_g$ with volume element $d\mu_g$ induced from g rather than $\Psi(g) = \int R(g) \, d\mu$ with fixed volume element $d\mu$, as e.g. in [3]. One then has (see II.2.2) at a critical flat metric

$$\text{(1)} \qquad d^2 \Phi(g_F) \cdot (\mathring{h}, \mathring{h}) = -\frac{1}{2}\int (\nabla \mathring{h})^2 \, d\mu_{g_F} + \frac{1}{2}\int (d \operatorname{tr} \mathring{h})^2 \, d\mu_{g_F}$$

rather than

$$\text{(2)} \qquad d^2 \Psi(g_F) \cdot (\mathring{h}, \mathring{h}) = -\frac{1}{2}\int (\nabla \mathring{h})^2 \, d\mu_{g_F} - \frac{1}{2}\int (d \operatorname{tr} \mathring{h})^2 \, d\mu_{g_F}.$$

There is now an important sign change in the second term; see also II.2.2, Remark 2. Brill [5] uses (1), together with the condition $\operatorname{tr} \mathring{h} =$ constant implied by the first order equations $DR(g) \cdot h = \Delta \operatorname{tr} \mathring{h} = 0$ to deduce $\nabla \mathring{h} = 0$ and hence his second order result.

To extend this result to a full neighborhood of the flat metrics, one may not use the first order condition $\operatorname{tr} \mathring{h} =$ constant. The indefinite sign in (1) now becomes a severe difficulty. We introduced the map Ψ with fixed volume element $d\mu$ since this difficulty is not present in (2).

Note that if the first order condition tr $\mathring{h}$ = constant is used, $d^2\Psi(g_F)\cdot(h,h) = d^2\Phi(g_F)\cdot(h,h) = -\frac{1}{2}\int(\nabla\mathring{h})^2\,d\mu_{g_F}$, so that the two treatments are then equivalent.

As a consequence of I.5.1, we have the following structure theorem for $\mathcal{M}_0^s$:

I.5.3. THEOREM. *Let* $s > n/2+1$, $\dim M \geqq 3$, *and if* $\dim M \geqq 4$, *assume* $\mathcal{F}^s \neq \varnothing$. *Then* $\mathcal{M}_0^s = (\mathcal{M}_0^s - \mathcal{F}^s) \cup \mathcal{F}^s$ *is the disjoint union of* C^∞ *closed submanifolds, and hence* $\mathcal{M}_0^s$ *is itself a* C^∞ *closed submanifold of* $\mathcal{M}^s$; *similarly,* $\mathcal{M}_0 = (\mathcal{M}_0 - \mathcal{F}) \cup \mathcal{F}$ *is a* C^∞ *closed submanifold of* $\mathcal{M}$.

Notes. 1. If $\dim M = 2$, $\mathcal{M}_0^s = \mathcal{F}^s$ is also a C^∞ closed submanifold.

2. We are allowing the possibility that $\mathcal{M}_0^s - \mathcal{F}^s$ is empty, and if $\dim M = 3$, we are also allowing the possibility that $\mathcal{F}^s$ is empty.

3. $\mathcal{M}_0^s$ is a manifold since we are allowing different components of a manifold to be modelled on different Hilbert spaces.

PROOF. If $\dim M = 3$, $\mathrm{Ric}(g) = 0 \Rightarrow g$ is flat. If $\dim M \geqq 4$ and $\mathcal{F}^s \neq \varnothing$, then from I.3.2 $\mathcal{F}^s = \mathcal{E}_0^s$. Thus in either case

$$\mathcal{M}_0^s = (\mathcal{M}_0^s - \mathcal{E}_0^s) \cup \mathcal{E}_0^s = (\mathcal{M}_0^s - \mathcal{F}^s) \cup \mathcal{F}^s$$

which from I.2.2 and I.3.2 is the disjoint union of C^∞ submanifolds, $\mathcal{F}^s$ closed.

Let $g_n \to g$ be a convergent sequence in $\mathcal{M}_0^s - \mathcal{F}^s$. Then $g \in \mathcal{M}_0^s$, and if $g \in \mathcal{F}^s$, there exists a neighborhood $U_g \subset \mathcal{M}^s$ such that $\bar{g} \in U_g$, $R(g) = 0 \Rightarrow g \in \mathcal{F}^s$. But then for n sufficiently large, $g_n \in U_g$, $R(g_n) = 0 \Rightarrow g_n \in \mathcal{F}^s$ contradicting $g_n \in \mathcal{M}_0^s - \mathcal{F}^s$. Thus $g \in \mathcal{M}_0^s - \mathcal{F}^s$, a closed set. □

REMARKS. 1. There are various topological conditions on M, $\dim M \geqq 4$, that imply a Ricci-flat metric is flat; see **[19]**, **[20]**. If we adopt any of these conditions, then we can drop the $\mathcal{F}^s \neq \varnothing$ assumption. Of course, if it turns out that Ricci-flat implies flat, then this assumption can be dropped.

2. If $\dim M \geqq 4$, $\mathcal{F}^s = \varnothing$, then $\mathcal{M}_0^s$ is still the union $\mathcal{M}_0^s = (\mathcal{M}_0^s - \mathcal{E}_0^s) \cup \mathcal{E}_0^s$, with $\mathcal{M}_0^s - \mathcal{E}_0^s$ a submanifold (I.2.2). But we do not know if $\mathcal{E}_0^s$, if nonempty, is a submanifold.

3. If $g(\lambda)$, $\lambda \in (-\delta, \delta)$, $\delta > 0$, is a continuous curve in $\mathcal{M}_0^s$, $g(0) \in \mathcal{F}^s$, then $g(\lambda) \in \mathcal{F}^s$, an interesting consequence of I.5.3.

If $g(\lambda)$ were a C^k function of λ, $1 \leqq k \leqq \infty$, then we could conclude from the method of I.4.2 that all k derivatives of $g(\lambda)$ at $\lambda = 0$ are parallel, $\nabla(g^{(k)}(0)) = 0$. Thus if $g(\lambda)$ is analytic, $g(\lambda) = \sum_{n=0}^\infty (\lambda^n/n!)g^{(n)}(0)$, and $\nabla(g^{(n)}(0)) = 0$ for all n, so that $g(\lambda) \in \mathcal{F}^s$, a conclusion from I.4.2 alone.

4. Finally we remark that although $\mathcal{M}_0^s$ is a manifold (under the hypothesis $\mathcal{F}^s \neq \varnothing$), the equation $R(g) = 0$ is not linearization stable at a flat solution, as we have seen from the example in I.4. Here the difficulty can be traced to the fact that $\mathcal{M}_0^s$ is a union of closed manifolds of different "dimensionalities", $\mathcal{F}^s$ being essentially finite-dimensional modulo the orbit directions; cf. Example 1 in §0.

II. Application to General Relativity

II.1. The general set-up and main idea. We now turn our attention to the Einstein empty-space field equations of general relativity. We apply the techniques used in the previous sections to prove that solutions to these equations are linearization stable if certain conditions are met; in certain exceptional cases, however, the solutions are not linearization stable.

Let ${}^{(4)}g$ be a smooth Lorentz metric (signature $-+++$) on a 4-manifold V. The Einstein empty-space field equations are that the Ricci tensor of ${}^{(4)}g$ vanish:

$$\text{(1)} \qquad \operatorname{Ric}({}^{(4)}g) = 0.$$

An *infinitesimal deformation* about a solution ${}^{(4)}g$ is then a solution ${}^{(4)}h \in S_2$ of the linearized equations

$$\text{(2)} \qquad D \operatorname{Ric}({}^{(4)}g) \cdot {}^{(4)}h = 0,$$

where $D \operatorname{Ric}({}^{(4)}g)$ is the derivative of the map $\operatorname{Ric}(\cdot)$ at ${}^{(4)}g$.

Assume that $(V, {}^{(4)}g)$ has a *compact* connected orientable spacelike hypersurface M so that $\dim M = 3$. Let g denote the induced riemannian metric on M and k the second fundamental form of M. Our conditions for linearization stability are as follows (see Note on p. 263):

$C_{\mathscr{H}}$: if $k = 0$, g is not flat;
C_δ : if $L_X g = 0$ and $L_X k = 0$, then $X = 0$;
C_{tr} : $\operatorname{tr} k =$ constant on M.

(The meaning of the subscripts $\mathscr{H}$ and δ will become clear.)

Our main result (II.5.1) is that if a solution $(V, {}^{(4)}g)$ of (1) has a compact spacelike hypersurface M whose induced metric g and second fundamental form k satisfy conditions $C_{\mathscr{H}}$, C_δ, and C_{tr}, then every solution ${}^{(4)}h$ of the linearized equations is tangent at ${}^{(4)}g$ to a curve ${}^{(4)}g(\lambda)$ of exact solutions of (1); i.e., there exist a tubular neighborhood V' of M and a curve ${}^{(4)}g(\lambda)$ of exact solutions of (1) on V' such that $({}^{(4)}g(0), {}^{(4)}g'(0)) = ({}^{(4)}g, {}^{(4)}h)$ on V'; see also **[16]**, **[17]**, **[39]** and **[40]**.

This conclusion asserts the linearization stability on a small piece of spacetime V' surrounding the Cauchy surface M. By standard arguments **[12]**, V' can be extended to a maximal common development of the spacetimes ${}^{(4)}g(\lambda)$, λ small, which approximates the maximal development of ${}^{(4)}g(0)$.

The case where V admits a noncompact spacelike hypersurface M is rather different. Here asymptotic conditions are necessary. For example, $k = 0$ and g the usual flat metric on $\boldsymbol{R}^3$ is not excluded. Thus the usual Minkowski metric on $\boldsymbol{R}^4$ is linearization stable in a tubular neighborhood of the hypersurface $M = \boldsymbol{R}^3$. This result was obtained independently using other methods by Choquet-Bruhat and Deser **[10]**, **[11]**. The treatment of the general noncompact case is in spirit similar, although there are certain technical difficulties associated with elliptic operators on noncompact manifolds which enter the problem in the nonflat case. We will present the noncompact case elsewhere. For the remainder of Part II, M will be compact.

It is convenient to introduce the supplementary variables (g, π) instead of (g, k), where $\pi = \pi' \otimes \mu_g \in S^2 \otimes \mu_g$ is a 2-covariant symmetric tensor density, $\pi' = ((\mathrm{tr}\, k)\, g - k)^\# \in S^2$ is the tensor part of π ($k^\#$ means the contravariant form of the tensor $k \in S_2$) and we write μ_g interchangeably with $d\mu_g$. In local coordinates

$$\pi = \pi^{ij} = ((\mathrm{tr}\, k)\, g^{ij} - k^{ij})(\det g_{ij})^{1/2}.$$

As is easy to see, $k = 0 \Leftrightarrow \pi = 0$, $\mathrm{tr}\ k = \text{constant} \Leftrightarrow \mathrm{tr}\ \pi' = \text{constant}$, and $L_X g = 0$, $L_X k = 0 \Leftrightarrow L_X g = 0$, $L_X \pi = 0$. Thus, in the conditions $C_{\mathscr{H}}$, C_δ, and C_{tr}, k can be replaced by π. Note that the divergence of X enters in the Lie derivative of a tensor density:

$$L_X \pi = (L_X \pi') \otimes \mu_g + \pi' \otimes L_X \mu_g = (L_X \pi') \otimes \mu_g + \pi' \otimes (\mathrm{div}\ X)\mu_g,$$

where $\mathrm{div}\ X = -\delta X$. In local coordinates,

$$L_X \pi = X^k \pi^{ij}{}_{|k} - \pi^{ik} X^j{}_{|k} - \pi^{jk} X^i{}_{|k} + X^k{}_{|k} \pi^{ij}.$$

As is well known every spacelike hypersurface in a Ricci-flat Lorentz manifold ([1], [14]) satisfies the constraint equations

$$\text{(C)} \qquad \begin{aligned} \mathscr{H}(g, \pi) &= \{\tfrac{1}{2}(\mathrm{tr}\ \pi')^2 - \pi' \cdot \pi' + R(g)\}\ \mu_g = 0, \\ \delta(g,\pi) &= \delta_g \pi = 0. \end{aligned}$$

In local coordinates,

$$\mathscr{H}(g, \pi) = (\det g_{ij})^{-1/2}\{\tfrac{1}{2}(g_{ab}\, \pi^{ab})^2 - \pi^{ab}\, \pi_{ab}\} + (\det g_{ij})^{1/2} R(g)$$

and

$$\delta(g, \pi) = \delta_g \pi = -\pi^{ij}{}_{|j} = 0$$

(so that δ_g now maps $S^2 \to \mathscr{X} \otimes \mu_g$).

We shall refer to $\mathscr{H}(g, \pi) = 0$ as the Hamiltonian constraint and $\delta_g \pi = 0$ as the divergence constraint.

Conversely, by means of existence of solutions to the evolution equations (see, e.g., [15]), every solution (g, π) to the constraint equations (C) generates a Ricci-flat spacetime in a tubular neighborhood of M. This spacetime is unique up to diffeomorphism of the neighborhood.

For the compact hypersurface M, we let $\mathscr{M}^s$ denote the H^s riemannian metrics on M; throughout, $s > n/2 + 1$. For $g \in \mathscr{M}^s$, we have, as in Part I, $T_g \mathscr{M}^s = S_2^s$, and we let

$$T_g^* \mathscr{M}^s = S_2^s \otimes \mu_g, \quad \text{the space of } H^s \text{ 2-covariant symmetric tensor densities,}$$

and

$$T^* \mathscr{M}^s = \bigcup_{g \in \mathscr{M}^s} T_g^* \mathscr{M}^s, \quad \text{the "cotangent" bundle of } \mathscr{M}^s.$$

Note that here we take the dual in the L_2 inner product but use only the closed subspace of such elements continuous in the H^s topology, so the dual of S_2^s is $S_s^2 \otimes \mu_g$; see also [14, p. 552].

The solutions (g, π) of the constraint equations may then be regarded as a certain subset $\mathscr{C}$ of $T^*\mathscr{M}^s$. We will, according to our general method, show that in a neighborhood of points (g, π) that satisfy conditions $C_{\mathscr{H}}$, C_δ and C_{tr}, $\mathscr{C}$ is a smooth submanifold of $T^*\mathscr{M}^s$ and that if $(h, \omega) \in T_{(g,\pi)}(T^*\mathscr{M}^s) \approx S_2^s \times (S_s^2 \otimes \mu_g)$ is tangent to $\mathscr{C}$, i.e., if (h, ω) is a solution to the linearized constraint equations, then there exist a $\delta > 0$ and a smooth curve $(g(\lambda), \pi(\lambda)) \in \mathscr{C}$, $-\delta < \lambda < \delta$, which is tangent to (h, ω) at (g, π).

Now suppose ${}^{(4)}h$ is a solution to the linearized equations, $D\,\mathrm{Ric}({}^{(4)}g) \cdot {}^{(4)}h = 0$. Then ${}^{(4)}h$ induces a solution (h, ω) to the linearized constraint equations about a solution (g, π) to the constraint equations. If (g, π) satisfies conditions $C_{\mathscr{H}}$, C_δ, and C_{tr}, then there exists a curve $(g(\lambda), \pi(\lambda))$ of solutions to the full constraint equations. By using the existence theory for the Einstein equations this curve of solutions to the constraint equations generates a curve ${}^{(4)}g(\lambda)$ of Ricci-flat Lorentz metrics (on a tubular neighborhood V' of M). After possible adjustment by a curve $\psi(\lambda)\colon V' \to V'$ of diffeomorphisms of V', ${}^{(4)}g(\lambda)$ will be tangent to ${}^{(4)}h$ (${}^{(4)}g(\lambda)$ will be an H^s spacetime and will be λ differentiable in H^{s-1}); see also the end of §0.

II.2. Solutions to the Hamiltonian constraint equation. For $s > n/2 + 1$, we consider $\mathscr{H}\colon T^*\mathscr{M}^s \to \Lambda^{s-2}$, $(g, \pi) \mapsto \mathscr{H}(g, \pi)$ as a map from $T^*\mathscr{M}^s$ to Λ^{s-2}, the H^{s-2} 3-forms on M. For $g \in \mathscr{M}^s$, Λ^{s-2} is isomorphic to $H^{s-2}(M; \boldsymbol{R}) \otimes \mu_g$ (by identifying $\lambda \in \Lambda^{s-2}$ with $f\mu_g$, $f \in H^{s-2}(M; \boldsymbol{R})$).

Let $\mathscr{C}^s_{\mathscr{H}} = \mathscr{H}^{-1}(0) = \{(g, \pi) \in T^*\mathscr{M}^s : \mathscr{H}(g, \pi) = 0\}$, the solution set to the Hamiltonian constraint. Note that the set $\mathscr{F}^s \times \{0\}$ is a subset of $\mathscr{C}^s_{\mathscr{H}}$, and that since $\dim M = 3$, $\mathscr{E}^s_0 = \mathscr{F}^s$ (without the assumption that $\mathscr{F}^s \neq \varnothing$). Thus, from I.3.3, $\mathscr{F}^s \times \{0\}$ is a smooth submanifold of $T^*\mathscr{M}$ (but is possibly empty). We will show that $\mathscr{F}^s \times \{0\}$ is the singular set on which $D\mathscr{H}(g, \pi)$ fails to be surjective. Thus

$$\mathscr{C}^s_{\mathscr{H}} = (\mathscr{C}^s_{\mathscr{H}} - \mathscr{F}^s \times \{0\}) \cup (\mathscr{F}^s \times \{0\})$$

is the disjoint union of submanifolds. Thus $\mathscr{C}^s_{\mathscr{H}}$ is somewhat similar to the structure of $\mathscr{M}^s_0$ (when $\mathscr{F}^s \neq \varnothing$); however, because of the kinetic terms involving the variable π, $\mathscr{F}^s \times \{0\}$ is not an isolated set of solutions of $\mathscr{H}(g, \pi) = 0$ (II.2.5); consequently, $\mathscr{C}^s_{\mathscr{H}} - \mathscr{F}^s \times \{0\}$ need not be closed, and $\mathscr{C}^s_{\mathscr{H}}$ itself need not be a manifold.

To prove our result, the basic argument of I.2.1 only has to be modified to take into account the kinetic terms:

II.2.1. THEOREM. *Let $(g, \pi) \in \mathscr{C}^s_{\mathscr{H}}$, $(g, \pi) \notin \mathscr{F}^s \times \{0\}$. Then in a neighborhood of (g, π), $\mathscr{C}^s_{\mathscr{H}}$ is a smooth submanifold of $T^*\mathscr{M}^s$, and $\mathscr{H}(g, \pi) = 0$ is linearization stable at (g, π).*

PROOF. We show that

$$D\mathscr{H}(g, \pi)\colon T^*_{(g,\pi)}(T^*\mathscr{M}^s) \approx S_2^s \times (S_2^s \otimes \mu_g) \to T_0\Lambda^{s-2} \approx \Lambda^{s-2} \approx H^{s-2} \otimes \mu_g$$

is surjective for $(g, \pi) \in \mathscr{C}^s_{\mathscr{H}} - \mathscr{F}^s \times \{0\}$, i.e., that $\mathscr{H}$ is a submersion at those (g, π) that satisfy condition $C_{\mathscr{H}}$.

By a straightforward computation, for $(h, \omega) \in S_2^s \times (S_s^2 \otimes \mu_g)$,

$$\begin{aligned}\gamma_{(g,\pi)}(h,\omega) &= D\mathscr{H}(g,\pi)\cdot(h,\omega)\\ &= \{\Delta \operatorname{tr} h + \delta\delta h - h\cdot \operatorname{Ric}(g) + \tfrac{1}{2}\operatorname{tr} h\, R(g)\\ &\qquad + 2(\tfrac{1}{2}(\operatorname{tr}\pi')\pi' - \pi'\times\pi')\cdot h - \tfrac{1}{2}(\tfrac{1}{2}(\operatorname{tr}\pi')^2 - \pi'\cdot\pi')\operatorname{tr} h\}\mu_g\\ &\qquad + 2(\tfrac{1}{2}(\operatorname{tr}\pi')\operatorname{tr}\omega - \pi'\cdot\omega),\end{aligned}$$

where $\pi' \times \pi' = \pi'^{ia}\,\pi'^{j}_{a}$ is the "product" of symmetric tensors.

Caution. $D\mathscr{H}(g,\pi)\cdot(h,\omega) = D_g\mathscr{H}(g,\pi)\cdot h + D_\pi\,\mathscr{H}(g,\pi)\cdot\omega$, so that one must take the derivative of $\mathscr{H}$ with respect to π (which is best done by using the coordinate expression involving $(\det g_{ij})^{-1/2}$ and not π', although the final expression is in terms of π').

The L_2-adjoint, $\gamma^*_{(g,\pi)}: H^s \otimes \mu_g \to S_2^{s-2} \times (S_s^2 \otimes \mu_g)$ is then given by

$$\begin{aligned}\gamma^*_{(g,\pi)}(N\mu_g) = (&g\Delta N + \operatorname{Hess} N - N\operatorname{Ric}(g) + \tfrac{1}{2}NR(g)g\\ &+ \{2(\tfrac{1}{2}(\operatorname{tr}\pi')\pi' - \pi'\times\pi')^\flat - \tfrac{1}{2}(\tfrac{1}{2}(\operatorname{tr}\pi')^2 - (\pi'\cdot\pi'))g\}N,\\ &\qquad\qquad 2(\tfrac{1}{2}(\operatorname{tr}\pi)g^\# - \pi)N).\end{aligned}$$

Since the symbol $\sigma_{\xi_x}(\gamma^*_{(g,\pi)}): \boldsymbol{R} \to (T^*_xM \otimes_{\text{sym}} T^*_xM) \times (T_xM \otimes_{\text{sym}} T_xM)$ is given by $\sigma_{\xi_x}(\gamma^*_{(g,\pi)}) = (\sigma_{\xi_x}(\gamma^*_g), 0)$ where $\gamma^*_g = (DR(g))^*$, for $\eta_x \neq 0$, $\sigma_{\xi_x}(\gamma^*_{(g,\pi)})$ is injective by the first factor alone. Thus we must show that $\gamma^*_{(g,\pi)}$ is injective. Thus, let $\gamma^*_{(g,\pi)}(N\mu_g) = 0$, so that

$$\text{(a)}\quad \begin{aligned}&g\Delta N + \operatorname{Hess} N - N^\flat \operatorname{Ric}(g) + \tfrac{1}{2}NR(g)g\\ &\quad + \{2(\tfrac{1}{2}(\operatorname{tr}\pi')\pi' - \pi'\times\pi')^\flat - \tfrac{1}{2}(\tfrac{1}{2}(\operatorname{tr}\pi')^2 - \pi'\cdot\pi')g\}N = 0;\end{aligned}$$

$$\text{(b)}\qquad 2(\tfrac{1}{2}(\operatorname{tr}\pi)g^\# - \pi)N = 0.$$

Taking the trace of (b) gives $(\operatorname{tr}\pi)N = 0$, and so again from (b), $N\pi = 0$. Since $\mathscr{H}(g,\pi) = 0$ and $N\pi = 0$, $NR(g) = 0$. Thus (a) simplifies to

$$\text{(c)}\qquad g\Delta N + \operatorname{Hess} N - N\operatorname{Ric}(g) = 0,$$

which, as in the proof of I.2.1, implies that $N = 0$ unless $\operatorname{Ric}(g) = 0 \Leftrightarrow g$ is flat for $\dim M = 3$. If g is flat, taking the trace of (c) gives $\Delta N = 0$ so that N is a constant. Thus from $N\pi = 0$, $N = 0$ unless $\pi = 0$. Thus, if g is flat and $\pi \neq 0$, $N = 0$, so that $\gamma^*_{(g,\pi)}$ is injective and $\gamma_{(g,\pi)}$ is surjective. □

Thus $\mathscr{H}$ is singular (in the sense that $D\mathscr{H}(g,\pi)$ is not surjective) on the set $\mathscr{F}^s \times \{0\}$, and on this set $\ker \gamma^*_{(g,\pi)} = \{\text{constant functions on } M\}$. Thus we expect the equation $\mathscr{H}(g,\pi) = 0$ to be linearization unstable in a neighborhood of a solution $(g_F, 0) \in \mathscr{F}^s \times \{0\}$. To find the extra condition implied by this instability, we introduce the integrated Hamiltonian density (= the Hamiltonian)

$$H: T^*\mathscr{M}^s \to R, \qquad (g,\pi) \mapsto \int \mathscr{H}(g,\pi) = \int \{\tfrac{1}{2}(\operatorname{tr}\pi')^2 - \pi'\cdot\pi' + R(g)\}\, d\mu_g,$$

the total kinetic energy,

$$K: T^*\mathscr{M}^s \to \boldsymbol{R}, \qquad (g,\pi) \mapsto \int (\tfrac{1}{2}(\operatorname{tr}\pi')^2 - \pi'\cdot\pi')\, d\mu_g,$$

and the integrated scalar curvature

$$\Phi: \mathscr{M}^s \to R, \qquad g \mapsto \int R(g)\, d\mu_g.$$

Thus $H(g, \pi) = K(g, \pi) + \Phi(g)$, and Φ serves as a potential for the Hamiltonian; see [14] for the geometrical consequences of this interpretation. First we consider the map Φ (see also [3]).

II.2.2. THEOREM. *Let* $\dim M \geqq 3$. *Then a metric* $g \in \mathcal{M}^s$ *is a critical point of* Φ *if and only if* $\mathrm{Ric}(g) = 0$. *At a critical point* $g \in \mathcal{M}^s$, *the Hessian of* Φ *is*

$$d^2\Phi(g)\cdot(h_1, h_2) = -\frac{1}{2}\int h_1\cdot\Delta_L h_2\, d\mu_g + \int \delta h_1\cdot\delta h_2\, d\mu_g + \frac{1}{2}\int \delta h_1\cdot d\,\mathrm{tr}\, h_2\, d\mu_g + \frac{1}{2}\int d\,\mathrm{tr}\, h_1\cdot\delta h_2\, d\mu_g + \frac{1}{2}\int d\,\mathrm{tr}\, h_1\cdot d\,\mathrm{tr}\, h_1\, d\mu_g.$$

If $g \in \mathcal{M}^{s+1}$ *so that* $h \in S_2^s$ *can be decomposed as* $h = \mathring{h} + L_X g$ (*where* $\mathring{h} \in \mathring{S}_2^s$ *is the divergence free part of* h), *then*

$$d^2\Phi(g)\cdot(h_1, h_2) = -\frac{1}{2}\int \mathring{h}_1\cdot\Delta_L\mathring{h}_2\, d\mu_g + \frac{1}{2}\int d\,\mathrm{tr}\,\mathring{h}_1\cdot d\,\mathrm{tr}\,\mathring{h}_2\, d\mu_g.$$

If $g_F \in \mathcal{F}^s$ (*so that* $h \in S_2^s$ *has a decomposition* $h = \mathring{h} + L_X g$) *then* $\Delta_L h = \bar{\Delta} h$, *and*

$$d^2\Phi(g_F)\cdot(h, h) = -\frac{1}{2}\int (\nabla\mathring{h})^2\, d\mu_g + \frac{1}{2}\int (d\,\mathrm{tr}\,\mathring{h})^2\, d\mu_g$$

for a flat riemannian metric g_F.

PROOF. First we find the critical points of Φ. We compute the derivative of Φ:

$$\begin{aligned} d\Phi(g)\cdot h = D\Phi(g)\cdot h &= \int DR(g)\cdot h\, d\mu_g + \int R(g) D(d\mu_g)\cdot h \\ &= \int (\Delta\,\mathrm{tr}\, h + \delta\delta h - h\cdot\mathrm{Ric}(g))\, d\mu_g + \int R(g)\,\tfrac{1}{2}\,\mathrm{tr}\, h\, d\mu_g \qquad (1)\\ &= -\int (\mathrm{Ric}(g) - \tfrac{1}{2} gR(g))\cdot h\, d\mu_g \end{aligned}$$

since $\int(\Delta\,\mathrm{tr}\, h + \delta\delta h)\, d\mu_g = 0$ for all (g, h) by Stokes' theorem. Thus $d\Phi(g)\cdot h = 0$ for all $h \in S_2^s \Leftrightarrow \mathrm{Ric}(g) - \frac{1}{2} gR(g) = 0$. Since $\dim M \geqq 3$, by considering the trace of this we see that it is equivalent to $\mathrm{Ric}(g) = 0$.

From (1), the second derivative of Φ is

$$\begin{aligned} d^2\Phi(g)\cdot(h, h) &= \int D^2R(g)\cdot(h, h)\, d\mu_g \\ &\quad + 2\int (DR(g)\cdot h)(D(d\mu_g)\cdot h) + \int R(g)\, D^2(d\mu_g)\cdot(h, h). \qquad (2) \end{aligned}$$

At a critical point g, $\mathrm{Ric}(g) = 0$, $R(g) = 0$, and $DR(g)\cdot h = \Delta\,\mathrm{tr}\, h + \delta\delta h$, and so from I.4.1, (2) becomes

$$\begin{aligned} d^2\Phi(g)\cdot(h, h) &= -\frac{1}{2}\int h\cdot\Delta_L h\, d\mu_g - \frac{1}{2}\int (d\,\mathrm{tr}\, h)^2\, d\mu_g + \int (\delta h)^2\, d\mu_g \\ &\quad + 2\int (\Delta\,\mathrm{tr}\, h + \delta\delta h)(\tfrac{1}{2}\,\mathrm{tr}\, h)\, d\mu_g \\ &= -\frac{1}{2}\int h\cdot\Delta_L h\, d\mu_g + \frac{1}{2}\int (d\,\mathrm{tr}\, h)^2\, d\mu_g + \int (\delta h)^2\, d\mu_g + \int d\,\mathrm{tr}\, h\cdot\delta h\, d\mu_g \end{aligned}$$

If $g \in \mathcal{M}^{s+1}$, $\mathrm{Ric}(g) = 0$, and $h = \mathring{h} + L_X g$, then $DR(g)\cdot h = DR(g)\cdot \mathring{h} = \Delta \mathrm{tr}\, \mathring{h}$, and so

$$\begin{aligned} d^2\Phi(g)\cdot(h,h) &= -\frac{1}{2}\int \mathring{h}\cdot\Delta_L \mathring{h}\, d\mu_g - \frac{1}{2}\int (d\,\mathrm{tr}\,\mathring{h})^2\, d\mu_g \\ &\quad + 2\int d\,\mathrm{tr}\,\mathring{h}\cdot d\delta X\, d\mu_g + \int (\Delta \mathrm{tr}\,\mathring{h})\,\mathrm{tr}\,h\, d\mu_g \\ &= -\frac{1}{2}\int \mathring{h}\cdot\Delta_L \mathring{h}\, d\mu_g - \frac{1}{2}\int (d\,\mathrm{tr}\,\mathring{h})^2\, d\mu_g \\ &\quad + \int d\,\mathrm{tr}\,\mathring{h}\cdot(2d\delta X + d\,\mathrm{tr}\,h)\, d\mu_g \\ &= -\frac{1}{2}\int \mathring{h}\cdot\Delta_L \mathring{h}\, d\mu_g + \frac{1}{2}\int (d\,\mathrm{tr}\,\mathring{h})^2\, d\mu_g \end{aligned}$$

since $\mathrm{tr}\,h = \mathrm{tr}\,\mathring{h} - 2\delta X$ so that $d\,\mathrm{tr}\,\mathring{h} = d\,\mathrm{tr}\,h + 2d\delta X$. □

REMARKS. 1. The change to a positive sign of the $\frac{1}{2}\int(d\,\mathrm{tr}\,\mathring{h})^2\, d\mu_g$ term in $d^2\Phi(g)\cdot(h,h) = d^2(\int R(g)\, d\mu_g)\cdot(h,h)$ as compared with $\int D^2R(g)\cdot(h,h)\, d\mu_g$ comes about because of the term $2\int(DR(g)\cdot h)(D(d\mu_g)\cdot h)$ involving the derivative of the volume element. Because of this sign change, a flat metric g_F is a saddle point for $d^2\Phi(g_F)$ (even within a slice), whereas $\int D^2 R(g)\cdot(h,h)\, d\mu_g \leqq 0$ on a slice at g_F. Thus because of this sign change, the behavior of the integrated scalar curvature $\Phi(g) = \int R(g)\, d\mu_g$ is somewhat different from the pointwise scalar curvature $R(g)$ at g_F.

2. That $d^2\Phi(g)\cdot(h,h) = d^2\Phi(g)\cdot(\mathring{h},\mathring{h})$ depends only on the divergence free part of h follows from the invariance of Φ by $\mathscr{D}$, so that $d^2\Phi(g)\cdot(L_X g, L_X g) = 0$, and by orthogonality of $\mathring{h}$ and $L_X g$, so that $d^2\Phi(g)\cdot(\mathring{h}, L_X g) = 0$.

We can now easily compute the critical points and the Hessian of the Hamiltonian $H\colon T^*\mathcal{M}^s \to \mathbf{R}$. From now on, $\dim M = 3$.

II.2.3. THEOREM. *A pair $(g,\pi) \in T^*\mathcal{M}^s$ is a critical point of $H\colon T^*\mathcal{M}^s \to \mathbf{R}$, $(g,\pi) \mapsto \int \mathscr{H}(g,\pi)$ if and only if $(g,\pi) \in \mathscr{F}^s \times \{0\}$.*

At a critical point $(g_F, 0)$, the Hessian of H is

$$\begin{aligned} d^2H(g_F,0)\cdot((h,\omega),(h,\omega)) &= -\frac{1}{2}\int (\nabla\mathring{h})^2\, d\mu_g + \frac{1}{2}\int (d\,\mathrm{tr}\,\mathring{h})^2\, d\mu_g \\ &\quad + 2\int \left(\tfrac{1}{2}(\mathrm{tr}\,\omega')^2 - \omega'\cdot\omega'\right) d\mu_g. \end{aligned}$$

PROOF. From the computation in the proof of II.2.1,

$$\begin{aligned} dH(g,\pi)\cdot(h,\omega) &= \int D\mathscr{H}(g,\pi)\cdot(h,\omega) \\ &= -\int(\mathrm{Ric}(g) - \tfrac{1}{2}gR(g))\cdot h\, d\mu_g + 2\int(\tfrac{1}{2}(\mathrm{tr}\,\pi')\pi' - \pi'\times\pi')\cdot h\, d\mu_g \\ &\quad - \frac{1}{2}\int(\tfrac{1}{2}(\mathrm{tr}\,\pi')^2 - \pi'\cdot\pi')\mathrm{tr}\,h\, d\mu_g + 2\int(\tfrac{1}{2}(\mathrm{tr}\,\pi')\,\mathrm{tr}\,\omega - \pi'\cdot\omega) = 0 \end{aligned}$$

for all $(h,\omega) \in S^2_s \times (S^2_s \otimes \mu_g)$. Thus

(a) $-(\mathrm{Ric}(g) - \tfrac{1}{2}gR(g)) + 2(\tfrac{1}{2}(\mathrm{tr}\,\pi')\pi' - \pi'\times\pi') - \tfrac{1}{2}(\tfrac{1}{2}(\mathrm{tr}\,\pi')^2 - \pi'\cdot\pi')g = 0,$

(b) $\tfrac{1}{2}(\mathrm{tr}\,\pi')\,g - \pi' = 0.$

Contracting (b) gives tr $\pi' = 0$, and thus from (b) again, $\pi' = 0$ (and so $\pi = 0$). Thus, from (a), $\mathrm{Ric}(g) - \frac{1}{2} g R(g) = 0 \Rightarrow \mathrm{Ric}(g) = 0 \Rightarrow g$ is flat.

In the computation for

$$d^2H(g, \pi)\cdot(h, h) = d^2\left(\int R(g)\, d\mu_g\right)\cdot(h, h)$$
$$+ d^2\left(\int (\tfrac{1}{2}(\mathrm{tr}\, \pi')^2 - \pi'\cdot\pi')\, d\mu_g\right)\cdot((h, \omega), (h, \omega)),$$

the terms due to the kinetic part of H are straightforward to compute; since $\pi = 0$ and the kinetic part is quadratic in π,

$$d^2K(g, 0)\cdot((h, \omega)(h, \omega)) = d^2\left(\int (\tfrac{1}{2}(\mathrm{tr}\, \pi')^2 - \pi'\cdot\pi')\, d\mu_g\right)\cdot((h, \omega), (h, \omega))$$
$$= 2\int (\tfrac{1}{2}(\mathrm{tr}\, \omega')^2 - \omega'\cdot\omega')\, d\mu_g.$$

The expression for $d^2(\int R(g)\, d\mu_g)\cdot(h, h)$ is given by II.2.2. □

Note that the critical points of H are exactly the set where $\mathscr{H}$ is singular; i.e. where $D\mathscr{H}(g, \pi)$ is not surjective. This "coincidence" follows from the fact that

$$dH(g, \pi)\cdot(h, \omega) = \int D\mathscr{H}(g, \pi)\cdot(h, \omega) = \int ((D\mathscr{H}(g, \pi))^*1)\cdot(h, \omega) = 0$$

for all $(h, \omega) \in T_{(g,\pi)}(T^*\mathscr{M}^s)$ iff $1 \in \ker(D\mathscr{H}(g, \pi))^*$ iff $D\mathscr{H}(g, \pi)$ is not surjective.

Around $(g_F, 0) \in \mathscr{F}^s \times \{0\}$, the equation $\mathscr{H}(g, \pi) = 0$ is linearization unstable. The extra condition that a first order deformation (h, ω) must satisfy for it to be tangent to a curve $(g(\lambda), \pi(\lambda))$ of exact solutions of $\mathscr{H}(g, \pi) = 0$ is given by the following:

II.2.4. THEOREM. *Let $(g(\lambda), \pi(\lambda)) \in \mathscr{C}^s_{\mathscr{H}} \subset T^*\mathscr{M}^s$, $\lambda \in (-\delta, \delta)$, $\delta > 0$, be a C^2 curve with $(g(0), \pi(0)) = (g_F, 0) \in \mathscr{F}^s \times \{0\}$, $(g'(0), \pi'(0)) = (h, \omega)$. Then (h, ω) must satisfy* tr $\mathring{h}$ = *constant and*

$$-\frac{1}{2}\int (\nabla\mathring{h})^2\, d\mu_{g_F} + 2\int (\tfrac{1}{2}(\mathrm{tr}\, \omega')^2 - \omega'\cdot\omega')\, d\mu_{g_F} = 0$$

where $\mathring{h}$ is the divergence free part of h.

PROOF. Differentiating $\mathscr{H}(g(\lambda), \pi(\lambda)) = 0$ twice and evaluating at $\lambda = 0$, we have

$$\text{(1)}\quad \begin{aligned} \frac{d\mathscr{H}}{d\lambda}(g(\lambda), \pi(\lambda))\Big|_{\lambda=0} &= D\mathscr{H}(g(\lambda), \pi(\lambda))\cdot(g'(\lambda), \pi'(\lambda))|_{\lambda=0} \\ &= D\mathscr{H}(g_F, 0)\cdot(h, \omega) = (\Delta \mathrm{tr}\, h + \delta\delta h)\, d\mu_g \\ &= (\Delta\, \mathrm{tr}\, \mathring{h})\, d\mu_g = 0 \end{aligned}$$

and

$$\text{(2)}\quad \begin{aligned} \frac{d^2\mathscr{H}}{d\lambda^2}(g(\lambda), \pi(\lambda))\Big|_{\lambda=0} &= D^2\mathscr{H}(g(\lambda), \pi(\lambda))\cdot((g'(\lambda), \pi'(\lambda)), (g'(\lambda), \pi'(\lambda)))|_{\lambda=0} \\ &\quad + D\mathscr{H}(g(\lambda), \pi(\lambda))\cdot(g''(\lambda), \pi''(\lambda))|_{\lambda=0} \\ &= D^2\mathscr{H}(g_F, 0)\cdot((h, \omega), (h, \omega)) \\ &\quad + D\mathscr{H}(g_F, 0)\cdot(g''(0), \pi''(0)) = 0. \end{aligned}$$

Integrating (2) over M gives

$$d^2H(g_F, 0)\cdot((h, \omega), (h, \omega)) = \int D^2\mathscr{H}(g_F, 0)\cdot((h, \omega), (h, \omega)) = 0 \tag{3}$$

since $(g_F, 0)$ is a critical point of H, so that

$$dH(g_F, 0)\cdot(g''(0), \pi''(0)) = \int D\mathscr{H}(g_F, 0)\cdot(g''(0), \pi''(0)) = 0.$$

From the first order condition (1), tr h = constant, so that from II.2.3, the second order condition (3) becomes

$$-\frac{1}{2}\int (\nabla\mathring{h})^2\, d\mu_g + 2\int \left(\tfrac{1}{2}(\mathrm{tr}\,\omega')^2 - \omega'\cdot\omega'\right) d\mu_g = 0. \quad \square$$

REMARKS. 1. The first order condition (1) does not give any restrictions on ω. However in the next section we shall see that the first order deformation (h, ω) of the $\delta_g\pi = 0$ constraint around $\pi = 0$ implies that $\delta\omega = 0$.

2. As an example of a nonintegrable deformation, let (T^3, g_F) and h be as in the example of I.4. Then if $\omega = 0$, $(h, 0)$ satisfies the linearized Hamiltonian constraint but not the second order condition, which for $\omega = 0$ reduces to $\nabla\mathring{h} = 0$. Thus $(h, 0)$ cannot be tangent to any curve $(g(\lambda), \pi(\lambda)) \in \mathscr{C}_{\mathscr{H}}$.

Even though g_F is flat, the integrated extra condition on a first order deformation (h, ω) cannot be converted to a pointwise condition as in I.4.2 or the above remark since the kinetic term $\int(\frac{1}{2}(\mathrm{tr}\,\omega')^2 - \omega'\cdot\omega')\, d\mu_g$ is not negative-definite, even if the condition $\delta\omega = 0$ is imposed. Not being able to convert to a pointwise extra condition signals that although there is linearization instability of $\mathscr{H}(g, \pi) = 0$ at $(g_F, 0) \in \mathscr{F} \times \{0\}$, these solutions are not isolated solutions. In fact, if we ignore the $\delta\pi = 0$ constraint, we can construct solutions algebraically to $\mathscr{H}(g, \pi) = 0$, $(g, \pi) \notin \mathscr{F}^s \times \{0\}$, which are arbitrarily close to a solution $(g_F, 0) \in \mathscr{F}^s \times \{0\}$.

This construction proceeds as follows: let $A^T \in S^2$ be any traceless tensor, tr $A^T = 0$, and let $\pi_\varepsilon = \varepsilon(A^T + ((2/3)A^T\cdot A^T)^{1/2}g_F)\,\mu_{g_F}$; here the trace and pointwise contraction "$\cdot$" are with respect to g_F. Then

$$\mathscr{H}(g_F, \pi_\varepsilon) = (R(g_F) + \tfrac{1}{2}(\mathrm{tr}\,\pi_\varepsilon')^2 - \pi_\varepsilon'\cdot\pi_\varepsilon')\,\mu_{g_F} = 0.$$

Thus for ε small, π_ε can be made arbitrarily close to 0.

In this construction, $\delta\pi_\varepsilon \neq 0$, and this situation cannot be remedied by choosing A^T to be transverse (i.e. $\delta A^T = 0$) as well as traceless, since $\psi = ((2/3)A^T\cdot A^T)^{1/2}$ need not be constant, so that $\pi = (A^T + \psi g_F)\,\mu_{g_F}$ need not be divergence free.

However, by being more subtle, we can still construct solutions to $\mathscr{H}(g, \pi) = 0$, $\delta_g\pi = 0$, $(g, \pi) \notin \mathscr{F}^s \times \{0\}$, which are arbitrarily close to the manifold of solutions $\mathscr{F}^s \times \{0\}$.

II.2.5. THEOREM. *Let $(g_F, 0) \in \mathscr{F}^s \times \{0\}$. Then in every neighborhood $U^s_{(g_F,0)} \subset T^*\mathscr{M}^s$ of $(g_F, 0)$, there exists a $(g, \pi) \in U^s_{(g_F,0)}$ such that $(g, \pi) \notin \mathscr{F}^s \times \{0\}$, $\mathscr{H}(g, \pi) = 0$, $\delta_g\pi = 0$, and* tr $\pi = c\mu_g$, c = *constant* $\neq 0$.

PROOF. We use a stability argument based on the Lichnerowicz [24] and Choquet-Bruhat [7], [8], [9] conformal method of constructing solutions to the constraint equations.

For $g_0 \in \mathcal{M}^s$, let $A^{TT} \in \mathring{S}^2_s$ be such that $\delta_{g_0} A^{TT} = 0$ and $\operatorname{tr} A^{TT} = 0$. Let $M = A^{TT} \cdot A^{TT}$ ("$\cdot$" is with respect to g_0) and let $c =$ constant. Then from [8], [9] (see also [33] for the case $c =$ constant $\neq 0$), if $M \neq 0$ ($\Leftrightarrow A^{TT} \neq 0$), $c \neq 0$, there exists a unique $\phi \in H^s(M; \boldsymbol{R})$, $\phi > 0$, that satisfies the Lichnerowicz equation

$$\text{(L)} \qquad 8\Delta\phi = -R(g)\,\phi + M\,\phi^{-7} - \tfrac{1}{6}c^2\,\phi^5.$$

Moreover, if $g = \phi^4 g_0$,

$$\pi = (\phi^{-4} A^{TT} + (c/3)\,\phi^2 g_0^{\#})\,\mu_{g_0} = (\phi^{-10} A^{TT} + (c/3)\,g^{\#})\,\mu_g$$

(where the last equality follows from $g^{\#} = \phi^{-4} g_0^{\#}$ and $\mu_g = \phi^6 \mu_{g_0}$), then $\mathcal{H}(g, \pi) = 0$, $\delta_g \pi = 0$, and $\operatorname{tr} \pi = c\mu_g$.

The stability theorem of [9], adapted to the case that $c \neq 0$, proves that solutions to (L) are stable with respect to g, M, and c^2, if $M \neq 0$, and $c \neq 0$; i.e. if we let

$$Y\colon T^*\mathcal{M}^s \to H^s(M; \boldsymbol{R}), \qquad (g, \pi) \mapsto \phi$$

be defined for those $(g, \pi) \in T^*\mathcal{M}^s$ such that π is of the form $\pi = (A^{TT} + (c/3)g)\mu_g$, $A^{TT} \neq 0$, $c \neq 0$, and let $\phi = Y(g, \pi) \in H^s(M; \boldsymbol{R})$, $\phi > 0$, be defined as the unique solution of (L), then Y is a continuous map.

From the uniqueness theorem for (L), if

$$-R(g) + M - \tfrac{1}{6}c^2 = 0, \qquad M \neq 0, c \neq 0,$$

then $\phi \equiv 1$ is the unique solution of (L). But then from stability of solutions to (L), it follows that if $-R(g) + M - (1/6)c^2$ ($M \neq 0$, $c \neq 0$) is H^{s-2} close to $0 \in H^{s-2}(M; \boldsymbol{R})$, then the unique solution $\phi > 0$ of (L) is H^s close to $1 \in H^s(M; \boldsymbol{R})$; i.e. if $U^s_1 \subset H^s(M; \boldsymbol{R})$ is a neighborhood of 1, there exists a neighborhood $U^{s-2}_0 \subset H^{s-2}(M; \boldsymbol{R})$ of 0 such that if $-R(g) + M - (1/6)c^2 \in U^{s-2}_0$, $M \neq 0$, $c \neq 0$, then the unique solution $\phi > 0$ of (L) is in U^s_1.

Now let

$$\begin{aligned} \bar{U}^s_{(g_F,0)} = \{(g, \pi) \in U^s_{(g_F,0)} &: \text{if } \pi = (A^{TT} + (c/3)\,g)\,\mu_g \text{ and } \phi \in U^s_1, \phi > 0, \\ &\text{then } (\phi^4 g, (\phi^{-4} A^{TT} + (c/3)\,\phi^2 g^{\#})\,\mu_g) \in U^s_{(g_F,0)}\}, \end{aligned}$$

and

$$\begin{aligned} \hat{U}^s_{(g_F,0)} = \{(g, \pi) \in U^s_{(g_F,0)} &: \text{if } \pi = (A^{TT} + (c/3)\,g)\,\mu_g, \\ &\text{then } -R(g) + M - \tfrac{1}{6}c^2 \in U^{s-2}_0\}, \end{aligned}$$

so that $\bar{U}^s_{(g_F,0)} \subset U^s_{(g_F,0)}$ and $\hat{U}^s_{(g_F,0)} \subset U^s_{(g_F,0)}$ are both neighborhoods of $(g_F, 0)$.

Now let $(g_0, \pi_0) \in \bar{U}^s_{(g_F,0)} \cap \hat{U}^s_{(g_F,0)}$ with π_0 of the form $\pi_0 = (A^{TT} + (c/3)\,g_0^{\#})\,\mu_{g_0}$, $A^{TT} \neq 0$, $c \neq 0$, and let $\phi > 0$ be the unique solution of (L) with coefficients (g_0, M, c). From the construction of $\hat{U}^s_{(g_F,0)}$ and U^{s-2}_0, $\phi \in U^s_1$, so from the construction of $\bar{U}^s_{(g_F,0)}$, if $g = \phi^4 g_0$, $\pi = (\phi^{-4} A^{TT} + (c/3)\,\phi^2 g_0^{\#})\,\mu_{g_0}$, $(g, \pi) \in U^s_{(g_F,0)}$, and $\mathcal{H}(g, \pi) = 0$, $\delta_g \pi = 0$, $\operatorname{tr} \pi = \phi^4 g_0 \cdot \pi = c\,\phi^6\,\mu_{g_0} = c\,\mu_g$, $c \neq 0$, and $(g, \pi) \notin \mathcal{F}^s \times \{0\}$. □

Thus the set $\mathscr{F}^s \times \{0\}$ is not isolated among the solutions of $\mathscr{H}(g, \pi) = 0$ and $\delta_g \pi = 0$. In fact, in [6], Brill-Deser show by example that a flat 3-torus and $\pi = 0$ is not an isolated solution of the constraint equations.

As shall be apparent from II.3.2, the divergence constraint also does not have any isolated solutions. Thus because $\mathscr{H}(g, \pi) = 0$ is linearization stable at all $(g, \pi) \notin \mathscr{F} \times \{0\}$, we can conclude:

There are no isolated solutions of the empty-space constraint equations of general relativity.

This result also holds for all physically reasonable stress-energy tensors.

Interestingly, *if we look for solutions to the constraint equations that also satisfy the condition* $\operatorname{tr} \pi = 0$, *then* $\mathscr{F}^s \times \{0\}$ *is an isolated manifold of solutions to the Hamiltonian constraint (and hence to both constraint equations).*

II.2.6. THEOREM. *Let* $(g_F, 0) \in \mathscr{F}^s \times \{0\}$. *Then there exists a neighborhood* $U_{(g_F,0)} \subset T^*\mathscr{M}^s$ *of* $(g_F, 0)$ *such that if* $(g, \pi) \in U_{(g_F,0)}$, $\mathscr{H}(g, \pi) = 0$, *and* $\operatorname{tr} \pi = 0$, *then* $(g, \pi) \in \mathscr{F}^s \times \{0\}$.

PROOF. From I.5.2 there exists a neighborhood $U_{g_F} \subset \mathscr{M}^s$ such that if $g \in U_{g_F}$, $R(g) \geqq 0$, then $g \in \mathscr{F}^s$. Let $U_{(g_F,0)} = T^*U_{g_F} \subset T^*\mathscr{M}^s$. Then if $(g, \pi) \in U_{(g_F,0)}$, $\mathscr{H}(g, \pi) = 0$, $\operatorname{tr} \pi = 0$, then $R(g) = \pi' \cdot \pi' \geqq 0$, and since $g \in U_{g_F}$, $g \in \mathscr{F}^s$, so $R(g) = 0$, and hence $\pi = 0$. □

REMARKS. 1. In particular, the solutions $\mathscr{F}^s \times \{0\}$ are isolated among the time-symmetric ($\pi = 0$) solutions to the Hamiltonian constraint.

2. The variation of the $\operatorname{tr} \pi = 0$ condition is

$$D(\operatorname{tr} \pi) \cdot (h, \omega) = h \cdot \pi + \operatorname{tr} \omega = 0.$$

Thus if $\pi = 0$, a deformation of $\operatorname{tr} \pi = 0$ must satisfy $\operatorname{tr} \omega = 0$. Using this condition, the second order condition of II.2.4 reduces to the pointwise condition $\nabla \mathring{h} = 0$ and $\omega = 0$. This is the basis of the "second order" version of II.2.6, proven in [5].

Although II.2.6 proves isolation in a full neighborhood of $\mathscr{F}^s \times \{0\}$, in light of II. 2.5, the isolation of this set is more a consequence of the $\operatorname{tr} \pi = 0$ condition than of the constraint equations.

II.3. The divergence constraint. Now let $\delta: T^*\mathscr{M}^s \to \mathscr{X}^{s-1} \times \mathscr{V}^s$, $(g, \pi) \mapsto \delta_g \pi = -\pi^{ij}{}_{|j}$, where $\mathscr{V}^s$ is the set of H^s volume elements on M. Let

$$\mathscr{C}_\delta^s = \delta^{-1}(0) = \{(g, \pi) \in T^*\mathscr{M}^s : \delta(g, \pi) = 0\}$$

denote the set of solutions to the divergence constraint.

II.3.1. THEOREM. *Let* $(g, \pi) \in \mathscr{C}_\delta^s$ *satisfy condition* C_δ: *if* $L_X g = 0$ *and* $L_X \pi = 0$, *then* $X = 0$. *Then in a neighborhood of* (g, π), $\mathscr{C}_\delta^s$ *is a smooth submanifold of* $T^*\mathscr{M}^s$, *and* $\delta_g \pi = 0$ *is linearization stable at* (g, π).

PROOF. $D\delta(g, \pi): S_2^s \times (S_s^2 \otimes \mu_g) \to \mathscr{X}^{s-1} \otimes \mu_g$ is computed to give

$$\beta_{(g,\pi)} \cdot (h, \omega) = D\delta(g, \pi) \cdot (h, \omega) = \delta\omega + \tfrac{1}{2} \pi^{lm} h_{lm}{}^{|i} - \pi^{lm} h^i{}_{l|m}$$

with L_2-adjoint $\beta^*_{(g,\pi)}: \mathscr{X}^s \otimes \mu_g \mapsto S_2^{s-1} \times (S^2_{s-1} \otimes \mu_g)$ given by

$$\beta^*_{(g,\pi)}(X\mu_g) = (-\tfrac{1}{2}(L_X\pi + X \otimes \delta\pi + \delta\pi \otimes X)'^{\flat}, \tfrac{1}{2}(L_Xg)^{\sharp}\,\mu_g)$$

($(L_X\pi)'$ means the tensor part of $L_X\pi$). Thus, since $\delta\pi = 0$, $\beta^*_{(g,\pi)}(X\mu_g) = 0 \Rightarrow L_X\pi = 0$ and $L_Xg = 0 \Rightarrow X = 0$, so $\beta^*_{(g,\pi)}$ is injective. Also, $\beta^*_{(g,\pi)}$ has injective symbol (which it inherits from the second factor alone), so that $\beta_{(g,\pi)} = D\delta(g, \pi)$ is surjective and $\delta(\cdot, \cdot)$ is a submersion at (g, π). □

For $(g, \pi) \in T^*\mathscr{M}^s$, let $I_g = \{\varphi \in \mathscr{D}^{s+1}: \varphi^*g = g\}$, the isometry group of g, and $I_\pi = \{\varphi \in \mathscr{D}^{s+1}: \varphi^*\pi = \pi\}$, the symmetry group of π (here $\varphi^*\pi = ((\varphi^{-1})_*\pi') \otimes (\varphi^*\mu_g)$ is the pullback of the contravariant tensor density π). I_g is a compact Lie group; I_π is closed in $\mathscr{D}^{s+1}$ but may be infinite-dimensional (e.g., if $\pi = 0$, $I_\pi = \mathscr{D}^{s+1}$). Let $I_{(g,\pi)} = I_g \cap I_\pi$, a compact Lie group, and let $\mathscr{I}_{(g,\pi)}$ denote its Lie algebra. Then $\ker \beta^*_{(g,\pi)} = \mathscr{I}_{(g,\pi)} = \{X \in \mathscr{X}^{s+1}: L_Xg = 0 \text{ and } L_X\pi = 0\}$.

If we consider the action $A: \mathscr{D}^{s+1} \times \mathscr{M}^s \to \mathscr{M}^s$ lifted to the cotangent bundle,

$$A': \mathscr{D}^{s+1} \times T^*\mathscr{M}^s \to T^*\mathscr{M}^s, \qquad (\varphi, (g, \pi)) \mapsto (\varphi^*g, \varphi^*\pi)$$

then the isotropy group of this action at a point $(g, \pi) \in T^*\mathscr{M}^s$ is $I_{(g,\pi)}$. Thus the map $\delta(g, \pi)$ is singular (i.e., fails to have surjective derivative) precisely where the action A' has isotropy group $I_{(g,\pi)}$ which is nondiscrete. At these (g, π), $\mathscr{D}^{s+1}$ does not act freely so that the quotient space $T^*\mathscr{M}^s/\mathscr{D}^{s+1}$ is singular (i.e., is not a manifold).

Note that if $(g, \pi) \in \mathscr{C}^s_\delta$, pulling back $\delta_g\pi = 0$ by $\varphi \in \mathscr{D}^{s+1}$ gives $\varphi^*(\delta_g\pi) = \delta_{\varphi^*g}(\varphi^*\pi) = 0$ so that $(\varphi^*g, \varphi^*\pi) \in \mathscr{C}^s_\delta$. Thus $\mathscr{C}^s_\delta$ is invariant under A' so that we have restricted action

$$B: \mathscr{D}^{s+1} \times \mathscr{C}^s_\delta \to \mathscr{C}^s_\delta.$$

Let $\mathscr{C}^s_\delta \,/\, \mathscr{D}^{s+1}$ denote the quotient space of $\mathscr{C}^s_\delta$ by this restricted action. Because the singular points (g, π) of $\mathscr{C}^s_\delta$ (as a manifold) correspond to singularities in the action B, and at these points (g, π), $\ker \beta^*_{(g,\pi)} = \mathscr{I}_{(g,\pi)}$, we conjecture that modulo the presence of discrete isotropy groups, $\mathscr{C}^s_\delta \,/\, \mathscr{D}^{s+1}$ is a smooth submanifold as a subset of the quotient space $T^*\mathscr{M}^s \,/\, \mathscr{D}^{s+1}$, the singular points of $\mathscr{C}^s_\delta$ precisely "cancelling out" the singularities in the quotient space $\mathscr{C}^s_\delta \,/\, \mathscr{D}^{s+1}$ due to the presence of nondiscrete isotropy. This possibility was pointed out to us by D. Ebin; cf. Marsden and Fischer **[28]**.

At those points $(g, \pi) \in \mathscr{C}^s_\delta$ for which $D\delta(g, \pi)$ is not surjective, there are extra second order conditions that must be satisfied for a deformation (h, ω) to be tangent to a curve in $\mathscr{C}^s_\delta$.

II.3.2. THEOREM. *Let $(g, \pi) \in \mathscr{C}^s_\delta$, $X \in \mathscr{X}^{s+1}$, $X \neq 0$ such that $L_Xg = 0$ and $L_X\pi = 0$. Let $(g(\lambda), \pi(\lambda)) \in \mathscr{C}^s_\delta$, $\lambda \in (-c, c)$, $c > 0$, be a C^2 curve with $(g(0), \pi(0)) = (g, \pi)$, and $(g'(0), \pi'(0)) = (h, \omega)$. Then (h, ω) satisfies*

$$D\delta(g, \pi) \cdot (h, \omega) = \delta\omega + \tfrac{1}{2}\pi^{lm} h_{lm}{}^{|i} - \pi^{lm} h^i{}_{l|m} = 0$$

and

$$\int h \cdot L_X\omega = 0.$$

PROOF. Differentiating $\delta(g(\lambda), \pi(\lambda)) = 0$ twice and evaluating at $\lambda = 0$ gives

$$\frac{d\delta}{d\lambda}(g(\lambda), \pi(\lambda))\Big|_{\lambda=0} = D\delta(g, \pi) \cdot (h, \omega) = 0,$$

and

$$\frac{d^2\delta}{d\lambda^2}(g(\lambda), \pi(\lambda))\Big|_{\lambda=0} = D^2\delta(g, \pi) \cdot ((h, \omega), (h, \omega)) + D\delta(g, \pi) \cdot (g''(0), \pi''(0)) = 0.$$

Thus, if $\ker \beta^*_{(g,\pi)} \neq 0$, then for each $X \in \ker \beta^*_{(g,\pi)}$,

$$\begin{aligned}\int X \cdot (D^2\delta(g, \pi) \cdot ((h,\omega), (h, \omega))) &+ \int X \cdot (\beta_{(g,\pi)} \cdot (g''(0), \pi''(0))) \\ &= \int X \cdot (D^2\delta(g, \pi) \cdot ((h, \omega), (h, \omega))) + \int (\beta^*_{(g,\pi)} X) \cdot (g''(0), \pi''(0)) \\ &= \int X \cdot (D^2\delta(g, \pi) \cdot ((h, \omega), (h, \omega))) = 0.\end{aligned}$$

A rather lengthy computation gives

$$D^2\delta(g, \pi) \cdot ((h, \omega), (h, \omega)) = 2\omega^{lm} \left(\tfrac{1}{2} h_{lm}{}^{|i} - h^i{}_{l|m}\right) - 2\pi^{lm} h^{ia} \left(\tfrac{1}{2} h_{lm|a} - h_{al|m}\right),$$

which, together with $D\delta(g, \pi) \cdot (h, \omega) = 0$, gives

$$\int X \cdot (D^2\delta(g, \pi) \cdot ((h, \omega), (h,\omega))) = - \int L_X\omega \cdot h = 0.$$

Thus, $\int L_X\omega \cdot h = 0$ is the necessary second order condition for each $X \in \ker \beta^*_{(g,\pi)}$ that must be satisfied for (h, ω) to be tangent to a curve in $\mathscr{C}^s_\delta$. □

Thus at points (g, π) of linearization instability of the equation $\delta_g\pi = 0$, there is an extra condition generated by each $X \in \ker (D\delta(g, \pi))^* = \{X \in \mathscr{X}^s\colon L_Xg = 0$ and $L_X\pi = 0\}$, so that the number of linearly independent extra conditions is equal to $\dim \ker (D\delta(g, \pi))^*$. For the Hamiltonian constraint there was one extra second order condition, corresponding to the fact that $\dim \ker (D\mathscr{H}(g_F, 0))^* = 1$.

II.4. The constraint manifold $\mathscr{C}^s$. We now consider the constraint set $\mathscr{C}^s = \mathscr{C}^s_{\mathscr{H}} \cap \mathscr{C}^s_\delta$. To show that $\mathscr{C}^s$ is a submanifold of $T^*\mathscr{M}^s$, we need additional restrictions in order to ensure that the intersection is transversal. At this point it is necessary to assume that $\operatorname{tr} \pi =$ constant. (See Note on p. 263.)

II.4.1. THEOREM. *Let $(g, \pi) \in \mathscr{C}^s = \mathscr{C}^s_{\mathscr{H}} \cap \mathscr{C}^s_\delta$ satisfy the following conditions:*

$C_{\mathscr{H}}$: *if $\pi = 0$, g is not flat;*
C_δ : *$L_Xg = 0$ and $L_X\pi = 0 \Rightarrow X = 0$;*
C_{tr} : *$\operatorname{tr} \pi' =$ constant.*

Then, in a neighborhood of (g, π), $\mathscr{C}^s$ is a C^∞ submanifold of $T^*\mathscr{M}^s$.

PROOF. Let $\Psi = (\mathscr{H}, \delta)\colon T^*\mathscr{M}^s \to \Lambda^{s-2} \times (\mathscr{X}^{s-1} \otimes \mathscr{V}^s)$, $(g, \pi) \mapsto (\mathscr{H}(g, \pi), \delta(g,\pi))$. Then

$$\begin{aligned}D\Psi(g, \pi) \cdot (h, \omega) &= (D\mathscr{H}(g, \pi) \cdot (h, \omega), D\delta(g, \pi) \cdot (h, \omega)) \\ &= (\gamma_{(g,\pi)} \cdot (h, \omega), \beta_{(g,\pi)} \cdot (h, \omega))\end{aligned}$$

and the L_2-adjoint of $D\Psi(g, \pi)$ is

$$(D\Psi(g, \pi))^* \cdot (N\mu_g, X \otimes \mu_g) = \gamma^*_{(g,\pi)}(N\mu_g) + \beta^*_{(g,\pi)}(X\mu_g).$$

Thus suppose $(D\Psi(g, \pi))^*(N\mu_g, X \otimes \mu_g) = 0$. Then

$$\text{(a)}\quad \begin{aligned} &g\Delta N + \operatorname{Hess} N - N \operatorname{Ric}(g) + \tfrac{1}{2} NR(g)g \\ &+ 2\left(\tfrac{1}{2}(\operatorname{tr}\pi')\,\pi' - \pi' \times \pi')^\flat - \tfrac{1}{2}(\tfrac{1}{2}(\operatorname{tr}\pi')^2 - \pi' \cdot \pi')g\right\} N \\ &- \tfrac{1}{2}(L_X\pi + X \otimes \delta\pi + \delta\pi \otimes X)'^\flat = 0; \end{aligned}$$

$$\text{(b)}\qquad 2(\tfrac{1}{2}(\operatorname{tr}\pi)\, g - \pi)\, N + \tfrac{1}{2}(L_X g)^\sharp\, \mu_g = 0.$$

Taking the trace of (a) and using $\mathscr{H}(g, \pi) = 0$ and $\delta\pi = 0$ gives

$$\text{(c)}\qquad 2\Delta N - \tfrac{1}{2}(X \cdot d \operatorname{tr}\pi' - \pi' \cdot L_X g - (\delta X)(\operatorname{tr}\pi')) = 0,$$

and from the trace of (b),

$$\text{(d)}\qquad \delta X = N \operatorname{tr}\pi'.$$

From (b),

$$\text{(e)}\qquad L_X g = 4(\pi - \tfrac{1}{2}(\operatorname{tr}\pi)g)N,$$

and, subsituting (d) and (e) into (c) gives

$$\text{(f)}\qquad 2\Delta N + 2(\pi' \cdot \pi' - \tfrac{1}{4}(\operatorname{tr}\pi')^2)N - \tfrac{1}{2}X \cdot d \operatorname{tr}\pi' = 0.$$

Since $P(\pi',\pi') \equiv \pi' \cdot \pi' - \frac{1}{4}(\operatorname{tr}\pi')^2 = (\pi' - \frac{1}{2}(\operatorname{tr}\pi')g) \cdot (\pi' - \frac{1}{2}(\operatorname{tr}\pi')g)$, the coefficient of N is positive-definite. Thus, if $\operatorname{tr}\pi' =$ constant,

$$2\Delta N + 2P(\pi', \pi')\, N = 0,$$

so that $N = 0$ unless $\pi' = 0$. If $N = 0$, from (a) $L_X\pi = 0$ and from (b) $L_X g = 0$ so $X = 0$.

If $\pi' = 0$, then $\Delta N = 0$ so N is constant and so from (a), $N(\operatorname{Ric}(g) - \frac{1}{2} gR(g)) = 0 \Rightarrow N = 0$, since $\operatorname{Ric}(g) \neq 0$ in the case that $\pi' = 0$. Then, again, $X = 0$. Thus, in either case, $(D\Psi(g, \pi))^*$ is injective.

The symbol $\sigma_{\xi_x}(D\Psi(g,\pi)^*)$ is given by

$$\begin{aligned}\sigma_{\xi_x} \cdot (s, Y) = \{&(-g\,\|\xi_x\|^2 + \xi_x \otimes \xi_x)\, s - \tfrac{1}{2}(-\pi_i'^k\, \xi_k\, Y_j - \pi_j'^k\, \xi_k\, Y_i + \pi'_{ij}\, \xi_k\, Y^k), \\ &\tfrac{1}{2}(Y \otimes \xi_x^\sharp + \xi_x^\sharp \otimes Y)\mu_g(x)\}.\end{aligned}$$

Thus if $\sigma_{\xi_x}(s, Y) = 0$, $\xi_x \neq 0$, from the second factor $Y = 0$ and so from the first factor $s = 0$. Thus σ_{ξ_x} is injective, $D\Psi(g, \pi)$ is surjective, and Ψ is a submersion at (g,π). □

It would be nice if the $\operatorname{tr}\pi' =$ constant condition could be dropped and we conjecture that it can. However, because of the coupling of equations (a) and (b) without the $\operatorname{tr}\pi' =$ const condition, it is possible that these equations have nonzero solutions (N, X) even at those (g, π) that satisfy conditions $C_{\mathscr{H}}$ and C_δ.

II.5. Integrating deformations of Ricci-flat spacetimes. As explained previously, we can use II.4.1 to prove the following result. We consider only the C^∞ case here.

II.5.1. THEOREM. *Let ${}^{(4)}g$ be a smooth Lorentz metric on a 4-manifold V satisfying*

Einstein's empty-space field equations $\mathrm{Ric}({}^{(4)}g) = 0$, *and let* ${}^{(4)}h$ *be a solution to the linearized equation*

$$D\ \mathrm{Ric}({}^{(4)}g) \cdot {}^{(4)}h = 0$$

about the solution ${}^{(4)}g$.

Assume that V *has a compact connected oriented spacelike hypersurface* M *with induced riemannian metric* g *and second fundamental form* k *that satisfy conditions* $C_{\mathscr{H}}$, C_δ, *and* C_{tr}. *Then there exist a tubular neighborhood* V' *of* M, *a* $\delta > 0$, *and a smooth curve* ${}^{(4)}g(\lambda)$, $-\delta < \lambda < \delta$, *of exact solutions to Einstein's equations defined on* V' *tangent to* ${}^{(4)}h$ *at* ${}^{(4)}g$, *i.e.*, ${}^{(4)}g(0) = {}^{(4)}g$, ${}^{(4)}g'(0) = {}^{(4)}h$, *and* $\mathrm{Ric}({}^{(4)}g(\lambda)) = 0$ *in a tubular neighborhood of* M.

PROOF. Let (g, π) be the variables on M induced by ${}^{(4)}g$. A deformation ${}^{(4)}h$ of $\mathrm{Ric}({}^{(4)}g) = 0$ induces a deformation (h, ω) of the linearized constraint equations, $D\mathscr{H}(g, \pi) \cdot (h, \omega) = 0$, $D\delta(g, \pi) \cdot (h, \omega) = 0$. Since (g, π) satisfies conditions $C_{\mathscr{H}}$, C_δ, and C_{tr}, by II. 4.1, $\mathscr{C}$ is a smooth submanifold with tangent space $T^{(g,\pi)}\mathscr{C} = \ker(D\mathscr{H}(g, \pi), D\delta(g, \pi))$. Since (h, ω) is tangent to $\mathscr{C}$, we can find a curve $(g(\lambda), \pi(\lambda)) \in \mathscr{C}$ tangent to (h, ω). By the evolution theory, this curve of solutions to the constraint equations gives us a curve ${}^{(4)}g(\lambda)$ of spacetimes defined in a tubular neighborhood V' of M. By a transformation of coordinates, ${}^{(4)}g(\lambda)$ can be made tangent to ${}^{(4)}h$. See **[16]** for details. □

Thus a solution of the linearized Einstein empty-space field equations actually approximates to first order a curve of exact solutions to the nonlinear equations in a tubular neighborhood of any compact spacelike hypersurface that satisfies conditions $C_{\mathscr{H}}$, C_δ, and C_{tr}. Because these conditions are so weak, presumably most spacetimes which have compact spacelike hypersurfaces have a hypersurface M satisfying these conditions, and thus is linearization stable in a tubular neighborhood of M. Moreover, by using standard arguments and by considering the maximal development (see **[12]**) of the Cauchy data of the curve of spacetimes ${}^{(4)}g(\lambda)$, there will be a maximal common development (which approximates the maximal development of ${}^{(4)}g(0)$) for which the spacetime is linearization stable.

REFERENCES

1. R. Arnowitt, S. Deser and C. W. Misner, *The dynamics of general relativity*, in *Gravitation: An Introduction to Current Research* (L. Witten, Ed.), Wiley, New York, 1962, pp. 227–265. MR **26** #1182.

2. M. Berger, *Sur les variétés d'Einstein compactes*, C. R. de la IIIe Réunion du Groupement des Mathématiciens d'Expression Latine (Namur, 1965), Librairie Universitaire, Louvain, 1966, pp. 35–55. MR **38** #6502.

3. ———, *Quelques formules de variation, pour une structure riemannienne*, Ann. Sci. École Norm. Sup. (4) **3** (1970), 285–294. MR **43** #3969.

4. M. Berger and D. Ebin, *Some decompositions of the space of symmetric tensors on a Riemannian manifold*, J. Differential Geometry **3** (1969), 379–392. MR **42** #993.

5. D. Brill, *Isolated solutions in general relativity*, in *Gravitation: Problems and Prospects*, "Naukova Dumka", Kiev, 1972, pp. 17–22.

6. D. Brill and S. Deser, *Instability of closed spaces in general relativity*, Comm. Math. Phys. **32** (1973), 291–304.

7. Y. Choquet-Bruhat, *New elliptic system and global solutions for the constraints equations in general relativity*, Comm. Math. Phys. **21** (1971), 211–218.

8. ———, *Global solutions of the equations of constraint in general relativity on closed manifolds*, Istituto Nazionale di Alta Matematica, Symposia Matematica, vol. XII, Bologna, Academic Press, New York, 1973.

9. ———, *Global solutions of the constraints equations on open and closed manifolds*, General Relativity and Gravitation **5** (1974), 49–60.

10. Y. Choquet-Bruhat and S. Deser, *On the stability of flat space*, Ann. Phys. **81** (1973), 165–178.

11. ———, *Stabilité initiale de l'espace temps de Minkowski*, C. R. Acad. Sci. Paris Sér. A-B **275** (1972), 1019–1021.

12. Y. Choquet-Bruhat and R. Geroch, *Global aspects of the Cauchy problem in general relativity*, Comm. Math. Phys. **14** (1969), 329–335. MR **40** #3872.

13. D. Ebin, *The manifold of Riemannian metrics*, Proc. Sympos. Pure Math., vol. 15, Amer. Math. Soc., Providence, R.I., 1970, pp. 11–40; see also: Bull. Amer. Math. Soc. **74** (1968), 1002–1004. MR **42** #2506.

14. A. Fischer and J. Marsden, *The Einstein equations of evolution—a geometric approach*, J. Mathematical Phys. **13** (1972), 546–568. MR **45** #8214.

15. ———, *The Einstein evolution equations as a first-order quasi-linear symmetric hyperbolic system*. I, Comm. Math. Phys. **28** (1972), 1–38. MR **46** #8616.

16. ———, *Linearization stability of the Einstein equations* (to appear); see also: Bull. Amer. Math. Soc. **79** (1973), 997–1003.

17. ———, *Global analysis and general relativity*, General Relativity and Gravitation **5** (1973), 73–77.

18. ———, *Deformations of the scalar curvature,* Duke Math. J. (to appear); see also: Bull. Amer. Math. Soc. **80** (1974), 479–484.

19. A. Fischer and J. A. Wolf, *The Calabi construction for compact Ricci-flat riemannian manifolds*, Bull. Amer. Math. Soc. **80** (1974), 92–97.

20. ———, *The structure of compact Ricci-flat riemannian manifolds*, J. Differential Geometry **10** (1975).

21. J. Kazdan and F. Warner, *Prescribing curvatures*, these PROCEEDINGS.

22. ———, *A direct approach to the determination of Gaussian and scalar curvature functions*, Invent. Math. (to appear).

23. S. Lang, *Differential manifolds*, Addison-Wesley, Reading, Mass., 1972.

24. A. Lichnerowicz, *L'intégration des équations de la gravitation relativiste et le problème des n corps*, J. Math. Pures Appl. (9) **23** (1944), 37–63. MR **7**, 266.

25. ———, *Propagateurs et commutateurs en relativité générale*, Inst. Hautes Études Sci. Publ. Math. No. 10 (1961), 56 pp. MR **28** #967.

26. ———, *Spineurs harmoniques*, C. R. Acad. Sci. Paris **257** (1963), 7–9. MR **27** #6218.

27. J. Marsden, D. Ebin and A. Fischer, *Diffeomeophism groups, hydrodynamics, and general relativity*, Proc. Thirteenth Biennial Seminar of the Canadian Math. Congress, Montreal, 1972, pp. 135–279.

28. J. Marsden and A. Fischer, *General relativity as a Hamiltonian system*, Istituto Nazionale di Alta Matematica, Symposia Matematica XIV, Bologna, Academic Press, New York, 1974, pp. 193–206.

29. E. Nelson, *Tensor analysis*, Princeton Univ. Press, Princeton, N.J., 1967.

30. L. Nirenberg and H. Walker, *The null spaces of elliptic partial differential operators in* R^n, J. Math. Anal. Appl. **42** (1973), 271–301.

31. M. Obata, *Certain conditions for a riemannian manifold to be isometric with a sphere*, J. Math. Soc. Japan **14** (1962), 333–340. MR **25** #5479.

32. ———, *Riemannian manifolds admitting a solution of a certain system of differential equations*, Proc. United States-Japan Seminar in Differential Geometry (Kyoto, 1965), Nippon Hyoronsha, Tokyo, 1966, pp. 101–114, MR **35** #7263.

33. N. O'Murchadha and J. W. York, *Existence and uniqueness of solutions of the Hamiltonian constraint of general relativity on compact manifolds*, J. Mathematical Phys. **14** (1973), 1551–1557.

34. K. Yano, *On harmonic killing vectors*, Ann. of Math. (2) **55** (1952), 38–45. MR **13**, 689.

35. J. A. Wolf, *Spaces of constant curvature*, 2nd ed., J. A. Wolf, Berkeley, Calif., 1972.

36. ———, *Local and global equivalence for flat affine manifolds with parallel geometric structures*, Geometriae Dedicata **2** (1973), 127–132.

37. J. P. Bourguignon, Thèse, Université de Paris VII, 1974; also, Compositio Math. (to appear).

38. Y. Muto, *On Einstein metrics*, J. Differential Geometry **9** (1974), 521–530.

39. V. Moncrief, *Spacetime symmetries and linearization stability of the Einstein equations*, J. Math. Phys. (to appear).

40. P. D'Eath, *Three perturbation problems in general relativity*, Thesis, Cambridge, 1974.

Note. In **[39]**, Moncrief shows that the map Ψ in II.4.1 is a submersion if and only if V, the spacetime, has no killing fields. Thus the conditions on p. 248 can be replaced by "V has no killing fields." For some applications to specific spacetimes, see **[40]**.

UNIVERSITY OF CALIFORNIA, SANTA CRUZ AND BERKELEY

Proceedings of Symposia in Pure Mathematics
Volume 27, 1975

THE SPECTRAL GEOMETRY OF REAL AND COMPLEX MANIFOLDS

PETER B. GILKEY*

Introduction. The relationship between the geometry of a smooth manifold and the spectrum of the Laplacian has been explored extensively recently. An analytic proof of the Atiyah-Singer index theorem has been discovered using spectral geometry ([1], [4], [5], [10], [11]). Patodi has proved that the spectral geometry of a real manifold determines whether it has constant scalar curvature, constant sectional curvature, flat curvature, or is Einstein [9]. We have proved that the spectral geometry of a complex Hermitian manifold determines whether the metric is Kaehler [6] and whether the metric has constant holomorphic sectional curvature [7].

The proof of these results depends upon certain asymptotic invariants of the spectrum of these operators. These invariants are computed by integrating local invariants of the metric tensor. We use the invariants of order 0,2, and 4 to prove the above results. There are conjectures regarding symmetric spaces which require the invariants of order 6.

In this paper, we compute the invariants of order 0,2,4 for differential operators of order 2 with leading symbol given by the metric tensor. This wide class of operators includes both the real and complex Laplacians. In § 1, we define the invariants and list their functorial properties. In § 2, we use invariance theory to determine all possible local invariants of order 2 and 4 for differential operators of this class. In § 3, we compute the asymptotic formulas explicitly. In § 4, we apply these formulas to obtain the geometric results for real manifolds noted above.

AMS (*MOS*) *subject classifications* (1970). Primary 53C55, 58G99.

*Research partially supported by NSF grant GP-34785X.

1. In this section, we describe the asymptotic invariants of the spectrum of certain elliptic operators. Although these invariants can be discussed in more generality, we restrict the discussion to differential operators of order 2 with leading symbol given by the metric tensor: Let M_m be a Riemannian manifold of dimension m. Let X_k be a system of local coordinates centered at x_0. The metric is given by

$$ds^2 = g_{ij}\, dX_i\, dX_j.$$

Throughout this paper we will sum over repeated indices unless otherwise indicated. Let V be a smooth r-dimensional vector bundle over M and let $s = (s_1,\cdots,s_r)$ be a local frame field for V. Let A be a square $r \times r$ matrix and let f be the r-tuple $f = (f_1,\cdots,f_r)$. We define

$$f \cdot s = f_j\, s_j, \qquad Af = (A_{1k}\, f_k,\cdots,\ A_{rk}\, f_k).$$

If $h(x)$ is any function on M, we use the following notation for the derivatives of h:

$$h_{/i_1 \cdots i_p} = d/dX_{i_2} \cdots d/dX_{i_p}(h).$$

Let D be a second order elliptic operator acting on the space of smooth sections to V. The leading order symbol of D is invariantly defined, but the lower order symbols depend upon the frame and coordinate system chosen. We restrict our attention to operators D with leading symbol given by the metric tensor, i.e. that D has the form:

$$D(f \cdot s) = D_s(f) \cdot s = -(g^{ij} f_{/ij} + A^X_{s,k}\, f_{/k} + B^X_s f) \cdot s.$$

The matrix g^{ij} is the inverse matrix of the matrix g_{ij}. The matrices $A^X_{s,k}$ and B^X_s express the lower order portion of the differential operator relative to the frame s and coordinate system X. When the identity of the coordinate system X or frame s under consideration is clear, the notational dependence upon X or s will be dropped.

One can use the calculus of pseudo-differential operators depending upon a complex parameter as developed by Seeley (**[12]**, **[13]**) to show that $\exp(-tD)$ is well defined for $t > 0$. It is an infinitely smoothing operator which is represented by the kernel function

$$K(t, x, y, D) \in \mathrm{HOM}(V_y, V_x).$$

$K(t, x, y, D)$ is a linear map from V_y to V_x such that

$$\exp(-tD)(f)(x) = \int_{M_m} K(t, x, y, D)\,(f)\, d\mathrm{vol}(y).$$

Let $K(t, x, D) = K(t, x, x, D)$ belong to $\mathrm{END}(V_x, V_x)$; it is an endomorphism of the fibre. $K(t, x, D)$ has a well-defined asymptotic expansion as $t \to 0^+$ of the form:

$$K(t, x, D) \sim \sum_{n=0}^{\infty} E_n(x, D)\, t^{(n-m)/2}.$$

Since D is a differential operator, the endomorphisms E_n for n odd can be shown to vanish identically.

A priori, the invariantly defined endomorphisms $E_n(x, D)$ are global in nature. Therefore, it is somewhat surprising that these endomorphisms are local invariants of the differential operator D. There is a functorially defined formula for E_n in the derivatives of the symbol of the operator. This formula is a noncommutative polynomial in these derivatives. Since the definition of these endomorphisms is independent of the frame s and coordinate system X, it follows that this formula is invariant under changes of frame and coordinates.

By using the universality of these formulas and dimensional analysis, we showed [3] that the formulas for E_n are homogeneous of order n in the following sense: define derivatives of degree k of

the metric having order k,

the matrices $A^X_{s,k}$, having order $k + 1$,

the matrix B^X_s having order $k + 2$.

To compute the total order of any monomial in these variables, we sum up the orders of each variable in the monomial. Thus, for example, $\mathrm{ord}(g_{11/2}\ A_{s,1/22}\ B^3_s) = 10$. According to this definition, $E_n(x, D)$ is composed of monomials which are all of order n in the derivatives of the symbol of D.

We construct scalar invariants of the operator by taking the trace of these endomorphisms. Let $B_n(x, D) = \mathrm{Tr}(E_n(x, D))$. If D is the Laplacian acting on functions, then $B_{2n}(x, D) = a_n$ is the Minakshisundaram coefficient. We have chosen to change this notation in order to index these invariants by their order and to avoid fractional indexing, if D is not a differential operator but is pseudo-differential.

If the vector bundle V is equipped with a hermitian inner product and if D is selfadjoint with respect to this inner product, let $\{\lambda_i, \phi_i\}_{i=1}^{\infty}$ be a spectral resolution of D into smooth orthonormal eigensections ϕ_i. We can define the kernel function by

$$K(t, x, y, D) = \sum_{i=1}^{\infty} \exp(-t\lambda_i)\, \phi_i(x) \otimes \bar{\phi}_i(y).$$

Consequently:

$$\sum_{n=0}^{\infty} E_n(x, D)\, t^{(n-m)/2} \sim \sum_{i=1}^{\infty} \exp(-t\lambda_i)\, \phi_i(x) \otimes \bar{\phi}_i(x),$$

$$\sum_{n=0}^{\infty} B_n(x, D)\, t^{(n-m)/2} \sim \sum_{i=1}^{\infty} \exp(-t\lambda_i) \cdot (\phi_i, \phi_i)(x).$$

The functions ϕ_i integrate to 1 since they are orthonormal. Therefore:

$$\sum_{n=0}^{\infty} \left[\int_{M_m} B_n(x, D)\, d\mathrm{vol}\right] t^{(n-m)/2} \sim \sum_{i=1}^{\infty} \exp(-t\lambda_i).$$

Consequently, if D is selfadjoint, the numbers $[\int_{M_m} B_n(x, D)\, d\mathrm{vol}]$ are determined by the spectrum of the operator and are said to be spectral invariants.

If D is either the complex or real Laplacian acting on a space of forms, the symbol of D is a functorial expression in the derivatives of the metric. Consequently, the invariants $B_n(x,D)$ are invariants of order n in the derivatives of the metric.

The integral of these invariants is determined by the spectral geometry of the manifold and can be used to deduce geometrical properties of the manifold.

However, we do not generally assume that D is selfadjoint or that the vector bundle has an inner product. We can use arbitrary matrices $A_{s,k}$ and B_s in defining D.

LEMMA 1.1. *Let $P(g_{i/i_1}\ldots, A_{s,j/j_1}\ldots, B_{s/k_1}\ldots,)$ be an arbitrary noncommutative matrix in the derivatives of the symbol of the operator. If P vanishes identically for every r, metric g, and matrices $A_{s,k}$, B_s, then P is the zero polynomial.*

PROOF. The proof in the commutative case is a direct application of Taylor's theorem; since the proof in the noncommutative case is only slightly more difficult, it is omitted.

The following functorial properties follow directly from the definition of the endomorphisms E_n:

THEOREM 1.2. (1) *Let c be constant. Then $E_n(x, D-c) = \sum_{k=0}^{n/2} c^k/k!E_{n-2k}(x,D)$.*

(2) *Let $\{D_i\}$ on $\Gamma(V_i)$ be a collection of differential operators with leading symbol given by the same metric tensor. Then $E_n(x, \oplus_i D_i) = \oplus_i E_n(x,D_i)$.*

(3) *Let (D_i, V_i, M_i) be given for $i=1,2$. The manifold $M = M_1 \times M_2$ inherits a natural Riemannian metric. Let $D=(D_1 \otimes 1+1 \otimes D_2)$ on $\Gamma(V_1 \otimes V_2)$ over M. Then*

$$E_n((x_1, x_2), D) = \sum_{p+q=n} E_p(x_1, D_1) \otimes E_q(x_2, D_2).$$

PROOF. The functional equations

(1)′ $\exp(-t(D-c)) = \exp(tc)\exp(-tD)$,

(2)′ $\exp(-t(\oplus_i D_i)) = \oplus_i \exp(-tD_i)$,

(3)′ $\exp(-tD) = \exp(-tD_1) \otimes \exp(-tD_2)$

imply that

(1)″ $K(t, x, D-c) = \exp(tc)\, K(t, x, D)$,

(2)″ $K(t, x, \oplus D_i) = \oplus K(t, x, D_i)$,

(3)″ $K(t, (x_1, x_2), D) = K(t, x_1, D_1) \otimes K(t, x_2, D_2)$.

We equate the coefficients of equal powers of t in the asymptotic expansions to prove (1), (2), (3) from (1)″, (2)″, (3)″.

2. Let $\mathscr{P}_n$ denote the space of all invariant maps of order n from the derivatives of the symbol of the operator D to endomorphisms. In this section, we will compute a basis for the linear spaces $\mathscr{P}_2$ and $\mathscr{P}_4$. These spaces depend upon the dimension of the manifold, but are independent of the dimension of the vector bundle. We will use Theorem 1.2(2) in §3 to prove that the polynomials representing $E_n(x, D)$ are independent of the dimension of the vector bundle and depend only upon n and m.

Let ∇ be a connection on the vector bundle V; $\nabla : \Gamma(V) \to \Gamma(V \otimes T^*M)$. Since M_m is a Riemannian manifold, there is a natural connection ∇_g^* on T^*M. We define the induced connection

$$(\nabla \otimes 1 + 1 \otimes \nabla_g^*) : \Gamma(V \otimes T^*M) \longrightarrow \Gamma(V \otimes T^*M \otimes T^*M).$$

We construct a unique second order elliptic operator D_∇ from the connection ∇

by the diagram:

$$\Gamma(V) \xrightarrow{\nabla} \Gamma(V\otimes T^* M) \xrightarrow{\nabla\otimes 1+1\otimes\nabla_g^*} \Gamma(V\otimes T^*M\otimes T^*M) \xrightarrow{-1\otimes G} \Gamma(V).$$

G is the map defined on $T^*M\otimes T^*M\to R$ induced by the Riemannian metric.

We compute the operator D_∇ in local coordinates as follows: Let

$$\nabla_g (d/dX_j) = \sum \Gamma_{ijk}\, d/dX_k\otimes dX_i.$$

We suppose that the coordinate system is normalized with respect to the metric so that $g_{ij}(x_0)=\delta_{ij}$ and $g_{ij/k}(x_0)=0$. Consequently, there is a formula of the form:

$$\Gamma_{ijk} = ag_{ij/k} + bg_{ik/j} + cg_{jk/i} + O(x^2).$$

We use the identities which define ∇_g uniquely:

$$\Gamma_{ijk}+\Gamma_{ikj}=g_{jk/i} \quad \text{(Riemannian)}, \quad \Gamma_{ijk}=\Gamma_{jik} \quad \text{(torsion free)}.$$

This implies that $a+b=0$, $b=c$, $2c=1$. Consequently

$$\Gamma_{ijk} = \tfrac{1}{2}(g_{jk/i} + g_{ik/j} - g_{ij/k}) + O(x^2).$$

We define ∇_g^* on T^*M so that $(\nabla_g d/dX_k, dX_j)+(d/dX_k, \nabla_g dX_j) = 0$. This implies $\Gamma^*_{ikj}+\Gamma_{ijk}=0$.

Let the connection ∇ on $\Gamma(V)$ have connection matrix w_k relative to the frame s so that $\nabla(fs)=(f_{/k}+w_k f)\, s\otimes dX_k$. We compute that:

$$\begin{aligned} D_\nabla(fs) &= -G(\nabla\otimes 1+1\otimes\nabla_g^*)(f_{/i}+w_i f)(s\otimes dX_i) \\ &= -G((f_{/i}+w_i f)_{/j}+w_j(f_{/i}+w_i f))(s\otimes dX_i\otimes dX_j) - G((f_{/i}+w_i f)\Gamma^*_{jik}) \\ &\quad \cdot(s\otimes dX_k\otimes dX_j) \\ &= -(g^{ij} f_{/ij}+(2g^{ij} w_j+g^{jk}\Gamma^*_{jik}) f_{/i}+(g^{ij}w_{i/j}+g^{ij}w_j w_i+g^{jk}w_i\Gamma^*_{jik})f)\cdot s. \end{aligned}$$

The operator $D-D_\nabla$ is an invariantly defined first order operator with leading symbol: $(2g^{ij} w_j+g^{jk}\Gamma^*_{jik}-A_{s,i})\, d/dX_i$. We define the connection ∇ uniquely by requiring that:

$$2g^{ij} w_j+g^{jk}\Gamma^*_{jik}-A_{s,i}=0, \quad i=1, \cdots, m.$$

This system of equations determines w_i uniquely. We compute that

$$w_i= \frac{1}{2}\left(A_{s,i}-\sum_{j=1}^{m} \Gamma^*_{jij}\right)+O(x^2).$$

(We suppose $A_{s,k}(x_0)=0$.) The connection defined above is the only connection we will consider for the remainder of this paper. We let $\bar{D}=D_\nabla$ for this connection.

We wish to normalize the choice of frames which we will consider. Let s' be another frame for V gnd let $s'_k=s_j U_{jk}$. Then $f's'=U(f')\cdot s$ and therefore $f=Uf'$. In the new frame, $Ds'(f')=U^{-1} D_s(Uf')$. Therefore:

$$\begin{aligned} D_{s'}(f') = -\,(g^{ij}f'_{/ij} &+ U^{-1}(A_{s,k} + 2g^{ik} U_{/i})f'_{/k} \\ &+ U^{-1}(B_s U + A_{s,k}U_{/k} + g^{ij} U_{/ij})\, f'). \end{aligned}$$

This yields the transformation equations:

$$A_{s',k} = U^{-1}(A_{s,k}U + 2g^{ik}U_{/i}), \qquad B_{s'} = U^{-1}(B_sU + A_{s,k}U_{/k} + g^{ij}U_{/ij}).$$

We have normalized the coordinate system with respect to the metric so that $g_{ij}(x_0) = \delta_{ij}$ and $g_{ij/k}(x_0) = 0$. We can change the coordinate system by orthogonal transformations as well as by transformations which vanish to second order. In such a coordinate system, the transformation equations for $A_{s,k}$ become

$$A_{s',k}(x_0) = A_{s,k}(x_0) + 2U_{/k}(x_0).$$

We may also normalize the local frame s so that $A_{s,k}(x_0)=0$ by a suitable change of frame. Henceforth, we will restrict our attention to such normalized coordinate systems and frames. We will only consider changes of frame such that $U(x_0)=I$ and $dU(x_0)=0$.

The canonical connection ∇ defines the operator $\bar{D}$ such that

$$\bar{D}(fs) = -(g^{ij}f_{/ij} + w_{k/k}f)\,s + O(x)(f)\cdot s.$$

The operator $\bar{D}-D$ is defined invariantly and is an operator of order zero. Consequently, it is an invariantly defined endomorphism E. We summarize these computations in

THEOREM 2.1.
(1) $\Gamma_{ijk} = \frac{1}{2}(g_{jk/i} + g_{ik/j} - g_{ij/k}) + O(x^2)$,
(2) $\Gamma^*_{ijk} = -\Gamma_{ikj}$,
(3) $w_k = \frac{1}{2}(A_{s,k} + \Gamma_{jjk}) + O(x^2)$,
(4) $\bar{D}(fs) = -(g^{ij}f_{/ij} + w_{k/k}f)\,s + O(x)(f)\cdot s$,
(5) $E = \bar{D} - D = B_s - w_{k/k} + O(x)$.

We will need the following technical lemma in our study of the spaces $\mathscr{P}_n$:

LEMMA 2.2. *Let $P \subset \mathscr{P}_n$ and let Q be any monomial of P. Let* $\deg_k(Q)$ *be the number of times that the index k appears in the monomial Q. Thus, for example,* $\deg_1(A^2_{s,1/11}\,g_{12/11}) = 9$. *Then* $\deg_k(Q)$ *is even for every k. Furthermore, the form of the polynomial P is unchanged by a coordinate permutation.*

PROOF. Both of these facts follow directly from the invariance of P under the action of $O(m)$ on the coordinate system. We consider the coordinate transformation which takes $X_k \to -X_k$, $X_j \to X_j$ for $j \neq k$. We decompose $P = \sum c_Q(P)\cdot Q$, where the sum ranges over all monic monomials and $c_Q(P)$ denotes the coefficient of Q in the polynomial P. In the new coordinate system,

$$P' = \sum (-1)^{\deg_k(Q)}\, c_Q(P)\cdot Q.$$

However, $P' = P$ since P is invariant. We proved in Lemma 1.1 that any relation among the derivatives of the symbol of the operator can be satisfied only if all the coefficients are zero. Therefore $(-1)^{\deg_k(Q)}c_Q(P) = c_Q(P)$. This implies that $\deg_k(Q)$ is even if Q is a monomial of P. Similarly, the form of P must be invariant under a coordinate permutation.

We use this lemma to compute the space $\mathscr{P}_2$:

THEOREM 2. 3. *Let $P \in \mathscr{P}_2$. Let $K(g)$ denote the scalar curvature of the metric g and let E denote the endomorphism $\tilde{D} - D$. Then there exist constants a, b such that $P = aK(g) + bE$.*

PROOF. The most general polynomial P which is of order 2 in the derivatives of the symbol of the operator has the form

$$P = P_2(g) + bB_s + cA_{s,k/k}.$$

Since $E = B_s$ + other terms, let $Q = P - bE = p_2'(g) + c'A_{s,k/k}$. We perform a change of frame:

$$A'_{s,k/k} = (U^{-1} A_{s,k} U + 2U^{-1} g^{ik} U_{/i})_{/k} = A_{s,k/k} + 2U_{/kk} + O(x).$$

Consequently, in the new frame,

$$Q'(x_0) = p_2'(g)(x_0) + c'A_{s,k/k}(x_0) + 2c'U_{/kk}(x_0).$$

Since Q is also invariant, $Q' = Q$ and therefore $2c'U_{/kk}(x_0) = 0$. This implies $c' = 0$ and therefore $Q = p_2'(g)$. Since Q is invariant, $p_2'(g)$ is an invariant polynomial of order 2 in the derivatives of the metric. It is well known that the only such polynomial is a multiple of the scalar curvature. Therefore $p_2'(g) = aK(g)$ for some a which completes the proof of the theorem.

There are more invariants of order 4 in the derivatives of the symbol. We can construct these invariants by using the connection we have defined. Let $\mathrm{END}(V) = V \otimes V^*$ be the bundle of all linear maps from V to V. The connection ∇ can be extended to $\mathrm{END}(V)$ by the equation $\nabla(\tilde{E})(fs) = \nabla(\tilde{E}fs) - \tilde{E}(\nabla fs)$. If w_k' denotes the induced connection matrix, then $w_k'(\tilde{E}) = w_k\tilde{E} - \tilde{E}w_k$. However, $w_k = \frac{1}{2}[A_{s,k} + \Gamma_{jjk}] + O(x^2)$ and the latter matrix commutes with any endomorphism E. Consequently:

$$w_k' \tfrac{1}{2} = \tfrac{1}{2}A_{s,k}\tilde{E} - \tfrac{1}{2}\tilde{E}A_{s,k} + O(x^2)\tilde{E}.$$

From the connection on $\mathrm{END}(V)$, we extend the differential operator $\tilde{D}$ to $\Gamma(\mathrm{END}(V))$.

We have the identity:

$$\begin{aligned} -\tilde{D}(E)(x_0) &= (g^{ij} E_{/ij} + w'_{k/k}(E)) \\ &= B_{s/kk} + \tfrac{1}{2}A_{s,k/k}B_s - \tfrac{1}{2}B_sA_{s,k/k} + \text{other terms not involving } B_s. \end{aligned}$$

Similarly, $E^2 = B_s^2 + \cdots$. Finally, we can construct one further invariant : let R_D be the curvature tensor defined by the connection ∇. $R_D \in \Gamma(\mathrm{END}(V) \otimes \Lambda^2 T^*M)$. By using matrix multiplication, $R_D^2 \in \Gamma(\mathrm{END}(V) \otimes \Lambda^2 T^*M \otimes \Lambda^2 T^*M)$. We use the Riemannian metric to map $\Gamma(\mathrm{END}(V) \otimes \Lambda^2 T^*M \otimes \Lambda^2 T^*M) \longrightarrow \mathrm{END}(V)$. We compute that:

$$\begin{aligned} R_D &= \sum_{j<k} (w_{k/j} - w_{j/k})\, dX_j \wedge dX_k + O(x) \\ &= \sum_{j<k} \tfrac{1}{2}(A_{s,k/j} - A_{s,j/k} + \text{terms in the metric}) + O(x), \\ G(R_D^2) &= \tfrac{1}{4}\sum_{j<k} (A_{s,k/j} - A_{s,j/k} + \text{terms in the metric})^2 + O(x). \end{aligned}$$

We summarize these computations in the following theorem which classifies $\mathscr{P}_4$:

THEOREM 2.4. *Let $P \in \mathscr{P}_4$; then there exist uniquely defined constants a, b, c, d such that $P = p_4(g) + aK(g)E + bE^2 + c\bar{D}E + dG(R_D^2)$ where $p_4(g)$ is an invariant of order 4 in the derivatives of the metric.*

PROOF. Since $E^2 = B_s^2 + \cdots$, $-\bar{D}E = B_{s/kk} + \cdots$, $G(R_D^2) = A_{s,k/j}A_{s,k/j}/4$, we can construct a new endomorphism Q which does not contain these three monomials by subtracting an appropriate multiple of these three invariants from P. Q must have the form:

$$Q = p_4'(g) + p_2'(g)B_s + aA_{s,k/k}B_s + a'B_sA_{s,kk} + b_{jk}A_{s,k/kjj} + c_{i,j,k,l}A_{s,i/j}A_{s,k/l} + d_{ij}(g)A_{s,i/j}$$

where $c_{ijij} = 0$. Let $s' = sU$ be a change of frame such that $U(x_0) = I$ and $dU(x_0) = 0$. We compute Q in the new frame in terms of the old frame. Since the form of Q is invariant, all the resulting terms involving the derivatives of U must cancel out. Terms in $U_{/kkjj}$ can only come from monomials of the form $A_{s,k/jj}$ and consequently $b_{j,k} + b_{k,j} = 0$. However, the form of Q is invariant under coordinate permutations so $b_{j,k} = b_{k,j}$. Therefore $c_{j,k} = 0$. Terms in $B_sU_{/kk}$ and $U_{/kk}B_s$ can only come from the monomials $A_{s,k/k}B_s$ and $B_sA_{s,k/k}$. This implies that $a = a' = 0$. Consequently, Q has the form:

$$Q = p_4'(g) + p_2'(g)B_s + c_{i,j,k,l}A_{s,i/j}A_{s,k/l} + d_{i,j}(g)A_{s,i/j}.$$

Let Y be a new coordinate system. Since the only term involving B_s is $p_2'(g)B_s$, $p_2'(g)$ must be invariant. This implies that $p_2'(g)$ is a multiple of the scalar curvature. We may assume without loss of generality by subtracting an appropriate multiple of $K(g)E$ that $p_2'(g) = 0$. Therefore to complete the proof of the theorem, it suffices to show that $d_{ij}(g) = c_{ijkl} = 0$.

We consider a change of frame and collect terms in $A_{s,i/j}U_{/ij}$ for $i \neq j$. These terms can only come from the monomials $A_{s,i/j}A_{s,i/j}$ and $A_{s,i/j}A_{s,j/i}$. Therefore $c_{ijij} + c_{ijji} = 0$. We have assumed $c_{ijij} = 0$ and therefore $c_{ijji} = 0$ as well. Finally, we collect terms in $A_{s,i/i}U_{/jj}$. These terms can only come from the monomial $A_{s,i/i} \cdot A_{s,j/j}$. This implies that $c_{iijj} = 0$. Since every index must appear an even number of times in every monomial, these are the only 3 terms of the form $A_{s,i/j}A_{s,k/l}$. This implies that $Q = p_4'(g) + d_{ij}^2(g)A_{s,i/j}$. By making a change of frame and collecting the terms in $U_{/ij}$, we conclude that $d_{ij}^2 + d_{ji}^2 = 0$. In geodesic polar coordinates, we can express the derivatives of the metric in terms of the curvature tensor R_{ijkl}. This implies $d_{ij}^2(g)$ must contain a monomial of the form $R_{i_1i_2i_3i_4}$ for some i_1, i_2, i_3, i_4. Every index appears with even multiplicity, and hence (i_1, i_2, i_3, i_4) is some permutation of (i, j, k, k) for some k. This implies that $d_{ij}^2(g) = cR_{ikjk} \cdot A_{s,i/j}$. The coefficient, however, does not change sign when i and j are permuted. Consequently $c = 0$ and this term does not appear as well. This completes the proof of the theorem.

3. In this section we compute the invariants $E_2(x, D)$ and $E_4(x, D)$. We use the following notation:

$\nabla_i = \nabla_{d/dX_i} : \Gamma(TM) \longrightarrow \Gamma(TM)$ is the Riemannian connection,

$R_{ijkl} = ((\nabla_i\nabla_j - \nabla_j\nabla_i)\, d/dX_k, d/dX_l)$ is the curvature tensor,

$\nabla_g^p : \Gamma(\Lambda^p T^*M) \longrightarrow \Gamma(\Lambda^p T^*M \otimes T^*M)$ is the induced connection,

R_g^p is the induced curvature tensor,

$K(g) = -\sum_{i<j} R_{ijij}$ is the scalar curvature,

$|R|^2 = \sum_{i,j,k,l} (R_{ijkl})^2$,

$|\mathrm{Ric}|^2 = \sum_{i,j} (\sum_k R_{ikjk})^2$,

$D_p^m = (d^*d + dd^*)$ on $\Gamma(\Lambda^p T^*M)$.

In §4, we will show that the connection induced by the operator D_p^m is the Riemannian connection. We will also show in Theorem 4.1 that:

$$\bar D_0^m = D_0^m, \quad R_{D_0^m} = 0, \qquad \mathrm{Tr}\,(\bar D_1^m - D_1^m) = -2K(g),$$
$$\mathrm{Tr}\,(\bar D_1^m - D_1^m)^2 = |\mathrm{Ric}|^2, \mathrm{Tr}\,(G(R_{D_1^m})^2) = -|R|^2/2.$$

It is well known [2], [8] that:

$$B_0(x, D_0^m) = (4\pi)^{-m/2}, \qquad B_2(x, D_0^m) = (4\pi)^{-m/2} K(g)/3,$$

$$B_4(x, D_0^m) = (4\pi)^{-m/2}(-D_0^m K(g)/15 + (10K(g)^2 - |\mathrm{Ric}|^2 + |R|^2)/180).$$

It should be noted that McKean-Singer [8] gave the coefficient of $|R|^2$ as 2/180 rather than the correct value of 1/180. McKean-Singer also showed that:

$$d/dt\,(\mathrm{Tr}\,(K(t, x, D_0^2) + K(t, x, D_2^2) - K(t, x, D_1^2))) = -D_0^2(\mathrm{Tr}\,K(t, x, D_0^2)).$$

By equating equal powers of t in the asymptotic expansion and by using Poincaré duality, we show that $2B_4(x, D_0^2) - B_4(x, D_1^2) = -D_0^2(B_2(t, D_0^2))$.

We apply these results to prove the following theorem:

THEOREM 3.1.

(1) $E_0(x, D) = (4\pi)^{-m/2} I$,

$B_0(x, D) = (4\pi)^{-m/2} r$ (*r is the dimension of the vector bundle*).

(2) $E_2(x, D) = B_2(x, D_0^m) I + (4\pi)^{-m/2}(E)$,

$B_2(x, D) = rB^2(x, D_0^m) + (4\pi)^{-m/2}\,\mathrm{Tr}\,(E)$ (*r is the dimension of V*).

(3) $E_4(x, D) = B_4(x, D_0^m) I + (4\pi)^{-m/2}$

$\cdot (K(g)E/3 + E^2/2 - \bar D E/6 + G(R_D^2)/6)$,

$B_4(x, D) = rB_4(x, D_0^m) + (4\pi)^{-m/2}$

$\cdot (K(g)\,\mathrm{Tr}\,(E)/3 + \mathrm{Tr}\,(E^2)/2 - D_0^m(E)/6 + \mathrm{Tr}\,(G(R_D^2)/6))$

where

$$B_2(x, D_0^m) = (4\pi)^{-m/2} K(g)/3 \quad \textit{and}$$
$$B_4(x, D_0^m) = (4\pi)^{-m/2}(-D_0^m K(g)/15 + (10K(g)^2 - |\mathrm{Ric}|^2 + |R|^2)/180).$$

PROOF. We deduce the formula for B_2 and B_4 from the formulas for E_2 and E_4.

The formulas (1) for E_0 are well known. We prove (2) as follows: By Theorem 2.3, there is a formula for E_2 of the form:

$$E_2(x, D) = a_{m,r} K(g) + b_{m,r}(\bar{D} - D).$$

By Theorem 1.2(2) applied to a sum of operators $E_2(x, \oplus_i D_i) = \oplus_i E_2(x, D_i)$. This shows that the constants $a_{m,r}$ and $b_{m,r}$ are independent of the dimension of the vector bundle involved. By Theorem 1.2(1),

$$\begin{aligned} E_2(x, D - c) &= E_2(x, D) + cE_0(x, D) = E_2(x, D) + (4\pi)^{-m/2} c \\ &= E_2(x, D) + cb_m. \end{aligned}$$

This implies that $b_m = (4\pi)^{-m/2}$. The endomorphism $E(D_0^m)$ vanishes, and hence $E_2(x, D_0^m) = B_2(x, D_0^m) = a_m K(g)$. This implies $a_m = (4\pi)^{-m/2} K(g)/3$ and completes the proof of (2).

We apply Theorem 2.4 to show that there is a formula for E_4 of the form:

$$E_4(x, D) = p_4(g) I + a_m K(g) E + b_m E^2 + c_m \bar{D}E + d_m G(R_D^2).$$

These constants are independent of the dimension of the vector bundle and depend only upon the dimension of the manifold. We apply this formula to the operator D_0^m. We have already noted that $E(D_0^m) = 0$ and consequently:

$$E_4(x, D_0^m) = B_4(x, D_0^m) = p_4(g).$$

To complete the proof of the theorem, we compute a_m, b_m, c_m, d_m.

By Theorem 1.2(1),

$$\begin{aligned} E_4(x, D - c) &= E_4(x, D) + cE_2(x, D) + c^2 / 2E_0(x, D) \\ &= E_4(x, D) + c(4\pi)^{-m/2}(K/3 + E) + c^2(4\pi)^{-m/2}/2 \\ &= E_4(x, D) + a_m cK(g) + b_m c^2 + c_m \bar{D}(c) + 2cb_m E. \end{aligned}$$

However, $\bar{D}(c) = 0$, and therefore $a_m = (4\pi)^{-m/2}/3$ and $b_m = (4\pi)^{-m/2}/2$. We complete the proof by computing c_m and d_m.

The invariants $\bar{D}(E)$ and $G(R_D^2)$ are linearly independent when restricted to a 2-dimensional manifold. If we take an arbitrary operator on M_2 and the flat operator on the flat torus T_{m-2}, we apply Theorem 1.2(3) to the operator $D \otimes 1 + 1 \otimes D_{\text{flat}}$ on $M_2 \otimes T_{m-2}$. This shows that

$$c_m \bar{D}E + d_m G(R_D^2) = (4\pi)^{-(m-2)/2}(c_2 \bar{D}E + d_2 G(R_D^2))$$

since all the higher order invariants of the flat operator vanish. We use the identity

$$2B_4(x, D_0^2) - B_4(x, D_1^2) = -D_2^0 B_2(x, D_0^2) = -(4\pi)^{-1} D_0^2 K(g)/3$$

to show that

$$\begin{aligned} &(4\pi)^{-1} \operatorname{Tr}(K(g) E(D_1^2)/3 + E(D_1^2)^2/2) \\ &\qquad + c_2 D_0^2 \operatorname{Tr}(E(D_1^2)) + d_2 \operatorname{Tr}(G(R_{D_1^2})) = (4\pi)^{-1} D_0^2 K(g)/3. \end{aligned}$$

We will show in §4 that $\operatorname{Tr}(E(D_1^2)) = -2K(g)$, $\operatorname{Tr}(E(D_1^2)^2) = |\operatorname{Ric}|^2 = 2K(g)^2$, and that $\mathrm{Tr}(G\,(R_{D_1^2})^2) = -\tfrac{1}{2}\,|R|^2 = -\,2K(g)^2$. Therefore:

$$-\,2K(g)^2\,((4\pi)^{-1}/\,3-(4\pi)^{-1}\,/\,2+d_2)\;-\;2c_2\,D_0^2\,K(g)=(4\pi)^{-1}\,D_0^2\,K(g)\,/\,3.$$

This implies that $d_2 = (4\pi)^{-1}/\,6$ and $c_2 = -(4\pi)^{-1}/\,6$. This completes the proof of Theorem 3.1.

4. In this section we will derive the combinatorial results which were used in §3. We will also determine the general formula for E_2, B_2, E_4, and B_4 for the operators D_p^m.

There is a canonical connection ∇_g^p which is induced by the Riemannian metric. There is also the connection $\nabla_{D_p^m}$ which is induced by the differential operator. The difference $\nabla_g^p - \nabla_{D_p^m}$ is an invariantly defined operator of order zero. We can compute it at x_0 functorially in terms of the first derivatives of the metric tensor. These derivatives vanish in a normalized coordinate system, and hence this linear map vanishes at x_0. Since it is invariantly defined, it must vanish identically and therefore $\nabla_g^p = \nabla_{D_p^m}$.

We will use the following notation: let

$$\begin{aligned} I &= (i_1,\cdots,i_p) \quad \text{for } i_1 < \cdots < i_p,\\ |I| &= p,\\ dX_I &= dX_{i_1} \wedge \cdots \wedge dX_{i_p},\\ ((\nabla_i^p\,\nabla_j^p &- \nabla_j^p\,\nabla_i^p)\,dX_I,\,dX_J) = \sum_{|J|=p} R_{ijIJ}. \end{aligned}$$

Define $R_{ijIJ} = 0$ for $|I| \neq |J|$. The curvature operator is expressed by $R_p(dX_I) = \sum_{i<j} R_{ijIJ}\,dX_J \otimes (dX_i \wedge dX_j)$. Therefore, $G\,(R_p^2)\,dX^I = \sum_{i<j} R_{ijIJ}\,R_{ijJK}\,dX_K$. Consequently:

$$\mathrm{Tr}\,(G\,(R^p)^2) = \sum_{i<j}\;\sum_{|I|=p}\;\sum_{|J|=p} -\,(R_{ijIJ})^2.$$

Let $\mathrm{Cliff}(T^*M)$ be the Clifford bundle on T^*M. It is defined from the universal tensor algebra on T^*M by the relation $dX_{ij} * dX + dX_j * dX_i = 2g_{ij}$. $\mathrm{Cliff}(T^*M)$ is an algebra; we will use the symbol "$*$" for the operation of multiplication. There is a functorial vector bundle isomorphism between $\mathrm{Cliff}(T^*M)$ and $\Lambda\,(T^*M)$ which is not an algebra map. We use this identification to define a new algebra structure on $\Lambda\,(T^*M)$. If $\xi \in T^*M$, we have the identity

$$\xi * \theta = (i\,(\xi) + e(\xi))\,(\theta).$$

$i\,(\xi)$ denotes interior multiplication; $e\,(\xi)$ denotes exterior multiplication.

We can construct a first order differential operator d_0, by using Clifford multiplication

$$d_0\colon \Gamma(\Lambda T^*M) \xrightarrow{\ \nabla\ } \Gamma(T^*M \otimes \Lambda T^*M) \xrightarrow{\ \text{Clifford multiplication}\ } \Gamma(\Lambda T^*M).$$

The leading order symbol of d_0 is Clifford multiplication. This is also the leading order symbol of the operator $(d + d^*)$. Consequently, their difference is a functorially defined endomorphism in the first derivatives of the metric. We have already shown that such an endomorphism must be zero and therefore $d_0 =$

$(d+d^*)$.Therefore, we may describe the de Rham complex by using Clifford multiplication.

We compute

$$(d + d^*)\,(f_I\, dX_I) = (f_{I/i}\, dX_i * dX_I + f_I\, \Gamma^p_{iIJ}\, dX_i * dX_J$$

where $\nabla_i\,(dX_I) = \Gamma^p_{iIJ}\, dX_J$. Consequently,

$$(d + d^*)^2\,(f_I\, dX_I) = f_I\, \Gamma^p_{iIJ/j}\, dX_j * dX_i * dX_J$$
$$+ \text{ terms involving derivatives of } f_I$$
$$+ \text{ terms which vanish in normal coordinates at } x_0.$$

This implies that $B_s\,(x_0)\,(f_I\, dX_I) = -\, f_I\, \Gamma^p_{iIJ/j}\, dX_j * dX_i * dX_J$. The connection defined by the operator agrees with the connection defined by the metric. Since $dX_i * dX_i = -1$ and $dX_j * dX_i = -dX_i * dX_j$ for $i \neq j$ at x_0, we conclude that:

$$(B_s - w_{k/k})\,(x_0)\,(dX_I) = \Gamma^p_{iIJ/i}\, dX_J - \Gamma^p_{iIJ/i}\, dX_J + \sum_{i<j} (\Gamma_{jIJ/i} - \Gamma_{iIJ/j})\, dX_j * dX_i * dX_J$$
$$= \sum_{i<j} R^p_{ijIJ}\, dX_j * dX_i * dX_J.$$

Since this must be an endomorphism from $\Lambda^p T^*M \longrightarrow \Lambda T^*M$, many of these terms vanish. This vanishing is due to the Bianchi identities.

We can now prove:

THEOREM 4.1. *Let* $E^m_p = E\,(D^m_p)$; *then*:

(1) $E^m_p\,(dX_I) = \sum_{i<j} R^p_{ijIJ}\, dX_j * dX_i * dX_J$,

(2) $G\,(R^2_p)\,(dX_I) = \sum_{i<j} R^p{}_{ijIJ}\, R_{ijJK}\, dX_K$,

(3) $E^m_0 = 0$, $R_0 = 0$,

(4) $E^m_1\,(dX_k) = \sum R_{ijik}\, dX_j$, $\mathrm{Tr}\,(E^m_1) = -\,2K\,(g)$, $\mathrm{Tr}\,(E^m_1)^2 = |\,\mathrm{Ric}\,|^2$,

(5) $G\,(R^2_1)\,(dX_k) = \sum_{i<j} R_{ijkl}\, R_{ijl\nu}\, dX_\nu$, $\mathrm{Tr}\,(G\,(R^2_1)) = -|\,R\,|^2/2$.

PROOF. We have previously computed (1) and (2). Since the connection on Λ^0 is flat, the curvature tensor vanishes. This proves (3) from (1) and (2). We compute $R^1_{ijkl} = \Gamma^1_{jkl/i} - \Gamma^1_{ikl/j} = -\Gamma_{jlk/i} + \Gamma_{ilk/j} = R_{jilk} = R_{ijkl}$. Consequently $R^1_{ijkl} = R_{ijkl}$. From (1), $E^m_1\,(dX_k) = \sum_{i<j} R_{ijkl} * dX_j * dX_i * dX_l$. This is an invariantly defined endomorphism of Λ^1. Consequently, we may ignore any terms which do not lie in $\Lambda^1\,(T^*M)$ since these extra terms must cancel off. Therefore $i = l$ or $j = l$. Consequently,

$$E^m_1(dX_k) = \sum_{i<j} -\, R_{ijki}\, dX_j + R_{ijkj}\, dX_i = \sum_{i,j} R_{ijik}\, dX_j.$$

This implies that

$$\mathrm{Tr}\,(E^m_1) = \sum_{i,j} R_{ijij} = -2K\,(g), \quad \mathrm{Tr}\,(E^m_1)^2 = \sum R_{ijik}\, R_{ikij} = |\mathrm{Ric}|^2.$$

Finally, $G\,(R^2_1)\,(dX_k) = \sum_{i<j} R_{ijkl}\, R_{ijl\nu}\, dX_\nu$ by (2). This implies

$$\mathrm{Tr}\,(G\,(R^2_1)) = \sum_{i<j} R_{ijkl}\, R_{ijlk} = -\,\tfrac{1}{2}\,|\,R\,|^2.$$

We combine Theorems 3.1 and 4.1 to prove

THEOREM 4.2.

(1) $E_0(x, D^m_p) = (4\pi)^{-m/2} I$,

(2) $E_2(x, D^m_p) = (4\pi)^{-m/2} (K(g) / 3I + E^m_p)$,

(3) $E_4(x, D^m_p) = (4\pi)^{-m/2} ((-D^m_0 K / 15 + (10K^2 - |\mathrm{Ric}|^2 + |R|^2) / 180) I + KE^m_p / 3 + (E^m_p)^2 / 2 - \bar{D}E^m_p / 6 + G(R^2_p) / 6)$.

We could take the trace of this formula to compute the corresponding invariants B_0, B_2, and B_4. Since this would involve rather lengthy computations, we use a simplified method. We proceed as follows:

$$B_4(x, D^m_1) = mB_4(x, D^m_0) + (4\pi)^{-m/2} (-2K^2/3 + |\mathrm{Ric}|^2/2 + D^m_0 K/3 - |R|^2/12).$$

If $m = 4$, the Euler class is defined by $C = (4\pi)^{-2} (2K^2 - 2|\mathrm{Ric}|^2 + |R|^2 / 2)$. We have shown previously in ([3], [4]):

$$\Sigma (-1)^p B_4(x, D^4_p) = C, \quad \Sigma (-1)^p B_2(x, D^2_p) = K/2\pi.$$

This proves that $B_4(x, D^4_2) = C - B_4(x, D^4_0) - B_4(x, D^4_4) + B_4(x, D^4_1) + B_4(x, D^4_3)$ and that $B_2(x, D^2_1) = B_2(x, D^2_0) + B_2(x, D^2_2) - K/2\pi$. By Poincaré duality, $B_n(x, D^m_p) = B_n(x, D^m_{m-p})$. We can use the results of Theorems 4.1 and 4.2 to compute $B_2(x, D^2_0)$, $B_4(x, D^4_0)$, and $B_4(x, D^4_1)$. We use these two relationships to compute $B_2(x, D^2_1)$ and $B_4(x, D^4_2)$, and use Poincaré duality to complete the computation of $B_2(x, D^2_p)$ and $B_4(x, D^4_p)$.

We can use these results to determine the invariants B_0, B_2, and $B_4(x, D^m_p)$. A basis for the set of invariants of order 4 in the derivatives of the metric is given by $\{D^m_0 K, K^2, |\mathrm{Ric}|^2, |R|^2\}$. We compute B_4 for the product manifold $M_4 \times T_{m-4}$, where T_{m-4} is the flat torus. We apply Theorem 1.2(2), (3) to the operator D^m_p to show that $B_4(x, D^m_p) = (4\pi)^{-(m-4)/2} \sum_q B_4(x, D^4_q)\binom{m-4}{p-q}$. Since the restriction of an invariant of order 4 to a manifold of dimension 4 is 1-1, this determines $B_4(x, D^m_p)$. Similarly, we compute $B_2(x, D^m_p) = (4\pi)^{-(m-2)/2} \sum_q B_2(x, D^2_q)\binom{m-2}{p-q}$. This proves

THEOREM 4.3.

(1) $B_2(x, D^2_0) = B_2(x, D^2_2) = (4\pi)^{-1} K/3$,

$B_2(x, D^2_1) = 2B_0(x, D^2_0) - K/2\pi = (4\pi)^{-1}(-2K/3)$,

$B_4(x, D^4_0) = B_4(x, D^4_4) = (4\pi)^{-2}(-12D^m_0 K + 10K^2 - |\mathrm{Ric}|^2 + |R|^2) / 180$,

$B_4(x, D^4_1) = B_4(x, D^4_3) = 4B_4(x, D^4_0) + (4\pi)^{-2}(-8K^2 + 6|\mathrm{Ric}|^2 + 4D^m_0 K - |R|^2)/12$,

$B_4(x, D^4_2) = C - 2B_4(x, D^4_0) + 2B_4(x, D^4_1)$

$\quad = 6B_4(x, D^4_0) + (4\pi)^{-2}(8K^2 - 12|\mathrm{Ric}|^2 + 4|R|^2 + 8D^m_0 K)/12$.

(2) $B_0(x, D^m_p) = (4\pi)^{-m/2}\binom{m}{p}$,

$B_2(x, D^m_p) = (4\pi)^{(2-m)/2} \sum \binom{m-2}{p-q} B_2(x, D^2_q)$,

$B_4(x, D^m_p) = (3\pi)^{(4-m)/2} \sum \binom{m-4}{p-q} B_4(x, D^4_q)$.

We apply this theorem to derive some geometric consequences from these computations for real manifolds. In the first section, we noted that

$$\int_{M_m} B_n(x, D_p^m)\, d\text{vol}$$

is a spectral invariant of the Laplacian D_p^m. The system of equations given in Theorem 4.3 is invertible. Consequently, we can express $B_4(x, D_q^4)$ for $q = 0, 1, 2$, in terms of the invariants $B_4(x, D_p^m)$ for $p = 0, 1, 2$. This proves that $\int_{M_m} B_4(x, D_q^4)$ is a spectral invariant. Let

$$(1) = \int 10K^2 - |\text{Ric}|^2 + |R|^2\, d\text{vol}\ \ (B_4(x, D_0^m)*180*(4\pi)^{m/2}),$$

$$(2) = \int -8K^2 + 6|\text{Ric}|^2 - |R|^2\, d\text{vol}\ \ ((B^4(x, D_1^4) - 4B_4(x, D_0^4))*12*(4\pi)^2),$$

$$(3) = \int 2K_2 - 2|\text{Ric}|^2 + |R|^2/2\, d\text{vol}$$

$$(C*32\pi = (2B^4(x, D_0^4) + B^4(x, D_2^4) - 2B^4(x, D_1^4))).$$

These are spectral invariants of the operators D_0^m, D_1^m, and D_2^m. The determinant of the matrix of coefficients is: $30 + 2 + 16 - 12 - 4 - 20 = 12 \neq 0$. Therefore the matrix is invertible. This proves $\int K^2\, d\text{vol}$, $\int |R|^2\, d\text{vol}$, and $\int |\text{Ric}|^2\, d\text{vol}$ are spectral invariants as well. We can compute $\int K d\text{vol}$ from $\int B_2(x, D_0^m)$ and $\text{vol}(M)$ from $\int B^0(x, D_0^m)$. This proves:

THEOREM 4.4. *The following are spectral invariants of D_0^m, D_1^m, and D_2^m:*

$$\text{vol}(M), \quad \int K(g) d\text{vol}, \int K^2(g) d\text{vol}, \int |\text{Ric}|^2\, d\text{vol}, \quad \int |R|^2\, d\text{vol}.$$

We can use this theorem to relate the spectrum of a manifold to its geometry. We suppose that $\dim(M) = m \geqq 4$. These results were first obtained by Patodi [9].

THEOREM 4.5. *From the spectrum of the Laplacian D_p^m for $p = 0, 1, 2$ it is possible to determine if a manifold M_m is* (a) *flat,* (b) *Einstein,* (c) *constant scalar curvature c,* (d) *constant sectional curvature c.*

PROOF. M is flat iff $\int |R|^2\, d\text{vol} = 0$. This proves (a). M_m is Einstein iff $R_{ijik} = g_{jk}K(g)/m$. Let H_{jk} be the tensor $R_{ijik} - g_{jk}K(g)/m$; M_m is Einstein iff $\int |H|^2\, d\text{vol} = 0$. However, $|H|^2$ is an invariant of order 4 in the derivatives of the metric, so by Theorem 4.4, $\int |H|^2\, d\text{vol}$ is a spectral invariant. This proves (b). M_m has constant scalar curvature c iff $\int (K - c)^2\, d\text{vol} = 0$. However, $\int (K - c)^2 = c^2 \text{vol}(M_m) - 2c \int K(g)\, d\text{vol} + \int K(g)^2\, d\text{vol}$ is a spectral invariant by Theorem 4.4. This proves (c). Finally, M_m has constant sectional curvature c provided that $R_{ijkl} = c(g_{ik}g_{jl} - g_{il}g_{kj})$. Consequently, M_m has constant sectional curvature c iff $\int |R - cg_{ik}\, g_{lj} + cg_{ij}\, g_{kl}|^2\, d\text{vol} = 0$. This is a spectral invariant by Theorem 4.4 and proves (d).

We apply the well-known result that the only simply connected manifold of constant positive curvature 1 is the standard sphere S^m to prove:

THEOREM 4.6. *Let S_m be the standard sphere, and let M_m be a Riemannian manifold. It is possible to determine if the universal covering space of M_m is S_m from the spectrum of the operators D^m_p for $p = 0, 1, 2$. Furthermore, if S_m is the universal Riemannian covering space of M_m, then $|\pi_1(M_m)|$ is a spectral invariant.*

The proof is very simple. M_m has universal covering space isometric to S_m iff M_m has constant sectional curvature 1. If S_m is the universal Riemannian covering space, then $|\pi_1(M_m)| = \mathrm{vol}(S_m) / \mathrm{vol}(N_m)$.

For such a manifold M_m, we can imbed $\pi_1(M_m) \subset O(m)$. We can use this observation to prove

THEOREM 4.7. *Let M_m be a Riemannian manifold of dimension $m \geqq 4$.*

(1) *Suppose that the spectrum of the operators D^m_p on M_m agrees with the corresponding spectrum on S_m. Then M_m is isometric to S_m.*

(2) *Suppose that the spectrum of the operators D^m_p on M_m agrees with the corresponding spectrum on RP_m. Then M_m is isometric to RP_m.*

PROOF. By Theorem 4.6 the universal covering space of M_m is S_m. If (1) holds, then $|\pi_1(M_m)| = |\pi_1(S_m)| = 1$ and hence M_m is isometric to S_m. If (2) holds, then $|\pi_1(M_m)| = |\pi_1(RP_m)| = 2$. Let $\mathcal{T} \in \pi_1(M_m)$ be the nonidentity element. $\mathcal{T}$ is an orthogonal matrix such that $\mathcal{T}^2 = 1$. Since $\mathcal{T}$ is a covering transformation, $\mathcal{T}$ acts without fixed points. This implies that $\mathcal{T} = -I$ and hence M_m is isometric to RP_m.

There are corresponding results for complex manifolds. Let

$$D^m_{p,q} = 2(\bar\partial\bar\partial^* + \bar\partial^*\bar\partial)$$

be the complex Laplacian acting on forms of type (p, q). We have shown **[6]** that the spectrum of the operators $D_{0,0}$, $D_{0,1}$, and $D_{1,0}$ is sufficient to determine if the manifold is Kaehler. We have also shown [7] that the spectrum of the operators $D^m_{p,q}$ is sufficient to determine if a Kaehler manifold has constant holomorphic sectional curvature c. This proves that the spectral geometry of a complex manifold is sufficient to determine if a manifold is complex holomorphic isometric to CP_m, complex projective space, since CP_m, is the only complex Kaehler manifold with constant holomorphic sectional curvature 1.

It is an unsolved question whether the spectral geometry suffices to determine if the manifold is symmetric. Such a theorem could follow from the computation of $B_6(x, D)$.

REFERENCES

1. M. Atiyah, R. Bott and V. K. Patodi, *On the heat equation and the index theorem*, Invent. Math. **19** (1953), 279–330.

2. M. Berger, *Sur les spectre d'une variété riemannienne*, C. R. Acad. Sci. Paris **263** (1963), 13–16.

3. P. Gilkey, *Curvature and the eigenvalues of the Laplacian for geometrical elliptic complexes*, Ph. D. Dissertation, Harvard University, Cambridge, Mass., 1972.

4. ———, *Curvature and the eigenvalues of the Laplacian for elliptic complexes*, Advances in Math. **10** (1973), 344–382.

5. P. Gilkey, *Curvature and the eigenvalues of the Dolbeault complex for Kaehler manifolds*, Advances in Math. **11** (1973), 311–325.

6. ———, *Spectral geometry and the Kaehler condition for complex manifolds*, Invent. Math. **26** (1974), 231–238.

7. P. Gilkey and J. Sacks, *Spectral geometry and manifolds of constant holomorphic sectional curvature*, these PROCEEDINGS.

8. H. P. McKean, Jr. and I. M. Singer, *Curvature and the eigenvalues of the Laplacian*, J. Differential Geometry **1** (1967), 43–69.

9. V. K. Patodi, *Curvature and the fundamental solution of the heat equation*, J. Indian Math. Soc. **34** (1970), 269–285.

10. ———, *Curvature and the eigenforms of the Laplace operator*, J. Differential Geometry **5** (1971), 233–249.

11. ———, *An analytic proof of the Riemann-Roch-Hirzebruch theorem for Kaehler manifolds*, J. Differential Geometry **5** (1971), 251–283.

12. R. T. Seeley, *Complex powers of an elliptic operator*, Proc. Sympos. Pure Math., vol. 10, Amer. Math. Soc., Providence, R. I., 1967, pp. 288–307.

13. ———, *Topics in pseudo-differential operators*, CIME, 1968, 167–305.

UNIVERSITY OF CALIFORNIA, BERKELEY

Proceedings of Symposia in Pure Mathematics
Volume 27, 1975

SPECTRAL GEOMETRY AND MANIFOLDS OF CONSTANT HOLOMORPHIC SECTIONAL CURVATURE

PETER B. GILKEY* AND JONATHAN SACKS

Let M_m be a compact analytic manifold of complex dimension m, and let G be a Hermitian metric for TM. Let $D^m_{p,q} = 2(\bar\partial\bar\partial^* + \bar\partial^*\bar\partial)$ be the complex Laplacian acting on $\Gamma(\Lambda^{p,q}(T^*M\otimes C))$. Let $\{\lambda^i_{p,q}\}^\infty_{i=1}$ be the eigenvalues of $D^m_{p,q}$ enumerated with multiplicity. The extent to which the spectrum of the Laplacian determines the geometry of the manifold has been studied extensively recently. An earlier paper by the first author **[5]** has shown that it is possible to determine if G is a Kaehler metric from the spectrum of the operators $D^m_{p,q}$. In this paper, we show that it is possible to determine whether the metric G has constant holomorphic sectional curvature from the spectrum of the operator $D^m_{p,q}$. This problem was originally proposed by M. Berger in 1968 **[1]**. The only manifold with a metric of constant positive holomorphic sectional curvature is CP_m, complex projective space with the standard metric. This will prove

THEOREM 1. *Let M_m be a compact analytic manifold. Suppose the spectrum of the operators $D^m_{p,q}$ on M_m agrees with the corresponding spectrum on CP_m. Then CP_m and M_m are isometric holomorphically.*

There is an analogous theorem for S^m and RP^m in the case of real manifolds. The proof of these and related results relies upon certain asymptotic invariants of the spectrum. We construct these invariants as follows: let $\{\lambda^i_{p,q}, \phi^i_{p,q}\}^\infty_{i=1}$ be

AMS (MOS) subject classifications (1970). Primary 32C10, 53C55, 58G99.

*Research partially supported by NSF grant GP-34785X.

a complete spectral decomposition of the selfadjoint operator $D^m_{p,q}$. By using the results of Seeley [7], we can define the series

$$f(t, x, D^m_{p,q}) = \sum_i \exp(-t\lambda_i)\, G(\phi^i_{p,q}, \phi^i_{p,q}) \quad \text{for } t > 0.$$

Furthermore, one can easily show that there is an asymptotic expansion as $t \to 0^+$ of the form:

$$f(t, x, D^m_{p,q}) \sim \sum_{n=0}^{\infty} B_n(x, D^m_{p,q})\, t^{(n-2m)/2}.$$

Because $D^m_{p,q}$ is a differential operator, the invariants B_n vanish for n odd. For n even, they are local invariants of order n in the derivatives of the Hermitian metric.

The functions $\{\phi^i_{p,q}\}$ form an orthonormal basis for $L^2(\Lambda^{p,q}T^*M)$. We integrate:

$$\sum_i \exp(-t\lambda^i_{p,q}) = \int f(t, x, D^m_{p,q})\, d\,\mathrm{vol} \sim \sum_{n=0}^{\infty} \left[\int B_n(x, D^m_{p,q})\, d\,\mathrm{vol}\right] t^{(n-2m)/2}.$$

Consequently, the number $\int B_n(x, D^m_{p,q})$ is a spectral invariant of the operator $D^m_{p,q}$.

Let D^m_0 be the real Laplacian $(dd^* + d^*d)$ acting on functions. McKean and Singer [6] proved that

$$B_0(x, D^m_0) = (4\pi)^{-m} \qquad (2m \text{ is the real dimension of } M_m),$$
$$B_2(x, D^m_0) = (4\pi)^{-m} K(g)/3 \quad (K(g) \text{ is the scalar curvature}).$$

The invariant $B_0(x, D^m_{p,q})$ is an invariant of order 0 in the derivatives of the Hermitian metric and is therefore a constant. It is well known that this constant is

$$B_0(x, D^m_{p,q}) = \dim(\Lambda^{p,q})\, B_0(x, D^m_0) = (4\pi)^{-m}\binom{m}{p}\binom{m}{q}.$$

Let M_m and N_n be two complex Hermitian manifolds, and let $(x, y) \in M_m \times N_n$. We give $M_m \times N_n$ the product metric. There is a natural isomorphism

$$\Lambda^{\bar p,\bar q} \cong \bigoplus [\Lambda^{p,q}(T^*M_m) \otimes \Lambda^{p',q'}(T^*N_n)]$$
$$(\text{summed over } p + p' = \bar p \text{ and } q + q' = \bar q).$$

We use this isomorphism to decompose

$$D^{m+n}_{\bar p,\bar q} = \bigoplus (D^m_{p,q} \otimes 1^n_{p',q'} + 1^m_{p,q} \otimes D^n_{p',q'})$$
$$(\text{summed over } p + p' = \bar p \text{ and } q + q' = \bar q).$$

This proves that

$$f(t, (x, y), D^{m+n}_{\bar p,\bar q}) = \sum f(t, x, D^m_{p,q}) f(t, y, D^n_{p',q'})$$
$$(\text{summed over } p + p' = \bar p \text{ and } q + q' = \bar q).$$

We equate coefficients of equal powers of t in the two asymptotic expansions to show:

LEMMA 2. $B_i((x, y), D^{m+n}_{\bar p,\bar q}) = \sum_{j+k=i} \sum_{p+p'=\bar p} \sum_{q+q'=\bar q} B_j(x, D^m_{p,q})\, B_k(y, D^n_{p',q'})$.

We will need the following lemma which characterizes the invariants of order 4 in the derivatives of a *Kaehler* metric:

LEMMA 3. *Let V^4_m denote the space of all invariants of order 4 in the derivatives of a Kaehler metric. If $m \neq 1$, this space is spanned by the invariants*:

$$K(g)^2, \quad D^m_0(K(g)), \quad *(c_1^2 \wedge \Omega^{m-2}), \quad *(c_2 \wedge \Omega^{m-2}).$$

Ω is the Kaehler form; $K(g)$ is the scalar curvature; c_i are the Chern classes; and $*$ is the Hodge operator.

PROOF. For $m \neq 1$, these four invariants are linearly independent. Therefore, it suffices to show that $\dim(V^4_m) \leqq 4$. Let R be the restriction map from $V^4_m \to V^4_{m-1}$ which restricts an invariant to a manifold of 1 lower complex dimension. We showed in an earlier paper [3] if $m > 2$, then R is 1-1. If $m = 2$, then the kernel of R is the span of the Chern classes. Consequently, $\dim(V^4_m) \leqq \dim(V^4_2)$ and $\dim(V^4_2) \leqq \dim(\mathrm{span}(c_1^2, c_2)) + \dim(V^4_1)$. However, the space of all invariants of order 4 on a Riemann surface has dimension 2. This proves $\dim(V^4_2) \leqq 4$ and proves the lemma.

As an immediate consequence of Lemma 3, we identify V^4_m and V^4_2. The restriction on any invariant of order 4 for a Kaehler manifold of dimension m to a manifold of dimension 2 is 1-1. Let M_2 be a 2-dimensional Kaehler manifold, and let T_{m-2} be the flat torus of complex dimension $m-2$. All the derivatives of the metric vanish on T_{m-2}. Consequently $B_k(x, D^{m-2}_{p,q}) = 0$ for $k > 0$ on T_{m-2}. By Lemma 2,

$$B_4(x, D^m_{p,q}) = \sum (4\pi)^{2-m} \binom{m-2}{p-p'}\binom{m-2}{q-q'} B_4(x, D^2_{p',q'}).$$

This determines $B_4(x, D^m_{p,q})$ in terms of the invariants $B_4(x, D^2_{p,q})$. This system of equations has a unique solution. Thus we may express the invariants $B_4(x, D^2_{p,q})$ in terms of the invariants $B_4(x, D^m_{p,q})$ for any m. Therefore $\int B_4(x, D^2_{p,q})\, d\,\mathrm{vol}$ is a spectral invariant.

Let M_2 be a Kaehler manifold. The local version of the index theorem proved by the first author ([2], [3]) shows

$$\sum (-1)^{p+q} B_4(x, D^2_{p,q}) = c_2 \qquad \text{(the Euler class integrand)},$$
$$\sum (-1)^q B_4(x, D^2_{0,q}) = (c_1^2 + c_2)/12 \qquad \text{(the Riemann-Roch integrand)}.$$

This implies that we can compute c_1^2 and c_2 from the invariants $B_4(x, D^2_{p,q})$. Therefore, $\int (c_1^2 \wedge \Omega^{m-2})$ and $\int (c_2 \wedge \Omega^{m-2})$ are spectral invariants of M_m.

LEMMA 4. *Let $P \in V^4_m$ and let M_m be Kaehler. Then $\int P\, d\,\mathrm{vol}$ is a spectral invariant.*

PROOF. By Lemma 3, we decompose $P = \alpha K(g)^2 + \beta D^m_0(K(g)) + \gamma *(c_1^2 \wedge \Omega^{m-2}) + \delta *(c_2 \wedge \Omega^{m-2})$. Therefore $\int P d\,\mathrm{vol} = \alpha \int K(g)^2\, d\,\mathrm{vol} + \gamma \int c_1^2 \wedge \Omega^{m-2} + \delta \int c_2 \wedge \Omega^{m-2}$. We can compute these last two terms as spectral invariants. Therefore, it suffices to show that $\int K(g)_2\, d\,\mathrm{vol}$ is a spectral invariant.

We decompose $B_4(x, D_0^m) = aK(g)^2 + \cdots$. To show that $\int K(g)^2$ is a spectral invariant, it suffices to show that $a \neq 0$. Let M_m be the complex manifold $S^2 \times T_{m-1}$. Since T_{m-1} is a flat torus, both c_1^2 and c_2 vanish identically on M_m. Since the scalar curvature is constant, $D_0^m(K(g))$ vanishes and $B_4(x, D_0^m) = aK(g)^2$. We can use the product formula to show that:

$$B_4(x, D_{0,0}^m) = (4\pi)^{1-m} B_4(x, D_{0,0}^1).$$

On a Riemann surface, McKean and Singer [6] showed that

$$B_4(x, D_{0,0}^1) = (4\pi)^{-1}(K(g)^2 + D_0^1(K(g)))/15.$$

Consequently $a = (4\pi)^{-m}/15 \neq 0$. This completes the proof of the lemma.

Goldberg [5] constructs the projective curvature tensor:

$$W^i{}_{jkl*} = R^i{}_{jkl*} + 1/(m+1)(R_{jl*}\,\delta^i_k + R_{kl*}\,\delta^i_j).$$

This tensor vanishes for a Kaehler manifold if and only if M_m has constant holomorphic sectional curvature. Let

$$|W|^2 = \sum |W^i_{jkl*}|^2 \in V_m^4.$$

THEOREM 5. *Let M_m be a Kaehler manifold. It is possible to determine if M_m has constant holomorphic sectional curvature c from the spectrum of the operators $D_{p,q}^m$.*

PROOF. $|W|^2 \in V_m^4$ so $\int |W|^2\, d$ vol is a spectral invariant. This vanishes if and only if M_m has constant holomorphic sectional curvature c. The scalar curvature $K(g)$ is a constant positive multiple of c. Since M_m is Kaehler, $D_{0,0}^m = D_0^m$ and

$$\int B_2(x, D_{0,0}^m) = (4\pi)^{-m} \int K(g)/3d \text{ vol},$$
$$\int B_0(x, D_{0,0}^m) = (4\pi)^{-m} \text{ volume}(M_m).$$

Consequently, $\int K(g)\, d$ vol/volume(M_m) is a spectral invariant, so c is a spectral invariant.

We combine these results to prove Theorem 1. Let M_m and CP_m have the same spectrum for the operators $D_{p,q}^m$. Since CP_m is Kaehler, so is M_m by [4]. Consequently, we may apply Theorem 5 to show that M_m has constant positive holomorphic sectional curvature. Since vol(M_m) = vol(CP_m) is a spectral invariant, M_m is isometric holomorphically to CP_m with the standard metric.

REFERENCES

1. M. Berger, *Eigenvalues of the Laplacian*, Proc. Sympos. Pure Math., vol. 16, Amer. Math. Soc., Providence, R.I., 1970, pp. 121–125. MR **41** #9141.

2. P. Gilkey, *Curvature and the eigenvalues of the Laplacian for elliptic complexes*, Advances in Math. **10** (1973), 344–382.

3. ———, *Curvatures and the eigenvalues of the Dolbeault complex for Kaehler manifolds*, **11** (1973), 311–325.

4. P. Gilkey, *Spectral geometry and the Kaehler condition for complex manifolds*, Inventiones **26** (1974), 231–258.

5. S. Goldberg, *Curvature and homology*, Pure and Appl. Math., vol. 11, Academic Press, New York, 1962. MR **25** #2537.

6. H. P. McKean, Jr. and I. M. Singer, *Curvature and the eigenvalues of the Laplacian*, J. Differential Geometry **5** (1971), 233–249.

7. R. T. Seeley, *Complex powers of an elliptic operator*, Proc. Sympos. Pure Math., vol. 10, Amer. Math. Soc., Providence, R.I., 1967, pp. 288–307. MR **38** #6220.

UNIVERSITY OF CALIFORNIA, BERKELEY

Proceedings of Symposia in Pure Mathematics
Volume 27, 1975

WHITNEY'S IMBEDDING THEOREM BY SOLUTIONS OF ELLIPTIC EQUATIONS AND GEOMETRIC CONSEQUENCES

R. E. GREENE AND H. WU*

The famous imbedding theorem of Whitney says that if M is a paracompact n-dimensional C^∞ manifold, then M has a C^∞ proper imbedding in R^{2n+1}. Here, a mapping's being *proper* means that the inverse image of each compact set is compact; this condition is equivalent to the continuity of the obvious extension of the mapping to the one-point compactifications of the domain and range, and it implies that the image of M in R^{2n+1} is a closed subset. We shall prove in this paper that, in case M is noncompact, a proper imbedding of M in R^{2n+1} could be achieved by using a much smaller class of C^∞ functions, which are of geometric interest. More precisely, we say that a linear differential operator P (always with real C^∞ coefficients) on M has the *UCP (unique continuation property) if any function u on M satisfying $P^*u = 0$ in a connected open set U must be identically zero on U if it is zero in some open subset of U. (Here $P^*=$ the adjoint of P relative to a Riemannian metric on M.) The main theorem of this paper then asserts:

Let M be an n-dimensional connected noncompact C^∞ manifold and let P be a linear elliptic operator with no zero order term on M which has the UCP. Then there is a proper C^∞ imbedding of M in R^{2n+1} whose component functions are solutions of $Pu = 0$ on M and a proper C^∞ immersion in R^{2n} whose component functions again satisfy $Pu = 0$ on M. (Here and throughout, all manifolds are to be assumed to be paracompact.)

AMS (MOS) subject classifications (1970). Primary 31C05, 35J30, 53C20; Secondary 32F10, 58G99.

*Research partially supported by National Science Foundation grants GP-27576 (first author) and GP-34785 (second author) and a Sloan Fellowship (second author).

It is by now well known that not every linear elliptic operator has the UCP. However, the important operators of differential geometry all have order at most two, and Aronszajn and Cordes ([**2**], [**7**], cf. also Protter [**23**]) have proved that every linear elliptic operator of second order has the UCP. In particular, the various Laplacians (see the following discussion) all have the *UCP and therefore every connected noncompact C^∞ manifold can be properly imbedded in euclidean space by harmonic functions (= functions annihilated by a Laplacian). This fact will be shown to have the following consequences: (1) Every connected noncompact Riemannian manifold has a nonnegative C^∞ proper function which is strictly subharmonic. (2) Every connected noncompact manifold admits a complete Riemannian metric of negative scalar curvature. (3) Every connected noncompact complex manifold has a C^∞ nonnegative proper function whose Levi form has at least one positive eigenvalue. The vanishing theorem of Malgrange and Siu ([**17**] and [**26**]) follows immediately from (3).

The question of whether a noncompact Riemannian manifold can be differentiably imbedded into some euclidean space by harmonic functions seems to have been explicitly asked for the first time in Chern's notes [**6**, p. 13]. Chern raised this question in the context of minimal submanifolds. However, we were led to this question via a completely different route and since we consider the interpretation of our theorem to be nearly as interesting as the theorem itself, we would like to discuss briefly imbedding theorems in general and the original motivation of our theorem in particular.

Given a class of *reasonable* topological spaces $\mathscr{T}$, a natural question is whether there exists any *reasonable* function on them, and if so, how many such functions. Generally, either one cannot produce any reasonable function or else one produces them in abundance. In the latter case, this abundance is usually epitomized by an imbedding theorem which says that there are enough of these reasonable functions to imbed any member of $\mathscr{T}$ into some standard model. Let us give some examples.

(I) $\mathscr{T}$ = normal topological spaces; then Reasonable Functions = continuous functions. Urysohn's lemma says that there are many continuous functions on a normal space and, with a minor twist, this lemma yields the fact that if X is normal and $\mathscr{D}$ = continuous maps into $I \equiv [0, 1]$, then the point evaluation map $X \to I^{\mathscr{D}}$ is a topological imbedding. (If one thinks of X as analogous to a vector space, then $\mathscr{D}$ is the analogue of the dual of X and $I^{\mathscr{D}}$ the analogue of the bi-dual.)

(II) $\mathscr{T}$ = C^∞ manifolds of real dimension n; then Reasonable Functions = C^∞ functions. The existence of partitions of unity guarantees a rich supply of C^∞ functions and the corresponding imbedding theorem is the Whitney imbedding theorem.

(III) $\mathscr{T}$ = real analytic manifolds of dimension n; then Reasonable Functions = real analytic functions. So far as we know, there is no simple proof of the existence of one nonconstant real analytic function on a real analytic manifold. Fortunately, such functions do exist and then of course there are enough of them to give an imbedding theorem. This is the imbedding theorem of Grauert-Morrey: Every n-dimensional real analytic manifold has a proper real analytic imbedding in $\boldsymbol{R}^{2n+1}$ ([**9**], [**19**]). The proof of this theorem is nonelementary and long, even if one assumes all the standard imbedding techniques of Whitney.

(IV) $\mathscr{T}$ = Stein manifolds of complex dimension n; then Reasonable Functions = holomorphic functions. The existence of nonconstant holomorphic functions in this case is built into the definition, but the corresponding imbedding theorem is still a deep result: Every n-dimensional Stein manifold has a proper holomorphic imbedding in $\boldsymbol{C}^{2n+1}$ (Remmert-Bishop-Narasimhan [**24**], [**4**], [**20**]). Chronologically this Stein manifold imbedding theorem preceded (and is needed in) the proof of the Grauert-Morrey result.

This last example is particularly pertinent to this paper. Noting that holomorphic functions on a complex manifold are just the (function) solutions of $\bar{\partial}u = 0$, one may paraphrase the Remmert-Bishop-Narasimhan theorem as asserting the existence of enough solutions of $\bar{\partial}u = 0$ on a Stein manifold to give a proper holomorphic imbedding of the manifold in complex euclidean space. The restriction of the manifold's being a Stein manifold is necessary for the validity of this imbedding theorem because there are complex manifolds homeomorphic to cells which have no nonconstant solutions of $\bar{\partial}u = 0$, a reflection of the fact that $\bar{\partial}u = 0$ is an overdetermined system for dimensions exceeding one. For some time now, we have been concerned with the general problem of producing nonconstant solutions of $\bar{\partial}u = 0$ on noncompact Kähler manifolds. In the course of our work on this problem, we found it natural as well as necessary to discuss the corresponding problem for noncompact Riemannian manifolds, where the obvious analogue of $\bar{\partial}$ is the usual Laplacian Δ ($= -\delta d =$ trace of the second covariant differential $=$ div grad). Since $\Delta u = 0$ is a single elliptic equation, it is reasonable to expect that nonconstant global solutions should exist on any noncompact M without further restrictions. Indeed, our main theorem coupled with the Aronszajn-Cordes theorem mentioned previously immediately implies that any n-dimensional noncompact Riemannian manifold M can be properly imbedded into $\boldsymbol{R}^{2n+1}$ by harmonic functions (solutions of $\Delta u = 0$). This special case was the original example of the general theorem; and it should be clear by now that one can interpret our theorem, not so much as a generalization of the Whitney imbedding theorem, but rather as a statement that on any noncompact n-dimensional Riemannian manifold M, there is such a rich supply of harmonic functions that one can extract $(2n + 1)$ of them which will simultaneously give local coordinates everywhere, separate the points of M, and carry the infinity of M to infinity.

Some geometric consequences of the main theorem will now be given. First we introduce two Laplacians, which are both linear second order differential elliptic operators without zero order terms. If M is a (real) C^∞ Riemannian manifold, then the classical Δ defined previously is such an operator. On the other hand, suppose M is a complex manifold with a C^∞ Hermitian metric G. Then define Δ_0 as follows: Let $G = \sum_{\alpha,\beta} G_{\alpha\beta} dz^\alpha \otimes d\bar{z}^\beta$ in local coordinate representation; then $\Delta_0 = \sum_{\alpha,\beta} G^{\alpha\beta} (\partial^2/\partial z^\alpha \partial \bar{z}^\beta)$. The operator Δ is defined on M by assigning to M the Riemannian metric associated to G as usual. To elucidate the relationship between Δ and Δ_0, we make the following observations: (A) If G is a Kähler metric, then $\Delta = 2\Delta_0$. (B) If G is not a Kähler metric, then Δ and $2\Delta_0$ will in general be distinct; in fact the first order terms of Δ and $2\Delta_0$ are, respectively, real but generically nonzero and identically zero. Since Δ_0 is not as well known as Δ, we add two more observations concerning Δ_0: (C) If D is the canonical Hermitian connection of G, then $\Delta_0 u = \text{trace } D^2 u$, where trace has to be taken in the nonstandard sense of $\sum_i D^2 u(Z_i, \bar{Z}_i)$, $\{Z_i\}$ being an orthonormal frame field of type (1, 0). (D) $\Delta_0 u = \text{trace } L_u = \text{trace } \partial\bar{\partial} u$, where the trace is taken relative to G and L_u is the Levi form of u.

We shall call a function u *Δ-harmonic* or *Δ_0-harmonic* if $\Delta u = 0$ or $\Delta_0 u = 0$ respectively. If no confusion is possible, we say just harmonic. We repeat the conclusion made previously: If M is a noncompact Riemannian manifold of real dimension n (respectively, noncompact Hermitian manifold of complex dimension m), then M can be properly imbedded into $\boldsymbol{R}^{2n+1}$ (respectively, $\boldsymbol{R}^{4m+1}$) by Δ-harmonic functions (respectively, Δ_0-harmonic functions).

Now let M be a noncompact Riemannian manifold of real dimension n and let $\{h_1, \cdots, h_{2n+1}\}$ be Δ-harmonic functions which give a proper imbedding in $\boldsymbol{R}^{2n+1}$. Then $\tau = \sum_i h_i^2$ is a C^∞ proper function on M which is strictly subharmonic in the sense that $\Delta\tau > 0$. Therefore we have:

COROLLARY 1. *Every connected noncompact Riemannian manifold admits a C^∞ strictly subharmonic nonnegative proper function.*

Proper functions are indispensable for the geometric study of noncompact manifolds, and those enjoying special properties are particularly desirable. Strict subharmonicity is one such special property. On the other hand, we have recently proved that *if a Riemannian manifold has a nonnegative continuous proper subharmonic function which is (uniformly) Lipschitzian, then the manifold has infinite volume* (cf. **[11]**). This result therefore puts Corollary 1 in the proper perspective.

The second geometric consequence of our main theorem is the following corollary, which was stated earlier. This result can be considered to be the higher dimensional Riemannian geometry analogue of the fact (deducible from the uniformization theorem, for instance) that any open Riemann surface admits a complete Hermitian metric of negative Gaussian curvature.

COROLLARY 2. *If M is a connected noncompact C^∞ manifold (of dimension $\geqq 2$), then M admits a complete Riemannian metric of negative scalar curvature.*

To deduce this result from the main theorem of this paper, first note that according to a theorem of M. L. Gromov **[12]** the manifold M admits a Riemannian metric g of negative sectional curvature; such a metric g cannot of course in general be chosen to be complete. The metric g necessarily has negative scalar curvature. It will now be shown how to obtain a complete metric of negative scalar curvature by multiplying g by an appropriate variable factor. The method to be given applies to any metric g of negative scalar curvature. Such metrics may also be obtained by applying a result of J. Kazdan and F. Warner **[14]** as an alternative to the use of Gromov's theorem: Their result is that if $f: M \to \boldsymbol{R}$ is a C^∞ function which is negative somewhere then there is a metric on M whose scalar curvature is f. In particular, f can be taken to be $\equiv -1$. However, Kazdan and Warner's result contains no information on obtaining a complete metric with specified scalar curvature.

For any C^∞ function $u: M \to \boldsymbol{R}$ and any C^∞ Riemannian metric on M (and in particular the metric g of negative scalar curvature at hand), the Riemannian metric $g' = e^{2u}g$ has scalar curvature R' given by the standard formula:

$$R' = e^{-2u}\{R - 2(n-1)(\Delta u) - (n-1)(n-2)\langle du, du\rangle\}.$$

Here $n =$ the dimension of M; $R =$ the scalar curvature of g; $\langle du, du\rangle$ is the inner product of the 1-form du with itself relative to the metric induced on 1-forms by g; and Δ is the Laplacian determined by g. Since R is everywhere negative according to the choice of g, R' will be everywhere negative provided that $\Delta u \geqq 0$ everywhere, i.e. provided that u is subharmonic. As noted (Corollary 1), the main theorem implies that M admits a proper C^∞ nonnegative (strictly) subharmonic function, say $\tau: M \to \boldsymbol{R}$. Now $e^{2\tau}g$ need not be complete. However, if $\chi: \boldsymbol{R} \to \boldsymbol{R}$ is a C^∞ monotone increasing convex function then $\chi \circ \tau: M \to \boldsymbol{R}$ is a C^∞ subharmonic function; and $\chi(x)$ may be chosen to grow so fast as $x \to +\infty$, $x \in \boldsymbol{R}$, that $e^{2(\chi\circ\tau)}g$ will be complete. That χ can be so chosen follows easily from the fact that τ is a nonnegative proper function (cf. [**22**]). The metric $e^{2(\chi\circ\tau)}g$ is then a complete metric of negative scalar curvature, as required.

It was only for convenience and brevity that the metric g needed to be chosen to have negative scalar curvature in the argument just given. In fact, using the fact that the function τ of Corollary 1 is strictly subharmonic, a very similar argument can be used to prove that if g is any Riemannian metric on M then there exists a C^∞ function $\chi: \boldsymbol{R} \to \boldsymbol{R}$ such that $e^{2(\chi\circ\tau)}g$ is a complete metric of everywhere negative scalar curvature. The required argument for this general case is the following: For any C^∞ function $\chi_1: \boldsymbol{R} \to \boldsymbol{R}$, the second derivative of $\chi_1\circ\tau$ along any C^∞ curve $c(t)$ in M (defined for t in a neighborhood of $t_0 \in \boldsymbol{R}$) is given by

$$\frac{d^2}{dt^2}(\chi_1\circ\tau)(c(t))\Big|_{t=t_0} = \chi_1''\big|_{\tau(c(t_0))}\cdot\left(\frac{d\tau(c(t))}{dt}\right)^2\Big|_{t_0} + \chi'\big|_{\tau(c(t_0))}\cdot\left(\frac{d^2}{dt^2}\tau(c(t))\right)\Big|_{t_0}.$$

Computing from this formula the Laplacian $\Delta(\chi_1\circ\tau)$ at the center p of a normal coordinate system $(x_1, \cdots, x_n)$ yields

$$\Delta(\chi_1\circ\tau)|_p = \chi_1''|_{\tau(p)}\cdot\sum_{i=1}^{n}\left(\frac{d\tau}{dx_i}\right)^2\Big|_p + \chi_1'|_{\tau(p)}\cdot(\Delta\tau)|_p.$$

If $\chi_1'' \geqq 0$ everywhere, then

$$\Delta(\chi_1\circ\tau)|_p \geqq \chi_1'|_{\tau(p)}\cdot(\Delta\tau)|_p.$$

From the facts that $\Delta\tau|_p > 0$ for all $p \in M$ and that τ is a proper nonnegative function, it follows that χ_1 may be taken to be convex and monotone increasing and to have $\chi_1'|_x$ going to $+\infty$ so rapidly as $x \to +\infty$, $x \in \boldsymbol{R}$, that, for every $p \in M$, $2(n-1)\Delta(\chi_1\circ\tau)|_p > R(p)$ where R is the scalar curvature of the metric g. Then the scalar curvature of the metric $e^{2(\chi_1\circ\tau)}g$ is everywhere negative by the previously given formula. Then by a previous argument applied to $e^{2(\chi_1\circ\tau)}g$ instead of to g itself, there exists a C^∞ function $\chi_2\colon \boldsymbol{R} \to \boldsymbol{R}$ such that $e^{2(\chi_2\circ\tau)}e^{2(\chi_1\circ\tau)}g$ is a complete metric of everywhere negative scalar curvature. Putting $\chi = \chi_1 + \chi_2$ yields then a function such that the metric $e^{2(\chi\circ\tau)}g$ is complete and has everywhere negative scalar curvature.

For the next corollary, let M be a noncompact complex manifold. Introduce in M an Hermitian metric and thus obtain the associated Laplacian Δ_0. Let then $\{h_1,\cdots, h_{4m+1}\}$ be Δ_0-harmonic functions which give a proper imbedding of M into $\boldsymbol{R}^{4m+1}$. As before, let $\tau_0 = \sum_i h_i^2$. Then τ_0 is C^∞ and proper and $\Delta_0\tau_0 > 0$. Since $\Delta_0\tau_0 =$ trace of the Levi form of τ_0, the positivity of $\Delta_0\tau_0$ implies that the Levi form of τ_0 has at least one positive eigenvalue. Thus, we have proved:

COROLLARY 3. *Any connected noncompact complex manifold admits a C^∞ nonnegative proper function whose Levi form has at least one positive eigenvalue.*

Now, Andreotti and Grauert have proved in [1] that if M is a complex manifold of dimension m and τ is a C^2 proper nonnegative function on M whose Levi form has at least q positive eigenvalues, then for any coherent sheaf $\mathscr{F}$ on M, $H^i(M, \mathscr{F}) = 0$ whenever $i = m - q + 1,\cdots, m$. Therefore, we have as an immediate consequence the following theorem of Malgrange and Siu ([17], [26]): If M is an m-dimensional noncompact complex manifold, then $H^m(M, \mathscr{F}) = 0$ for any coherent sheaf $\mathscr{F}$ on M. We note that Siu has further extended this theorem to complex spaces in [27].

For further comments about the relationship of the main theorem of this paper with the older special cases of the Grauert-Morrey imbedding theorem as proved by Bochner [5], Malgrange [18] and Royden [25], the reader is referred to the introduction of our paper [10].

We note that the main theorem is not valid for compact manifolds. Indeed, since the maximum principal is valid for harmonic functions, the only harmonic functions on compact manifolds are constants.

We now give the proof of the main theorem. The special case where P is the Laplacian Δ of a Riemannian metric is treated in great detail in [10], and the proof of the general theorem differs from that of this special case only in a few technical points. Thus the following proof has been purposely written in a way that will complement [10]; wherever the technical details are given in full in [10], we shall give only the conceptual framework here, and wherever the proof in [10] is inadequate for the present general case, we shall supply the details needed.

It will be sufficient to prove the imbedding statement. The immersion statement follows from the imbedding statement by standard projection arguments plus a technical observation about the particular proper maps one can obtain to insure that properness is preserved under projection; see §3 of [10]. The kernel of the proof is the following approximation theorem of Runge type of Lax [15] and Malgrange [16]. To state this theorem precisely, we define: A function u_0 is said to satisfy $Pu = 0$ on a compact set K if there is a solution of $Pu = 0$ defined on an open set containing K which agrees with u_0 when restricted to K.

THEOREM OF LAX-MALGRANGE. *If M is a noncompact manifold and P is a linear elliptic operator which has the *UCP, and if K is a compact set in M such that $M - K$ has no component with compact closure, then any function $v: K \to \boldsymbol{R}$ which satisfies $Pu = 0$ on K is the uniform limit of functions $u: M \to \boldsymbol{R}$ which satisfy $Pu = 0$ everywhere on M.*

In the following discussion, we shall let $\mathscr{P}$ stand for the linear space of all (global) solutions u of $Pu = 0$ on M. By the regularity theorem of elliptic operators, $\mathscr{P} \subseteq \mathscr{C}^\infty(M)$, the algebra of C^∞ functions on M. We have to show that $(2n + 1)$ members of $\mathscr{P}$ suffice to give a proper imbedding of the n-dimensional manifold M into $\boldsymbol{R}^{2n+1}$. It follows immediately from the Lax-Malgrange result that $\mathscr{P}$ separates points of M. Indeed, if p, q are distinct points of M, let $v: \{p, q\} \to \boldsymbol{R}$ be $v(p) = 0$, $v(q) = 1$. Since the function u which is 0 in a neighborhood of p and 1 in a disjoint neighborhood of q satisfies $Pu = 0$ on its domain, we see that v satisfies $Pv = 0$ on $\{p, q\}$. Then any member of $\mathscr{P}$ that (closely) approximates v on $\{p, q\}$ will separate p and q. To show the functions in $\mathscr{P}$ give local coordinates at each point, we appeal to the following existence result.

LEMMA. *Let P be a linear elliptic operator on $\boldsymbol{R}^n$. Then there exist n local solutions of $Pu = 0$ around the origin 0 which give local coordinates at 0.*

This is proved in [3, p. 228], using the explicit expression of a fundamental solution of elliptic operators with constant coefficients. We give a brief proof here, assuming the conventional existence theorem of local solutions (without specifying any initial conditions). It is easy to see that it suffices to prove: Given a unit tangent vector t at 0, there exists a v such that locally at 0, $Pv = 0$ and $tv \neq 0$. By rotating $\boldsymbol{R}^n$ if necessary, we may assume $t = (\partial/\partial x_n)(0)$. Let $\delta > 0$ be fixed and let φ be a C^∞ function with support in the ball B_δ of radius δ, $0 \leqq \varphi \leqq 1$, and $\varphi \equiv 1$ in $B_{\delta/2}$. For an $a \in \boldsymbol{R}$, let $u_0(x) = \varphi(x)(ax_n^m + x_n)$, where $m =$ order P, and let $f = Pu_0$. Then support $f \subseteq B_\delta$ and $tu_0 \neq 0$. Furthermore, if

$$P = \alpha_n(x)\,(\partial^m/\partial x_n^m) + \alpha_1(x)(\partial/\partial x_n) + \text{other terms},$$

then $\alpha_n(x)$ is nowhere zero because P is elliptic. Since at 0, $f(0) = P(ax_n^m + x_n)(0) = a\alpha_n(0)m! + \alpha_1(0)$, we may assume (upon suitably adjusting a) that $f(0) = 0$. We shall now produce a u such that $Pu = f$ in B_δ and $|tu|$ is arbitrarily small if δ is sufficiently small. Then $v = u - u_0$ will provide the sought-for solution.

To find this u, first recall a weak form of the Sobolev inequality [8, p. 537]: In B_δ, every function g which is N-time continuously differentiable for $N > n/2$ satisfies:

$$|g(0)| \leqq \sum_{|\alpha|=0}^{N} a_i\, \delta^{|\alpha|-n/2} \|D^\alpha g\|_{2,\delta},$$

where α is a multi-index, $\|\ \|_{2,\delta}$ is the L^2 norm over B_δ, and the $\{a_i\}$ are constants independent of δ and g. Now since all the $D^\alpha g$ are continuous, each $\delta^{-n/2}\|D^\alpha g\|_{2,\delta} \to$

some constant c_α as $\delta \to 0$. Thus

$$|g(0)| \leqq a_0\delta^{-n/2}\|g\|_{2,\delta} + o(\delta).$$

Now let $g = \partial u/\partial x_n$, where u is the local solution of $Pu = f$ given by Hörmander in [**13**, p. 174]. Recall that u has the property

$$\|D^\alpha u\|_2 \leqq A\|f\|_2 = A\|f\|_{2,\delta},$$

for $|\alpha| \leqq m$, A is a constant independent of f, and the last equality is because support $f \subseteq B_\delta$. In particular,

$$|tu| \leqq a_0 A\, \delta^{-n/2}\|f\|_{2,\delta} + o(\delta).$$

Now, $\delta^{-n/2}\|f\|_{2,\delta} \to 0$ as $\delta \to 0$ because in B_δ, $|f| \leqq |P(ax_n^m + x_n)| \leqq$ (constant independent of δ) $\cdot\,\delta$, so that $\|f\|_{2,\delta} = o(\delta^{(n+1)/2})$. Thus the right side $\to 0$ as $\delta \to 0$. This proves our claim.

To return to the proof of the main theorem, the lemma guarantees that given $p \in M$, there exist n functions $\{v_1,\cdots, v_n\}$ defined in a compact coordinate ball around p which satisfy $Pu = 0$ and give local coordinates at p. Using the Lax-Malgrange theorem again, let $\{u_1,\cdots, u_n\} \subseteq \mathscr{P}$ such that u_i approximates v_i sufficiently closely for each i in this coordinate ball. Since for solutions of elliptic equations, L^2 convergence on compact sets is identical with uniform convergence of functional values and all derivatives on compact sets ([**16**] or [**21**]), the approximation of the v's by the u's means that the first partial derivatives of u_i converge uniformly to the corresponding first partial derivatives of v_i for each i. In particular, the Jacobian of $\{u_i\}$ at p converges to the Jacobian of $\{v_i\}$ at p. The latter is nonzero, and thus $\{u_i\}$ gives local coordinates at p.

Equip $\mathscr{C}^\infty(M)$ with the standard topology of uniform convergence on compact subsets of M of functional values and derivatives of all orders. $\mathscr{P}$ is then a closed subspace of $\mathscr{C}^\infty(M)$. Now any closed subspace $\mathscr{H}$ of $\mathscr{C}^\infty(M)$ which separates points and gives local coordinates at each point of M has the property that the imbeddings of M in $\boldsymbol{R}^{2n+1}$ whose component functions are in $\mathscr{H}$ form a dense subset of $\mathscr{H} \times \cdots \times \mathscr{H}$ ($2n + 1$ factors). Similarly, the immersions of M in $\boldsymbol{R}^{2n}$ by functions in $\mathscr{H}$ form a dense subset of $\mathscr{H} \times \cdots \times \mathscr{H}$ ($2n$ factors). These statements are the essential content of the Whitney projection technique of imbedding. Since $\mathscr{P}$ enjoys both properties by the previous remarks, one sees that the imbeddings of M in $\boldsymbol{R}^{2n+1}$ whose component functions are in $\mathscr{P}$ form a dense subset of $\mathscr{P} \times \cdots \times \mathscr{P}$ ($2n + 1$ factors) and that the immersions of M in $\boldsymbol{R}^{2n}$ whose component functions are in $\mathscr{P}$ form a dense subset of $\mathscr{P} \times \cdots \times \mathscr{P}$ ($2n$ factors). It should be noted that to prove these density properties of imbeddings and immersions it is not in fact necessary to assume that the operator P has no term of order zero. Minor modifications of the proof just given establish the same density properties for the case where P is an arbitrary C^∞ real linear elliptic differential operator with the unique continuation property. Specifically, the only use made in the prevoius argument of the fact that the constants were assumed to be solutions of $Pu = 0$ was in the argument concerning separation of points by functions in $\mathscr{P}$. In the general case, the constants' being solutions can be replaced in that argument by a second use of the initial value specification statement in [3, p. 228], which implies that for any C^∞ linear elliptic differential operator the value at a point of a (local) solution in a neighborhood of that point can be specified arbitrarily. However, the argument to be given now to establish the existence of proper mappings into some $\boldsymbol{R}^k$ by functions in $\mathscr{P}$ does depend on knowing that any constant λ satisfies $P\lambda = 0$, and it is for this reason that the assumption that P is without zero order term is made in the main theorem and thus also in the following argument.

The only essential step remaining in the proof of the main theorem is to produce a proper map of M into some $\boldsymbol{R}^k$ by members of $\mathscr{P}$. Once this step is carried out we shall have at least a proper imbedding in $\boldsymbol{R}^{2n+1+k}$. The proof in §3 of [**10**] carries over verbatim to show that in fact k can be taken to be $(n + 1)$. Moreover, a careful analysis also given therein shows how to project from $\boldsymbol{R}^{2n+1+k}$ to $\boldsymbol{R}^{2n+1}$ while preserving both the imbedding property of the mapping and the properness, provided that the imbedding and the proper map were chosen correctly in the first place. In general, this procedure becomes quite complicated, but we can give a simple example which will clearly illustrate the main ideas behind the proof in [**10**] of the existence of a proper map into some $\boldsymbol{R}^k$ by elements of $\mathscr{P}$.

Let P be defined on $\boldsymbol{R}^2$. We shall show how to produce a proper map $\boldsymbol{R}^2 \to \boldsymbol{R}^4$ using elements of $\mathscr{P}$. The construction depends on a careful choice of a multi-sequence of compact sets. Let

$$\begin{aligned} B_n &= \text{closed disc of radius } n, \\ \boldsymbol{R}^2_+ &= \text{closed upper half-plane}, \\ \boldsymbol{R}^2_- &= \text{closed lower half-plane}. \end{aligned}$$

Now define

$$\begin{aligned} S_1 &= \textstyle\bigcup_{n=0}^\infty (\text{closure}\,(B_{2n+1} - B_{2n}) \bigcap \boldsymbol{R}^2_+), \\ S_2 &= \textstyle\bigcup_{n=0}^\infty (\text{closure}\,(B_{2n+1} - B_{2n}) \bigcap \boldsymbol{R}^2_-), \\ S_3 &= \textstyle\bigcup_{n=1}^\infty (\text{closure}\,(B_{2n} - B_{2n-1}) \bigcap \boldsymbol{R}^2_+), \\ S_4 &= \textstyle\bigcup_{n=1}^\infty (\text{closure}\,(B_{2n} - B_{2n-1}) \bigcap \boldsymbol{R}^2_-). \end{aligned}$$

Clearly $S_1 \cup S_2 \cup S_3 \cup S_4 = \boldsymbol{R}_2$ and each component of S_i $(i = 1, 2, 3, 4)$ is itself compact and has a connected unbounded complement. Also, each compact set meets only a finite number of the components of the sets $S_1, \cdots, S_4$. Thus on each component of $S_1, \cdots, S_4$, every solution of $Pu = 0$ is the uniform limit of elements of $\mathscr{P}$ (by the Lax-Malgrange theorem, as usual). By an iteration process, we can produce for any increasing sequence of positive integers $\{\alpha_1, \alpha_2, \cdots\}$ four elements $f_1, \cdots, f_4 \in \mathscr{P}$ such that $f_i \,|\, S_i^j > \alpha_j$, where we have written each $S_i = \bigcup_{j=1}^\infty S_i^j$ as the union of its components. Thus $(f_1, \cdots, f_4)$ give a proper map of $\boldsymbol{R}^2$ into $\boldsymbol{R}^4$, as claimed.

References

1. A. Andreotti and H. Grauert, *Théorèmes de finitude pour la cohomologie des espaces complexes*, Bull. Soc. Math. France **90** (1962), 193–259. MR **27** #343.

2. N. Aronszajn, *A unique continuation theorem for solutions of elliptic partial differential equations or inequalities of second order*, J. Math. Pures Appl. (9) **36** (1957), 235–249. MR **19**, 1056.

3. L. Bers, F. John and M. Schechter, *Partial differential equations*, Proc. Summer Seminar (Boulder, Col., 1957), Lectures in Appl. Math., Vol. III, Interscience, New York, 1964. MR **29** #346.

4. E. Bishop, *Partially analytic spaces*, Amer. J. Math. **83** (1961), 669–692. MR **25** #5191.

5. S. Bochner, *Analytic mappings of compact Riemannian spaces into Euclidean space*, Duke Math. J. **3** (1937), 339–354.

6. S. S. Chern, *Minimal submanifolds in a Riemannian manifold*, Technical Report 19, Department of Mathematics, University of Kansas, Lawrence, Kansas, 1968. MR **40** #1899.

7. H. O. Cordes, *Über die eindeutige Bestimmtheit der Lösungen elliptischer Differentialgleichungen durch Anfangsvorgaben*, Nachr. Akad. Wiss. Göttingen. Math.-Phys. Kl. IIa **1956**, 239–258. MR **19**, 148.

8. H. Federer, *Geometric measure theory*, Die Grundlehren der math. Wissenschaften, Band 153, Springer-Verlag, New York, 1969. MR **41** #1976.

9. H. Grauert, *On Levi's problem and the imbedding of real-analytic manifolds*, Ann. of Math. (2) **68** (1958), 460–472. MR **20** #5299.

10. R. E. Greene and H. Wu, *Embeddings of open Riemannian manifolds by harmonic functions* (to appear).

11. ———, *Integrals of subharmonic function on manifolds of nonnegative curvature* (to appear).

12. M. L. Gromov, *Stable mappings of foliations into manifolds*, Izv. Akad. Nauk SSSR Ser. Mat. **33** (1969), 707–734 = Math. USSR Izv. **3** (1969), 671–694. MR **41** #7708.

13. L. Hörmander, *Linear partial differential operators*, Springer-Verlag, Berlin and New York, 1963. MR **28** #4221.

14. J. L. Kazdan and F. W. Warner, *Prescribing curvature*, these Proceedings.

15. P. D. Lax, *A stability theorem for solutions of abstract differential equations, and its application to the study of the local behavior of solutions of elliptic equations*, Comm. Pure Appl. Math. **9** (1956), 747–766. MR **19**, 281.

16. B. Malgrange, *Existence et approximation des solutions des équations aux dérivées partielles et des équations de convolution*, Ann. Inst. Fourier (Grenoble) **6** (1955/56), 271–355. MR **19**, 280.

17. ———, *Faisceaux sur des variétés analytiques réeles*, Bull. Soc. Math. France **85** (1957), 231–237. MR **20** #1340.

18. ———, *Plongement des variétés analytiques-réeles*, Bull. Soc. Math. France **85** (1957), 101–112. MR **20** #1338.

19. C. B. Morrey, *The analytic imbedding of abstract real-analytic manifolds*, Ann. of Math. (2) **68** (1958), 159–201. MR **20** #5504.

20. R. Narasimhan, *Imbeddings of holomorphically complete complex spaces*, Amer. J. Math. **82** (1960), 917–934. MR **26** #6438.

21. ———, *Analysis on real and complex manifolds*, Advanced Studies in Pure Math., Vol. 1, Masson, Paris; North-Holland, Amsterdam, 1968. MR **40** #4972.

22. K. Nomizu and H. Ozeki, *The existence of complete Riemannian metrics*, Proc. Amer. Math. Soc. **12** (1961), 889–891. MR **24** #A3610.

23. M. H. Protter, *Unique continuation for elliptic equations*, Trans. Amer. Math. Soc. **95** (1960), 81–91. MR **22** #3871.

24. R. Remmert, *Sur les espaces analytiques holomorphiquement séparables et holomorphiquement convexes*, C. R. Acad. Sci. Paris **243** (1956), 118–121. MR **18**, 149.

25. H. L. Royden, *The analytic approximation of differentiable mappings*, Math. Ann. **139** (1960), 171–179. MR **22** #4067.

26. Y.-T. Siu, *Analytic sheaf cohomology groups of dimension n of n-dimensional noncompact complex manifolds*, Pacific J. Math. **28** (1969), 407–411. MR **39** #4440.

27. ———, *Analytic sheaf cohomology groups of dimension n of n-dimensional complex spaces*, Trans. Amer. Math. Soc. **143** (1969), 77–94. MR **40** 5902.

University of California, Los Angeles

University of California, Berkeley

Proceedings of Symposia in Pure Mathematics
Volume 27, 1975

FOURIER INTEGRAL OPERATORS FROM THE RADON TRANSFORM POINT OF VIEW

V. GUILLEMIN AND D. SCHAEFFER

The purpose of these notes is not to discuss any new results but to describe a point of view toward some existing results: i.e. Hörmander's *Fourier integral operators* [1] and a related series of papers by Donald Ludwig, especially [3]. The reasons for this are partly pedagogical, but also our approach relates well to recent developments in the theory of generalized functions. (See the concluding paragraph below.)

We recall that distributions on manifolds enjoy two types of functoriality, "push-forward" and "pull-back". "Push-forward" is simpler to describe. For a *proper* map $f\colon X \to Y, f_*$ on distributions is the dual operation to f^* on compactly supported C^∞ functions. "Pull-back" is only defined for submersions, $f\colon X \to Y$. Locally a submersion is just a projection map, for example the standard projection $\boldsymbol{R}^k \times \boldsymbol{R}^l \to \boldsymbol{R}^k$. Given a C_0^∞ density $\rho(x, y)\, dx\, dy$ on $\boldsymbol{R}^k \times \boldsymbol{R}^l$ we define its push-forward as $(\int \rho(x, y)\, dy)\, dx$. The dual operation is the pull-back operation on distributional densities. The point of view toward Fourier integrals which we want to describe here is one that makes maximum use of these two functors.

We will begin with "wave-front" sets. Given a distribution μ on a manifold X, Hörmander attaches to it a subset $WF(\mu)$ of the cotangent bundle T^*X called its wave-front set.[1] Intuitively (x_0, ξ_0) is not in $WF(\mu)$ if μ is smooth at x_0 in the direction ξ_0. Hörmander's precise definition is $(x_0, \xi_0) \notin WF(\mu) \Leftrightarrow \widehat{\rho\mu}(\xi)$ is rapidly decreasing in a conical nbhd of ξ_0, where ^ is the Fourier transform and ρ a bump function at x_0.

We will now give an alternative definition of $WF(\mu)$.

AMS (MOS) subject classifications (1970). Primary 47G05; Secondary 58G15.

[1]There are various wave-front sets. For simplicity we will use the projective wave front set: $(x, \xi) \in WF(\mu) > (x, \lambda \xi) \in WF(\mu)$ for $\lambda \neq 0$.

ROUGH DEFINITION. $(x_0, \xi_0) \notin WF(\mu)$ if for every function $f\colon X \to \boldsymbol{R}$ with $(df)_{x_0} = \xi_0$ the push-forward, $f_*\rho\mu$, is smooth, ρ being, as above, a bump function at x_0. The workable definition adds: if f depends smoothly on parameters so does $f_*\rho\mu$.

We will show that the two definitions agree. First however let us use the second definition to compute $WF(\mu)$ for a particularly simple class of distributions.

DEFINITION. A distribution, μ, on X is *wave-like* if it is of the form $\mu = f^*\nu$ where ν is a distribution on $\boldsymbol{R}$ and $f\colon X \to \boldsymbol{R}$ a submersion.

LEMMA 1. *For a wave-like distribution, $f^*\nu$, the wave-front set is the set of normal vectors to the surfaces $f = c$, c being in the singular support of ν.*

PROOF. Suppose $g\colon X \to \boldsymbol{R}$ and $(dg)_{x_0} \neq (df)_{x_0}$. Then we can choose coordinates $x_1, \cdots, x_n$ centered at x_0 such that $f = x_1$ and $g = x_2$. Let ρ be a bump function at x_0. Then

$$g_*\rho f^*\nu = \int\left(\int \nu(x_1)\,\rho(x_1, \cdots, x_n)\,dx_1\right) dx_3\,dx_4 \cdots dx_n$$

which is smooth.

We will also need the Radon inversion formula. For simplicity we will state this in odd dimensions.

LEMMA 2. *Given $\omega \in S^{n-1}$, let $\phi_\omega\colon \boldsymbol{R}^n \to \boldsymbol{R}$ be the map $x \to x\cdot\omega$ and let $D = (-1)^{-1/2}d/dt$ on $\boldsymbol{R}$. Then for a compactly supported distribution, μ, on $\boldsymbol{R}^n$ one has*

$$\text{(i)} \qquad \int \phi_\omega^*\, D^{n-1}(\phi_\omega)_*\mu\, d\omega = \frac{1}{2(2\pi)^n}\mu.$$

PROOF. See Fritz John [2].

Finally we will need a trivial identity:

LEMMA 3. $\widehat{(\phi_\omega)^*\mu}(t) = \hat{\mu}(t\omega)$ *for* $\mu \in \mathscr{E}'(\boldsymbol{R}^n)$.

To prove the equivalence of our two definitions we note that by Lemma 3 if $\hat{\mu}$ is rapidly decreasing in the direction ω_0, $(\phi\omega_0)_*\mu$ is smooth. Suppose now that ω_0 is not in $WF(\mu)$ at x_0 (in the sense of the first definition). Then the integrand in (i) is smooth for ω near ω_0 by Lemma 3 and its push-forward in the direction of ω_0 is smooth for $\omega \neq \omega_0$ by Lemma 1. Q.E.D.

REMARK. The functorial definition seems to make some of the standard facts about WF easier to understand. For example the pull-back of distributions with respect to a map $f\colon X \to Y$ is usually defined only if f is a submersion. One can try to define it for more general maps by the "method of continuity" i.e. let μ_i be a sequence of smooth functions converging to μ in the distribution sense and define $f^*\mu = \lim f^*\mu_i$ if it exists. Hörmander shows that the method of continuity works if f satisfies:

$$\text{(ii)} \qquad (df)_x^t\,\xi \neq 0 \quad \text{whenever } (f(x), \xi) \in WF(\mu).$$

To prove this observe first that the thorem is true

(a) when μ is smooth. Then $f^*\mu$ can be defined as the usual pull-back of functions.

(b) when μ is wave-like. If $\mu = g^*\nu$ for some $g\colon Y \to \boldsymbol{R}$ then the transversality condition above just says that $g \cdot f$ is a submersion, so we can define $f^*\mu = (g \circ f)^*\nu$.

For the general case just break μ into a sum of (a) and (b) using the Radon transform.

We will now describe Hörmander's Fourier integrals from the "Radon" point of view. This approach requires first of all a familiarity with the classical homogeneous distributions on the real line namely:

(iii) $$\delta(x), \quad \delta^{(\nu)}(x), \quad x^{-n}, \quad x_+^\lambda, \quad (x+io)^\lambda, \quad \text{etc.}$$

Consider the pull-backs of these distributions with respect to submersions $f\colon Z \to \boldsymbol{R}$. These are the classical "couches" or "boundary layers". By Lemma 1 their WF set is the normal bundle to the hypersurface $f = 0$.

Finally given a submersion $\pi\colon Z \to X$, consider distributions of the form

(iv) $$\pi_* \rho f^* \mu$$

μ being a distribution on the list (iii).

PROPOSITION. *The Fourier integrals of Hörmander are just those distributions on X which can be approximated to arbitrary order of smoothness by finite sums of distributions of the type* (iv).

We will now prove a theorem of Ludwig concerning the wave-front sets of distributions of the form (iv).

THEOREM. *Let $Z = X \times Y$ and let $\pi\colon Z \to X$ be the usual projection. For fixed $y \in Y$ let $S(y)$ be the hypersurface in X consisting of the points*: $x \in X$, $f(x, y) = 0$. (*Assume for simplicity that $S(y)$ has no singular points.*) *Let $\Lambda \subset T_X^*$ be the normal bundle of the envelope of these hypersurfaces. Then $WF(\mu) \subset \Lambda$.*

PROOF. We first prove

LEMMA 4. *Let $\pi\colon Z \to X$ be a submersion and μ a distribution on Z. Then $WF(\pi_*\mu)$ is contained in the set of (x, η) with the property: $\exists z \in Z$, $\pi(z) = x$ and $(d\pi)_z^t \eta \in WF(\mu)$.*

PROOF. Let $f\colon X \to \boldsymbol{R}$ with $(df)_x = \eta$ and observe that $f_*(\pi_*\mu)$ smooth $\Leftrightarrow (f \circ \pi)_* \mu$ smooth.

To prove the theorem we note that, as was already observed, if $f\colon X \to \boldsymbol{R}$ and ν is on the list (iii), then $WF(f^*\nu)$ is just the normal bundle to $f = 0$. By Lemma 4 the WF of the push-forward is the set of vectors x, $\partial f(x, y)/\partial x$ where $f(x, y) = \partial f(x, y)/\partial y = 0$ which, by classical surface theory (see Struik **[4]**) is the equation for the normal bundle to the envelope.

As an illustration of Ludwig's result let X be $\boldsymbol{R}^n$ and let Y be a hypersurface in $\boldsymbol{R}^n$. Let $f\colon \boldsymbol{R}^n \times Y \to \boldsymbol{R}$ be the function $(x, y) \to |x - y|^2 - c^2$ and let ν be the delta function. For fixed $y \in Y$, $f_y^* \nu$ is the delta function of the sphere of radius c

about y in $\boldsymbol{R}^n$. "Summing" these $f_y^*\nu$'s, we get, by Ludwig, a distribution on $\boldsymbol{R}^n$ whose wave-front set lies on the envelope of the spheres $|x - y| = c$, $y \in Y$, i.e. the singularities not on the envelope cancel each other out by "interference" (corroborating Huygens' principle!). See Figure 1.

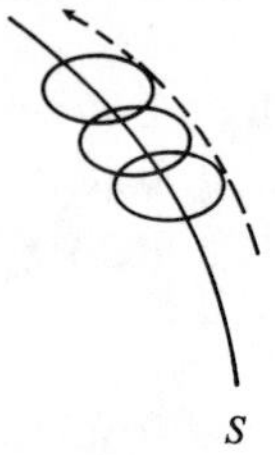

FIGURE 1

The above results convey, we hope, some of the flavor of the "Radon" approach to Fourier integrals.

To conclude we note that there have been two recent developments in the theory of generalized functions having many analogies with the work of Hörmander. One is the recent work of Sato on hyperfunctions and the other the work of Maslov-Leray on "asymptotic functions". In both cases there are analogues of wave-front sets, Fourier integrals, etc. There is some evidence that this is due to the existence of analogous push-forwards and pull-backs.

We have learned recently that W. Ambrose has independently proposed an approach to Fourier integral operators using the Radon transform (private communication).

REFERENCES

1. L. Hörmander, *Fourier integral operators*. I, Acta Math. **127** (1971), 79–183.

2. F. John, *Plane waves and spherical means applied to partial differential equations*, Interscience, New York, 1955. MR **17**, 746.

3. D. Ludwig, *Exact and asymptotic solutions of the Cauchy problem*, NYU Report, 1959; Comm. Pure Appl. Math. **13** (1960), 473–508. MR **22** #5816.

4. D. Struik, *Lectures on classical differential geometry*, Addison-Wesley, Reading, Mass., 1950. MR **12**, 127.

MASSACHUSETTS INSTITUTE OF TECHNOLOGY

Proceedings of Symposia in Pure Mathematics
Volume 27, 1975

A HIERARCHY OF NONSOLVABILITY EXAMPLES

C. DENSON HILL*

1. Introduction. Let P be an $l \times k$ system of linear partial differential operators with C^∞ coefficients. We want to consider the determined/overdetermined/underdetermined system of inhomogeneous equations

$$Pu = f. \tag{1}$$

We shall consider only those f which satisfy the compatibility conditions

$$Qf = 0. \tag{2}$$

Here Q is a similar $m \times l$ system of operators such that $QP = 0$ gives a complete set of consistency conditions for the formal solvability of (1). Thus we have a complex

$$\mathscr{E}^k \xrightarrow{P} \mathscr{E}^l \xrightarrow{Q} \mathscr{E}^m, \tag{3}$$

where $\mathscr{E}$ denotes C^∞ functions, and our assumption is that for smooth f satisfying (2) there is no obstruction to the existence of formal power series solutions of (1).

DEFINITION. (a) $Pu = f$ is *locally solvable* at a point x_0 (or a Poincaré lemma for (1) is valid at x_0) means: Given any open neighborhood ω of x_0, there exists an open neighborhood ω' of x_0 with $\omega' \subset \omega$, such that given any f which is C^∞ and satisfies (2) in ω, there exists a u that is C^∞ and satisfies (1) in ω'.

(b) $Pu = f$ is *not locally solvable* at x_0 (or (1) does not have a Poincaré lemma at x_0) means: The point x_0 has a fundamental sequence of neighborhoods $\{\omega\}$ such

AMS (MOS) subject classifications (1970). Primary 35N10, 32F10; Secondary 35N15.

* Supported by an Alfred P. Sloan Research Fellowship.

that for most (say, in the sense of Baire category) f which are C^∞ and satisfy (2) in ω, there is no C^∞ u which solves (1) in *any* neighborhood ω' of x_0.

Note that in case (b) the complex (3) cannot be exact at x_0 in the sense of sheaves.

In addition to the algebraic consistency conditions (2) there are (as is well known in the determined case where $Q = 0$ and $Pu = f$ consists of a single equation) also conditions of a geometric nature which may obstruct the actual solvability of (1). The purpose of this note is to call attention to a hierarchy of examples of systems (1) in which it is possible to explain just how certain geometric conditions involving curvature determine the solvability or nonsolvability. The hierarchy is especially interesting in that it contains a sequence of nonsolvability examples (some overdetermined and some underdetermined) with the original example **[5]** of Hans Lewy occurring as the first term in the sequence. We give a single geometric proof of nonsolvability for all of these examples; in particular, it provides a new proof of the nonsolvability for the Lewy example.

These examples were obtained in the joint work **[4]** of A. Andreotti and the author. We refer the reader to **[2]**, **[3]**, **[4]** for detailed proofs of some arguments which will be only sketched here. Other related aspects of this work have been summarized in **[1]** and **[6]**.

2. The tangential Cauchy-Riemann complex. Let S be a C^∞ hypersurface in $\boldsymbol{C}^n$ and let x_0 be a point on S; we may assume that, near x_0, S is given by $S = \{\rho = 0\}$ where ρ is a C^∞ real-valued function with $d\rho \neq 0$. The specific example of (3) which we consider is the jth position

$$\cdots \longrightarrow \mathscr{C}^{j-1} \xrightarrow{\bar\partial_s} \mathscr{C}^{j} \xrightarrow{\bar\partial_s} \mathscr{C}^{j+1} \longrightarrow \cdots \tag{4}$$

in the tangential Cauchy-Riemann complex to S near x_0. Here $\mathscr{C}^j$ is the C^∞ Cauchy data on S with respect to $\bar\partial$ acting on $\mathscr{E}^{0j}$ (C^∞ forms of type $(0, j)$) and $\bar\partial_s$ is the tangential part of $\bar\partial$, so $\bar\partial_s^2 = 0$. Alternately: $\mathscr{C}^j = \mathscr{E}^{0j} \,/\, \mathscr{I}^{0j}$ where $\mathscr{I}^{0j}$ denotes those C^∞ $(0, j)$ forms which can be written as $\rho\alpha + \bar\partial\rho \wedge \beta$, and (4) is defined in the obvious way as a quotient complex. For complete details see **[3]**.

The signature sig $\mathscr{L}(x_0)$ of the Levi form of S at x_0 is the signature of the quadratic form

$$\sum_{j,k} \frac{\partial^2 \rho}{\partial z_j \partial \bar z_k}(x_0)\, w_j \bar w_k$$

restricted to the analytic tangent space $\sum_j (\partial\rho/\partial z_j(x_0)) w_j = 0$. This signature at x_0 is a biholomorphic invariant and is independent of the particular function ρ used to define S.

Note that (4) is a nonelliptic complex.

3. Positive results. The following theorem is obtained by combining the results of Theorem 2 of **[3]** and Theorems 1–3 of **[4]**.

THEOREM 1. *Assume* sig $\mathscr{L}(x_0)$ *has p positive, q negative, and r zero eigenvalues. Then* (4) *is locally solvable at x_0 provided* $j \notin [p, p + r] \cup [q, q + r]$.

COROLLARY. *Assume the Levi form of S is nondegenerate with p positive and q negative eigenvalues at x_0. Then* (4) *is locally solvable at x_0 provided $j \neq p, q$.*

REMARK. Actually much more is proved in **[4]** than is indicated by Theorem 1 above: For example, if $S = S \cap U$ where U is an open domain of holomorphy, if $j > 0$ and if $j \notin [\min(p, q), \max(p, q) + r]$, then, under the hypotheses of Theorem 1, it is shown that $H^j(S) = 0$ (there is no boundary cohomology). Also precise ranges are given in which one can solve the Cauchy and Riemann-Hilbert problems in cohomology.

4. Negative results. As a companion to the above corollary we have

THEOREM 2. *Assume the Levi form of S is nondegenerate with p positive and q negative eigenvalues at x_0. Then for $j = p$ and $j = q$ the system in* (4) *is not locally solvable at x_0.*

EXAMPLE. In $\boldsymbol{C}^2$ consider $S = \{\operatorname{Im} z_2 = |z_1|^2\}$ and pick any point $x_0 \in S$. Then the Levi form has only one eigenvalue and it is not zero. Hence the theorem says that there is a nonsolvability example in (4) for $j = 1$. The complex (4) in that case is just

$$C^\infty(S) \xrightarrow{L} C^\infty(S) \longrightarrow 0$$

where $L = \frac{1}{2}(\partial/\partial x_1 + i\partial/\partial y_1) - i(x_1 + iy_1)\,\partial/\partial x_2$ is the operator of Hans Lewy. Thus we recover his example. See **[2]** and **[3]** for details.

5. Sketch of the proof of Theorem 2. If a solution u of (1) is sought in the same region $\omega' = \omega$ where f is prescribed, then nonsolvability means nontriviality of cohomology. So we introduce the boundary cohomology

$$H^j(S) = \frac{\ker \bar{\partial}_s : \mathscr{C}^j(S) \longrightarrow \mathscr{C}^{j+1}(S)}{\operatorname{im} \bar{\partial}_s : \mathscr{C}^{j-1}(S) \longrightarrow \mathscr{C}^j(S)}$$

on S. The following lemma gives us a handle on $H^j(S)$.

LEMMA A. *Let $S = S \cap U$, where U is an open domain of holomorphy, and set $U^+ = U \cap \{\rho \geqq 0\}$ and $U^- = U \cap \{\rho \leqq 0\}$. Then*

$$H^j(U^+) \oplus H^j(U^-) \xrightarrow[\text{jump}]{\sim} H^j(S),$$

where $H^j(U^\pm)$ denotes Dolbeault cohomology which is C^∞ up to the boundary S.

Lemma A is a consequence of the Mayer-Vietoris sequence of [3].

Thus $H^j(S)$ will be nontrivial, or infinite dimensional, if either $H^j(U^+)$ or $H^j(U^-)$ is.

Next consider the Bochner-Martinelli forms

$$\psi_\alpha = \frac{\sum_j (-1)^j \bar{z}_j^{\alpha_j}\, d\bar{z}_1^{\alpha_1} \wedge \cdots \wedge \widehat{d\bar{z}_j^{\alpha_j}} \wedge \cdots \wedge d\bar{z}_n^{\alpha_n}}{(\sum_j z_j^{\alpha_j} \bar{z}_j^{\alpha_j})^n}$$

where $\alpha = (\alpha_1, \cdots, \alpha_n)$ and $\alpha + 1 = (\alpha_1 + 1, \cdots, \alpha_n + 1)$.

LEMMA B. *Let U be an open set in $\boldsymbol{C}^n - \{0\}$ $(n \geqq 2)$ which contains a closed half-sphere $\{z \in \boldsymbol{C}^n \mid \sum_j | z_j |^2 = \varepsilon, \operatorname{Re} z_1 \geqq 0\}$. Then for $\alpha \in \boldsymbol{N}^n$ the forms $\psi_{\alpha+1}$ represent linearly independent classes of $H^{n-1}(U, \mathcal{O})$ over $\boldsymbol{C}$.*

Lemma B is proved in **[4]**. It uses Stokes' theorem and a Runge approximation argument.

LEMMA C. *Assume the hypotheses of Theorem 2. Then for any sufficiently small neighborhood U of x_0 which is a domain of holomorphy we have, setting $S = S \cap U$, that the groups $H^p(U^+)$, $H^q(U^-)$, $H^p(S)$, $H^q(S)$ are all infinite dimensional over $\boldsymbol{C}$.*

Lemma C is also proved in **[4]**. It involves a local biholomorphic change of variables near x_0, Lemma A, and Lemma B applied to lower dimensional slices of $\boldsymbol{C}^n$.

REMAINDER OF THE PROOF OF THEOREM 2. We present the rest of the proof in detail, since the argument as written in **[4]** is not very easy to follow. Let $\{U_k\}$ $(k = 0, 1, 2, \cdots)$ be a fundamental sequence of neighborhoods of x_0 which are domains of holomorphy, and set $S_k = S \cap U_k$ and $S = S_0$. We have

$$S \supset S_1 \supset S_2 \supset \cdots, \quad x_0 = \bigcap_k S_k.$$

We will treat the case $j = p$, the case $j = q$ being the same.

Consider the diagram

$$\begin{array}{ccc} Z^p(S) & \xrightarrow{\ r_k\ } & Z^p(S_k) \\ & & \uparrow \bar{\partial}_s \\ & & \mathcal{C}^{p-1}(S_k) \end{array}$$

where r_k is the restriction map and

$$Z_p(S_k) = \ker \bar{\partial}_s : \mathcal{C}^p(S_k) \longrightarrow \mathcal{C}^{p+1}(S_k).$$

Let $\omega_1, \cdots, \omega_{n-1}, \bar{\partial}\rho$ be a basis for C^∞ (0, 1) forms near x_0. Then $\mathcal{C}^p(S_k) = \{\sum_\alpha a_\alpha \omega_\alpha \mid a_\alpha \in C^\infty(S_k)\}$ where $\omega_\alpha = \omega_{\alpha_1} \wedge \cdots \wedge \omega_{\alpha_p}$. Thus $\mathcal{C}^p(S_k)$ has the topology of the Fréchet-Schwartz space $[\mathcal{E}(S_k)]^m$ with $m = \binom{n-1}{p}$. The maps r_k, $\bar{\partial}_s$ are continuous; $Z^p(S_k)$ and

$$E_k = \{(f, u) \in Z^p(S) \times \mathcal{C}^{p-1}(S_k) \mid \bar{\partial}_s u = r_k f \text{ on } S_k\}$$

are Fréchet-Schwartz spaces. Let

$$\operatorname{Im} E_k = \text{projection of } E_k \text{ onto } Z^p(S);$$

it is the continuous image of a Fréchet-Schwartz space. Hence according to a theorem of Banach there are only two possibilities: Either

1. $\operatorname{Im} E_k = Z^p(S)$, or
2. $\operatorname{Im} E_k$ is of the first category in $Z^p(S)$. But Lemma B exhibits explicit non-

trivial cohomology classes; so possibility 1 is ruled out. Hence possibility 2 prevails and it follows that $\bigcup_{k=1}^{\infty} \operatorname{Im} E_k$ is of the first category in $Z^p(S)$.

Thus for most (in the sense of second category) $f \in Z^p(S)$ there can be no u which is C^∞ and satisfies $\bar{\partial}_s u = f$ in any neighborhood of x_0.

References

1. A. Andreotti, *E. E. Levi convexity and the Hans Lewy problem*, Proc. Internat. Congress Math. (Nice, 1970), vol. 2, Gauthier-Villars, Paris, 1971, pp. 607–611.

2. A. Andreotti and C. D. Hill, *Complex characteristic coordinates and tangential Cauchy-Riemann equations*, Ann. Scuola Norm. Sup. Pisa **26** (1972), 299–324.

3. ———, *E. E. Levi convexity and the Hans Lewy problem*. I: *Reduction to vanishing theorems*, Ann. Scuola Norm. Sup. Pisa **26** (1972), 325–363.

4. ———, *E. E. Levi convexity and the Hans Lewy problem*. II: *Vanishing theorems*, Ann. Scuola Norm. Sup. Pisa **26** (1972), 747–806.

5. H. Lewy, *An example of a smooth linear partial differential equation without solution*, Ann. of Math. (2) **66** (1957), 155–158. MR **19**, 551.

6. C. D. Hill, *The Cauchy problem for $\bar{\partial}$*, Proc. Sympos. Pure Math., vol. 23, Amer. Math. Soc., Providence, R. I., 1973, pp. 135–144.

STANFORD UNIVERSITY

Proceedings of Symposia in Pure Mathematics
Volume 27, 1975

EXTENDING ISOMETRIC EMBEDDINGS

H. JACOBOWITZ*

Let H be a submanifold of a Riemannian manifold U. Assume $f: H \to E^N$ is an isometric embedding into some Euclidean space. When can f be extended to an isometric embedding of U into the same Euclidean space? In this paper the local problem is considered, i.e. find conditions on f and H at a point $p \in H$ which guarantee the existence of such an extension of f to an open neighborhood containing p. When everything is real analytic these conditions are implicit in the proofs of the Cartan-Janet theorem and the main goal of this paper is to obtain such extension results in the nonanalytic case. The approach is basically that of Nash. One inverts a linearized operator and tries to construct the solutions of a nonlinear problem. The difficulty occurs in preserving the initial conditions.

Let $L: T_pH \times T_pH \to \boldsymbol{R}^1$ be the second fundamental form at p of H in U and $\bar{L}$: $T_pH \times T_pH \to \boldsymbol{R}^{N-\dim H}$ the second fundamental form at p of H in E^N. One writes $\bar{L} > L$ if there exists some $\nu \in T_{f(p)}E^N$ with $\langle \nu, \bar{L}(x, y)\rangle = L(x, y)$, $\langle \nu, f_x(x)\rangle = 0$ for x and y in T_pH and $\|\nu\| < 1$. If only $\|\nu\| \leqq 1$, then one writes $\bar{L} \geqq L$.

THEOREM. *Let U be a Riemannian manifold of dimension n, H a codimension one submanifold, $f: H \to E^N$, $N \geqq \frac{1}{2}n(n+3)$ a nondegenerate isometric embedding and p a point of H at which $\bar{L} > L$. If U, H and f are of class C^∞, then there exists a C^∞ isometric embedding $\hat{f}$ of some open set in U containing p with $\hat{f}|_H = f$.*

A version of this theorem is also true for finite differentiability. Note that $\bar{L} \geqq L$ is always necessary for the existence of an isometric extension.

RICE UNIVERSITY

AMS (MOS) subject classifications (1970). Primary 53A05, 53C40; Secondary 53B25, 35A10.
*This work was supported in part by NSF grant GP-18961 and will be published in J. Differential Equations.

Proceedings of Symposia in Pure Mathematics
Volume 27, 1975

PRESCRIBING CURVATURES

JERRY L. KAZDAN* AND F. W. WARNER*

For many years mathematicians have investigated the global constraints that the topology of a differentiable manifold imposes on its differential geometric structures. One of the most elementary and appealing of these constraints is expressed by the Gauss-Bonnet theorem, which among other things, gives the following necessary sign condition on the Gaussian curvature K of a compact 2-manifold M in terms of its Euler characteristic, $\chi(M)$:

if $\chi(M) > 0$, then K must be positive somewhere,
(A) if $\chi(M) = 0$, then K must change sign unless it is identically zero,
if $\chi(M) < 0$, then K must be negative somewhere.

We have recently shown **[18]** that this sign condition is also sufficient for a given function K to be the curvature of some metric. After briefly summarizing this work and related work on scalar curvature functions, we present some new results concerning scalar curvature; in particular, we show that there are topological obstructions to manifolds admitting metrics of zero scalar curvature. One intriguing aspect of scalar curvature (as well as Ricci and sectional curvature) is that current knowledge of topological obstructions is quite fragmentary. § 5 contains several open questions.

Gaussian Curvature

1. Summary. The main result here is quite simply stated. Let M be a compact connected 2-manifold.

AMS (MOS) subject classifications (1970). Primary 53C20, 53A30, 35J60; Secondary 47H15, 41A30, 58G99.

*Supported in part by NSF grants GP 28976X and GP 29258.

THEOREM 1.1. *$K \in C^\infty(M)$ is the Gaussian curvature of a metric on M if and only if K satisfies the sign condition* (A).

Various partial results were previously obtained in **[5]**, **[11]**, **[15]**, **[20]**, by a variety of methods. In our approach, we fix a metric g on M and seek the metric g_1 with curvature K as conformally equivalent to g, so $\varphi^*(g_1) = e^{2u}g$, where φ is a diffeomorphism. If Δ and k are the Laplacian and curvature of g, then u satisfies the nonlinear elliptic equation

$$(1.2) \qquad \Delta u = k - (K \circ \varphi)e^{2u}.$$

Thus, we seek a diffeomorphism φ and a function u such that (1.2) is satisfied. Our previous work **[15]** showed that in many natural cases, a solution of

$$(1.3) \qquad \Delta u = k - Ke^{2u}$$

may not exist. Thus we were led to content ourselves with a modest existence theorem for (1.3) and place more emphasis on finding diffeomorphisms φ such that equation (1.2) is solvable. In this way, we obtain the following theorem, of which Theorem 1.1 is an immediate consequence. A proof is sketched in the next section.

THEOREM 1.4. *Let (M, g) be a compact connected Riemannian 2-manifold with constant Gaussian curvature. Then $K \in C^\infty(M)$ is the Gaussian curvature of a metric conformally equivalent to g if and only if K satisfies the sign condition* (A).

On the basis of this result, we were also able to show that if M is an open manifold obtained by deleting an arbitrary closed set from a compact manifold, then any $K \in C^\infty(M)$ is the Gaussian curvature of some metric **[16]**. By an entirely different method, we also characterized those functions $K \in C^\infty(R^2)$ that are Gaussian curvatures of *complete* metrics **[16]**. It would be interesting if one could extend this result on complete metrics to open manifolds other than just R^2.

2. Proof of Theorem 1.4. Rewrite (1.2) as

$$(2.1) \qquad T(u) \equiv -e^{-2u}(\Delta u - k) = K \circ \varphi.$$

The problem is to find a function u and a diffeomorphism φ such that $T(u) = K \circ \varphi$. There are three key ingredients:

(i) INVERSE FUNCTION THEOREM. *Let $T'(u_0)v$ denote the linearization of T about u_0, so*

$$T'(u_0)\,v = \frac{d}{dt} T(u_0 + tv)\Big|_{t=0} = -e^{-2u_0}\,[\Delta v - 2(\Delta u_0 - k)v].$$

Thus $T'(u_0)$ is a linear elliptic operator. Let $f \in C^\infty(M)$. If $T'(u_0)$ is invertible, then, as one might expect, there is an $\eta > 0$ such that if $\|f - T(u_0)\|_p < \eta$ then there is a $u \in C^\infty(M)$ such that $T(u) = f$. Here $\|\ \ \|_p$ is the $L_p(M)$ norm and we take $p >$ dim M.

(ii) APPROXIMATION THEOREM. *Given $f \in C(M)$, we find which functions $h \in C(M)$*

we can approximate arbitrarily closely in $L_p(M)$ by $f \circ \varphi$, where φ is a diffeomorphism of M. In other words, we wish to describe the L_p closure of the orbit of f under the group of diffeomorphisms. The result is that one can approximate h arbitrarily closely if and only if the range of h is contained in the range of f, that is, $\min f \leqq h(x) \leqq \max f$ *for all $x \in M$. One can readily extend this to open M and $h \in L_p(M)$.*

(iii) PERTURBATION THEOREM. *Since in many common cases the linearization $T'(u_0)$ is not invertible, we also need to know that the set of u's such that $T'(u_0)$ is invertible is an open dense set in, say, $C^2(M)$.*

Given these three ingredients, which we do not prove here, it is easy to prove Theorem 1.4. To illustrate the proof by a special case, let $M = T^2$, the torus, with the standard flat metric, so $k = 0$. Other manifolds M are treated similarly. Assuming K changes sign, we shall solve (2.1). Now $T'(0)v = -\Delta v$, which is not invertible (ker $T'(0)$ is the constant s). We are saved by the Perturbation Theorem which guarantees that for any $\varepsilon > 0$ there is a $u_0 \in C^\infty$ so close to zero that $T'(u_0)$ is invertible and $\max |T(u_0)| < \varepsilon$. Since K changes sign we can pick $\varepsilon > 0$ so small that

$$\min K < T(u_0) < \max K. \tag{2.2}$$

By the Inverse Function Theorem there is an $\eta > 0$ such that one can solve $T(u) = f$ if $\|f - T(u_0)\|_p < \eta$. In view of (2.2) the Approximation Theorem yields a diffeomorphism φ of M such that $\|K \circ \varphi - T(u_0)\|_p < \eta$. Thus, one can solve $T(u) = K \circ \varphi$, just as desired.

REMARK. While at this conference we learned of the work of Fischer-Marsden (this same volume) which led us to a slightly different proof of Theorem 1.1 and Theorem 4.3 below. It is simpler in that it avoids more subtle consideration of conformal deformations, but consequently it cannot yield the facts concerning conformal deformation. This approach directly solves the differential equation $F(g) = K$, where F is the map from metrics to Gaussian (or scalar) curvature functions. The new proof follows our steps (i), (ii), (iii) as before, with step (i) replaced by the appropriate implicit function theorem.

Note, however, that neither this direct proof of Theorem 1.1 (without Theorem 1.4) nor the proof just sketched of Theorem 1.4 yield the deeper assertions of **[15]**, **[17]**, **[20]** concerning curvatures of pointwise conformally equivalent metrics, that is, the more delicate information on the existence or nonexistence of solutions to (1.3) and (3.1) below.

SCALAR CURVATURE

3. The functional $\lambda_1(g)$. A simple way to deform a metric g is pointwise conformally, $g_1 = u^{4/(n-2)}g$, where $u > 0$ and $n = \dim M$ (one could write, say, $g_1 = e^{2v}g$, but the other form turns out to be more convenient). If Δ and k are the Laplacian and scalar curvature, respectively, of g, then the scalar curvature K of g_1 is found from

(3.1) $$L_g u \equiv -\alpha \Delta u + ku = Ku^a,$$

where $\alpha = 4(n-1)/(n-2)$ and $a = (n+2)/(n-2)$. Assume that M is compact and let $\lambda_1(g)$ be the lowest eigenvalue of the linear elliptic operator L_g, so $L_g \phi = \lambda_1(g)\phi$. Note that the corresponding eigenfunction ϕ does not change sign so we can assume that $\phi > 0$.

This functional $\lambda_1(g)$ has a number of striking properties showing its intimate relation with the geometry of M. Let us collect them.

LEMMA 3.2. *Let M be compact and connected, with* dim $M = n \geqq 3$.

(a) *The sign of $\lambda_1(g)$ is a conformal invariant.*

(b) *M admits a metric of positive* (*resp. zero, negative*) *scalar curvature pointwise conformal to g if and only if $\lambda_1(g) > 0$* (*resp. $\lambda_1(g) = 0$, $\lambda_1(g) < 0$*).

(c) *There are topological obstructions to a manifold admitting a metric with $\lambda_1(g) > 0$, and with $\lambda_1(g) = 0$.*

(d) *On any M, there are metrics with $\lambda_1(g) < 0$.*

(e) *If M admits a metric g_+ with $\lambda_1(g_+) > 0$, then it admits a metric with $\lambda_1(g) = 0$.*

(f) *Critical points of $\lambda_1(g)$ among all metrics with* Vol$(M, g) = 1$ *are Einstein metrics.*

PROOF. Part (a) of this is Theorem 3.2 of **[17]**. For the "only if" half of (b), if K is the scalar curvature of $g_1 = u^{a-1}g$, then $L_g u = Ku^a$, so $\langle \phi, Ku^a \rangle = \langle \phi, L_g u \rangle = \lambda_1(g)\langle \phi, u \rangle$; the assertion is clear. The second half of (b) follows from the observation that $L_g \phi = \lambda_1 \phi = \lambda_1 \phi^{1-a}\phi^a$, so by (3.1) the metric $g_1 = \phi^{a-1}g$ has scalar curvature $K = \lambda_1 \phi^{1-a}$. To prove (c), note that, by (b), M has a metric g with $\lambda_1(g) > 0$ (resp. $\lambda_1(g) = 0$) if and only if it has a metric with positive (resp. zero) scalar curvature. However Lichnerowicz has shown **[19]** (see also **[12]**) that certain spin manifolds do not admit metrics with positve scalar curvature, while in Theorem 5.1 below we prove that certain manifolds do not admit metrics with zero scalar curvature. Part (d) is a consequence of a result of Elíasson **[9]** and Aubin **[1]**, who show that one can always find a metric g whose scalar curvature satisfies $\int k\, dV < 0$. The variational characterization of $\lambda_1(g)$ shows that $\lambda_1(g) < 0$ (cf. Remark 2.4 in **[17]**). For (e), use (d) to find g_- with $\lambda_1(g_-) < 0$ and let $g_t = tg_- + (1-t)g_+$. Then $\lambda_1(g_t)$ depends continuously on t. Since $\lambda_1(g_-) < 0$ and $\lambda_1(g_+) > 0$, the result is clear.

To prove (f), we use the formula (3.5) below for the first variation of λ_1. Then at a critical point of λ_1 among metrics with fixed volume, using Lagrange multipliers one finds that $\hat{S}\phi^2 = \nu g$ for some constant ν. Eliminating g in favor of $\hat{g}$, this reads $\hat{S} = \nu\phi^{-2n/(n-2)}\hat{g}$. Thus, in the standard way using the Bianchi identities, $\phi \equiv$ constant and $\hat{g}$ is an Einstein metric. But $\phi \equiv$ const then implies that $\hat{g} =$ const g and $\hat{S} =$ const S, so g is an Einstein metric. Q.E.D.

It is important to calculate the first and second variation of λ_1. For this purpose, for a given metric g on M let S^2T denote the space of bilinear symmetric forms, let $TZ = \{h \in S^2T\colon \operatorname{tr} h = h^i_i = 0\}$, let $N = \{h \in S^2T\colon \delta h = 0\}$, where we define (in classical tensor notation) $(\delta h)_j = -h^i_{j;i}$, and define Lichnerowicz's Laplacian on symmetric bilinear forms by $\Delta_L h = \Delta h - Rh$, where R is the operator

$$(3.3)\qquad (Rh)_{ij} = S_{it}\, h^t_j + S_{jt}\, h^t_i - 2R_{isjt}\, h^{st}$$

with $(\Delta h)_{ij} = g^{ks}\, h_{ij;ks}$, S_{ij}, and R_{ijkl} the Laplace-Beltrami operator on $S^2\, T$, and the Ricci and Riemannian curvature tensors, respectively, of g. (Caution—many authors use different sign conventions on the Laplacians.) We let $(\ ,\ )$ and $\langle\ ,\ \rangle = \int(\ ,\ )$ denote the local and global inner products induced by g on $S^2\, T$. Also, we let $g(t)$ denote a family of metrics depending analytically on a parameter t and let $h = (dg/dt)|_{t=0}$. For convenience, normalize $g(0)$ (but not variations of g) so that $\mathrm{Vol}(M, g(0)) = 1$.

LEMMA 3.4. *Let $h \in S^2 T$ and let $\psi > 0$ be the (normalized) first eigenfunction of L_g. Then*

$$(3.5)\qquad \text{(a)}\qquad \lambda_1' = \frac{d}{dt}\lambda_1(g(t))\Big|_{t=0} = -\langle h, \psi^2 \hat{S}\rangle,$$

where $\hat{S}$ is the Ricci curvature of $\hat{g} = \psi^{4/(n-2)} g$. In particular, if g has constant scalar curvature (so $\psi = 1$ and $\hat{S} = S$), then

$$(3.6)\qquad \lambda_1' = -\langle h, S\rangle.$$

(b) *If g is Ricci flat ($S = 0$), then*

$$(3.7)\qquad \begin{aligned} \lambda_1'' = \frac{d^2}{dt^2}\lambda_1(g(t))\Big|_{t=0} &= \tfrac{1}{2}\langle h, \Delta_L h\rangle \quad \text{if } h \in N \cap TZ,\\ &= 0 \qquad\qquad\quad\ \text{if } h \in (N \cap TZ)^{\perp}. \end{aligned}$$

The proof of this is contained in the Appendix (§ 6).

4. Existence of certain metrics. Just as in the two dimensional case discussed in § 1, one can attempt to prove the existence of a metric g_1 with prescribed scalar curvature K on a compact manifold M ($\dim M = n \geqq 3$) by using conformal deformations, so here $\varphi^*(g_1) = u^{4/(n-2)}\, g$ for a diffeomorphism φ of M, and, as in (3.1) (cf. (1.2))

$$(4.1)\qquad L_g\, u \equiv -\alpha\Delta u + ku = (K \circ \varphi)\, u^a, \qquad u > 0.$$

In view of Lemma 3.2(a), the sign of $\lambda_1(g)$ will be significant (it plays a role similar to that of the Euler characteristic in the two dimensional case). Let $CE(g)$ denote the set of C^∞ functions that are scalar curvatures of metrics conformal to g. The next theorem is proved by the procedure of § 2.

THEOREM 4.2. (a) *If $\lambda_1(g) < 0$, then $CE(g)$ is precisely the set of C^∞ functions that are negative somewhere.*

(b) *If $\lambda_1(g) = 0$, then $CE(g)$ is precisely the set of C^∞ functions that either change sign or are identically zero.*

(c) *If $\lambda_1(g) > 0$, then $CE(g)$ contains the C^∞ functions K for which there is a constant $c > 0$ such that* $\min K < ck < \max K$. *Moreover, if there is a constant in $CE(g)$, then $CE(g)$ is precisely the set of C^∞ functions that are positive somewhere.*

REMARKS. For part (c), it is not known if $\lambda_1(g) > 0$ implies there is a constant in $CE(g)$. This is the only part of Yamabe's question **[23]** still unresolved (see **[9]**, **[17]** and **[21]**), since the remaining cases are covered by (a) and (b).

Theorem 4.2, along with Lemma 3.2(d)—(e), implies the existence of certain metrics with prescribed scalar curvature.

THEOREM 4.3. (a) *Any function that is negative somewhere is the scalar curvature of some metric.*

(b) *All elements of $C^\infty(M)$ are scalar curvatures if and only if M admits a metric of constant positive scalar curvature.*

In particular, on S^n ($n \geqq 3$), given any $K \in C^\infty$, there is a metric having K as the scalar curvature. Both Theorems 4.2 and 4.3 are proved in **[18]**. One can also use Theorem 4.2 to prove that if M is an open manifold obtained by deleting an arbitrary closed set from a compact manifold, then any function on M is the scalar curvature of some metric. Nothing appears to be known about scalar curvatures of *complete* Riemannian manifolds: For instance, which functions on $\boldsymbol{R}^3$ are—or are not—scalar curvatures of complete metrics? Is $K \equiv 1$ allowed for example?

5. Zero scalar, Ricci, and sectional curvatures. The previously mentioned result of Lichnerowicz gives a topological obstruction to the existence of metrics with positive scalar curvature, but leaves open the possibility of zero scalar curvature. We will discuss this now. The following theorem gives an elementary example of a topological obstruction to zero scalar curvature. For a stronger version see Remark 5.4. Here $\hat{A}(M)$ denotes the Hirzebruch genus, $\beta_1(M)$ the first Betti number, and all manifolds are assumed compact and connected.

THEOREM 5.1. *Let M be a spin manifold with $\hat{A}(M) \neq 0$ and $\beta_1(M) = \dim M$. Then M does not admit a metric of zero scalar curvature.*

The proof of this requires two lemmas. The first is due to Bourguignon **[6]**. We give a different somewhat simpler proof.

LEMMA 5.2. *Let M be a compact manifold that does not admit a metric of positive scalar curvature. Then any metric with zero scalar curvature must have zero Ricci curvature.*

PROOF. Let g be a metric with zero scalar curvature so $\lambda_1(g) = 0$. If the Ricci curvature S of g is not zero, then by Lemma 3.4(a), $\lambda_1(g - tS) > 0$ for all $t > 0$ sufficiently small (since, if $h = -S$, then $\lambda_1'(0) > 0$). Hence, by Lemma 3.2(b), there is a metric with positive scalar curvature. Q.E.D.

The second lemma is due to Bochner **[24]**. We repeat the short proof.

LEMMA 5.3. *A Riemannian manifold (M, g) with zero Ricci curvature and $\beta_1(M) = \dim M$ is flat.*

PROOF. Since (M, g) is Ricci flat, the Hodge-Laplacian equals the Laplace-Beltrami operator on 1-forms so every harmonic 1-form has zero covariant derivative. Since $\beta_1(M) = \dim M = n$, the Hodge-de Rham theorem asserts there are n such linearly independent harmonic 1-forms with zero covariant derivative. But then (M, g) must be flat. Q.E.D.

PROOF OF THEOREM 5.1. If M is a spin manifold with $\hat{A}(M) \neq 0$, then M does not admit a flat metric, since $\hat{A}(M)$ is expressed in terms of the Pontrjagin numbers, which are zero for a flat metric. Moreover, M does not admit a metric with positive scalar curvature, by the result of Lichnerowicz [**19**]. In view of Lemmas 5.2—5.3, the assertion is clear. Q.E.D.

REMARK 5.4. As N. Hitchin pointed out to us, a simple example for Theorem 5.1 is if M is the connected sum of a 4-dimensional torus with a 4-dimensional spin manifold M_1, having $\hat{A}(M_1) \neq 0$ (such as $M_1 = K3$ surface). He also observed that one can strengthen Theorem 5.1 to $\beta_1(M) \neq 0$ by replacing Lemma 5.3 with the assertion that "$\beta_1 \neq 0$ and Ricci flat implies $\hat{A} = 0$".

In [**7**], Calabi conjectured the existence of certain metrics. A special case of this conjecture asserts that if M is a Kähler manifold with $c_1(M) = 0$, then there is a Kähler metric with zero Ricci curvature. This conjecture is still open. However, if one could find a Kähler manifold with $\hat{A}(M) \neq 0$, $\beta_1(M) \neq 0$, and $c_1(M) = 0$ (no such example occurs in complex dimension 2), then, in view of Theorem 5.1 and the above remark, this would be a counterexample. If Calabi's conjecture were true, there would be many manifolds which are Ricci flat (i.e. zero Ricci curvature) but which are not flat. In the case $\dim M = 3$, it is easy to see that Ricci flat $\Rightarrow$ flat by purely local algebraic considerations. However, in $\dim M = 4$, Willmore [**22**] has an example of a Riemannian metric, and Calabi [**8**] has an example of a Kähler metric, on an open set in $\boldsymbol{R}^4$ which is Ricci flat but not flat. In this connection we remind the reader of a long standing problem.

QUESTION 1. Is there an example of a compact Riemannian manifold which is Ricci flat but not flat? What if one replaces compactness by asking for a complete metric?

Fischer and Wolf have observed [**10**] that if a compact M admits a flat metric then any Ricci flat metric on M is flat.

It is easy to give an example of a metric with scalar curvature zero but Ricci curvature nonzero. Indeed, by Lemma 3.2(b) and (e), S^3 has a metric with zero scalar curvature, but this metric cannot have zero Ricci curvature since zero Ricci curvature implies flat in three dimensions.

Theorem 5.1 shows that certain manifolds that do not admit positive scalar curvature metrics also do not admit zero scalar curvature metrics.

QUESTION 2. (a) If a compact manifold M has a metric with zero scalar curvature, does it also admit a metric with positive scalar curvature?

(b) In particular, does the 3-torus, T^3, admit a positive scalar curvature metric?

Incidentally, an affirmative answer to Question 2(a) would also yield a counterexample to Calabi's conjecture, say for M a $K3$ surface since $K3$ surfaces have $\hat{A}(M) \neq 0$ and hence do not admit metrics with positive scalar curvature.

We have some fragmentary information on Question 2. The answer to 2(a) is "yes" if the Ricci curvature of this metric is not zero, by Lemma 3.4(a), and

Lemma 3.2(b), since then for all $t > 0$ sufficiently small, $\lambda_1(g - tS) > 0$ (in particular, if M does not admit an Einstein metric, the answer is "yes"). On the other hand, if the Ricci curvature of this metric is zero, then $\lambda_1'(0) = 0$ for all perturbations h so we must turn to $\lambda_1''(0)$ to see how $\lambda_1(g + th)$ varies. In view of Lemma 3.4, this reduces to investigating the negativity of the operator Δ_L (see Remark 6.16). For the special case of the flat metric g_0 on T^3 (or a flat metric on any M, for that matter), we find that $\lambda_1' = 0$ and $\lambda_1'' \leqq 0$ for all variations h. Thus, up to second order variations there are no positive scalar curvature metrics near a flat metric.

6. Appendix. We will now prove Lemma 3.4 and obtain formulas for the first and second variations of $\lambda_1(g)$. Write $g_t = g(t)$ and L_t for L_{g_t}. Let $\phi = \phi(t) > 0$ be the first eigenfunction and $\lambda_1(t) = \lambda_1(g_t)$ be the first eigenvalue of L_t, so

$$L_t\,\phi(t) = \lambda_1(t)\,\phi(t). \tag{6.1}$$

Normalize ϕ so that $\langle\phi(t), \phi(0)\rangle = 1$, where the inner product is in $L_2(M, g)$. Since L_t is analytic in t, so are $\phi(t)$ and $\lambda_1(t)$ for all $|t|$ sufficiently small [**14**, VII, § 6, especially 6.5]. Let $'$ denote the derivative with respect to t. Then from (6.1) we find

$$L_t'\phi + L_t\phi' = \lambda_1'\,\phi + \lambda_1\phi', \tag{6.2}$$

$$L_t''\,\phi + 2L_t'\phi' + L_t\phi'' = \lambda_1''\,\phi + 2\lambda_1'\phi' + \lambda_1\phi''. \tag{6.3}$$

Taking the inner product of these with ϕ and setting $t = 0$, we find

$$\lambda_1' = \lambda_1'(0) = \langle\phi(0), L_0'\phi(0)\rangle, \tag{6.4}$$

and, if $\lambda_1'(0) = 0$,

$$\lambda_1'' = \lambda_1''(0) = \langle\phi(0), L_0''\phi(0)\rangle + 2\langle\phi(0), L_0'\phi'(0)\rangle, \tag{6.5}$$

where in both (6.4) and (6.5), we used

$$\langle\phi(0), (L_0 - \lambda_1(0))f\rangle = \langle(L_0 - \lambda_1(0))\,\phi(0), f\rangle = 0$$

for any $f \in C^2$.

(a) To compute (6.4), we need only evaluate
$L_0' = -\alpha\Delta' + k'$, where

$$\Delta' u = \frac{d}{dt}\Delta_{g_t}u\Big|_{t=0} = -h^{ij}u_{;ij} + (\tfrac{1}{2}h_i^{i;j} - h^{ij}_{;i})u_{;j} \tag{6.6}$$

for any $u \in C^2$ and

$$k' = \frac{d}{dt}k_{g_t}\Big|_{t=0} = -\Delta h_i^i + h^{ij}_{;ij} - h^{ij}S^i_j. \tag{6.7}$$

Writing $\phi = \phi(0)$ from now on, we see that

$$L_0'\phi = -\alpha[-h^{ij}\phi_{;ij} + (\tfrac{1}{2}h_i^{i;j} - h^{ij}_{;i})\phi_{;j}] + [-\Delta h_i^i + h^{ij}_{;ij} - h^{ij}S_{ij}]\,\phi. \tag{6.8}$$

Substitute this into (6.4) and integrate by parts, taking all derivatives off of the h, to obtain a formula of the form $\lambda_1' = \langle h, \text{something}\rangle$. One finds that "something" $= -\varphi^2 \hat{S}_{ij}$, where $\hat{S}_{ij}$ is the Ricci curvature of the metric $\hat{g} = \varphi^{4/(n-2)}g$. Then, as claimed, $\lambda_1' = -\langle h, \varphi^2 \hat{S}\rangle$.

(b) Here we assume that g is Ricci flat. In this case $k = 0$, so $L_0 = -\alpha\Delta$ and the first eigenfunction $\varphi(0) = \varphi$ of $L\varphi = \lambda_1(g)\varphi$ is a constant, $\varphi = [\mathrm{Vol}(M, g)]^{-1/2} = 1$, while $\lambda_1(g) = 0$. Because $\varphi = 1$, we have $L_0''\varphi = k''$. Thus (6.5) reads

$$\lambda_1'' = \langle 1, k''\rangle + 2\langle 1, -\alpha\Delta'\varphi' + k'\varphi'\rangle. \tag{6.9}$$

Also, from (6.2), $\varphi' = \varphi'(0)$ is the unique solution of $\alpha\Delta\varphi' = k'$ such that $\langle 1, \varphi'\rangle = 0$ (this last condition follows by differentiating $\langle\varphi(0), \varphi(t)\rangle = 1$).

The reduction of (6.9) to (3.7) is a bit tedious, so we will only indicate the main steps. First of all

$$\langle 1, k''\rangle = \frac{1}{2}\int [h^{ij;l}(2h_{jl;i} - h_{ij;l}) - h^i_{i;l}h_j^{j;l}]\, dV_g, \tag{6.10}$$

as one finds from a straightforward computation taking the derivative of (6.7)—before evaluating (6.7) at $t = 0$, of course. Then integrate (6.10) by parts, taking a derivative off the left factors. Using the Ricci identities on the resulting first term, since $S = 0$ we find

$$\langle 1, k''\rangle = \frac{1}{2}\int h^{ij}[-2h^l_{j;li} + (\Delta_L h)_{ij} + g_{ij}\Delta h^l_l]\, dV_g. \tag{6.11}$$

For the second term in (6.9), it is useful to define the function r as the unique solution r (such that $\langle r, 1\rangle = 0$) of $\Delta r = h^{ij}_{;ij}$, which we write symbolically as $r = Q(h)$. Then, using (6.7), the equation defining φ' reads $\alpha\Delta\varphi' = k' = \Delta(r - h^i_i)$. Thus $\alpha\varphi' = r - h^i_i + \text{const}$. From this formula for φ', (6.6), and several integrations by parts we find that

$$-2\alpha\langle 1, \Delta'\varphi'\rangle = \int h^{ij}[h^l_{l;ij} - g_{ij}\Delta h^l_l]\, dV_g. \tag{6.12}$$

From (6.7) and the formula for φ' in terms of r, as well as several integrations by parts, we similarly obtain

$$\langle 1, \alpha k'\varphi'\rangle = \int h^{ij}[(Qh)_{;ij} - 2h^l_{l;ij} + g_{ij}\Delta h^l_l]\, dV_g. \tag{6.13}$$

Substituting (6.11)—(6.13) into (6.9), we conclude that

$$2\lambda_1'' = \langle h, Ph\rangle, \tag{6.14}$$

where the linear differential operator $P: S^2T \to S^2T$ is the unique formally selfadjoint operator satisfying (6.14), namely

$$\begin{aligned}(Ph)_{ij} = &-h^l_{j;li} - h^l_{i;lj} + (\Delta_L h)_{ij} + (n-1)^{-1}(n-2)(Qh)_{;ij}\\ &+(n-1)^{-1}(-g_{ij}\Delta h^l_l + g_{ij}h^{ls}_{;ls} + h^l_{l;ij}).\end{aligned} \tag{6.15}$$

Our remaining task is to decipher (6.14)—(6.15) (this was the most difficult part for us). Berger-Ebin [**4**, §§ 3 and 4] have shown that every $h \in N^\perp$ can uniquely be written in the form $h_{ij} = \frac{1}{2}(\xi_{i;j} + \xi_{j;i})$ for some 1-form ξ_i. Armed with this, a straightforward computation using the Ricci and Bianchi identities shows that if $h \in N^\perp$ then $Ph = 0$ (to find $Q(h)$, note that here $\Delta r = \Delta \xi_i^{;i}$ so $Q(h) = r = \xi_i^{;i}$).

On the other hand, if $h \in (TZ)^\perp$, that is, if $h = ug$ for some function $u \in C^\infty$, it is easy to see that $Ph = 0$ (note here $\Delta r = \Delta u$ so $Q(h) = u +$ const).

These last two facts show that if $h \in (N \cap TZ)^\perp = N^\perp + TZ^\perp$ then $Ph = 0$. But if $h \in N \cap TZ$, then $Qh = 0$ and $Ph = \Delta_L h \in N \cap TZ$. This proves the assertion of part (b). Q.E.D.

REMARK 6.16. To examine the nature of critical points of $\lambda_1(g)$, it is important to determine if, for a Ricci flat metric, the operator Δ_L is nonpositive. This is an open question. However, one can prove Δ_L is nonpositive on $N^\perp + TZ^\perp$. To see this, note that by a computation using the Ricci and Bianchi identities, $\delta\Delta_L = \Delta\delta$ (here $\Delta\xi_i = \xi_{i;j}^{;j}$ for a 1-form). Therefore $\Delta_L : N \to N$ and $\Delta_L : N^\perp \to N^\perp$. Also $\mathrm{tr} \circ \Delta_L = \Delta \circ \mathrm{tr}$ so $\Delta_L : TZ^\perp \to TZ^\perp$. Let $\Phi \in S^2T$ be an eigentensor of Δ_L with eigenvalue γ. If $\Phi \in N^\perp$, then $\delta\Phi \neq 0$ and $\Delta(\delta\Phi) = \delta\Delta_L\Phi = \gamma(\delta\Phi)$ so $\gamma \leqq 0$ since it is an eigenvalue of Δ. Similarly, if $\Phi \in TZ^\perp$, then $\Delta(\mathrm{tr}\,\Phi) = \mathrm{tr}(\Delta_L\Phi) = \gamma(\mathrm{tr}\,\Phi)$ and again $\gamma \leqq 0$. In a slightly different direction, if we define $Bh = \Delta h - \beta Rh$ (cf. (3.3) and [**4**, § 6]) then B is nonpositive for $-1 \leqq \beta \leqq \frac{1}{2}$, since it is then possible to write B as $B = -D^*D$ for some operator D. Our case, however, is $\beta = 1$. In personal correspondence, Marcel Berger has pointed out that there are examples of non Ricci flat manifolds where B is indefinite for all $\beta > \frac{1}{2}$.

BIBLIOGRAPHY

1. T. Aubin, *Metriques riemanniennes et courbure*, J. Differential Geometry **4** (1970), 383–424. MR **43** #5452.

2. Marcel Berger, *Quelques formules de variation pour une structure riemannienne*, Ann. Sci. Ecole Norm Sup. (4) **3** (1970), 285–294. MR **43** #3969.

3. ———, *Sur les premiers valeurs propres des variétés Riemanniennes*, Composito Math. **26** (1973), 129–149.

4. Marcel Berger and D. Ebin, *Some decompositions of the space of symmetric tensors on a Riemannian manifold*, J. Differential Geometry **3** (1969), 379–392. MR **42** #993.

5. Melvyn Berger, *Riemannian structures of prescribed Gaussian curvature for compact 2-manifolds*, J. Differential Geometry **5** (1971), 325–332. MR **45** #4329.

6. J.-P. Bourguignon, personal communication.

7. E. Calabi, *On Kähler manifolds with vanishing canonical class*, Algebraic Geometry and Topology (A Sympos. in Honor of S. Lefschetz), Princeton Univ. Press, Princeton, N. J., 1957, pp. 78–89. MR **19**, 62.

8. ———, *A construction of nonhomogeneous Einstein metrics*, these PROCEEDINGS.

9. H. Eliasson, *On variations of metrics*, Math. Scand. **29** (1971), 317–327.

10. A. Fischer and J. Wolf, *The structure of compact Ricci-flat Riemannian manifolds*, J. Differential Geometry **10** (1975) (to appear).

11. H. Gluck, *The generalized Minkowski problem in differential geometry in the large*, Ann. of Math. (2) **96** (1972), 245–276. MR **46** #8132.

12. N. Hitchin, *The space of harmonic spinors* (to appear).

13. G. Jensen, *The scalar curvature of left-invariant Riemannian metrics*, Indiana Univ. Math. J. **20** (1970/71), 1125–1144. MR **44** #6914.

14. T. Kato, *Perturbation theory for linear operators*, Die Grundlehren der math. Wissenschaften, Band 132, Springer-Verlag, New York, 1966. MR **34** #3324.

15. Jerry L. Kazdan and F. W. Warner, *Curvature functions for compact 2-manifolds*, Ann. of Math. (2) **99** (1974), 14–47.

16. ———, *Curvature functions for open 2-manifolds*, Ann. of Math. **99** (1974), 203–219.

17. ———, *Scalar curvature and conformal deformation of Riemannian structure*, J. Differential Geometry **10** (1975) (to appear).

18. ———, *Existence and conformal deformation of metrics with prescribed Gaussian and scalar curvature*, Ann. of Math. (to appear).

19. A. Lichnerowicz, *Spineurs harmoniques*, C. R. Acad. Sci. Paris **257** (1963), 7–9. MR **27** #6218.

20. J. Moser, *On a nonlinear problem in differential geometry*, Dynamical Systems (M. Peixoto, Editor), Academic Press, New York, 1973.

21. N. S. Trudinger, *Remarks concerning the conformal deformation of Riemannian structures on compact manifolds*, Ann. Scuola Norm. Sup. Pisa (3) **22** (1968), 265–274. MR **39** #2093.

22. T. J. Willmore, *On compact Riemannian manifolds with zero Ricci curvature*, Proc. Edinburgh Math. Soc. (2) **10** (1956), 131–133. MR **17**, 783.

23. H. Yamabe, *On a deformation of Riemannian structures on compact manifolds*, Osaka Math. J. **12** (1960), 21–37. MR **23** #A2847.

24. K. Yano and S. Bochner, *Curvature and Betti numbers*, Ann. of Math. Studies, no. 32, Princeton Univ. Press, Princeton, N. J., 1953. MR **15**, 989.

UNIVERSITY OF PENNSYLVANIA

Proceedings of Symposia in Pure Mathematics
Volume 27, 1975

ON SYMPLECTIC RELATIONS IN PARTIAL DIFFERENTIAL EQUATIONS

BOHDAN LAWRUK*

This is a report on part of a paper, *Special symplectic spaces* by B.Lawruk, J. Śniatycki and W.Tulczyjew* which will be submitted for publication elsewhere.

A symplectic interpretation of procedures used in studying boundary value problems for partial differential equations is given. This interpretation leads to certain criteria for the choice of spaces of functions and of boundary conditions. It can also be used to obtain solutions of boundary value problems for large and nonsimply connected domains by reduction to corresponding problems for small and simply connected parts of these domains. The method is illustrated for the simple case of a second order equation with a formally selfadjoint differential operator and Dirichlet and Neumann boundary conditons. The same approach can also be adopted for a system of equations with a formally selfadjoint matrix differential operator of any order and a much richer class of boundary conditions.

Let Ω be a bounded domain in $\boldsymbol{R}^n$ with a piecewise $\boldsymbol{C}^\infty$ boundary $\partial\Omega$. We denote by ν the unit normal to $\partial\Omega$ directed towards the exterior of Ω, and by dS the induced surface element in $\partial\Omega$.

Let A be a second order linear formally selfadjoint differential operator on $\bar{\Omega}$. For each $u \in \boldsymbol{C}^2(\bar{\Omega})$,

$$Au = \sum_{i,j=1}^{n} \frac{\partial}{\partial x_i}\left(a_{ij}\frac{\partial u}{\partial x_j}\right) + au$$

AMS (MOS) subject classifications (1970). Primary 53C15; Secondary 35A99.

*This research was supported in part by operating grants of the National Research Council of Canada, and was begun at a 1972 Summer Research Institute of the Canadian Mathematical Congress.

where a_{ij}, $a \in C^\infty(\bar{\Omega})$, $a_{ij} = a_{ji}$, $i, j = 1, \cdots, n$. Then for any u and u' we have the following Green's formula:

$$\int_{\partial\Omega} \sum_{i,j=1}^{n} \left(u' a_{ij} \frac{\partial u}{\partial x_j} - u a_{ij} \frac{\partial u'}{\partial x_j} \right) \nu_i \, dS - \int_{\Omega} (u' Au - uAu') \, dx = 0,$$

where ν_i denotes the ith component of the unit normal ν.

We do not assume that A is of any definite type.

Let P be a linear space of triplets (u, f, φ), where u and f are functions on $\bar{\Omega}$ and φ is a function on $\partial\bar{\Omega}$. The actual choice of the space P should depend on the operator A. We have not made any assumptions about A, and therefore we do not specify P either. Since the space of C^∞ functions is dense in most spaces used in partial differential equations, one can think of P as consisting of triplets of C^∞ functions.

Let ω be an antisymmetric bilinear form on P defined by

$$\omega((u, f, \varphi), (u', f', \varphi')) = \int_{\partial\Omega} (u'\varphi - u\varphi') \, dS - \int_{\Omega} (uf' - u'f) \, dx.$$

It is nondegenerate in all spaces of interest here. Thus (P, ω) is a symplectic space in the weak sense. In the sequel the adjective "weak" is dropped for simplicity.

Let N be the subspace of P defined by $(u, f, \varphi) \in N$ if and only if $Au = f$ and

$$\sum_{i,j=1}^{n} a_{ij} \frac{\partial u}{\partial x_j} \nu_i \Big|_{\partial\Omega} = \varphi.$$

From the Green formula it follows that ω restricted to N is identically zero. Such a subspace of a symplectic space is called isotropic. Among isotropic subspaces of special interest are maximal isotropic and Lagrangian subspaces. A Lagrangian subspace is an isotropic subspace possessing an isotropic complement. Every Lagrangian subspace is maximal isotropic, and the converse holds if P is a Hilbert space. This fact and the definition of a Lagrangian subspace in the infinite-dimensional case are both due to A. Weinstein **[1]**, **[2]**.

The subspace of P consisting of elements of the form $(0, f, \varphi)$ is clearly isotropic and it is also a complement of N in P. Hence, N is Lagrangian.

Let M be the subspace of P consisting of elements of the form $(u, 0, \varphi)$. The intersection of M and N is the space of solutions of the equation $Au = 0$. The restriction ω_M of ω to M is no longer nondegenerate. Its characteristic space R consists of all elements $(u, 0, \varphi)$ of M such that $u/\partial\Omega = 0$ and $\varphi = 0$. Let P^r denote the quotient space M/R and let $\rho : M \to P^r$ be the canonical projection. The form ω_M induces an antisymmetric nondegenerate bilinear form ω^r in P^r. The elements of P^r written as (ψ, φ) can be interpreted as Cauchy data[1] on $\partial\Omega$, where ψ represents the Dirichlet and φ the Neumann part of the data. For any (ψ, φ) and (ψ', φ') in P^r,

[1]The term "Cauchy data" is used here to mean the collection of
$u/\partial\Omega$ and $\sum_{i,j=1}^{n} a_{ij} (\partial u/\partial x_j) \nu_i |_{\partial\Omega}$,
also in the case when $\partial\Omega$ has characteristic direction.

$$\omega^r((\psi, \varphi), (\psi', \varphi')) = \int_{\partial\Omega} (\psi'\varphi - \psi\varphi')\, dS.$$

Let $N^r = \rho(N \cap M)$ be the projection of $N \cap M$ to P^r. It consists of all the Cauchy data on $\partial\Omega$ which can be extended to solutions u of the equation $Au = 0$. From the Green formula it follows that N is an isotropic subspace of P^r.

In general, N^r need not be a Lagrangian subspace or even maximal isotropic; however, if every Dirichlet (or Neumann) problem for the operator A is always solvable, then N^r is a Lagrangian subspace. For an elliptic operator A and for P consisting of C^∞ functions, N^r is maximal isotropic. If P is a suitable Hilbert space (Sobolev space) then N^r is Lagrangian [**1**].

We shall assume that the space P has been properly chosen for the differential operator A if N^r is a maximal isotropic subspace.

The space P^r has a product structure: $P^r = Q \times \tilde{Q}$ where Q is the space of Dirichlet data and $\tilde{Q}$ is the space of Neumann data. Let $\pi : P^r \to Q$ denote the projection onto the first factor, i.e. $\pi(\psi, \varphi) = \psi$ and let θ be the bilinear from on P^r defined by

$$\theta((\psi, \varphi), (\psi', \varphi')) = \int_{\partial\Omega} \psi\varphi'\, dS.$$

Then $\omega^r = \theta^T - \theta$. The quadruplet (P^r, Q, π, θ) is called a special symplectic space. Here Q has been chosen to represent the Dirichlet data. Note that $Q \times 0$ is a Lagrangian subspace of the space P^r, and so is $0 \times \tilde{Q}$; however, we could choose another type of boundary data and obtain a special symplectic structure corresponding to this choice, provided the chosen data form a Lagrangian subspace of P^r.

Consider now the projection $\pi(N^r)$ of N^r to Q. This is the set of all Dirichlet data which can be extended to a solution u of $Au = 0$ in Ω. Further, the restriction of θ to N^r induces on $\pi(N^r)$ a bilinear form γ given as follows: for any (ψ, φ) and (ψ', φ') in N^r,

$$\gamma(\psi, \psi') = \theta((\psi, \varphi), (\psi', \varphi')).$$

It can be easily seen that γ is well defined and is symmetric. Using the explicit form of θ we have $\gamma(\psi, \psi') = \int_{\partial\Omega} \psi\varphi'\, dS$. If N^r is maximal isotropic, then it is uniquely determined by $\pi(N^r)$ and γ. Conversely, if N^r is uniquely determined by $\pi(N^r)$ and γ is defined as above, then it is maximal isotropic. That is why we consider the space P to be properly chosen for the differential operator A if N^r is maximal isotropic. In this case we say that N^r is generated by γ and that γ is a generating form of N^r.

The symplectic formalism developed here can be used to obtain solutions of boundary value problems for large and nonsimply connected domains by reducing them to corresponding boundary value problems for small and simply connected parts of these domains.

By a solution of the equation $Au = 0$ in Ω, we mean the knowledge of its Cauchy data on $\partial\Omega$, hence an element of N^r; however, this element may be extended to a solution u of the equation $Au = 0$ in Ω not uniquely (the space $N \cap R$ consists of all triplets $(u, 0, 0)$ such that $Au = 0$ in Ω, $u/\partial\Omega = 0$ and

$$\sum_{i,j=1}^{n} a_{ij} \frac{\partial u}{\partial x_j} \nu_i / \partial\Omega = 0,$$

and may not reduce to $(0, 0, 0)$).

Let $\Omega = \operatorname{Int}(\bar{\Omega}_1 \cup \bar{\Omega}_2)$ where Ω_1 and Ω_2 are two domains such that $\Omega_1 \cap \Omega_2 = \varnothing$ and $\partial\Omega_1 \cap \partial\Omega_2$ is a piecewise $\boldsymbol{C}^\infty$ $(n-1)$-dimensional submanifold of $\boldsymbol{R}^n$. Let A be a differential operator in $\bar{\Omega}$ as before, and A_1 and A_2 be the restrictions of A to $\bar{\Omega}_1$ and $\bar{\Omega}_2$, respectively. Following the same reasoning as before we obtain special symplectic spaces $(P_1^r, Q_1^r, \pi_1, \theta_1)$ and $(P_2^r, Q_2^r, \pi_2, \theta_2)$ corresponding to the choice of Dirichlet data on $\partial\Omega_1$ and $\partial\Omega_2$. Similarly we obtain a special symplectic space (P^r, Q^r, π, θ) corresponding to the choice of Dirichlet data on $\partial\Omega$. In each of our symplectic spaces, we have spaces of solutions of the corresponding differential equations. We assume that these spaces form maximal isotropic subspaces: $N_1^r \subset P_1^r$, $N_2^r \subset P_2^r$ and $N^r \subset P^r$. Let γ_1, γ_2 and γ be the generating forms of N_1^r, N_2^r and N^r, respectively. We would like to relate $\pi(N^r)$ and γ to $\pi_1(N_1^r)$, $\pi_2(N_2^r)$, γ_1 and γ_2. Note that $\pi(N^r)$ are Dirichlet data on $\partial\Omega$ which extend to a solution of $Au = 0$ in Ω, and similar interpretation holds for $\pi_1(N_1^r)$ and $\pi_2(N_2^r)$. Then $\phi \in \pi(N^r)$ if and only if ϕ extends to a function $\tilde{\phi}$ on $\partial\Omega_1 \cup \partial\Omega_2$ and such that $\tilde{\phi}/\partial\Omega_1 \in \pi_1(N_1^r)$, $\tilde{\phi}/\partial\Omega_2 \in \pi_2(N_2^r)$ and $\gamma_1(\tilde{\phi}/\partial\Omega_1, \tilde{\phi}/\partial\Omega_1) + \gamma_2(\tilde{\phi}/\partial\Omega_2, \tilde{\phi}/\partial\Omega_2)$ is stationary with respect to a variation of $\tilde{\phi}/\partial\Omega_1 \cap \partial\Omega_2$. The value of γ on ϕ is given by

$$\gamma(\phi, \phi) = \gamma_1(\tilde{\phi}/\partial\Omega_1, \tilde{\phi}/\partial\Omega_1) + \gamma_2(\tilde{\phi}/\partial\Omega_2, \tilde{\phi}/\partial\Omega_2),$$

where ϕ satisfies the conditions above.

This holds under some additional technical assumptions which are satisfied if all the spaces considered here are Hilbert spaces and all the subspaces are closed.

As an example let us consider the wave equation in $\boldsymbol{R}^2$. Suppose that Ω is the interior of the square in $\boldsymbol{R}^2$ with vertices $(0, 0)$, $(0, 1)$, $(1, 0)$, and $(1, 1)$, and let $Au = \partial^2 u/\partial x^2 - \partial^2 u/\partial y^2$. The boundary of Ω consists of four segments of straight lines. Using the d'Alambert formula for the solution of the wave equation, one can verify that N^r is a maximal isotropic subspace of P^r with the projection $\pi(N^r)$ uniquely characterized by the condition: $\phi \in \pi(N^r)$ if and only if for each $z \in [0, 1]$, $\phi(z, 0) + \phi(1-z, 0) - \phi(0, z) - \phi(1, 1-z) = 0$. The generating form of N^r can be computed to give

$$\gamma(\phi, \phi) = \int_0^1 dz\, \{\phi(0, z)\, \phi_x(z, 0) - \phi(z, 0)\, \phi_y(0, z) + \phi(z, 1)\, \phi_y(1, z) - \phi(1, z)\phi_x(z, 1)\},$$

where ϕ_x and ϕ_y denote the derivatives of ϕ with respect to the coordinates x and y, respectively.

It should be noted that if we took Ω to be the interior of a characteristic square, then N^r would not be a maximal isotropic subspace of P^r.

Examples of this type were considered also by R. A. Aleksandrian and S. L. Sobolev; cf. [3], [4] and [5].

REFERENCES

1. A. Weinstein, *Symplectic manifolds and their Lagrangian submanifolds*, Advances in Math. **6** (1971), 329–346, MR **44** #3351.

2. ———, *Symplectic structures on Banach manifolds*, Bull. Amer. Math. Soc. **75** (1969), 1040–1041. MR **39** #6364.

3. R. A. Aleksandrian, Dissertation, Moscow, 1949.

4. ———, *On Dirichlet's problem for the equation of a chord and on the completeness of a system of functions in the circle*, Dokl. Akad. Nauk SSSR **73** (1950), 869–872. (Russian) MR **12**, 615.

5. S. L. Sobolev, *An instance of a correct boundary value problem for the equation of string vibration with the conditions given all over the boundary*, Dokl. Akad. Nauk SSSR **109** (1956), 707–709. (Russian) MR **18**, 215.

McGILL UNIVERSITY

Proceedings of Symposia in Pure Mathematics
Volume 27, 1975

ON PERIODS OF SOLUTIONS OF A CERTAIN NONLINEAR DIFFERENTIAL EQUATION AND THE RIEMANNIAN MANIFOLD O_n^2

TOMINOSUKE OTSUKI

1. Introduction. Our object here is to investigate certain properties of periodic solutions of the nonlinear differential equation of order 2:

$$\text{(E)} \qquad nx(1-x^2)\frac{d^2x}{dt^2}+\left(\frac{dx}{dt}\right)^2+(1-x^2)(nx^2-1)=0,$$

which appeared first in [4], relating to some kind of minimal hypersurfaces in the $(n+1)$-dimensional unit sphere S^{n+1}.

We recall the following results.

THEOREM A. *Let M be a complete minimal hypersurface immersed in S^{n+1} with two principal curvatures; then*

(I) *in case of their multiplicities m and $n-m \geqq 2$, M is congruent to the generalized Clifford torus: $S^m((m/n)^{1/2}) \times S^{n-m}(((n-m)/n)^{1/2}) \subset S^{n+1}$ in R^{n+2}, where $S^m(r)$ denotes the m-dimensional sphere of radius r, and*

(II) *in case of one of their multiplicities equal to one, M is constructed by the following method:*

(i) *C is a plane curve in R^2 $(= C)$ given by*

AMS (MOS) subject classifications (1970). Primary 34C25, 34C40, 53A05, 53A10, 53C25.

$$q(t) = \exp i(t - \pi/2)\cdot\{x(t) + ix'(t)\},$$

where $x(t)$ *is a solution of* (E) *and the support function of the tangent direction angle* t *for* C.

(ii) $M \ni p = (1 - x(t)^2 - x'(t)^2)^{1/2} e_n + q(t)$, *where* $e_n \in R^n$, $\|e_n\| = 1$ *and* $S^{n+1} \subset R^n \times R^2$.

THEOREM B. *Any solution* $x(t)$ *of* (E) *such that* $x^2 + (dx/dt)^2 < 1$ *is periodic and its period* T *is given by*

$$T = 2\int_{a_0}^{a_1} [1 - x^2 - C(x^{-2} - 1)^\alpha]^{-1/2}\, dx, \tag{1}$$

where

$$C = (a_0^2)^\alpha(1 - a_0^2)^{1-\alpha} = (a_1^2)^\alpha (1 - a_1^2)^{1-\alpha} \tag{2}$$

$(0 < a_0 < \alpha^{1/2} < a_1 < 1, \alpha = 1/n)$ *is the integral constant of* (E) *and* $0 < C < A = \alpha^\alpha(1 - \alpha)^{1-\alpha}$.

Regarding T as a function of C, the following is known in [**4**]:

(i) T is differentiable and $T > \pi$,

(ii) $\lim_{C\to 0} T = \pi$ and $\lim_{C\to A} T = 2^{1/2}\pi$.

By these facts, there are countably and infinitely many compact minimal hypersurfaces immersed but not imbedded in S^{n+1}. $M = S^{n-1}(((n - 1)/n)^{1/2}) \times S^1(n^{-1/2})$ corresponds to the constant solution $x(t) = n^{-1/2}$ of (E) and is minimally imbedded in S^{n+1}.

We have been interested in the problem on the existence of imbedded compact minimal hypersurfaces of this type other than the Clifford tori $S^m((m/n)^{1/2}) \times S^{n-m}(((n - m)/n)^{1/2})$, $m = 1,\cdots, n - 1$, from the point of view of differential geometry and from that of the theory of nonlinear differential equations, because the existence of such hypersurfaces in S^{n+1} is equivalent to the existence of solutions of (E) with period 2π.

The answer to this problem was negative, for expecting the existence of such hypersurfaces. First, S. Furuya gave the inequality

$$T < (1 - \alpha)^{1/2}\cdot 2\pi \tag{3}$$

in [**2**] and then we have proved a little sharper inequality

$$T < (2^{-1/2} + (1 - \alpha)^{1/2})\cdot\pi \tag{4}$$

in [**5**], for any integer $n \geqq 2$.

On the other hand, by means of a numerical analysis and observation about (E), M. Urabe made a conjecture:

$$T < 2^{1/2}\pi. \tag{U}$$

(See [**5**] and [**12**].) Equation (E) however may be considered for any real number n.

We shall discuss the inequality (U) and show that it is true for any real number $n > 1$.

2. A geometrical meaning of (E). Let n be a real number and consider a metric on the uv-plane given by

$$(5)\qquad ds^2 = (1 - u^2 - v^2)^{n-2}\,\{(1 - v^2)\,du^2 + 2uv\,du\,dv + (1 - u^2)\,dv^2\}.$$

We denote the 2-dimensional Riemannian manifold with the metric (5) in the unit disk D^2: $u^2 + v^2 < 1$ by O_n^2. Then, we have easily

THEOREM 1. *For any geodesic of O_n^2 not passing through the origin of D^2, its support function $x(t)$ satisfies the differential equation* (E).

Therefore, (E) may be considered as the equation of geodesics of O_n^2. The equation of geodesics of O_n^2 in the coordinates u and v is

$$(6)\quad (1 - u^2 - v^2)\frac{d^2v}{du^2} = n\left(-v + u\frac{dv}{du}\right)\left\{1 - v^2 + 2uv\frac{dv}{du} + (1 - u^2)\left(\frac{dv}{du}\right)^2\right\},$$

which can be written in the polar coordinate r, θ in the uv-plane as

$$(7)\qquad r(1 - r^2)\frac{d^2r}{d\theta^2} + \{(n + 2)r^2 - 2\}\left(\frac{dr}{d\theta}\right)^2 + r^2(1 - r^2)(nr^2 - 1) = 0.$$

Since Theorem B holds for any real number $n > 1$, any solution of (7) is periodic with respect to the argument angle θ. Its period is given by

$$(8)\qquad \Theta = 2\int_{r_0}^{r_1} [C_1 r^4(1 - r^2)^n - r^2(1 - r^2)]^{-1/2}\,dr,$$

where

$$(9)\qquad C_1 = \frac{1}{r_0^2(1 - r_0^2)^{n-1}} = \frac{1}{r_1^2(1 - r_1^2)^{n-1}} \qquad (0 < r_0 < \alpha^{1/2} < r_1 < 1)$$

is the integral constant of (7). Furthermore, if we set $C_1 = 1/C^n$, then $T = \Theta$. Setting $r^2 = R$, (8) can be written as

$$= C^{n/2}\int_{R_0}^{R_1} \frac{dR}{R((1 - R)\{R(1 - R)^{n-1} - C^n\})^{1/2}},$$

which may be considered as a generalized elliptic integral when n is an integer. (See [6].)

Thus, a compact minimal hypersurface in S^{n+1} in case (II) of Theorem A corresponds to a closed geodesic in O_n^2 and especially an imbedded one to a simple closed geodesic.

Remarking that the metric (5) is invariant under any rotation about the origin of D^2, O_n^2 can be considered as a surface of revolution around the zw-axis plane in the Lorentzian 4-space L^4 with the metric

$$ds^2 = dx^2 + dy^2 + dz^2 - dw^2,$$

whose profile curve $C = C_0 \cup C_1$ is given by

$$C_0: \begin{cases} x = r(1-r^2)^{(n-1)/2}, \quad y = 0, \\ z = \int_0^r t(1-t^2)^{(n-3)/2}(2n-1-n^2t^2)^{1/2}\,dt, \quad w = 0, \end{cases}$$
$$(0 \leqq r \leqq (2n-1)^{1/2}/n);$$

$$C_1: \begin{cases} x = r(1-r^2)^{(n-1)/2}, \quad y = 0, \quad z = b, \\ w = \int_{(2n-1)^{1/2}/n}^r t(1-t^2)^{(n-3)/2}(n^2t^2-2n+1)^{1/2}dt, \quad ((2n-1)^{1/2}/n \leqq r < 1), \end{cases}$$

where b is $z|r=(2n-1)^{1/2}/n$ for C_0. (See **[8]**.) For $n > 1$, O_n^2 is the surface punctured at a singular point from a closed surface in L^4.

3. Properties of T. Replacing nx^2 and nC by x and C respectively, the period T given by (1) can be written as

$$T = T_n(x_0) := \int_{x_0}^{x_1} \frac{dx}{(x(n-x) - Cx^{1-\alpha}(n-x)^{\alpha})^{1/2}}, \tag{10}$$

where

$$C = x_0^{\alpha}(n-x_0)^{1-\alpha} = x_1^{\alpha}(n-x_1)^{1-\alpha} \quad (0 < x_0 < 1 < x_1 < n). \tag{11}$$

We give first two lemmas as follows [10]:

LEMMA 1. *The function* $\varphi(x) := x^{\alpha}(n-x)^{1-\alpha}$ $(0 \leqq x \leqq n)$ *is monotone increasing in* $[0, 1]$ *and decreasing in* $[1, n]$ *and*

$$\varphi'(x) = ((1-x)/x(n-x))\,\varphi(x). \tag{12}$$

LEMMA 2. *The function*

$$F(x) := \frac{x(n-x)}{(1-x)^2} \cdot \frac{B-\varphi(x)}{\varphi(x)} \quad (0 \leqq x \leqq n,\ x \neq 1)$$
$$:= 1/2 \quad (x = 1),$$

is smooth and positive in $(0, n)$, *where* $B = \varphi(1) = nA$.

Now, define a function $X = X_n(x)$ $(0 \leqq x \leqq 1)$ by

$$x(n-x)^{n-1} = X(n-X)^{n-1}, \quad 1 \leqq X \leqq n. \tag{13}$$

Noting (12) and the equality

$$\int_{x_0}^1 \frac{d\varphi(x)}{((B-\varphi(x))(\varphi(x)-C))^{1/2}} = \pi,$$

we can write the right-hand side of (10) as follows:

$$T = \int_{x_0}^1 \{(F(x))^{1/2} + (F(X_n(x)))^{1/2}\} \frac{d\varphi(x)}{((B-\varphi(x))(\varphi(x)-C))^{1/2}}. \tag{14}$$

Therefore, it suffices to prove (U) that we can prove the inequality

$$(F(x))^{1/2} + (F(X_n(x)))^{1/2} < 2^{1/2} \quad \text{for } 0 < x < 1. \tag{15}$$

THEOREM 2. *$T_n(x_0) = T_m(y_0)$, where $m = n/(n-1)$, $y_0 = m - (m-1)x_1$, $x_1 = X_n(x_0)$.*

PROOF. Supposing $n > 1$, we have $m = n/(n-1) > 1$. Changing integral parameter x into $y = m - (m-1)x$, we have

$$T_n(x_0) = \int_{y_0}^{y_1} \frac{dy}{(y(m-y) - (m-1)Cy^{1-1/m}(m-y)^{1/m})^{1/2}} = T_m(y_0),$$

where $y_0 = m - (m-1)x_1$, $y_1 = m - (m-1)x_0$. We have easily

$$(m-1)C = y_0^{1/m}(m-y_0)^{1-1/m} = y_1^{1/m}(m-y_1)^{1-1/m}, \qquad 0 < y_0 < 1 < y_1 < m.$$

Therefore, in order to prove the inequality (U) for all real numbers $n > 1$, it is sufficient to prove it for $n \geqq 2$.

Now, as for the function $F(x)$ we have

$$((F(x))^{1/2})' = \frac{\{n + (n-2)x\}\{B - \varphi(x)\} - B(1-x)^2}{2(1-x)^2(x(n-x)\varphi(x))^{1/2}\{B - \varphi(x)\}^{1/2}}, \tag{16}$$

where $\{B - \varphi(x)\}^{1/2}$ denotes the function

$$\{B - \varphi(x)\}^{1/2} = (1-x)\left(\frac{B - \varphi(x)}{(x-1)^2}\right)^{1/2}.$$

4. Some constants depending on n (> 2).

LEMMA 3. *The function*

$$g_0(x) := \frac{x(n-x)^{n-1}\{n + (n-2)x\}^n}{(n - 1 + nx - x^2)^n}$$

is monotone increasing in $[0, n/2]$ and decreasing in $[n/2, n]$ and $g_0(1) = (n-1)^{n-1}$.

(See **[10]**.) By the Lemma 3, we define a constant $\Lambda = \Lambda_n$ for $n > 2$ by

$$g_0(\Lambda) = (n-1)^{n-1} \quad \text{and} \quad 1 < \Lambda < n. \tag{17}$$

As for this constant Λ_n, we can prove the following lemmas (**[10]**, **[11]**).

LEMMA 4. *$n/2 < \Lambda_n < n - 1$ for $n > 2$.*

Writing the condition in (17) in detail and comparing it with the right-hand side of (16), we can prove

LEMMA 5. *The function $F(x)$ is monotone increasing in $(0, \Lambda]$ and decreasing in $[\Lambda, n)$, and $F(1) = 1/2$, $F'(1) = (n-2)/6(n-1)$.*

Making use of the fact the function $x(n-x)/(n-1+nx-x^2)$ is monotone decreasing in $[n/2, n]$ and Lemma 4, we have

$$\begin{aligned} F(x) \leqq F(\Lambda) &= \frac{\Lambda(n-\Lambda)}{(\Lambda-1)^2} \cdot \frac{B - \varphi(\Lambda)}{\varphi(\Lambda)} \\ &= \frac{\Lambda(n-\Lambda)}{n - 1 + n\Lambda - \Lambda^2} < \frac{n^2}{n^2 + 4n - 4} \end{aligned}$$

from which we obtain a little sharper inequality than (4), since we have $n/(n^2 + 4n - 4)^{1/2} < (n - 1)/n$ (**[5]**, **[10]**).

THEOREM 3. *$T < (2^{-1/2} + (1 + 4\alpha - 4\alpha^2)^{-1/2})\cdot\pi$ $(n \geqq 2)$.*

Now, we have, in $0 < x < 1$,

$$((F(x))^{1/2})' + ((F(X_n(x)))^{1/2})' = \frac{1 - x}{2x(n-x)\,(\varphi(x)\,\{B - \varphi(x)\})^{1/2}}\cdot\{f(x) - f(X_n(x))\},$$

where

$$(18)\qquad f(x) := \begin{cases} \dfrac{(x(n-x))^{1/2}}{(1-x)^3}[\{n + (n-2)\,x\}\,\{B-\varphi(x)\} - B(1-x)^2] & (x \neq 1), \\ \dfrac{(n-2)B}{6(n-1)^{1/2}} & (x = 1). \end{cases}$$

LEMMA 6. *$f(x) > 0$ in $(0, \Lambda)$ and $f(x) < 0$ in (Λ, n).*

LEMMA 7. *$(F(x))^{1/2} + (F(X_n(x)))^{1/2}$ is increasing at x $(0 < x < 1)$, if and only if $f(x) > f(X_n(x))$.*

As for $f(x)$, we have

$$(19)\qquad \begin{aligned} f'(x) = \tfrac{1}{2}(1-x)^{-4}(x(n-x))^{-1/2} \\ \cdot[\{n(n+2) + 2(4n^2 - 5n - 2)x + (3n^2 - 16n + 16)x^2\} \\ \cdot\{B - \phi(x)\} - 3B(1-x)^2\,\{n + (n-2)x\}]. \end{aligned}$$

For simplicity, setting

$$(20)\qquad P(x) := n(n+2) + 2(4n^2 - 5n - 2)x + (3n^2 - 16n + 16)x^2,$$

we can say that

LEMMA 8. *$f(x)$ is decreasing at x $(0 < x < n)$, if and only if*

$$[P(x) - 3(1-x)^2\,\{n + (n-2)x\}]\,B < P(x)\,\varphi\,(x).$$

Since $P(x) - 3(1-x)^2\,\{n + (n-2)x\} > 0$ in $[0, n]$, we define an auxiliary function

$$g(x) := \frac{P(x)\,\varphi(x)}{P(x) - 3(1-x)^2\,\{n + (n-2)x\}},$$

for which we can prove the following **[10]**.

LEMMA 9. *$g'(x) = 0$ $(0 < x < n)$ has unique roots $\gamma = \gamma_n$ in $(0, 1)$ and $\bar{\gamma} = \bar{\gamma}_n$ in $(1, n)$ and $n/2 < \bar{\gamma} < n$, which are given by the quadratic equation of order 2:*

$$n^2(n+2) - n(9n^2 - 2n + 8)x + 4(3n^2 - 2n + 2)x^2 = 0.$$

LEMMA 10. *$g(x)$ is monotone increasing in $(0, \gamma]$ and $[1, \bar{\gamma}]$ and decreasing in $[\gamma, 1]$ and $[\bar{\gamma}, n)$.*

Since $g(1) = \varphi(1) = B$, this lemma implies that $g(x) = B$ has a unique solution in $(0, 1)$ and $(1, n)$ respectively. We denote them by $\sigma = \sigma_n$ and $\bar{\sigma} = \bar{\sigma}_n$ respectively. Then we have

$$0 < \sigma_n < \gamma_n \quad \text{and} \quad \bar{\gamma}_n < \bar{\sigma}_n < n. \tag{21}$$

LEMMA 11. *The function $f(x)$ is monotone decreasing in $(\sigma, \bar{\sigma})$ and increasing in $(0, \sigma]$ and $[\bar{\sigma}, n)$ and $f(\sigma) \geqq f(x) \geqq f(\bar{\sigma})$.*

THEOREM 4. *The function $(F(x))^{1/2} + (F(X_n(x)))^{1/2}$ is monotone increasing and less than $2^{1/2}$ in $[\sigma, 1)$.*

PROOF. By Lemma 6 and Lemma 11, we have

$$f(\Lambda) = 0 > f(\bar{\sigma}) \quad \text{and} \quad f(x) > f(X_n(x)) \quad \text{for } \sigma \leqq x < 1.$$

Hence, by Lemma 7, $(F(x))^{1/2} + (F(X_n(x)))^{1/2}$ is monotone increasing in $[\sigma, 1)$. Since $F(1) = 1/2$, we get

$$(F(x))^{1/2} + (F(X_n(x)))^{1/2} < 2^{1/2} \quad \text{for } \sigma \leqq x < 1.$$

By this theorem, the only thing we have to do is to prove (15) for $0 < x \leqq \sigma$. For such x, by Lemma 4 and Lemma 5 we have

$$F(x) \leqq F(\sigma), \qquad F(X_n(x)) \leqq F(\Lambda) < n^2/(n^2 + 4n - 4),$$

hence

$$(F(x))^{1/2} + (F(X_n(x)))^{1/2} < (F(\sigma))^{1/2} + n/(n^2 + 4n - 4)^{1/2} \quad \text{for } 0 < x \leqq \sigma. \tag{22}$$

5. Proof of $T < 2^{1/2}\pi$ for $3 \leqq n \leqq 14$. Since we can prove that $\gamma_n < 1/5$ for $n \geqq 3$, we get from (21) and Lemma 5 that $\sigma = \sigma_n < 1/5$ for $n \geqq 3$ and so

$$F(\sigma) < F(1/5) = (1/16)\,\{5(n-1)\,((5n-1)/(n-1))^{1/n} - (5n-1)\}.$$

Hence we have the inequality

$$\begin{aligned}(F(x))^{1/2} + (F(X_n(x)))^{1/2} &< \frac{1}{4}\left(5(n-1)\left(\frac{5n-1}{n-1}\right)^{1/n} - (5n-1)\right)^{1/2} \\ &\quad + \frac{n}{(n^2+4n-4)^{1/2}} \quad \text{for } 0 < x \leqq \sigma.\end{aligned} \tag{23}$$

Now, setting $n = 1/t$ and $x = 1/a$ in $F(x)$, we get

$$F(x) = \frac{a}{(a-1)}\left[\left(\frac{a-t}{1-t}\right)^t\left(\frac{1}{t} - 1\right) - \frac{1}{t} + \frac{1}{a}\right]$$

and we shall investigate the auxiliary function of t:

(24) $$G_a(t) := \left(\frac{a-t}{1-t}\right)^t\left(\frac{1}{t}-1\right)-\frac{1}{t} \qquad (0 < t < 1 < a).$$

As for $G_a(t)$, we can prove the following fact [**10**].

LEMMA 12. *$G_5(t)$ is monotone increasing in* $(0, 1/3]$.

THEOREM 5. *When* $3 \leqq n \leqq 14$, *we have*

$$(F(x))^{1/2} + (F(X_n(x)))^{1/2} < 2^{1/2} \quad \text{for } 0 < x < 1.$$

PROOF. It is sufficient to prove this inequality for $0 < x \leqq \sigma$. By Lemma 12 we can do the following estimations:

When $3 \leqq n \leqq 4$,

$$\tfrac{1}{4}\left(5(n-1)\left(\frac{5n-1}{n-1}\right)^{1/n} - (5n-1)\right)^{1/2} \leqq \tfrac{1}{4}(5\cdot 2\cdot 7^{1/3} - 14)^{1/2} \doteqdot 0.56620,$$
$$n/(n^2+4n-4)^{1/2} \leqq 4/28^{1/2} = 2/7^{1/2} \doteqdot 0.75593.$$

When $4 \leqq n \leqq 10$,

$$\tfrac{1}{4}\left(5(n-1)\left(\frac{5n-1}{n-1}\right)^{1/n} - (5n-1)\right)^{1/2} \leqq \tfrac{1}{4}(5\cdot 3\cdot (19/3)^{1/4} - 19)^{1/2} \doteqdot 0.54748,$$
$$n/(n^2+4n-4)^{1/2} \leqq 10/136^{1/2} = 5/34^{1/2} \doteqdot 0.85749.$$

When $8 \leqq n \leqq 14$,

$$\tfrac{1}{4}\left(5(n-1)\left(\frac{5n-1}{n-1}\right)^{1/n} - (5n-1)\right)^{1/2} \leqq \tfrac{1}{4}(5\cdot 7\cdot (39/7)^{1/8} - 39)^{1/2} \doteqdot 0.52336,$$
$$n/(n^2+4n-4)^{1/2} \leqq 14/248^{1/2} = 7/62^{1/2} \doteqdot 0.88900.$$

By virtue of these estimations, we obtain the inequality

$$\tfrac{1}{4}\left(5(n-1)\left(\frac{5n-1}{n-1}\right)^{1/n} - (5n-1)\right)^{1/2} + n/(n^2+4n-4)^{1/2} < 2^{1/2} \quad \text{for } 3 \leqq n \leqq 14.$$

By Theorem 5 and (14), (U) is true for $3 \leqq n \leqq 14$.

6. Proof of $T < 2^{1/2}\pi$ for $n \geqq 14$. In this section, first of all we shall show that $\sigma_n < 1/11$ for $n \geqq 14$. This is equivalent to $g(1/11) > B$ by Lemma 10. Since we have

$$g\left(\frac{1}{11}\right) = \frac{(53n^2+29n-7)(11n-1)^{1-1/n}}{583n^2-581n+73},$$

$g(1/11) > B = (n-1)^{1-1/n}$ is equivalent to

(25) $$\left(\frac{11-t}{1-t}\right)^{1-t} > \frac{583-581t+73t^2}{53+29t-7t^2},$$

where $t = 1/n$. (25) also can be written as

$$(1 - t)\log\frac{11 - t}{1 - t} > \log\frac{583 - 581t + 73t^2}{53 + 29t - 7t^2}.$$

Since we have

$$\frac{11 - t}{1 - t} = 11\left\{1 + \frac{10t}{11(1 - t)}\right\} \quad \text{and} \quad 0 < \frac{10t}{11(1 - t)} \leqq \frac{1}{11}$$

for $0 < t \leqq 1/11$, we have

$$\begin{aligned}(1 - t)\log\frac{11 - t}{1 - t} &= (1 - t)\log 11 + 10t/11 - \tfrac{1}{2}(10/11)^2 t^2/(1 - t) \\ &\quad + \cdots + (-1)^{m-1} m^{-1}(10/11)^m t^m/(1 - t)^{m-1} + \cdots. \end{aligned} \tag{26}$$

Next, we have

$$\frac{583 - 581t + 73t^2}{53 + 29t - 7t^2} = 11(1 - Q), \quad \text{where } Q = \frac{150t(6 - t)}{11(53 - 29t + 7t^2)}$$

and

$$\begin{aligned} &0 < Q \leqq \frac{150 \cdot 65}{11^3 \cdot 53} < 1 \quad \text{for } 0 < t \leqq \frac{1}{11}, \\ &\log\frac{583 - 581t + 73t^2}{53 + 29t - 7t^2} = \log 11 - Q - \frac{Q^2}{2} - \frac{Q^3}{3} - \cdots - \frac{Q^n}{n} - \cdots. \end{aligned} \tag{27}$$

Comparing (26) with (27) and using the properties of the function Q of t, we can prove the following fact **[10]**.

THEOREM 6. $0 < \sigma_n < 1/11$ *for* $n \geqq 14$.

By Theorem 6 and Lemma 5, we have

$$F(\sigma) < F(1/11) = \frac{1}{100}\left\{11(n - 1)\left(\frac{11n - 1}{n - 1}\right)^{1/n} - (11n - 1)\right\}$$

and hence

$$\begin{aligned}(F(x))^{1/2} + (F(X_n(x)))^{1/2} &< \frac{1}{10}\left(11(n - 1)\left(\frac{11n - 1}{n - 1}\right)^{1/n} - (11n - 1)\right)^{1/2} \\ &\quad + \frac{n}{(n^2 + 4n - 4)^{1/2}} \quad \text{for } 0 < x \leqq \sigma. \end{aligned} \tag{28}$$

As in the previous case, we can prove the following fact **[10]**.

LEMMA 13. $G_{11}(t)$ *is monotone increasing in* $(0, 1/11]$.

THEOREM 7. *When* $n \geqq 14$, *we have*

$$(F(x))^{1/2} + (F(X_n(x)))^{1/2} < 2^{1/2} \quad \textit{for } 0 < x < 1.$$

PROOF. It is sufficient to prove this inequality for $0 < x \leqq \sigma$. By Lemma 13 and (24) we can do the following estimations:

When $14 \leqq n \leqq 21$,

$$\frac{1}{10}\left(11(n-1)\left(\frac{11n-1}{n-1}\right)^{1/n} - (11n-1)\right)^{1/2} + \frac{n}{(n^2+4n-4)^{1/2}}$$
$$< \frac{1}{10}\left(11\cdot 13\cdot\left(\frac{153}{13}\right)^{1/14} - 153\right)^{1/2} + \frac{21}{521^{1/2}}$$
$$\doteqdot 0.41877 + 0.92003 = 1.33880 < 2^{1/2};$$

When $n \geqq 21$,

$$\frac{1}{10}\left(11(n-1)\left(\frac{11n-1}{n-1}\right)^{1/n} - (11n-1)\right)^{1/2} + \frac{n}{(n^2+4n-4)^{1/2}}$$
$$< \frac{1}{10}(220(23/2)^{1/21} - 230)^{1/2} + 1 \doteqdot 0.41393 + 1 < 2^{1/2}.$$

Hence, by means of (22) we obtain the inequality in this theorem for $0 < x \leqq \sigma$. Thus we can show that (U) is true for $n > 14$.

7. Proof of $T < 2^{1/2}\pi$ for $2 \leqq n \leqq 3$. Since we can prove easily that $\gamma_n < 1/4$ for $2 \leqq n \leqq 3$, we get from (21) and Lemma 5 that $\sigma = \sigma_n < 1/4$ for $2 \leqq n \leqq 3$ and so

$$F(\sigma) < F(\tfrac{1}{4}) = \tfrac{1}{9}\left\{4(n-1)\left(\frac{4n-1}{n-1}\right)^{1/n} - (4n-1)\right\}.$$

Hence we have the inequality

$$(F(x))^{1/2} + (F(X_n(x))^{1/2} > \frac{1}{3}\left(4(n-1)\left(\frac{4n-1}{n-1}\right)^{1/n} - (4n-1)\right)^{1/2} + \frac{n}{(n^2+4n-4)^{1/2}} \quad \text{for } 0 < x \leqq \sigma. \tag{29}$$

Since we have

$$\frac{n}{(n^2+4n-4)^{1/2}} \leqq \frac{3}{17^{1/2}} \quad \text{for } 2 \leqq n \leqq 3,$$

in order to prove that $(F(x))^{1/2} + (F(X_n(x)))^{1/2} < 2^{1/2}$ for $0 < x \leqq \sigma$, it is sufficient to prove that

$$\frac{1}{3}\left(4(n-1)\left(\frac{4n-1}{n-1}\right)^{1/n} - (4n-1)\right)^{1/2} < 2^{1/2} - \frac{3}{17^{1/2}},$$

which is equivalent to

$$4(n-1)\left(\frac{4n-1}{n-1}\right)^{1/n} - (4n-1) < \frac{9(43 - 6\cdot 34^{1/2})}{17}.$$

Setting $n = 2 + u$ $(0 \leqq u \leqq 1)$, the above inequality can be written as

$$(1+u)\left(\frac{7+4u}{1+u}\right)^{1/(2+u)} - u < \frac{253 - 27\cdot 34^{1/2}}{34} \doteqdot 2.81071.$$

As for the left-hand side, we can prove the following facts [**11**].

LEMMA 14. $(1+u)((7+4u)/(1+u))^{1/(2+u)} - u < 7^{1/2}$ *for* $0 < u < 1$.

THEOREM 8. *When* $2 \leqq n \leqq 3$, *we have* $(F(x))^{1/2} + (F(X_n(x)))^{1/2} < 2^{1/2}$ *for* $0 < x < 1$.

REMARK. As is shown in § 1, the differential equation (E) was originally treated only for integers $n \geqq 2$. Under this restriction for n, the proof of the inequality (U) was given first by S. Furuya when $n = 3$. (The case of $n = 2$ is included in (3) and (4).) We have tried to extend his method to the general case of any integer $n > 3$ but could not succeed. (See [**9**].) Quite recently, K. Tandai succeeded in proving (U) for any integer n, by devising ingeniously a modification of our method in [**9**].

REFERENCES

1. S. S. Chern, M. do Carmo and S. Kobayashi, *Minimal submanifolds of a sphere with second fundamental form of constant length*, Functional Analysis and Related Fields, Springer-Verlag, Berlin and New York, 1970, pp. 60–75.

2. S. Furuya, *On periods of periodic solutions of a certain nonlinear differential equation*, Japan-United States Seminar on Ordinary Differential and Functional Equations, Springer-Verlag, Berlin and New York, 1971, pp. 320–323.

3. W. Y. Hsiang and H. B. Lawson, Jr., *Minimal submanifolds of low cohomogeneity*, J. Differential Geometry **5** (1970), 1–38.

4. T. Otsuki, *Minimal hypersurfaces in a Riemannian manifold of constant curvature*, Amer. J. Math. **92** (1970), 145–173.

5. ———, *On integral inequalities related with a certain nonlinear differential equation*, Proc. Japan Acad. **48** (1972), 9–12.

6. ———, *On a 2-dimensional Riemannian manifold*, Differential Geometry (In Honor of K. Yano, Kinokuniya), Tokyo, 1972, pp. 401–414.

7. ———, *On a family of Riemannian manifolds defined on an m-disk*, Math. J. Okayama Univ, **16** (1973), 85–87.

8. T. Otsuki and M. Maeda, *Models of the Riemannian manifolds* O_n^2 *in the Lorentzian 4-space*, J. Differential Geometry **9** (1974), 97–108.

9. T. Otsuki, *On a differential equation appeared in differential geometry*, Sūgaku **25** (1973), 97–109. (Japanese)

10. ———, *On a bound for periods of solutions of a certain nonlinear differential equation.* I, J. Math. Soc. Japan **26** (1974), 206–233.

11. ———, *On a bound for periods of solutions of a certain nonlinear differential equation.* II, Funkcialj Ekvacioj **17** (1974) (to appear).

12. M. Urabe, *Computations of periods of certain nonlinear autonomous oscillations*, Study of Algorithms of Numerical Computations **149** (1972), 111–129. (Japanese)

TOKYO INSTITUTE OF TECHNOLOGY

Proceedings of Symposia in Pure Mathematics
Volume 27, 1975

SINGULARITIES AND THE OBSTACLE PROBLEM

DAVID G. SCHAEFFER

One of the classical problems of mathematical physics is to determine the equilibrium position of an elastic membrane stretched over a rigid boundary and subjected to a force per unit area f. It is well known [1] that the displacement function u is a solution of Poisson's equation

$$\Delta u = -f \quad in\ \Omega, \qquad u = 0 \quad on\ \partial\Omega.$$

(For simplicity we assume the boundary to be planar.) Suppose however that the membrane is displaced by inserting a blunt instrument partially through the boundary ring. If the height of the obstacle is $\psi(x)$, where $\psi < 0$ on $\partial\Omega$, the displacement function must satisfy $u \geqq \psi$. Presumably there will be a region I in the interior of Ω where the membrane and the obstacle are in contact, and elsewhere u will be a solution of Laplace's equation, as there is no external force acting. We are thus led to the following free boundary value problem, the so-called obstacle problem: Find a function $u \in C^1(\Omega)$ and a closed set $I \subset \Omega$ such that

$$u \geqq \psi \quad and \quad I = \{x : u(x) = \psi(x)\},$$
$$\Delta u = 0 \quad on\ \Omega \sim I, \qquad u = 0 \quad on\ \partial\Omega.$$

A weak solution of the obstacle problem may be obtained variationally by minimizing the Dirichlet integral $\int_\Omega (\nabla u)^2\, dx$ in the class of functions $\{u : u \geqq \psi$ and $u = 0$ on $\partial\Omega\}$; as this set is convex, existence and uniqueness follow from the remark that a closed convex set in Hilbert space contains a unique element of least norm.

AMS (MOS) subject classifications (1970). Primary 35J65; Secondary 35J20.

Unfortunately the variational formulation of the problem sheds no light on the nature of the free boundary. Indeed, it is easy to construct examples where ∂I is extremely unpleasant. Suppose for example that ψ assumes its maximum (which is positive) on a set $K \subset \Omega$ with nonempty interior. Then the solution u equals ψ on K; that is $K \subset I$. However, consider the problem with the obstacle function $\psi' = \psi - \phi$ where ϕ is a smooth function with $\phi \geqq 0$ and supp $\phi \subset \operatorname{Int}(K)$. Both problems have the same solution u, but in the second problem, the region of contact is

$$I' = I \sim \{x: \phi(x) > 0\}.$$

Thus for any open set $\mathcal{O}$ with $\operatorname{Cl} \mathcal{O} \subset \operatorname{Int} K$, there is an obstacle problem which has the region of contact $I \sim \mathcal{O}$, since there is a smooth nonnegative function ϕ such that $\mathcal{O} = \{x: \phi(x) > 0\}$. In particular, we could arrange for the free boundary to have positive measure.

Even if the obstacle function satisfies $\Delta\psi < 0$, singularities can still occur in the free boundary. Let Ω be an open set in $\boldsymbol{R}^2$ containing the square $\{(x, y): |x|, |y| \leqq 1\}$ such that Ω is invariant under the transformation $y \mapsto -y$. Suppose

$$(*) \qquad \psi(x, y) = \{1-(x/\varepsilon)^2\} w(y) - 1$$

where $w \in C^\infty(\boldsymbol{R})$; if $w > 0$, then $\Delta\psi < 0$, providing we require that ε is sufficiently small. Choose a nonnegative even function $\phi \in C^\infty(\boldsymbol{R})$ with $\phi(0) = 1$ and supp $\phi \subset (-\frac{1}{3}, \frac{1}{3})$ such that ϕ is monotonically decreasing for $y > 0$. If $a \in \boldsymbol{R}$ let

$$w_a(y) = \tfrac{1}{2} + \phi(y - a) + \phi(y + a).$$

Now if $a = \frac{1}{2}$ and ψ is given by $(*)$, then $\{(x, y) : \psi > 0\}$ has two components. But the region of contact must be contained in $\{(x, y) : \psi > 0\}$, and since the problem is symmetric under the transformation $y \mapsto -y$ the region of contact must also have two components. On the other hand, if $a = 0$, the region of contact will have only one component. For some value of $a \in (0, \frac{1}{2})$ a singularity will occur in the free boundary when the number of components changes.

In spite of the above examples, we conjecture that generically the free boundary is a manifold of class C^∞. (By generic we mean for data ψ in an open dense subset of $C^\infty(\bar{\Omega})$.) It would then follow that generically u would be piecewise C^∞. We feel that this conjecture should follow from the Thom transversality theorem, but have been unable to prove this.

In the lecture we went on to discuss [2]. This paper contains an analogous result for a problem which in some ways resembles the obstacle problem but is two degrees of differentiability smoother.

References

1. R. Courant and D. Hilbert, *Methods of mathematical physics*, Vol. II: *Partial differential equations*, Interscience, New York, 1962. MR **25** #4216.

2. D. Schaeffer, *An example of generic regularity for a nonlinear elliptic equation*, Arch. Rational Mech. Anal. (to appear).

Massachusetts Institute of Technology

Proceedings of Symposia in Pure Mathematics
Volume 27, 1975

HARMONIC MAPPINGS OF SPHERES

R. T. SMITH

Let N and M be compact Riemannian manifolds. The basic existence result of Eells and Sampson is

THEOREM 1 [2]. *If all sectional curvatures of M are nonpositive, there is a harmonic representative of every homotopy class of maps $N \to M$.*

Little is known if M has positive curvatures. By direct construction methods we have some examples:

THEOREM 2 [3]. *$\pi_n(S^n) = Z$ is represented by harmonic maps for $n = 1, \cdots, 7$.*

More generally, if $f\colon S^n \to S^m$ is a harmonic polynomial map of homogeneity $k > ((n-1)/2)(2^{1/2}-1)$, one can produce six harmonic suspensions of f. Examples of such polynomial maps include the Hopf maps $S^{2n-1} \to S^n$ for $n = 2, 4$. These are the only topologically nontrivial maps which arise from

OBSERVATION 3 ([1], [3]). *Let G be a transitive group of isometries of S_n. Each irreducible representation V of G on the spherical harmonics determines a harmonic polynomial map of S^n to the unit sphere in V.*

However, other topologically interesting harmonic polynomial maps (related to Cartan's theory of isoparametric hypersurfaces) have been found by R. Wood (unpublished).

AMS (MOS) subject classifications (1970). Primary 58E15, 53C20.

References

1. M. do Carmo and N. R. Wallach, *Representations of compact groups and minimal immersions into spheres*, J. Differential Geometry **4** (1970), 91–104.

2. J. Eells and J.H. Sampson, *Harmonic mappings of Riemannian manifolds*, Amer. J. Math. **86**(1964), 109–160.

3. R.T. Smith, *Harmonic mappings of spheres*, Amer. J. Math. (to appear).

Columbia University

Proceedings of Symposia in Pure Mathematics
Volume 27, 1975

HOLOMORPHIC *R*-TORSION FOR LIE GROUPS

NANCY K. STANTON

1. Introduction. Ray and Singer [3] have defined the holomorphic R-torsion $T_0(M, \chi)$ of a compact complex Hermitian manifold M and a character χ of $\pi_1(M)$. Let $L(\chi)$ be the line bundle associated to χ. Let $\mathscr{D}^{0,q}(M, L(\chi))$ denote the $(0, q)$ forms on M with values in $L(\chi)$. Let $\Delta_{0,q}$ denote the $\bar\partial$ Laplacian on $\mathscr{D}^{0,q}(M, L(\chi))$, and let $P_{0,q}$ denote the orthogonal projection onto $\operatorname{Ker} \Delta_{0,q}$. Then

$$\zeta_{0,q}(s, \chi) = \frac{1}{\Gamma(s)} \int_0^\infty t^{s-1} \operatorname{tr}(\exp(-t\Delta_{0,q}) - P_{0,q})dt$$

is an analytic function of s for Re s large and it extends to a meromorphic function in the s-plane which is analytic at $s = 0$. The holomorphic R-torsion is defined as the positive root of

$$\log T_0(M, \chi) = \frac{1}{2} \sum_q (-1)^q q\zeta'_{0,q}(0, \chi).$$

$T_0(M, \chi)$ is an invariant of the complex Hermitian structure on M. Furthermore, if χ_1 and χ_2 are characters of $\pi_1(M)$ such that the corresponding complexes are acyclic, then $T_0(M, \chi_1)/T_0(M, \chi_2)$ is independent of the choice of metric.

Let G be a compact even-dimensional Lie group and T a maximal torus of G. Then G and the left coset space G/T can be given complex structures [4]. Also, G is a holomorphic principal bundle over G/T. Let χ be a character of $\pi_1(G)$ and χ_0 the induced character of $\pi_1(T)$. Our main theorem is

THEOREM 1.1. $T_0(G, \chi) = T_0(T, \chi_0)$.

AMS (MOS) subject classifications (1970). Primary 58G05, 32M10, 35P20.

(For details, see [5].)

The torsion $T_0(T, \chi_0)$ has been computed by Ray and Singer [3]. If $\dim_{\boldsymbol{R}} T > 2$, $T_0(T, \chi_0) = 1$. If $\dim_{\boldsymbol{R}} T = 2$, $T_0(T, \chi_0)$ can be expressed in terms of theta functions and the Dedekind η-function.

EXAMPLE. Let G be the Hopf surface $S^1 \times S^3$. Then

$$S^3 = \{(z_1, z_2) \in \boldsymbol{C}^2 : |z_1|^2 + |z_2|^2 = 1\}.$$

Let $T' = \{(z, 0) : |z|^2 = 1\} \subset S^3$. Then $T = S^1 \times T'$ is a maximal torus of G. Let γ be a generator of $\pi_1(S^1 \times S^3) = \boldsymbol{Z}$. If χ is a character of $\pi_1(S^1 \times S^3)$, then for some $0 < u < 1$, $\chi(\gamma) = e^{2\pi iu}$. The induced character χ_0 of $\pi_1(T)$ is

$$\chi_0(m\gamma_1 + n\gamma_2) = e^{2\pi imu},$$

where γ_1 is the generator of $\pi_1(S^1)$ corresponding to γ and γ_2 is the generator of $\pi_1(T')$. Then Theorem 1.1 becomes

$$T_0(S^1 \times S^3, \chi) = T_0(T, \chi_0) = |\theta_1(u, i)/\eta(i)|.$$

This has been computed directly by Ray and Singer [3].

The proof of Theorem 1.1 involves two steps. The first is to adapt the proof of the product formula for torsion to the fibre bundle $G \to G/T$. This gives the theorem if rank $G > 2$. If rank $G = 2$, it gives a formula for

$$\sum_q (-1)^q q \operatorname{tr}(\exp(-t\Delta_{0,q})).$$

The second step is to use the action of the Weyl group on $H^*(G/T)$ to simplify the formula for rank 2 groups. Before describing these steps more explicitly, we discuss the product formula.

2. Product formula. Let M_1 and M_2 be compact complex manifolds and let χ_i be a character of $\pi_1(M_i)$, $i = 1, 2$. Then $\chi = \chi_1 \chi_2$ is a character of $\pi_1(M_1 \times M_2)$. Give $M = M_1 \times M_2$ the product metric. Let $\Delta^{(i)}$ denote the Laplacian on $L(\chi_i)$ and let $\chi(M_i, L(\chi_i))$ be the arithmetic genus of M_i with values in $L(\chi_i)$, i.e.,

$$\chi(M_i, L(\chi_i)) = \sum_q (-1)^q \dim(\operatorname{Ker} \Delta_{0,q}^{(i)}).$$

Then, if $\chi(M_1, L(\chi_1)) = 0$,

$$\log T_0(M, \chi) = \chi(M_2, L(\chi_2)) \log T_0(M_1, \chi_1). \tag{2.1}$$

(See [3] for the proof.)

REMARK. If M_1 is a torus, $\chi(M_1, L(\chi_1)) = 0$, since if χ_1 is nontrivial, the corresponding $\bar\partial$ complex is acyclic, and if χ_1 is trivial and $\dim_{\boldsymbol{C}} M_1 = N$, then

$$\chi(M_1, L(\chi_1)) = \sum_q (-1)^q \binom{N}{q} = 0.$$

We sketch the proof of (2.1). Let f_i be a form on M_i and let $f_1 \otimes f_2$ denote the wedge product of f_1 and f_2 lifted to M. Since M has the product metric,

$$\Delta(f_1 \otimes f_2) = (\Delta^{(1)} f_1) \otimes f_2 + f_1 \otimes (\Delta^{(2)} f_2). \tag{2.2}$$

If we think of M as a fibre bundle with fibre M_1 and base M_2, then f_1 is vertical, f_2 is horizontal, and (2.2) says that Δ is the sum of a vertical Laplacian $\Delta^{(1)} \otimes I$ and a horizontal Laplacian $I \otimes \Delta^{(2)}$, where I is the identity operator. If f_i is an eigenform of $\Delta^{(i)}$ with eigenvalue λ_i, then $f_1 \otimes f_2$ is an eigenform with eigenvalue $\lambda_1 + \lambda_2$, and every eigenform of Δ is a finite linear combination $\sum_{j=1}^{n} f_{1_j} \otimes f_{2_j}$, where f_{i_j} is an eigenform of $\Delta^{(i)}$. Thus

$$\exp(-t\Delta_{0,q}) = \sum_{q_1+q_2=q} \exp(-t\Delta^{(1)}_{0,q_1}) \otimes \exp(-t\Delta^{(2)}_{0,q_2}). \tag{2.3}$$

Formula (2.1) now follows by a straightforward computation from (2.3) and the observation that

$$\sum_q (-1)^q \operatorname{tr}(\exp(-t\Delta^{(i)}_{0,q})) = \chi(M_i, L(\chi_i)).$$

3. Adapting the product formula. The keys to the proof of (2.1) are the splitting of Δ into a vertical Laplacian and a horizontal Laplacian and the completeness of products of eigenforms on M_1 and eigenforms on M_2. We now explain the analogues for the fibre bundle $G \to G/T$. Since the metric on G is not a product of the metrics on G/T and T, one would not expect a simple formula for the Laplacian on G in terms of the Laplacians on G/T and T. However, we will introduce a family of line bundles $\boldsymbol{V}^\lambda$ over G/T. Then we will obtain a family of splittings of the Laplacian on G into the sum of a (horizontal) Laplacian on $\boldsymbol{V}^\lambda$ and a (vertical) Laplacian on $L(\chi_0)$.

Let $\tilde{G}$ be the universal covering group of G, p the covering map, and $\tilde{T} = p^{-1}(T)$. Then $L(\chi_0) = \tilde{T} \times_\chi \boldsymbol{C}$, and $\tilde{T}$ acts on sections of $L(\chi_0)$. In fact, $\tilde{T}$ acts as a group of unitary operators, so eigenspaces of Δ are invariant under the action of $\tilde{T}$. Let L be the μ-eigenspace. Then, since $\tilde{T}$ is a commuting family of operators on L, L splits into a direct sum of one-dimensional invariant subspaces, $L = \bigoplus_{\lambda=\mu} V^\lambda$. $\tilde{G}$ is a principal bundles over G/T with fibre $\tilde{T}$. Hence, if ρ is the representation of $\tilde{T}$ on V^λ, we can form a Hermitian line bundle $\boldsymbol{V}^\lambda$ over G/T, $\boldsymbol{V}^\lambda = \tilde{G} \times_\rho V^\lambda$.

If $\pi : G \to G/T$ is the projection, then $\pi^* \boldsymbol{V}^\lambda = L(\chi)$. Hence, a section s of $\boldsymbol{V}^\lambda$ can be lifted to a section $\tilde{s}$ of $L(\chi)$, and the restriction of $\tilde{s}$ to a coset of T is an eigensection of Δ with eigenvalue λ. There is a connection on G which gives rise to a covariant differential D_λ on $\boldsymbol{V}^\lambda$. The $(0,1)$ component D''_λ of D_λ satisfies $(D''_\lambda)^2 = 0$ and the associated Laplacian $\Delta_\lambda = D''_\lambda (D''_\lambda)^* + (D''_\lambda)^* D''_\lambda$ satisfies

$$\Delta\tilde{s} = \lambda\tilde{s} + \widetilde{\Delta_\lambda s}. \tag{3.1}$$

Thus, on lifts of sections of $\boldsymbol{V}^\lambda$, the Laplacian splits into a vertical Laplacian, multiplication by λ, and a horizontal Laplacian, Δ_λ.

In this context, the appropriate completeness theorem is

PROPOSITION 3.2. *Finite linear combinations of lifts of eigensections of Δ_λ on $\boldsymbol{V}^\lambda$, λ varies, are dense in $\mathscr{D}^{0,0}(G, L(\chi))$.*

The method of lifting forms on G/T with values in $\boldsymbol{V}^\lambda$ to forms on G with values in $L(\chi)$ is more complicated, but (3.1) and a modification of (3.2) are true for forms as well as sections.

This completeness result and (3.1) can be used to imitate the proof of the product formula and prove the following theorem.

THEOREM 3.3. *If* rank $G > 2$,

$$T_0(G, \chi) = 1 = T_0(T, \chi_0).$$

If rank $G = 2$,

$$\sum_q (-1)^q q \operatorname{tr}(\exp(-t\Delta_{0,q})) = -\sum_\lambda e^{-t\lambda} \chi(G/T, V^\lambda),$$

where the sum on the right is over eigenvalues of Δ *on* $\mathscr{D}^{0,0}(T, L(\chi_0))$ *counted with multiplicity and* $\chi(G/T, V^\lambda)$ *is the arithmetic genus of* G/T *with values in* V^λ.

4. Action of the Weyl group. To complete the proof of Theorem 1.1 for rank 2 groups, we must compute $S = \sum_\lambda e^{-t\lambda} \chi(G/T, V^\lambda)$. By the Riemann-Roch theorem,

$$\chi(G/T, V^\lambda) = \int_{G/T} \operatorname{ch} V^\lambda \wedge \mathscr{T}(G/T),$$

where ch V^λ is the Chern character of V^λ and $\mathscr{T}(G/T)$ is the Todd class of G/T. The contribution of the μ-eigenspace to the sum S is

$$e^{-t\mu} \sum_{\lambda=\mu} \chi(G/T, V^\lambda) = e^{-t\mu} \int_{G/T} \Big(\sum_{\lambda=\mu} \operatorname{ch} V^\lambda\Big) \wedge \mathscr{T}(G/T). \tag{4.1}$$

The Weyl group is the key to simplifying (4.1). Let W be the Weyl group of G, i.e., $W = N(T)/T$ where $N(T)$ is the normalizer of T. Then W acts on G/T by right translation, and hence it acts on $H^*(G/T)$. By a lemma of Leray [**1**, § 27], the action of W on $H^*(G/T)$ is equivalent to the regular representation of W. Thus, the subspace of $H^*(G/T)$ invariant under W is $H^0(G/T)$.

We will sketch a proof that $a = \sum_{\lambda=\mu} \operatorname{ch} V^\lambda$ is invariant under the action of W. It then follows that $a \in H^0(G/T)$ and thus $a = M(\mu)$, the multiplicity of μ. Since $\int_{G/T} \mathscr{T}(G/T) = 1$ (see [**2**]), (4.1) becomes

$$e^{-t\mu} \sum_{\lambda=\mu} \chi(G/T, V^\lambda) = e^{-t\mu} M(\mu) = \sum_{\lambda=\mu} e^{-t\lambda}.$$

Combining this with Theorem 3.3 gives

$$\sum_q (-1)^q q \operatorname{tr}(\exp(-t\Delta_{0,q})) = -\sum_\lambda e^{-t\lambda}.$$

It follows from the definition of torsion that $T_0(G, \chi) = T_0(T, \chi_0)$. The proof that a is invariant under the action of W relies on the fact that W acts as a group of isometries of T. Thus, W preserves eigenspaces of Δ. Also, if $\sigma \in W$, then σ permutes the one-dimensional invariant subspaces of the μ eigenspace. Hence, if $\sigma(V^\lambda) = V^{\sigma(\lambda)}$, there is a bundle map

$$\begin{array}{ccc} V^{\lambda} & \xrightarrow{\sigma} & V^{\sigma(\lambda)} \\ \downarrow & & \downarrow \\ G/T & \xrightarrow{\sigma} & G/T \end{array}$$

By naturality of Chern classes, $\sigma^*(\mathrm{ch}\ V^{\sigma(\lambda)}) = \mathrm{ch}\ V^{\lambda}$ and $\sum_{\lambda=\mu} \mathrm{ch}\ V^{\lambda}$ is invariant under W.

REFERENCES

1. A. Borel, *Sur la cohomologie des espaces fibrés principaux et des espaces homogènes de groupes de Lie compacts*, Ann. of Math. (2) **57** (1953), 115–207. MR **14**, 490.

2. A. Borel and F. Hirzebruch, *Characteristic classes and homogeneous spaces*. I, II, III, Amer. J. Math. **80** (1958), 458–538; ibid. **81** (1959), 315–382; **82** (1960), 491–504. MR **21** #1586; **22** #988; #11413.

3. D. B. Ray and I. M. Singer, *Analytic torsion for complex manifolds*, Ann. of Math. (2) **98** (1973), 154–177.

4. H. Samelson, *A class of complex-analytic manifolds*, Portugal. Math. **12** (1953), 129–132. MR **15**, 505.

5. N. K. Stanton, *Holomorphic R-torsion for Lie groups*, Thesis, M. I. T., Cambridge, Mass., 1973.

MASSACHUSETTS INSTITUTE OF TECHNOLOGY

HOMOGENEOUS SPACES

Proceedings of Symposia in Pure Mathematics
Volume 27, 1975

THE FIRST EIGENVALUE OF THE LAPLACIAN ON MANIFOLDS OF NONNEGATIVE CURVATURE

I. CHAVEL AND E. FELDMAN

Our purpose in this note is to announce the following results:

THEOREM 1. *Let M be a compact orientable Riemannian 2-manifold, diffeomorphic to the 2-sphere, having Gauss curvature $K(p)$, $p \in M$, which satisfies for all $p \in M$*

$$0 \leqq K(p) \leqq \kappa \tag{1}$$

where κ is some positive constant. Then the first eigenvalue, λ_1, of the Laplacian on M (acting on C^∞ functions on M) satisfies

$$\lambda_1 \leqq 2\kappa, \tag{2}$$

with equality if and only if the metric on M has constant curvature κ.

THEOREM 2. *Let M be a Riemannian manifold diffeomorphic to the real projective plane and assume the Gaussian curvature of M satisfies* (1) *at all $p \in M$. Let $\rho =$* arc sin $1/3^{1/2} \in [0, \pi/2]$. *Then*

$$\lambda_1 \leqq (\pi^2/4\rho^2)\kappa < (3\pi^2/4)\kappa. \tag{3}$$

THEOREM 3. *Let M be a compact manifold without boundary of dimension $n \geqq 2$ and with Riemannian metric g. Assume that the Ricci curvature of g on M is everywhere positive semidefinite and that there exist points $p_-, p_+ \in M$ with distance to one another $\geqq d$ (d is a given positive number) and each having injectivity radius $> d/2$. Then*

$$\lambda_1 \leqq 4j^2_{n/2-1}/d^2 \tag{4}$$

AMS (MOS) subject classifications (1970). Primary 35P15, 53C20; Secondary 34B25.

where $j_{n/2-1}$ *is the first zero of the* $(n/2-1)$*st Bessel function.*

By W. Klingenberg's result we immediately obtain

THEOREM 4. *Let M be a compact manifold without boundary of even dimension* $\geqq 2$ *and with Riemannian metric g. Assume that for every 2-section,* σ, *tangent to M the Riemann sectional curvature of* σ, $K(\sigma)$, *satisfies*

$$0 < K(\sigma) \leqq \kappa. \tag{5}$$

Then

$$\lambda_1 \leqq 4\kappa j^2_{n/2-1}/\pi^2. \tag{6}$$

We note that we do not know whether inequalities (3), (4), and (6) are sharp; indeed we would expect the sharp upper bound for (3) to be 6κ, which at the same time would characterize the real projective plane of constant curvature κ.

Also note that if the Ricci curvature of M is bounded below by a positive constant then one does have the appropriate sharp inequality to estimate λ_1 from below. The result was obtained by S. Bochner, A. Lichnerowicz, and M. Obata—for a detailed exposition with references cf. [**1**, pp. 179–185].

It is classical [**6**, p. 486] that $j^2_{n/2-1} \leqq n(n/2+2)$. Thus, Theorem 3 implies

$$\lambda_1 \leqq 4n(n/2+2)/d^2. \tag{4'}$$

If in particular we choose d to be the diameter of M our result invites comparison with J. Cheeger's theorem [**2**, pp. 1, 101–102, 188–196] that

$$\lambda_1 \leqq 16(n+1)^2(n+2)/d^2$$

without the assumption on the injectivity radii of points in M.

We finally remark that Theorem 1 has already been proven by J. Hersch [**3**], via conformal mapping and the standard imbedding of the 2-sphere in Euclidean 3-space. Our ability to obtain by our method the sharp inequality (2) is due to the existence of a simple closed geodesic for any Riemannian metric on the 2-sphere [**4**]. Indeed the method of Theorem 1 immediately generalizes to

THEOREM 5. *Let M be a compact oriented n-dimensional Riemannian manifold,* $n \geqq 2$, *without boundary and with Riemannian sectional curvatures* $K(\sigma)$ *satisfying* (5) *for all 2-sections* σ *tangent to M. Furthermore assume M has an oriented* $n-1$ *dimensional, totally geodesic submanifold N which bisects M. Then* $\lambda_1 \leqq n\kappa$ *with equality if and only if M is isometric to the n-sphere of constant curvature* κ.

In all three applications our methods are essentially the same: we use Rayleigh's principle to estimate λ_1 from above via calculus of variations; the calculus of variations leads to two mixed boundary-value problems in two open pairwise disjoint sets in M each having appropriate geodesic coordinates; the volume elements are averaged over the field of geodesics to yield a one-dimensional Sturm-Liouville problem. The following comparison theorem of W. T. Reid [**5**] is then invoked

THEOREM 6. *Let*

$$Lu = (\Phi u')' + \lambda \Phi u, \qquad Mv = (\Psi v')' + \mu \Psi v$$

be two ordinary differential operators on $[\alpha, \beta]$, *where* Φ, Ψ *are* C^∞ *on* $[\alpha, \beta]$ *and both positive on* (α, β). *Let* u *and* v *be respective smooth solutions of* $Lu = 0$, $Mv = 0$ *satisfying the boundary conditions*

$$u(\alpha) = v(\alpha) = u'(\beta) = v'(\beta) = 0;$$

and furthermore assume that neither u *nor* v *vanishes on* (α, β). *If* $(\Phi/\Psi)' \geqq 0$ *on all of* (α, β) *then* $\lambda \leqq \mu$ *with equality if and only if* $\Phi/\Psi \equiv \text{const}$ *on* (α, β).

Detailed proofs will appear elsewhere.

REFERENCES

1. M. Berger, P. Gauduchon and E. Mazet, *Le spectre d'une variété Riemanniene*, Lecture Notes in Math., vol. 194, Springer-Verlag, Berlin and New York, 1971. MR **43** #8025.

2. J. Cheeger, *The relation between the Laplacian and the diameter for manifolds of non-negative curvature*, Arch. Math. **19** (1968), 558–560. MR **38** #6503.

3. J. Hersch, *Quatre propriétés isopérimétriques de membranes sphinque homogènes*, C. R. Acad. Sci. Paris **270** (1970), 1645–1648.

4. H. Poincaré, *Sur les lignes géodésiques des surfaces convexes*, Trans. Amer. Math. Soc. **5** (1905), 237–274.

5. W. T. Reid, *A comparison theorem for self-adjoint differential equations of second order*, Ann. of Math. (2) **65** (1957), 197–202. MR **19**, 1052.

6. G. N. Watson, *A treatise on the theory of Bessel functions*, Cambridge Univ. Press, London; Macmillan, New York, 1944. MR **6**, 64.

CITY COLLEGE OF THE CITY UNIVERSITY OF NEW YORK

GRADUATE CENTER OF THE CITY UNIVERSITY OF NEW YORK

Proceedings of Symposia in Pure Mathematics
Volume 27, 1975

THE GENERALIZED GEODESIC FLOW

LEON W. GREEN

Let M be a simply-connected, complete Riemannian manifold with $\frac{1}{4}$-pinched negative sectional curvature. The classical Busemann construction of horospheres is amplified to find horocycle fields, $H(\eta)$, in the principal bundle $\boldsymbol{F}$, parametrized by unit vectors η orthogonal to a fixed unit vector ξ in the ambient Euclidean space. If B is the basic horizontal vector field corresponding to ξ, its one parameter group of diffeomorphisms of $\boldsymbol{F}$ is called a generalized geodesic flow in $\boldsymbol{F}$. Then $[B, H(\eta)] = -H(L\eta)$, where L is a linear operator associated with the second fundamental form of the horospheres. This suggests an analogy with the Iwasawa decomposition for symmetric (noncompact) spaces of rank one, with B corresponding to the abelian part and the horocycle fields spanning what should be the nilpotent part.

By constructing similar "negatively directed" horocycles, one obtains an analogue of the Bruhat decomposition. These analogies lead to a proof, essentially by Mautner's group representation method, of the fact that the generalized geodesic flow is weakly mixing for the principal bundle of a compact manifold covered by M. (Details are in Duke Math. J. **41** (1974), 115–126; cf. also the Correction, ibid., June 1975.)

It was also claimed that these analogies with symmetric spaces lead to helpful interpretations of the boundary, $H(\infty)$, in the sense of B. O'Neill and P. Eberlein, of spaces like M. Namely, the projection of the above dynamical picture to the usual geodesic flow in the unit tangent bundle SM leads naturally to considering SM as a bundle over $H(\infty)$ with fibers as the stable manifolds of the flow. This suggests that the correct measure class on the boundary is that which, when multiplied by a (suitably normalized) flow invariant measure on the fibers, yields the usual invariant measure in SM.

AMS (MOS) subject classifications (1970). Primary 53C20, 58F05, 28A65; Secondary 54H20, 34C35.

Proceedings of Symposia in Pure Mathematics
Volume 27, 1975

THE EIGENFUNCTIONS OF THE LAPLACIAN ON A TWO-POINT HOMOGENEOUS SPACE: INTEGRAL REPRESENTATIONS AND IRREDUCIBILITY

SIGURDUR HELGASON

As stated in [3(a)], with a proof sketched there and in [3(c)], the eigenfunctions of the Laplace-Beltrami operator on the non-Euclidean disk D are precisely the functions

$$z \to \int_B P(z, b)^\mu \, dT(b) \qquad (z \in D)$$

where P is the Poisson kernel, μ any complex number and T any analytic functional on the boundary B of D. Here we state without proofs extensions of this result to noncompact two-point homogeneous spaces X, obtained by similar methods, sharpening an earlier result [3(a), p.137]. We also state irreducibility criteria for the corresponding eigenspace representations. The spaces X are the noncompact symmetric spaces of rank one and the Euclidean spaces. We treat these two cases separately.

I. $X = G/K$, a symmetric space of rank one. Here we take G to be the largest connected group of isometries of X. Then G is semisimple and K is a maximal compact subgroup. Let $\mathfrak{g} = \mathfrak{k} + \mathfrak{p}$ be the corresponding Cartan decomposition of the Lie algebra $\mathfrak{g}$ of G, $\mathfrak{g} = \mathfrak{k} + \mathfrak{a} + \mathfrak{n}$ an Iwasawa decomposition with $\mathfrak{a}$ a maximal abelian subspace of $\mathfrak{p}$ and let M be the centralizer of $\mathfrak{a}$ in K. The metric d on X is assumed induced by the Killing form $\langle \ , \ \rangle$ of $\mathfrak{g}$. Let Δ denote the corresponding Laplace-Beltrami operator. Given $x \in X$, $b \in B$ $(= K/M)$ let $A(x, b) \in \mathfrak{a}$

AMS (MOS) subject classifications (1970). Primary 22E45, 22E30, 33A30, 35C15, 35J05.

denote log of the complex distance from the origin $o = \{K\}$ in X to the horocycle in X passing through x with normal b [3(a), p. 9]. Let $\mathfrak{a}^*$ be the dual of $\mathfrak{a}$, $\mathfrak{a}_c^* = \boldsymbol{C} \otimes \mathfrak{a}^*$. If $\lambda \in \mathfrak{a}_c^*$ and $2\rho \in \mathfrak{a}^*$ denotes the sum of the roots of $(\mathfrak{g}, \mathfrak{a})$ with multiplicity, positive with respect to the ordering given by $\mathfrak{n}$, let $\mathscr{E}_\lambda$ denote the eigenspace

$$\mathscr{E}_\lambda = \{f \in C^\infty(X) : \Delta f = -(\langle \lambda, \lambda \rangle + \langle \rho, \rho \rangle f\}$$

with the topology induced by that of $C^\infty(X)$.

THEOREM 1. (i) *The eigenfunctions of Δ on X with eigenvalue $\geqq -\langle \rho, \rho \rangle$ are precisely the functions*

$$f(x) = \int_B e^{(i\lambda+\rho)(A(x,b))}\, dT(b) \tag{1}$$

where $\lambda \in i\mathfrak{a}^*$ and T is an analytic functional on B.

(ii) *The natural representation of G on $\mathscr{E}_\lambda$ is irreducible if and only if*

$$\boldsymbol{e}(\lambda)\, \boldsymbol{e}(-\lambda) \neq 0,$$

where $\boldsymbol{e}(\lambda)^{-1}$ is the denominator of Harish-Chandra's function $\boldsymbol{c}(\lambda)$ (cf. [3(b)]).

(iii) *If $\operatorname{Re}\langle i\lambda, \rho \rangle > 0$ then as $x \to \infty$ we have formally*

$$e^{-d(0,x)r} f(x) \to \boldsymbol{c}(\lambda)\, T$$

where, in terms of the notation below, $r = l[2(m_\alpha + 4m_{2\alpha})]^{-1/2}$.

(iv) *If X is a real hyperbolic space the functions (1), with $\lambda \in \mathfrak{a}_c^*$ and T an analytic functional, constitute all the eigenfunctions of Δ.*

While (ii) is proved in [3(a), p. 143], part (i) is given there, p. 137, only in a weaker form. The sharper form above is based on the study of the K-finite eigenfunctions of Δ which in [3(a), p. 133] are expressed by means of the integral

$$\Phi_{\lambda,\delta}(x) = \int_{K/M} e^{(i\lambda+\rho)(A(x,kM))} \langle v, \delta(k)v \rangle\, dk_M \tag{2}$$

where δ is an irreducible unitary representation of K for which $\delta(M)$ has a fixed unit vector v, $\langle\ ,\ \rangle$ denotes the inner product, and dk_M the normalized invariant measure on K/M. The integral (2) is an Eisenstein integral in the sense of [1]. The set $\hat{K}_0$ of representations δ is parametrized in Kostant [5] by a pair of integers, but for us the parametrization in Johnson and Wallach [4] is more convenient. If α, and possibly 2α, are the positive roots, m_α and $m_{2\alpha}$ their multiplicities, and $\boldsymbol{Z}^+$ the set of nonnegative integers, the parametrization has the following properties:

(1) If $m_{2\alpha} = 0$, $\hat{K}_0$ corresponds to the set of pairs (p, q) where $p \in \boldsymbol{Z}^+$ and $q = 0$.

(2) If $m_{2\alpha} = 1$, $\hat{K}_0$ corresponds to the set of pairs $(p, q) \in \boldsymbol{Z}^+ \times \boldsymbol{Z}$ where $p \pm q \in 2\boldsymbol{Z}^+$.

(3) If $m_{2\alpha} = 3$ or 7, $\hat{K}_0$ corresponds to the set of pairs $(p, q) \in \boldsymbol{Z}^+ \times \boldsymbol{Z}^+$ with $p - q \in 2\boldsymbol{Z}^+$.

THEOREM 2. *Select $H \in \mathfrak{a}$ such that $\alpha(H) = 1$ and put $h_t = \exp tH$, $\alpha_0 = \alpha/\langle\alpha, \alpha\rangle$. Then if $\delta \in \hat{K}_0$ corresponds to (p, q) the Eisenstein integral $\Phi_{\lambda,\delta}$ is given in terms of the hypergeometric function F by*

$$\Phi_{\lambda,\delta}(h_t \cdot o) = c_{\lambda,\delta} \tanh^p t \cosh^l t$$

$$\cdot F\left(\frac{-l+p+q}{2}, \frac{-l+p-q+1-m_{2\alpha}}{2}, p+\frac{m_\alpha+m_{2\alpha}+1}{2}, \tanh^2 t\right) \tag{3}$$

where $l = \langle i\lambda - \rho, \alpha_0\rangle$ and

$$c_{\lambda,\delta} = \frac{\Gamma(\langle\rho,\alpha_0\rangle + \frac{1}{2}(l+p+q))\,\Gamma(\frac{1}{2}(m_\alpha+m_{2\alpha}+1+l+p-q))\,\Gamma(\frac{1}{2}(m_\alpha+m_{2\alpha}+1))}{\Gamma(\langle\rho,\alpha_0\rangle+\frac{1}{2}l)\,\Gamma(\frac{1}{2}(m_\alpha+m_{2\alpha}+1+l))\,\Gamma(\frac{1}{2}(m_\alpha+m_{2\alpha}+1)+p)}. \tag{4}$$

REMARK. According to Lemma 6.2 in [3(d)] the Eisenstein integrals $\Phi_{\lambda,\delta}$ and $\Phi_{-\lambda,\delta}$ are related by

$$\frac{\Phi_{-\lambda,\delta}}{\Phi_{\lambda,\delta}} = \frac{\boldsymbol{C}_s(-\lambda)}{\boldsymbol{c}(\lambda)},$$

where $\boldsymbol{C}_s$ is the generalized $\boldsymbol{c}$-function. Letting $t \to 0$ we therefore obtain

$$\boldsymbol{C}_s(-\lambda) = \boldsymbol{c}(\lambda)\frac{\boldsymbol{C}_{-\lambda,\delta}}{\boldsymbol{C}_{\lambda,\delta}}, \tag{5}$$

which upon using (4) is in agreement with Theorem 3.1 in Johnson and Wallach [4].

II. $X = \boldsymbol{R}^n$, a Euclidean space. Here the analogous questions can be dealt with by the same methods but are of course much simpler. Let Δ_0 denote the Laplacian on $\boldsymbol{R}^n$ and if $\lambda \in \boldsymbol{C}$ let

$$\mathscr{E}_\lambda = \{f \in \boldsymbol{C}^\infty(\boldsymbol{R}^n) : \Delta_0 f = -\lambda^2 f\}.$$

Let (,) denote the inner product on $\boldsymbol{R}^n$ and S^{n-1} the unit sphere $(x, x) = 1$. Then we have

THEOREM 3. (i) *The group of isometries of $\boldsymbol{R}^n$ acts irreducibly on $\mathscr{E}_\lambda$ if and only if $\lambda \neq 0$.*

(ii) *The eigenfunctions in $\mathscr{E}_\lambda$ $(\lambda \neq 0)$ are*

$$\int_{S^{n-1}} e^{i\lambda(x,\omega)}\, dT(\omega)$$

where T runs through certain "functionals" on $\boldsymbol{S}^{n-1}$.

(iii) *For $n = 2$ the functionals T are the elements in the dual of the inductive limit*

$$E = \bigcup_{a>0,b>0} E_{a,b},$$

where $E_{a,b}$ is the Banach space of restrictions to $\boldsymbol{S}^1$ of holomorphic functions $f(z)$ in $\boldsymbol{C} - \{0\}$ satisfying

$$\sup_z |f(z)| e^{-a|z|-b|z|^{-1}} < \infty.$$

REMARK. As I found out after the lecture, Part (ii) of this theorem is also contained in [2].

ADDED IN PROOF. Another proof of Theorem 1(iv) and (i) for the hermitian space has been found by K. Minemura (preprint).

REFERENCES

1. Harish-Chandra, *On the theory of the Eisenstein integral*, Proc. Internat. Conf. on Harmonic Analysis (Univ. of Maryland, 1971), Lecture Notes in Math., vol. 266, Springer-Verlag, Berlin and New York, 1972.

2. M. Hashizume, A. Kowata, K. Minemura and K. Okamoto, *An integral representation of an eigenfunction of the Laplacian on the Euclidean space*, Hiroshima Math. J. **2** (1972), 535–545.

3. S. Helgason, (a) *A duality for symmetric spaces with applications to group representations*, Advances in Math. **5** (1970), 1–154. MR **41** #8587.

(b) *Group representations and symmetric spaces*, Proc. Internat. Congress Math. (Nice, 1970), vol. II, Gauthier-Villars, Paris, 1971, pp. 313–319.

(c) *Harmonic analysis in the non-Euclidean disk*, Proc. Internat. Conf. on Harmonic Analysis (Univ. of Maryland, 1971), Lecture Notes in Math., no. 266, Springer-Verlag, Berlin and New York, 1972.

(d) *The surjectivity of invariant differential operators on symmetric spaces*. I, Ann. of Math. **98** (1973), 451–479.

4. K. Johnson and N. Wallach, *Composition series and intertwining operators for the spherical principal series*, Bull. Amer. Math. Soc. **78** (1972), 1053–1059. MR **46** #9238.

5. B. Kostant, *On the existence and irreducibility of certain series of representations*, Bull. Amer. Math. Soc. **75** (1969), 627–642. MR **39** #7031.

MASSACHUSETTS INSTITUTE OF TECHNOLOGY

Proceedings of Symposia in Pure Mathematics
Volume 27, 1975

THE COHOMOLOGY RING OF $SO(2n+2)/SO(2)\times SO(2n)$ AND SOME GEOMETRICAL APPLICATIONS

HON-FEI LAI

The homogeneous space $\boldsymbol{R}_{2n,2} \equiv SO(2n+2)/SO(2) \times SO(2n)$ can be interpreted as the Grassmannian manifold of oriented $2n$-planes in Euclidean space $\boldsymbol{R}^{2n+2}$, and as such there are a natural $2n$-plane bundle and a natural 2-plane bundle over it, with Euler classes denoted by Ω, $\tilde{\Omega}$, respectively; thus $\Omega \in H^{2n}(\boldsymbol{R}_{2n,2})$, $\tilde{\Omega} \in H^2(\boldsymbol{R}_{2n,2})$. By introducing a fixed complex structure on $\boldsymbol{R}^{2n+2}$, we also get an embedded n-dimensional complex projective space K, which is the set of all complex n-dimensional subspaces of $\boldsymbol{C}^{n+1} = \boldsymbol{R}^{2n+2}$. Let $\kappa \in H^{2n}(\boldsymbol{R}_{2n,2})$ be its Poincaré dual in $\boldsymbol{R}_{2n,2}$. Then the (integral) cohomology ring of $\boldsymbol{R}_{2n,2}$ is generated by $\tilde{\Omega}$ and κ with the relation $\tilde{\Omega}^{n+1} = 2\kappa \cup \tilde{\Omega}$. We also have $\Omega + \tilde{\Omega}^n = 2\kappa$, $\kappa \cup \tilde{\Omega}^n = (-1)^n$ and $\kappa \cup \Omega = 1$. These facts are proved in [3] using characteristic class arguments, and in [1] using Schubert varieties [4]. Some applications: Let M be a $2n$-dimensional compact orientable manifold with Euler characteristic χ.

(1) *Real-complex singularities*. Let $i:M \to \boldsymbol{C}^{n+1}$ be an embedding. A point $x \in M$ is called an $\boldsymbol{RC}$-singular point if the tangent space to $i(M)$ at $i(x)$ is complex. Such points are generically isolated. We can assign an index $\nu(x) = \pm 1$ to each $\boldsymbol{RC}$-singular point, and the sum of these indices is equal to χ.

(2) *Gauss-Kronecker curvature*. Let i: $M \to \boldsymbol{R}^{2n+1}$ be an embedding. For any $2n$-plane Y in $\boldsymbol{R}^{2n+1}$, there are at least $|1-\chi| + 1$ points on M where the tangent plane is parallel to Y, and where the Gauss-Kronecker curvature is $\geqq 0$ or $\leqq 0$, according to whether $\chi \geqq 0$ or $\chi \leqq 0$.

AMS (MOS) subject classifications (1970). Primary 53C40, 57F15.

(3) *Parallel tangents*. For any embedding of M in $\boldsymbol{R}^{2n+2}$, there are at least $\chi^2/4$ pairs of points where the tangent planes are parallel and oppositely oriented.

References

1. H.-F. Lai, *Characteristic classes of real manifolds immersed in complex manifolds*, Trans. Amer. Math. Soc. **172** (1972), 1-33. MR **47** #2618.

2. ——, *On parallel tangents of embeddings of codimension* 2 *in Euclidean spaces*, Indiana Univ. Math. J. **22** (1973), 1171–1181.

3. ——, *On the topology of the even-dimensional complex quadrics*, Proc. Amer. Math. Soc. **46** (1974), 419–425.

4. W. T. Wu, *Sur les classes caractéristiques des structures fibrées sphériques*, Actualités Sci. Indust., no. 1183, Hermann, Paris, 1952. MR **14**, 1112.

Tulane University

Proceedings of Symposia in Pure Mathematics
Volume 27, 1975

REPRESENTATIONS OF LINEAR FUNCTIONALS ON H^p SPACES OVER BOUNDED HOMOGENEOUS DOMAINS IN C^N $(N > 1)$

JOSEPHINE MITCHELL*

1.1. Introduction. The paper represents joint work with Kyong T. Hahn. It generalizes some results of Duren, Romberg and Shields [2]. We keep the domains as general as possible.

D is a *domain of type* A if

(1) D is a bounded, homogeneous domain in complex vector space C^N $(N > 1)$, which is star-shaped and circular with respect to $0 \in D$.

(2) The group Γ of holomorphic automorphisms of D extends continuously to the topological boundary ∂D.

Thus D has a unique Bergman-Šilov (B-Š) boundary b which is also circular and invariant under Γ [13].

(3) The isotropy group $\Gamma_0 = \{\gamma \in \Gamma : \gamma(0) = 0\}$ is transitive on b. D is of *type* A′ if it is of type A and

(4) b has a unique normalized Γ-invariant measure $d\mu_t(z) = P(z, t)ds_t$, $z \in D$, $t \in b$, $P(z, t)$ the Poisson kernel of D and ds_t the circularly invariant measure at $t \in b$.

REMARK. The bounded symmetric domains are domains of type A′. R.-QP. Lu's nonsymmetric homogeneous domains $S_{p,m,n}$, suitably modified, may be domains of type A or A′ [14].

1.2. NOTATION. For $p > 0$ the *Hardy H^p space* on D is defined by

AMS (MOS) subject classifications (1970). Primary 32M10, 32M15, 32A30; Secondary 30A78.
*Supported in part by NSF grants.

$$H^p \equiv H^p(D) = \left\{f: f \text{ holomorphic on } D \text{ and } \sup_{0<r<1} M_p(r,f) < \infty\right\},$$

where

$$M_p(r,f) = \left(\frac{1}{V}\int_b |f(rt)|^p \, ds_t\right)^{1/p},$$

$$\|f\|_p = \sup_{0<r<1} M_p(r,f) \quad (\| \ \|_p \text{ is not a norm for } 0 < p < 1),$$

$$(f,g) = \frac{1}{V}\int_b f(t)\bar{g}(t)\, ds_t, \qquad f, g \in L^1(b).$$

Let $f: D \to C \cdot f_r$ be the *slice function* of f defined by $f_r(z) = f(rz)$, $z \in D \cdot D_r = \{rz : z \in D\}$. Similarly for b_r.

We give a summary of the results in the paper. Proofs will appear elsewhere.

2.1. Representation of bounded linear functionals on H^p. Take $p > 0$. Let T be a bounded linear functional on H^p, that is, $T \in (H^p)^*$, the dual space of H^p. The first theorem gives a representation for T.

THEOREM 1. (i) *Let D be a domain of type* A *and $T \in (H^p)^*$, $p > 0$. Then there exists a unique function G, holomorphic on D, such that*

$$T(f) = \lim_{r\to 1} (f_{r\rho^{-1}}, G_\rho), \qquad 0 < r < \rho < 1, \forall f \in H^p. \tag{1}$$

(ii) *If D is a domain of type* A′, *G is holomorphic on D and the limit in* (1) *exists and equals $T(f)$ for all $f \in H^p$, then $T \in (H^p)^*$.*

OUTLINE OF PROOF. Since D is of type A there exists a complete system $\{\varphi_{k\nu}\}$ ($k = 0,1,2,\cdots; \nu = 1,\cdots, m_k < \infty$) of homogeneous orthogonal polynomials which are orthonormal on b **[11]**. A function f, holomorphic on D, has a series expansion **[8]**

$$\begin{aligned} f(z) &= \sum_{k,\nu} a_{k\nu}(f)\varphi_{k\nu}(z), \\ a_{k\nu}(f) &= \lim_{r\to 1}(f_r, \varphi_{k\nu}) \qquad (z \in D), \\ \sum_{k,\nu} &= \sum_{k=0}^{\infty}\sum_{\nu=1}^{m_k}, \end{aligned} \tag{2}$$

where convergence is uniform on compact subsets of D. If f and g are holomorphic on D and $0 < r < 1$, $r < \rho < 1$, then

$$(f_{r\rho^{-1}}, g_\rho) = \sum_{k,\nu} a_{k\nu}(f)\overline{a_{k\nu}}(g) r^k. \tag{3}$$

(3) follows easily from (2).

Let $T \in (H^p)^*$. By (2), $a_{j\mu}(\varphi_{k\nu}) = \delta_{kj}\delta_{\nu\mu}$ (Kronecker δ). Thus by (3) for any g, holomorphic on D, $(\varphi_{k\nu, r\rho^{-1}}, g_\rho) = \bar{a}_{k\nu}(g) = \bar{T}(\varphi_{k\nu})$ if (1) holds. Define

$$G(z) = \sum_{k,\nu} \bar{T}(\varphi_{k\nu})\varphi_{k\nu}(z) \quad \text{(unique)}. \tag{4}$$

Series (4) converges uniformly on compact subsets of D. This follows from the inequality $|T(\varphi_{k\nu})| \leqq \|T\| \, \|\varphi_{k\nu}\|_p$ and estimates for $\|\varphi_{k\nu}\| \, |\varphi_{k\nu}(z)|$ obtained from the maximum principle on b and inequality $|S(z, t^*)| \leqq V^{-1}(1-r)^{-N}$, $z \in \bar{D}_r$, $r < 1$, $S(z, t^*)$ the Szegö kernel of D **[12]**. Here we use (3) in definition of domain of type A. Thus G is holomorphic on D.

Let $f \in H^p$. By calculation $T(f_r) = (f_{r\rho-1}, G_\rho)$. Since D is circular and star-shaped there exists $\tilde{f} \in L^p(b)$ such that $\|f_r - \tilde{f}\|_p \to 0$ as $r \to 1$ **[1]**. Also $\|f_r - \tilde{f}\|_p = \|f_r - f\|_p$ (**[7]**, **[9]**). By continuity of T, $T(f_r) \to T(f)$ as $r \to 1$. This proves (i).

(ii) Set $T_r(f) = (f_{r\rho-1}, G_\rho)$. It is defined on [0, 1) by (3) and is a linear functional on H^p. Prove T_r bounded by using the inequality $|f_{r\rho^{-1}}(t)| \leqq \|f\|_p (1 - r\rho^{-1})^{-2N/p}$ and the fact that G_ρ is holomorphic on $\bar{D}$ (**[7]**, **[9]**). Here we use (4) of definition of domain of type A′. By hypothesis $\lim_{r\to 1} T_r(f)$ exists for each $f \in H^p$, but the limit equals $\sup_{0 \leqq r<1} T_r(f) < \infty$, since bounded and continuous are the same for linear functionals. By uniform boundedness principle $\sup_{0<r<1} \|T_r\| = B < \infty$. Thus $|T_r(f)| \leqq B\|f\|_p$. By continuity of T_r in [0, 1], $|T(f)| \leqq B\|f\|_p$ and $T \in (H^p)^*$.

2.2. CASE $1 < p < \infty$.

THEOREM 2. *Let D be a domain of type* A′.

(i) $(H^p)^*$ *is isometrically isomorphic to* $L^q/(H^p)^\perp$ *where*

$$(5) \qquad (H^p)^\perp = \{g \in L^q : (g, \varphi_{k\nu}) = 0 \ \ for \ \ k \geqq 0\}$$

and $1/p + 1/q = 1$.

(ii) $T(f) = (f, g)$ *for every* $f \in H^p$ *and* g *is the unique function in cosets of* $L^q/(H^p)^\perp$ *whose Fourier coefficients* $a_{k\nu}(g) = 0$ *for* $k \geqq 0$.

(iii) *If* $p = 2$, $T(f) = (f, h)$, *where* h *is the "analytic part" of* g.

OUTLINE OF PROOF. Use the Riesz representation theorem on L^p $(1 \leqq p < \infty)$. Then prove that the annihilator

$$(H^p)^\perp = \{T \in (H^p)^* : T(f) = 0 \ \forall f \in H^p\}$$

equals (5) by using results in **[8]**, **[9]**. Let $\{r_n\} \uparrow 1$, $f_n(t) = f(r_n t)$. Then $\{f_n\}$ is a bounded sequence in H^p since $\|f_r\|_p$ is monotone **[7]**. This implies $\{f_n(z)\}$ is bounded independently of n and z on compact subsets of D, which implies that $\{f_n\}$ converges weakly to f in H^p for $p > 1$ **[8]**, that is, $(f, g) = 0$ for every $g \in L^q$ which annihilates f in H^p. (5) follows from this.

REMARK. The second reference to **[8]** uses a result of Weyl **[17]**, which assumes that the group of automorphisms of b is a compact Lie group but Γ_0 acts by unitary transformations which form a compact Lie group **[15**, p. 223].

For proof of (ii) see **[4**, Theorem 7.1]. For (iii) use Hilbert space techniques and the representation $g(t) \sim \sum_{-\infty}^{\infty} a_{k\nu}(g)\varphi_{k\nu}(t)$ for $g \in L^2(b)$ **[8]**.

3. Representation theorems for bounded symmetric domains, $0 < p < 1$. For the bounded symmetric domains R_j (j = I, II, III, IV) **[11]** we get more explicit results on the function G of Theorem 1.

3.1. ORDER PROPERTIES OF THE SZEGÖ KERNEL, $S(z, t^*)$. For the domains R_j,

$S(z, t^*) = Q^{-s}(z, \bar{t})$, Q a polynomial in z and $\bar{t}$ and s a positive integer or half-integer, $z \in D$, $t \in b$.

We have

THEOREM 3. *If D is a bounded symmetric domain, there exists p_0, $0 < p_0 \leqq 1$, such that $\|S_t\|_p \leqq C_{pN}$, for $0 < p < p_0$, C_{pN} a constant depending only on p and N; $p_0 = 1$ for $R_{\mathrm{I}}(1, N)$ (the ball), $p_0 = 1/s$ for $R_{\mathrm{I}}(m, n)$ $(1 < m \leqq n)$, and R_{II}, $p_0 = 1/2s$ for R_{III} and $2/N$ for R_{IV}. p_0 is sharp for the ball, $R_{\mathrm{I}}(2, 2)$ and R_{II} with $n = 2$.*

The proof of Theorem 3 uses classical analytic techniques due to Hua, Hua and Look, Rauch, Mitchell (**[11]**, **[12]**, **[16]**).

3.2. BOUNDEDNESS OF G FOR DOMAINS R_j.

THEOREM 4. *Let $T \in (H^p)^*$, $0 < p < p_0$, where $D = R_j(\mathrm{I}, \cdots, \mathrm{IV})$. Then $G \in H^\infty$ and*

$$(1) \qquad T(f) = \lim_{r \to 1} (f_r, \tilde{G}),$$

$\tilde{G}$ the boundary value of G.

PROOF. Let $\zeta \in D$, $r < \rho < 1$. By Theorem 1(i),

$$T(S_{\zeta^*}) = \lim_{r \to 1} (S_{r\rho^{-1}, \zeta^*}, G_\rho) = \lim_{r \to 1} \int_b S(r\rho^{-1} t, \zeta^*) \bar{G}_\rho(t)\, ds_t$$
$$= \lim_{r \to 1} \int_b \overline{S(r\rho^{-1}\zeta, t^*) G_\rho(t)}\, ds_t = \lim_{r \to 1} \bar{G}_\rho(r\rho^{-1}\zeta) = \bar{G}(\zeta).$$

Here we used the symmetry properties of $S(z, t^*)$, the Cauchy integral formula for G_ρ on D and holomorphy of G on D. Thus $|G(\zeta)| \leqq \|T\| \; \|S_{\zeta_1^*}\|_p = O(1)$ for $0 < p < p_0$. Therefore $G \in H^\infty$. (1) follows by the Lebesgue dominated convergence theorem.

4. The properties of derivatives of class H^p, $0 < p < 1$.

4.1. We can prove some results on the derivatives of G and Lipschitz classes Λ_α but first we need some properties of derivatives of class H^p.

DEFINITION. Let D be a domain of type A, f be holomorphic on D and continuous on $D \cup b$. $f \in$ *Lipschitz class* Λ_α $(0 < \alpha \leqq 1)$ if it has a boundary function $\tilde{f}$ on b and

$$(1) \qquad |\tilde{f}(t') - \tilde{f}(t'')| \leqq K \|t' - t''\|^\alpha \quad \text{as } t' - t'' \to 0,$$

$t', t'' \in b$, K a constant independent of t', t'', and $\| \;\|$ euclidean distance between t' and t''. (For case $N = 1$ see **[2]**.) $f \in$ class Λ_* if

$$|\tilde{f}(t+h) - 2\tilde{f}(t) + \tilde{f}(t-h)| = O(h), \qquad t+h, t-h, t \in b.$$

For $N = 1$ we have two theorems of Hardy and Littlewood **[10]**.

THEOREM a. *Let f be analytic on $|z| < 1$. Then $f \in \Lambda_\alpha$ $(0 < \alpha \leqq 1)$ if and only if $f'(z) = O((1 - r)^{\alpha - 1})$, $|z| = r$.*

THEOREM b. *If $f' \in H^p$ for some $p < 1$, then $f \in H^{p_1}$, $p_1 = p/(1-p)$. For each p, p_1 is the best possible index.*

Theorem a is generalized as follows:

THEOREM 5(a). *Let D be a bounded domain with B-Š boundary b, star-shaped and circular with respect to 0, $0 \in D$. If $f \in \Lambda_\alpha$ $(0 < \alpha \leqq 1)$, then*

$$df_t(w)/dw = O((1-r)^{\alpha-1}), \qquad |w| = r, w \in C^1.$$

The proof uses slice function techniques and the same method of proof as in the case $N = 1$.

THEOREM 5(b). *Let f be holomorphic on a bounded symmetric domain D. If*

$$|\partial f(z) / \partial z_j| \leqq C(1-r)^{\alpha-1} \tag{2}$$

$(0 < \alpha \leqq 1)$ for $1 \leqq j \leqq N$, $z \in \bar{D}_r$ and C a constant independent of z and r, then $f \in \Lambda_\alpha$.

OUTLINE OF PROOF. We need the extra hypothesis on D in order to use the Dirichlet problem for f to get the continuity of f on $D \cup b$ (**[11]**, **[12]**).

(2) gives

$$|f(r't) - f(r''t)| < \varepsilon \quad \text{for } r' - r'' \text{ sufficiently small,} \tag{3}$$

where $t \in b$ and $r't$, $r''t \in D$, $0 < r < 1$, since D is star-shaped. Let $\{r_n\} \uparrow 1$. Then $\{f(r_n t)\}$ is a Cauchy sequence and has a limit. Call the limit $\tilde{f}(t)$. It is independent of the sequence $\{r_n\}$.

Let $f \in \Lambda_\alpha$ and set $h = 1 - \|t' - t''\| > 0$ (without loss of generality). Then

$$\begin{aligned} |\tilde{f}(t') - \tilde{f}(t'')| &= \lim_{\rho\to1} |f(\rho t') - f(\rho t'')| \\ &\leqq \lim_{\rho\to1} \{|f(\rho t') - f(ht')| + |f(ht'') - f(\rho t'')|\} + |f(ht') - f(ht'')| \\ &\leqq \text{const} \| t' - t'' \|^\alpha + |f(ht') - f(ht'')| \end{aligned}$$

by a variation of (3). On the last term on right use the mean value theorem for functions of several variables, convexity of D and hence of D_h and the Schwarz inequality to give

$$|f(ht') - f(ht'')| \leqq \text{const} \|t' - \|t''.$$

Thus $\tilde{f}$ is continuous on b and $\in \Lambda_\alpha$. The above inequality also implies that $f \in H^p$. Thus by **[11]**, **[12]**, f has a Poisson integral representation $f(z) = \int_b P(z, t)\tilde{f}(t)\, ds$. Since $\tilde{f}$ is continuous on b, the Poisson integral represents a harmonic function on $\bar{D}$, which implies that f is continuous on $D \cup b$. Hence $f \in \Lambda_\alpha$.

4.2. Theorem b is generalized in Theorem 7. This is done by means of a lemma whose proof uses a procedure due to Flett in **[5]** and Flett uses the Marcinkiewicz interpolation theorem **[18]**.

LEMMA. *Let D be of type A, $\tilde{f} \in L^{p'}(b)$ and $u(\rho t) = \int_b P(\rho t, v)\tilde{f}(v)\, ds_v$.*

(i) *If* $1 \leqq p' \leqq q' \leqq +\infty$, $\alpha' = 1/p' - 1/q'$, $1 \leqq r' \leqq +\infty$, $1/q' - 1/r' - 1 \geqq 0$, *then*

$$M_{q'}(\rho, u) \leqq \text{const}\ (1-\rho)^{-N\alpha'}\ \|\tilde{f}\|_{p'}.$$

(ii) *If* $1 < p' < q' \leqq +\infty$, $\alpha' = 1/p' - 1/q'$, $p' \leqq k' < +\infty$, *then*

$$\left(\int_0^1 (1-\rho)^{Nk'\alpha'-1} M_{q'}^{k'}(\rho, u)d\rho\right)^{1/k'} \leqq C_{p'}^{k'}\ \|\tilde{f}\|_{p'}.$$

OUTLINE OF PROOF. Use the Hölder inequality twice on the integrand $P(\rho t, v)\tilde{f}(v)$, getting an inequality for $u(\rho t)$. Raise both sides to power q' and integrate over b. This gives an inequality for $M_{q'}^{q'}(u, \rho)$ which may be further refined to give (i).

To prove (ii) we show that Tf defined by $Tf(\rho) = (V^{-1}2^N\rho^{-N})^{1/q'} M_{q'}(u, 1-\rho)$ on $0 < \rho < 1$ satisfies the hypotheses of the Marcinkiewicz interpolation theorem and from this theorem and (i) the result follows.

THEOREM 6. *Let D be a domain of type* A. *Suppose* $p < q \leqq +\infty$, $p \leqq k < \infty$, $\alpha = 1/p - 1/q$ *and* $f \in H^p$. *Then*

$$\left(\int_0^1 (1-\rho)^{Nk\alpha-1} M_q^k(\rho, f)d\rho\right)^{1/k} \leqq C(p,q,k,N)\ \|\tilde{f}\|_p.$$

(Note that the constants in the Lemma and Theorem 6 are independent of f.)

This theorem follows easily from the Lemma and properties of plurisubharmonic functions.

We can now generalize Theorem b. Notationally it is simpler to use fractional derivatives and integrals.

DEFINITION. (1) The *αth fractional derivative* of f is

$$f^{[\alpha]}(z) = \sum_{k,\nu} \frac{\Gamma(k+\alpha+1)}{\Gamma(k+1)} a_{k\nu}(f)\varphi_{k\nu}(z).$$

(2) The *αth fractional integral* of f is

$$f_{[\alpha]}(z) = \sum_{k,\nu} \frac{\Gamma(k+1)}{\Gamma(k+\alpha+1)} a_{k\nu}(f)\varphi_{k\nu}(z).$$

$f^{[\alpha]}$ and $f_{[\alpha]}$ are holomorphic on D.

Formulas connecting fractional and ordinary derivatives and integrals are

$$f^{[1]}(rt) = \frac{\partial}{\partial r}[rf(rt)], \qquad rf_{[1]}(rt) = \int_0^r f_t(\rho)d\rho,$$

$$r^q f_{[q]}(rt) = \int_0^r \rho^{q-1} f_{t,\,[q-1]}(\rho)d\rho.$$

THEOREM 7. *Let D be of type* A *and* $f \in H^p$ *for some* $p < 1$. *Then* $f_{[1]} \in H^q$ *for* $q = pN/(N-p)$.

We follow the method of proof of Theorem 5.12 of **[4]**. The proof depends on Theorem 6 and Bochner's generalization **[1]** of the Hardy-Littlewood maximal theorem for circular domains.

COROLLARY. *If* $f_{[q-1]} \in H^p$, *then* $f_{[q]} \in H^{p_1}$, $p_1 = Np/(N-p)$.

5. Derivatives of G.

5.1. THEOREM 8. *Let D be of type A′, and $N/(N+q) < p < N/(N+q-1)$ and $G^{[q-1]} \in \Lambda_\alpha$, $\alpha = N(1/p-1) - q + 1$, $q = 1, 2, 3, \cdots$; then $\lim_{r\to 1} V^{-1}(f_{r\rho-1}, G_\rho)$ exists for all $f \in H^p$ and defines a functional $T \in (H^p)^*$.*

If $p = N/(N+q)$ and $G^{[q-1]} \in \Lambda_*$, the conclusion also holds.

OUTLINE OF PROOF. Given $G(z) = \sum_{k,\nu} b_{k\nu}\varphi_{k\nu}(z)$ with $G^{[q-1]} \in \Lambda_\alpha$. From the definition of fractional integral $G^{[0]}(z) = G(z)$ is holomorphic on D. Let $f \in H^p$. By (1.3),

$$\psi(r) = (f_{r\rho-1}, G_\rho) = \sum_{k,\nu} a_{k\nu}(f)\,\bar{b}_{k\nu}\, r^k.$$

Show that $\lim_{r\to 1}\psi(r)$ exists. Then by Theorem 1(ii) $T \in (H^p)^*$. Since $|\psi(r)| \leqq \int_0^1 |\psi'(\rho)|d\rho + |\psi(0)|$, we show that $\int_0^1 |\psi'(\rho)|d\rho < \infty$. This follows from the formula

$$\psi'(r^2) = \int_b f_{[q-1]}(rt)\frac{\partial G^{[q-1]}}{\partial r}(rt)\, ds_t.$$

By Theorem 7, $f \in H^p$ implies $f_{[q-1]} \in H^{p_{q-1}}$, $p_{q-1} = pN/(N-(q-1)p)$. By Theorem 5(a), $G^{[q-1]} \in \Lambda_\alpha$ implies $(\partial G^{[q-1]}/\partial r)(rt) = O((1-r)^{\alpha-1})$. Multiplying these functions together and by r and integrating over $[0, 1]$ gives the desired result.

5.2. The converse to Theorem 8 holds for the ball.

THEOREM 9. *For the ball if $N/(N+q) < p < N/(N+q-1)$ and $\lim_{r\to 1}(f_{r\rho^{-1}}, G_\rho) = T(f)\ \forall f \in H^p$, then $G^{[q-1]} \in \Lambda_\alpha$, $\alpha = N(1/p-1) - q + 1$, $0 < \alpha \leqq 1$, $q = 1, 2, \cdots$. If $p = N/(N+q)$, then $G^{[q-1]} \in \Lambda_*$.*

OUTLINE OF PROOF. Let $N/(N+q) < p < N/(N+q-1)$. Take $\zeta \in \bar{D}_\rho$, $\rho < 1$, $z \in D$. As in the proof of Theorem 3 we need only evaluate the order of the integral

$$\left(\int_b \left|\frac{\partial}{\partial \bar{\zeta}_j}(S^{[q-1]}(\rho t, \zeta^*))\right|^p ds_t\right)^{1/p} = O((1-\rho)^{N(1/p-1)-q}) \qquad (1 \leqq j \leqq N).$$

Then $\partial S^{[q-1]}_{\zeta^*}/\partial\bar{\zeta}_j \in H^p$.

By a similar calculation to that in Theorem 4

$$T\left(\frac{\partial}{\partial\bar{\zeta}_j} S^{[q-1]}_{\zeta^*}\right) = \frac{\partial}{\partial\bar{\zeta}_j}\bar{G}^{[q-1]}(\zeta).$$

Theorem 5(b) gives $G^{[q-1]} \in \Lambda_\alpha$ where $\alpha = N(1/p-1) - q + 1$, $0 < \alpha \leqq 1$.

REMARK. A similar result to Theorem 9 may be proved for the other bounded symmetric domains but in these cases the exponents of Theorems 8 and 9 do not check so that the next theorem is not valid for these domains.

5.3. THEOREM 10. *If $N/(N+q) < p < N/(N+q-1)$, then the Banach spaces $(H^p)^*$ and Λ_α^{q-1} with $\alpha = N(1/p-1) - q + 1$ are equivalent for the ball. If $p = N/(N+q)$, then $(H^p)^*$ is equivalent to Λ_*^{q-1}.*

DEFINITION. $\Lambda_\alpha^q = \{f : f, f^{[1]}, \cdots, f^{[q]}$ are holomorphic on D and continuous on $\bar{D}$ and $f^{[q]} \in \Lambda_\alpha$ with

$$\|f\| = \|f\|_\infty + \sup_{t', t'' \in b; t' \neq t''} \left\{ \frac{|f^{[q]}(t') - f^{[q]}(t'')|}{\|t' - t''\|^\alpha} \right\}.$$

Λ_*^q is defined similarly [2]. Λ_α^q and Λ_*^q are Banach spaces with this norm.

Two Banach spaces are *equivalent* if there is a one-one linear mapping L of X onto Y such that both L and L^{-1} are bounded. By the open mapping theorem it is sufficient that L be bounded.

It follows easily from Theorems 1 and 9 that $L: T \to G$ is a bounded linear functional from $(H^p)^*$ to Λ_α^{q-1}.

6. B^p spaces. Let D be a domain of type A.

DEFINITION. Fix p, $0 < p < 1$. Then $B^p = \{f : f$ holomorphic on D and

(1) $$\|f\|_{B^p} = \int_0^1 (1-\rho)^{N(1/p-1)-1} M_1(\rho, f)\{d\rho < \infty\}.$$

Duren and Shields introduced B^p spaces for $N = 1$ ([2], [10]). B^p is a metric space.

THEOREM 11. *B^p with norm (1) is a Banach space. Also*

(i) $|f(z)| \leqq C_{pN} \|f\|_{B^p} (1-\rho)^{-N/p}$ *for* $f \in B^p$, $z \in \bar{D}_\rho$ *and* $f(z) = o((1-\rho)^{-N/p})$. C_{pN} *independent of* f.

(ii) $\forall f \in B^p$, $f_\rho \to f$ *in* B^p *norm as* $\rho \to 1$.

(iii) H^p *is a dense subset of* B^p.

(iv) $\|f\|_{B^p} \leqq C_{pN} \|f\|_p$ *if* $f \in H^p$.

OUTLINE OF PROOF. By the monotonicity of $M_1(\rho, f)$ ([7], [9]),

$$\begin{aligned}\|f\|_{B^p} &\geqq M_1(R, f) \int_R^1 (1-\rho)^{N(1/p-1)-1} d\rho \\ &= (1-R)^{N(1/p-1)} M_1(R, f)(N(1/p-1))^{-1}.\end{aligned}$$

This gives a bound for $M_1(R, f)$. Since f is holomorphic on $\bar{D}_R$ Cauchy integral formula gives the bound $|f(z)| \leqq$ const $(1-\rho)^{-N} M_1(R, f)$ for $z \in \bar{D}_\rho$ and $\notin D_r$ for $r < \rho$ and $R = \frac{1}{2}(1+\rho)$ [11]. This gives (i).

The "o" inequality, completeness of B^p and (ii) follow as in [2]. $H^p \subset B^p$ and (iv) follow from Theorem 6 with $k = q = 1$ and $\alpha = 1/p - 1$. Also H^p contains all functions holomorphic in D_R where $R > 1$ and such functions are dense in B^p by (ii).

OPEN QUESTIONS. 1. Characterize $(H^p)^*$ spaces for other bounded symmetric domains, using B^p spaces. Arlene P. Frazier [6] has done this for the polydisc.

2. Study further properties of H^p $(0 < p < 1)$ and its enveloping spaces such as B^p ([2], [3], Yanagihara).

References

1. S. Bochner, *Classes of holomorphic functions of several variables in circular domains*, Proc. Nat. Acad. Sci. U.S.A. **46** (1960), 721–723. MR **22** #11144.

2. P. L. Duren, B. W. Romberg and A. L. Shields, *Linear functionals on H^p spaces with* $0 < p < 1$, J. Reine Angew. Math. **238** (1969), 32–60. MR **41** #4217.

3. P. L. Duren and A. L. Shields, *Properties of H^p $(0 < p < 1)$ and its containing Banach space*, Trans. Amer. Math. Soc. **141** (1969), 255–262. MR **39** #6065.

4. P. L. Duren, *Theory of H^p spaces*, Pure and Appl. Math., no. 38, Academic Press, New York and London, 1970. MR **42** #3552.

5. T. M. Flett, *On the rate of growth of mean values of holomorphic and harmonic functions*, Proc. London Math. Soc. (3) **20** (1970), 749–768. MR **42** #3286.

6. Arlene P. Frazier, *The dual space of H^p of the polydisc for* $0 < p < 1$, Duke Math. J. **39** (1972), 369–379. MR **45** #2198.

7. Kyong T. Hahn and Josephine Mitchell, *H^p spaces on bounded symmetric domains*, Trans. Amer. Math. Soc. **146** (1969), 521–531. MR **40** #6247.

8. ———, *H^p spaces on bounded symmetric domains*, Ann. Polon. Math. **28** (1973), 89–95.

9. Kyong T. Hahn, *Properties of holomorphic functions of bounded characteristic on star-shaped circular domains*, J. Reine Angew. Math. **254** (1972), 33–40.

10. G. H. Hardy and J. E. Littlewood, *Some properties of fractional integrals*. II, Math. Z. **34** (1932), 403–439.

11. L. K. Hua, *Harmonic analysis of functions of several complex variables in classical domains*, Science Press, Peking, 1958; English transl., Transl. Math. Monographs, vol. 6, Amer. Math. Soc., Providence, R. I., 1963. MR **30** #2162.

12. L. K. Hua and K. H. Look, *Theory of harmonic functions in classical domains*, Sci. Sinica (1958), 1031–1094. MR **22** #11148.

13. A. Korányi and J. A. Wolf, *Realization of hermitian symmetric spaces as generalized half-planes*, Ann. of Math. (2) **81** (1965), 265–288. MR **36** #4980.

14. R.-Q. Lu, *Harmonic functions in a class of nonsymmetric transitive domains*, Acta Math. Sinica **15** (1965), 614–650= Chinese Math. Acta **7** (1965), 339–377. MR **33** #5940.

15. Y. Matsushima, *Differentiable manifolds*, Marcel Dekker, New York, 1972.

16. J. Mitchell, *Summability methods on matrix spaces*, Canad. J. Math. **13** (1961), 63–77. MR **27** #504.

17. H. Weyl, *Harmonics on homogeneous manifolds*, Ann. of Math. **35** (1934), 486–499. MR **27** #504.

18. A. Zygmund, *Trigonometric series*. Vol. II, 2nd rev. ed., Cambridge Univ. Press, New York, 1959. MR **21** #6498.

University of Michigan

State University of New York at Buffalo

Proceedings of Symposia in Pure Mathematics
Volume 27, 1975

NONCOMPACT RIEMANNIAN MANIFOLDS ADMITTING A TRANSITIVE GROUP OF CONFORMORPHISMS*

MORIO OBATA

The following two theorems are proved.

THEOREM 1. *Let M be a connected noncompact riemannian n-manifold, $n \geqq 3$. If the largest connected group $C_0(M)$ of conformorphisms of M is essential and transitive, then M is conformorphic to a euclidean n-space E^n with standard metric.*

THEOREM 2. *Let M be a connected riemannian n-manifold, $n \geqq 3$. If the largest connected group $I_0(M)$ of isometries of M is transitive, and if $C_0(M) \neq I_0(M)$, then M is isometric to either a euclidean n-space E^n or a euclidean n-sphere S^n with standard metric.*

Theorem 2 has been proved by Goldberg and Kobayashi in the compact case.

TOKYO METROPOLITAN UNIVERSITY

AMS (MOS) subject classifications (1970). Primary 53Axx, 53A30.
*The details will appear in Tôhoku Math. J., Vol. 25

Proceedings of Symposia in Pure Mathematics
Volume 27, 1975

CRITICAL SETS OF ISOMETRIES

V. OZOLS

Let M be a connected C^∞ manifold with Riemannian inner product $\langle \cdots, \cdots \rangle$ and distance ρ. Let $I(M)$ be the full isometry group, and for each $x \in M$ let $I_x(M)$ be the isotropy subgroup of $I(M)$ at x.

For each $g \in I(M)$ one has the following associated "critical" subsets of M:

(i) $\mathrm{Fix}(g) = \{x | gx = x\}$,

(ii) $\mathrm{Crit}(g) = \{x | x$ is a critical point of $x \mapsto \rho^2(x, gx)\}$,

(iii) $\mathrm{Pres}(g) = \{x | g$ preserves some minimizing geodesic from x to $gx\}$.

("Preserves" means there is a number a such that $g\gamma(t) = \gamma(t + a)$ for all t, where t is proportional to arc length.) Similarly, if X is a Killing vector field on M, one has:

(i) $\mathrm{Zero}(X) = \{x | X_x = 0\}$.

(ii) $\mathrm{Crit}(|X|^2) = \{x | x$ is a critical point of $x \mapsto \langle X_x, X_x \rangle\}$.

Zero(X) has been studied by a number of people (see [1], [2], [6]). Their main results are relations between characteristic numbers of M and those of Zero(X). Some of these results are also valid for Fix(g).

Crit(g) is only defined when the function $x \mapsto \rho^2(x, gx)$ is differentiable, and this is only known if gx does not lie in the cut locus of x for any $x \in M$ (one needs the squares in $\rho^2(x, gx)$ resp. $|X|^2$ to get differentiability on Fix(g) resp., Zero(X)). If g satisfies this cut locus condition then it is called "of small displacement," and then one can use the first variation formula to prove that Crit(g) = Pres(g). Similarly, one can easily show that Crit($|X|^2$) is the set of points lying on geodesic orbits of the 1-parameter group g_t associated to X. Then it is evident that Crit ($|X|^2$) $\subset$ Pres(g_t) for all t.

AMS (MOS) subject classifications (1970). Primary 53C35, 53C30; Secondary 53C70.

By using the period bounding lemma of ordinary differential equations, one can prove:

THEOREM [8]. *Let M be a compact Riemannian manifold with a Killing vector field X and its associated 1-parameter group g_t. Then there is a number $a > 0$ such that* $\mathrm{Crit}(|X|^2) = \mathrm{Pres}(g_t)$ *for all* $0 < |t| < a$.

This equality is false in general for large t, and the geometric significance of the largest bound a for which equality holds is not known.

Suppose now that $M = G/K$ is a normal Riemannian homogeneous space, $\boldsymbol{g}, \boldsymbol{k}$ are the Lie algebras of G, K resp., and $\boldsymbol{g} = \boldsymbol{k} + \boldsymbol{m}$ is an $\mathrm{ad}(K)$-invariant splitting of g. Identify $\boldsymbol{m}$ with T_xM (where $x \equiv e \cdot K$) as usual, giving $\boldsymbol{m}$ the $\mathrm{ad}(K)$-invariant inner product coming from $\langle \cdots, \cdots \rangle$ on T_xM. Then:

PROPOSITION [9]. *Let $g \in G$. Then $x \in \mathrm{Pres}(g)$ if and only if there is $Y \in \boldsymbol{m}$ of minimal length such that:*

(i) $g = (\exp Y) \cdot k$ *for some* $k \in K$.

(ii) $\mathrm{ad}(k) Y = Y$.

COROLLARY. $\mathrm{Pres}(g) = \{h^{-1} x \mid hgh^{-1} = (\exp Y_h) k_h$ *as above*$\}$.

In this generality no further results are known. One must assume M is a symmetric space. Let M be a compact symmetric space, $s_x \in I_x(M)$ the geodesic symmetry at x, and $\sigma = \mathrm{ad}(s_x)$ the involution of G and $\boldsymbol{g}$. Since $I(M)$ is a compact Lie group, it has certain abelian subgroups $S \subset I(M)$ characterized by the properties:

(i) there is $g \in S$ whose powers are dense in S (such a g is called a *generator* of S).

(ii) S has finite index in its normalizer.

These subgroups need not be either connected or of maximal rank if $I(M)$ is not connected, but they include all the maximal tori of $I(M)$. J. de Siebenthal [11] denotes these groups by $T^{(h)}$ so we will call them $T^{(h)}$-subgroups.

LEMMA [9]. *Let M be a compact symmetric space, and $g \in I(M)$. Suppose there is $Y \in \boldsymbol{m}$ of minimal length such that $g \in (\exp Y)k$ for some $k \in K$ such that* $\mathrm{ad}(k)Y = Y$. *Then there is a $T^{(h)}$-subgroup $S \subset I(M)$ such that:*

(a) $g \in S$,

(b) $\sigma S^0 = S^0$,

(c) $S \cdot x = S^0 \cdot x$.

Using this lemma we can prove the following

PROPOSITION. *Let M be a compact symmetric space, and $g \in I(M)$. Then there is a set $\{g_\alpha\}$ of generators of the $T^{(h)}$-subgroups of $I(M)$ containing g such that* $\mathrm{Pres}(g) \subset \bigcup_\alpha \mathrm{Pres}(g_\alpha)$.

THEOREM. *Let M be a symmetric space of compact type, and $g \in I(M)$ a generator of a $T^{(h)}$-subgroup S. Then* $\mathrm{Pres}(g) = \bigcup_{i=1}^n S^0 \cdot x_i$, *where $\{x_i\}_{i=1}^n$ is a set of representatives of the components of* $\mathrm{Pres}(g)$, *and S^0 is the identity component of S.*

The proof makes essential use of both the finite index and the abelian conditions on S. Combining these results, one obtains:

THEOREM. *Let M be a symmetric space of compact type, and $g \in I(M)$. Then* $\operatorname{Pres}(g) = \bigcup_{i=1}^{n} Z^0(g) \cdot x_i$, *where $Z^0(g)$ is the identity component of the centralizer of g, and $\{x_i\}_{i=1}^{n}$ is a set of representatives of the components of* $\operatorname{Pres}(g)$.

By somewhat different and simpler methods on the Lie algebra level one obtains the same results for symmetric spaces of noncompact and Euclidean types.

Application to Clifford translations. A Clifford translation of a metric space is an isometry g which moves each point the same distance as every other point. An argument involving the first variation of arc length allows one to prove the following geodesic characterization of Clifford translations (in arbitrary Riemannian manifolds):

THEOREM [9]. *Let M be a C^∞ Riemannian manifold and $g \in I(M)$. The following are equivalent:*

(i) *g is a Clifford translation;*

(ii) *for each $x \in M$ there is a minimizing geodesic from x to gx preserved by g;*

(iii) *for each $x \in M$, every minimizing geodesic from x to gx is preserved by g.*

Thus, g is a Clifford translation exactly when $M = \operatorname{Pres}(g)$. The statement of the theorem makes sense in the G-spaces of Busemann and it would be interesting to know whether it is also true in that generality.

In symmetric spaces of compact type we see from the above results that if g is Clifford then $M = \bigcup Z^0(g) \cdot x_i$, so in fact $M = Z^0(g) \cdot x$ for any $x \in M$. The converse is easy to prove, so we have:

THEOREM [9]. *Let M be a symmetric space of compact type. Then $g \in I(M)$ is a Clifford translation if and only if $Z^0(g)$ is transitive on M.*

There are two previous proofs of this result ([3], [14]), both using ad hoc arguments involving the classification of symmetric spaces. In fact, J. A. Wolf in [14] proved the following general result:

THEOREM. *Let Γ be a properly discontinuous group of isometries acting freely on a simply connected symmetric space M. Then the following are equivalent:*

(i) *Γ consists of Clifford translations;*

(ii) *M/Γ is homogeneous;*

(iii) *$Z(\Gamma)$ is transitive on M.*

So far it has not been possible to extend the methods of our proof beyond the case where Γ is cyclic.

Finally, it should be remarked that the arguments in [7] suffice to prove that if M is symmetric of noncompact type, then $\operatorname{Crit}(g)$ is always defined and equals $Z^0(g) \cdot x$. The proof shows, in fact, that if g is a Clifford translation then $\exp \boldsymbol{m} \subset Z^0(g)$, so the centralizer of any set of Clifford translations is transitive. There are, however, no nontrivial Clifford translations of symmetric spaces of noncompact type, so this result is vacuous.

REFERENCES

1. P. Baum and J. Cheeger, *Infinitesimal isometries and Pontryagin numbers*, Topology **8** (1969), 173–193. MR **38** #6627.

2. R. Bott, *Vector fields and characteristic numbers*, Michigan Math. J. **14** (1967), 231–244. MR **35** #2297.

3. H. Freudenthal, *Clifford-Wolf-Isometrien symmetrischer Räume*, Math. Ann. **150** (1963), 136–149. MR **27** #693.

4. R. Hermann, C-W *cell decompositions of symmetric homogeneous spaces*, Bull. Amer. Math. Soc. **66** (1960), 126–128. MR **22** #12163.

5. ———, *Totally geodesic orbits of groups of isometries*, Nederl. Akad. Wetensch. Proc. Ser. A **65** = Indag. Math. **24** (1962), 291–298. MR **25** #2554.

6. S. Kobayashi, *Fixed points of isometries*, Nagoya Math. J. **13** (1958), 63–68. MR **21** #2276.

7. V. Ozols, *Critical points of the displacement function of an isometry*, J. Differential Geometry **3** (1969), 411–432. MR **42** #1010.

8. ———, *Critical points of the length of a Killing vector field*, J. Differential Geometry **7** (1972), 143–148.

9. ———, *Clifford translations of symmetric spaces*, Proc. Amer. Math. Soc. **44** (1974), 169–175.

10. G. Segal, *The representation ring of a compact Lie group*, Inst. Hautes Études Sci. Publ. Math. No. 34 (1968), 113–128. MR **40** #1529.

11. J. de Siebenthal, *Sur les groupes de Lie compacts non connexes*, Comment. Math. Helv. **31** (1956), 41–89. MR **20** #926.

12. G. Vincent, *Les groupes linéaires finis sans points fixes*, Comment. Math. Helv. **20** (1947), 117–171. MR **9**, 131.

13. J. A. Wolf, *Vincent's conjecture on Clifford translations of the sphere*, Comment. Math. Helv. **36** (1961), 33–41. MR **25** #532.

14. ———, *Locally symmetric homogeneous spaces*, Comment. Math. Helv. **37** (1962/63), 65–101. MR **26** #5522.

15. ———, *Spaces of constant curvature*, 2nd ed., 1972.

UNIVERSITY OF WASHINGTON

Proceedings of Symposia in Pure Mathematics
Volume 27, 1975

PARTIAL SPIN STRUCTURES AND INDUCED REPRESENTATIONS OF LIE GROUPS

JOSEPH A. WOLF

In this paper I will sketch a geometric setting for induced representations. So far, the setting has been useful in the study of unitary representations of reductive Lie groups. The square integrable representations of those groups can be obtained by examining geometrically defined differential operators on appropriate vector bundles. The other representations involved in harmonic analysis on the group are then constructed by inducing square integrable representations of certain subgroups. The setting can be viewed as a sort of measurable version of variation of certain types of geometric structures.

1. Measurable families. We start with the notion of *measurable family of C^∞ manifolds*. This consists of (i) a locally trivial Borel fibre space $p: X \to Z$ where X and Z are analytic Borel spaces, (ii) the structure of C^∞ n-manifold on each fibre $Y_z = p^{-1}(z)$, and the compatibility condition (iii) X induces the intrinsic Borel structure on each Y_z. Then *partially C^∞ bundle* over X means a locally trivial Borel fibre space $\mathscr{W} \to X$ such that the $\mathscr{W}|_{Y_z} \to Y_z$ all are C^∞ bundles with the same structure group, and *partially C^∞ section* means a Borel section C^∞ over each Y_z.

EXAMPLE 1. *Partially C^∞ function* on X means a Borel function C^∞ on each Y_z.

EXAMPLE 2. The *partial tangent bundle* $\mathscr{T} \to X$ is the union of the tangent bundles $\mathscr{T}_z \to Y_z$, with Borel structure as follows. A partially C^∞ function $f: X \to R$ has *differential* $df: \mathscr{T} \to R$ given by $(df)|_{\mathscr{T}_z} = d(f|_{\mathscr{T}_z})$, and $\mathscr{T}$ has Borel structure defined by all these df. Then $\mathscr{T} \to X$ is a partially C^∞ vector bundle; its partially C^∞

AMS (*MOS*) *subject classifications* (1970). Primary 22E30, 43A65, 43A85, 53C35; Secondary 10D20, 15A66, 17B20, 53C99.

sections are the Borel families of C^∞ vector fields on the Y_z.

Let $\mathfrak{S}$ be a geometric structure, e.g. riemannian. By *measurable family of $\mathfrak{S}$-manifolds* we mean a measurable family $p: X \to Z$ of C^∞ manifolds, and a collection of partially C^∞ bundles and sections over X whose restrictions to each Y_z define an $\mathfrak{S}$-structure there. For example, a measurable family of riemannian manifolds comes from the partial tangent bundle $\mathscr{T} \to X$ and an appropriate partially C^∞ section of $\mathscr{T}^* \otimes \mathscr{T}^*$.

2. Example: partially harmonic spinors. Fix a measurable family $p: X \to Z$ of oriented riemannian n-manifolds and a Lie group homomorphism $\alpha: U \to SO(n)$ that factors $U \to^{\tilde{\alpha}} \mathrm{Spin}(n) \to SO(n)$. As above, differentials of partially C^∞ functions specify the *partial oriented orthonormal frame bundle* $\pi: \mathscr{F} \to X$. Suppose that we have a partially C^∞ principal U-bundle $\pi_U: \mathscr{F}_U \to X$, a bundle map $\bar{\alpha}: \mathscr{F}_U \to \mathscr{F}$ given by α on each fibre, and a partially C^∞ connection on $\mathscr{F}_U$ whose Y_z-restrictions map under $\bar{\alpha}$ to the riemannian connections on the $\mathscr{F}|_{Y_z}$.

Let μ be a finite-dimensional unitary representation of U, say on V_μ. We have the associated partially C^∞ vector bundle $\mathscr{V}_\mu = \mathscr{F}_U \times_U V_\mu \to X$. If s is the spin representation of Spin(n) this gives the *partial spin bundle* $\mathscr{S} = \mathscr{V}_{s\cdot\alpha} \to X$ and the bundle $\mathscr{S} \otimes \mathscr{V}_\mu \to X$ of $\mathscr{V}_\mu$-valued partial spinors. These bundles are hermitian.

Let dz be a positive σ-finite Borel measure on Z. If γ, δ are Borel sections of $\mathscr{S} \otimes \mathscr{V}_\mu$ their global inner product is $\langle\gamma, \delta\rangle = \int_Z(\int_{Y_z}\langle\gamma_y, \delta_y\rangle\, dy)\, dz$. This defines a Hilbert space $L_2(\mathscr{S} \otimes \mathscr{V}_\mu) = \{\gamma$ Borel section of $\mathscr{S} \otimes \mathscr{V}_\mu: \langle\gamma, \gamma\rangle < \infty\}$, the *square integrable $\mathscr{V}_\mu$-valued partial spinors* on X. Evidently it is the direct integral $\int_Z L_2((\mathscr{S} \otimes \mathscr{V}_\mu)|_{Y_z})\, dz$ of the spaces of square integrable sections over the Y_z.

If n is even, $\mathscr{S}$ splits as direct sum of the partial half-spin bundles $\mathscr{S}^\pm$, and we have the usual Dirac operators $D_z^\pm: C^\infty(\mathscr{S}^\pm \otimes \mathscr{V}_\mu|_{Y_z}) \to C^\infty(\mathscr{S}^\mp \otimes \mathscr{V}_\mu|_{Y_z})$ on sections C^∞ over Y_z. These fit together to form the *partial Dirac operators* $D^\pm$: $C^\infty(\mathscr{S}^\pm \otimes \mathscr{V}_\mu) \to C^\infty(\mathscr{S}^\mp \otimes \mathscr{V}_\mu)$ on partially C^∞ $\mathscr{V}_\mu$-valued spinors. If the riemannian manifold Y_z is complete then $D_z = D_z^+ \oplus D_z^-$ is essentially selfadjoint on $L_2(\mathscr{S} \otimes \mathscr{V}_\mu|_{Y_z})$ from domain consisting of the compactly supported C^∞ sections. Thus, if Y_z is complete a.e. (Z, dz), then $D = D^+ \oplus D^-$ is essentially selfadjoint, and we have Hilbert spaces

$$H_2(\mathscr{V}_\mu) = H_2^+(\mathscr{V}_\mu) \oplus H_2^-(\mathscr{V}_\mu)$$

where $H_2^\pm(\mathscr{V}_\mu) = \{\omega \in L_2(\mathscr{S}^\pm \otimes \mathscr{V}_\mu): D^\pm\omega = 0\} = \int_Z H_2^\pm(\mathscr{V}_\mu|_{Y_z})\, dz$. These spaces are closed in L_2 and consist of partially C^∞ sections. We refer to the elements of $H_2(\mathscr{V}_\mu)$ as the *square integrable $\mathscr{V}_\mu$-valued partially harmonic spinors* on X.

3. Application to symmetric spaces. Let G be a reductive Lie group, $\mathfrak{g}$ its Lie algebra. We assume that ad(g) is an inner automorphism on $\mathfrak{g}_C$ whenever $g \in G$. G^0 denotes the identity component of G, $Z_G(G^0)$ its G-centralizer, and $G^\dagger = Z_G(G^0)G^0$. We assume that $Z_G(G^0)$ has a closed abelian subgroup Z with G/ZG^0 finite.

Let H be a Cartan subgroup of G, θ a Cartan involution with $\theta(H) = H$, and K the fixed point set of θ. Split $H = T \times A$ where $T = H \cap K$. Choose a positive

$\mathfrak{a}$-root system $\Sigma_\mathfrak{a}^+$ on $\mathfrak{g}$, let $\mathfrak{n}$ be the sum of the negative $\mathfrak{a}$-root spaces, and define $N = \exp(\mathfrak{n})$. Then the normalizer of N in G is a cuspidal parabolic subgroup $P = MAN$ of G, where $Z_G(A) = M \times A$. The H-series of unitary equivalence classes of representations of G consists of the $[\pi_{\eta,\sigma}] = [\mathrm{Ind}_{P\uparrow G}(\eta \otimes e^{i\sigma})]$ where $[\eta]$ is a square integrable representation class of M and $\sigma \in \mathfrak{a}^*$. Plancherel measure on the unitary dual of G is concentrated on the union of the various H-series.

Choose a positive $\mathfrak{t}_C$-root system $\Sigma_\mathfrak{t}^+$ on $\mathfrak{m}_C$ and let $2\rho_\mathfrak{t} = \sum_{\beta\in\Sigma^+}\beta$. If $\nu \in i\mathfrak{t}^*$ is integral and $\mathfrak{m}_C$-regular let $[\eta_\nu]$ denote the corresponding square integrable representation class of M^0; its restriction to $Z_{M^0} = Z_M(M^0) \cap M^0$ is $e^{\nu-\rho_\mathfrak{t}}$. If χ is an irreducible unitary representation of $Z_M(M^0)$ with same Z_{M^0}-restriction then $\chi \otimes \eta$ is a square integrable representation of $M = Z_M(M^0)M^0$, and $\eta_{\chi,\nu} = \mathrm{Ind}_{M^\dagger\uparrow M}(\chi \otimes \eta_\nu)$ is an irreducible square integrable representation of M. All the unitary equivalence classes of irreducible square integrable representations of M are of that form $[\eta_{\chi,\nu}]$, so the H-series of G consists of the $[\pi_{\chi,\nu,\sigma}] = [\pi_{\eta_{\chi,\nu}\sigma}]$.

Retain the notation above, define $U = K \cap M^\dagger$, and consider $X = G/UAN \to G/M^\dagger AN = Z$. Replacing G by a Z_2-extension if necessary, the linear isotropy action of U on the tangent space of $M^\dagger/U$ factors through $\mathrm{Spin}(n)$, $n = \dim M^\dagger/U$, and we have the setup of § 2 with $\mathscr{F}_U = G/AN \to G/UAN = X$. If μ is a finite-dimensional unitary representation of U, say on V_μ, we view $\mathscr{V}_\mu \to G/UAN = X$ as the G-homogeneous bundle for the representation of UAN given by $uan \to e^{\rho_\mathfrak{a}+i\sigma}(a)\mu(u)$ where $2\rho_\mathfrak{a} = \sum_{\alpha\in\Sigma_\mathfrak{a}^+}(\dim \mathfrak{g}^\alpha)\alpha$. Then G acts on $H_{\frac{1}{2}}^{\pm}(\mathscr{V}_\mu)$ by a unitary representation $\pi_{\mu,\sigma}^{\pm}$.

T is a Cartan subgroup of U; let $\rho_{\mathfrak{t},\mathfrak{u}}$ be half the sum of the positive $\mathfrak{t}_C$-roots of μ_C. $U = Z_M(M^0)U^0$. If μ is irreducible now $\mu = \chi \otimes \mu^0$ and we express the highest weight of μ^0 as $\nu - \rho_\mathfrak{t} + \rho_{\mathfrak{t},\mathfrak{u}}$. Suppose that $\nu + \rho_\mathfrak{t}$ is $\mathfrak{m}$-regular and $\sigma \in \mathfrak{a}^*$. Then the distribution characters of $\pi_{\mu,\sigma}^{\pm}$ and $\pi_{\chi,\nu+\rho_\mathfrak{t},\sigma}$ are related by a difference formula

$$\Theta_{\pi_{\mu,\sigma}^+} - \Theta_{\pi_{\mu,\sigma}^-} = \Theta_{\pi_{\chi,\nu+\rho_\mathfrak{t},\sigma}}.$$

In particular, $\pi_{\chi,\nu+\rho_\mathfrak{t},\sigma}$ occurs as a subrepresentation in one of $\pi_{\mu,\sigma}^{\pm}$. Incidentally, every H-series class is of this form $[\pi_{\chi,\nu+\rho_\mathfrak{t},\sigma}]$. Further, under a mild additional condition on $\nu + \rho_\mathfrak{t}$, one of the Hilbert spaces $H_{\frac{1}{2}}^{\pm}(\mathscr{V}_\mu)$ is reduced to zero, and so the other $[\pi_{\mu,\sigma}^{\mp}]$ is equal to $[\pi_{\chi,\nu+\rho_\mathfrak{t}}]$.

In summary, the action of G on the Hilbert spaces $H_{\frac{1}{2}}^{\pm}(\mathscr{V}_\mu)$ of square integrable partially harmonic spinors provides geometric realizations for a family of unitary representation classes that carries the Plancherel measure of G.

UNIVERSITY OF CALIFORNIA, BERKELEY

RELATIVITY

Proceedings of Symposia in Pure Mathematics
Volume 27, 1975

EINSTEIN-MAXWELL THEORY AND THE STRUCTURE EQUATIONS

GEORGE DEBNEY

1. Introduction. Einstein's theory of gravitation contains the field equations[1]

$$G_{\mu\nu} = -4\pi T_{\mu\nu} \tag{1}$$

where $G_{\mu\nu} \equiv R_{\mu\nu} - \frac{1}{2}g_{\mu\nu} R$ is the Einstein tensor and T is the stress-energy tensor for matter and electromagnetic fields present in the region of space-time under consideration. If $T = 0$ the equations reduce to $\mathrm{Ric}(g) = 0$, corresponding to $\nabla^2\phi = 0$ in Newtonian potential theory; the analog for $T \neq 0$ is $\nabla^2\phi = 4\pi\rho$, where ρ is a density. Usually one has additional equations on the functions represented in T so as to provide enough information to describe the problem physically and, at least in principle, to arrive at a quantitative prediction based on the metrics $\{g\}$ found as solutions. Such additional equations might be equations of state for matter or Maxwell's equations for electromagnetic contributions.

Einstein-Maxwell theory is the special case of equation (1) where there exists an electromagnetic field tensor $F_{\mu\nu} = -F_{\nu\mu}$ giving rise to the trace-free stress-energy tensor

$$T \equiv (4\pi)^{-1}(F_{\mu\sigma} F_{\nu}{}^{\sigma} - \tfrac{1}{4}g_{\mu\nu} F_{\alpha\beta} F^{\alpha\beta})\, dx^{\mu}dx^{\nu}.$$

Since $\mathrm{tr}\,(T) = 0$, equation (1) becomes

$$\mathrm{Ric}(g) = -4\pi T \tag{2}$$

AMS (MOS) subject classifications (1970). Primary 83C05, 83C20, 83C50; Secondary 53B30.

[1] $R_{\mu\nu\rho\sigma}$, $R_{\mu\nu}$, R are the components of the Riemann curvature tensor, Ricci tensor, and Ricci scalar, respectively. Greek indices refer to a natural basis (although the basis is unimportant at this stage).

and must be consistent with Maxwell's equations

(3) $\nabla_\mu F^{\mu\nu} = -4\pi J^\nu$ (J is the current 4-vector),

(4) $\nabla_{[\sigma}F_{\mu\nu]} = 0.$

In terms of the exterior derivative d and co-derivative δ these equations are $\delta F = -4\pi J$ and $dF = 0$, respectively. At this point we restrict the subsequent discussion to the "source-free" case, $J = 0$.

An example of a solution for this system possessing simple properties is the Reissner-Nordström solution [1] for a point mass $\boldsymbol{m}$ with charge $\boldsymbol{e}$. The gravitational field is

(*) $$g = (1 - 2m/r + e^2/r^2)^{-1}dr^2 - (1 - 2m/r + e^2/r^2)dt^2 + r^2 d\Omega^2,$$

where r, t are radius and time coordinates and $d\Omega^2$ is the metric for the 2-sphere S^2. This obviously reduces to the well-known Schwarzschild solution ($R_{\mu\nu} = 0$) whenever the charge $\boldsymbol{e} = 0$. The electromagnetic field tensor is essentially $F = (e/r^2)\, dr \wedge dt$.

Another well-known solution which is a generalization of that above is the so-called "charged Kerr" [2], [3]. In essence this metric is interpreted to be that of a mass $\boldsymbol{m}$ with charge $\boldsymbol{e}$, specific angular momentum $\boldsymbol{a}$, and magnetic dipole moment $\boldsymbol{ea}$:

(**) $$\begin{aligned} g = {} & \rho^2\Delta^{-1}dr^2 + \rho^2 d\theta^2 + [r^2 + a^2 + (2\,mr - e^2)\rho^{-2}a^2 \sin^2\theta] \sin^2\theta\, d\phi^2 \\ & + 2a(2mr - e^2)\rho^{-2}\sin^2\theta\, d\phi\, dt - [1 - (2mr - e^2)\rho^{-2}]\, dt^2, \end{aligned}$$

where $\Delta \equiv r^2 - 2mr + e^2 + a^2$ and $\rho^2 \equiv r^2 + a^2\cos^2\theta$. This geometry does not have the space-like spherical symmetry of (*) but does possess two Killing vectors ∂_ϕ and ∂_t, giving rise to the term "axial symmetry." The electromagnetic field is $F = d\alpha$, where $\alpha \equiv er\rho^{-2}(dt + a\sin^2\theta\, d\phi)$ is the vector potential for the electromagnetic field.

2. Geodesic and shearfree null vector fields. An important concept in general relativity over the past several years has been the study of conditions, both necessary and sufficient, for the existence of a null vector field k ($k\cdot k = 0$) which is both a geodesic and is shearfree; its existence has certain consequences in the Petrov algebraic classification of the conformal tensor. The *shear* of a null vector field $k = k_\mu\, dx^\mu$ is defined by

$$|\sigma| \equiv [\tfrac{1}{2}\nabla_{(\mu}k_{\nu)}\nabla^\mu k^\nu - (\tfrac{1}{2}\nabla_\mu k^\mu)^2]^{1/2}.$$

The best physical description of this quantity (one of Sach's "optical scalars" [4]) is in the context of some of the other optical scalars such as *divergence* and *twist*. For example, if one places two screens at different places along a geodesic null congruence so that a circle drawn on the first defines an image via the flow on the second, the image will be distorted in a number of ways, corresponding to those terms which are obtained from $\nabla_\nu k_\mu$. The shear is one of these contributions to the distortion and has a close analogy with similar quantities in fluid mechanics, where one looks at the symmetric shear matrix of an element of matter.

The Petrov classification of conformally related space-times is based on an eigenvalue problem for the conformal tensor C of Weyl. Any four-dimensional Lorentz manifold, no matter what the field equations, possesses four null eigenvector fields for the conformal tensor in the generic case. The algebraic types arise from the multiplicities possible in case of degeneracies. If multiplicities are indicated by 2,3,4 one may list the combinations:

Type I	[1,1,1,1]	nondegenerate C
	....	
Type II	[1,1,2]	
Type D	[2,2]	degenerate C
		or
Type III	[1,3]	"algebraically special"
Type N	[4]	
	....	
Conformally flat	[–]	indeterminate since $C = 0$

A number of important theorems tie the existence of a geodesic and shearfree null vector field to the algebraic classification of C and to the null eigenvectors of C. Theorem 1 below holds for the most general form of Einstein's equations.

THEOREM 1 (ROBINSON AND SCHILD [5], OTHERS). *If k is a geodesic and shearfree null vector field then it is a null eigenvector for C.*

If we consider only Ricci-flat space-times we obtain the following well-known theorem.

THEOREM 2 (GOLDBERG AND SACHS [6]). *Let* $\mathrm{Ric}(g) = 0$ ($\Rightarrow C(g) \equiv \mathrm{Riem}(g)$). *Then C is algebraically degenerate if and only if it admits a geodesic and shearfree null vector field. This field is a degenerate null eigenvector for C.*

If we now consider what sort of similar theorems we might obtain in the context of the source-free ($J = 0$) Einstein-Maxwell system equations (2) – (4) we must first examine the two-form F. In the generic case an eigenvector k for F satisfies $F_{\mu\nu} k^\nu = \alpha k_\mu$ where $\alpha \neq 0$. Hence, $k_\mu k^\mu = 0$, and two principal null directions exist in this case. If $\alpha = 0$ one finds that a single principal null direction exists for F, plus an entire null hypersurface of space-like eigendirections. The case $\alpha \neq 0$ is that for which F is *nondegenerate*; $\alpha = 0$ defines a *degenerate* F. A result of Ruse and Synge [7] states that F is degenerate if and only if $(F + iF^*)\colon F = 0$.[2] The degenerate fields have been studied extensively in this context and represent electromagnetic radiation. Certainly the system of equations (2) – (4) is considerably simplified, particularly due to the following results.

THEOREM 3 (MARIOT/ROBINSON [8]). *In any space-time admitting the source-free Maxwell equations $\delta F = 0 = dF$, the condition $(F + iF^*)\colon F = 0$ implies the (single) null eigenvector for F is geodesic and shearfree.*

THEOREM 4. *In an Einstein-Maxwell space-time admitting $\delta F = 0 = dF$, if k is a*

[2] "*" on a pair of skew indices refers to the Hodge dual (or "adjoint").

geodesic and shearfree null eigenvector for F, then k is a degenerate null eigenvector for C. (*See* [9].)

An obvious corollary to Theorems 3 and 4 tells us that all degenerate Maxwell fields F give rise to degenerate space-times through C.

COROLLARY. *Let $\delta F = 0 = dF$ and let the Einstein-Maxwell equations hold. Then $(F + iF^*): F = 0$ implies C is degenerate; furthermore F and C have a degenerate null eigenvector in common.*

One would like to have theorems for generally nonradiative Maxwell fields and their corresponding space-times. Such nondegenerate F are interpreted physically as representing a charge and/or a magnetic moment. The Reissner-Nordström and "charged-Kerr" metrics are well-known examples in this category. One problem confronting this endeavor is that Theorem 4 does not admit a general converse; i.e., a null vector k might be geodesic and shearfree and might also go to define a degenerate C, but nothing in the field equations forces it to be a null eigenvector for F as well. It was seen in one particular class of metrics [3], that to produce some solvable cases, one had to impose a certain alignment of a special given null vector k with the field F plus a geodesic condition on k; only then did a kind of "Goldberg-Sachs" situation arise to simplify the field equations.

3. The Einstein-Maxwell equations with a Killing vector field. The Reissner-Nordström and "charged-Kerr" metrics both possess two degenerate null eigenvectors for C, implying they are both type [2,2]. These same null vectors are eigenvectors for F, but are nondegenerate with respect to F since $(F + iF^*): F \neq 0$. Furthermore these vector fields are geodesic and shearfree. However, both metrics above possess other common symmetry properties as defined by Killing vector fields. The "charged-Kerr" solution (and therefore its special case, the Reissner-Nordström solution) admits two commuting Killing vector fields: one space-like, the other time-like.

Recall that a Killing vector field $\xi = \xi_\mu dx^\mu$ satisfies

$$\nabla_\nu \xi_\mu + \nabla_\mu \xi_\nu = 0 \tag{5}$$

so that $d\xi$ is a two-form. Furthermore, an integrability condition for equation (5) is

$$\nabla_\mu \nabla_\nu k_\rho = R_{\rho\nu\mu\sigma} k^\sigma. \tag{6}$$

A special kind of converse to Theorem 4, involving a nondegenerate F in an Einstein-Maxwell space-time, is possible with such an additional structure.

THEOREM 5. *Suppose F is nondegenerate and satisfies the Einstein-Maxwell equations. Let k be a null eigenvector for F, $d\xi$, and C. Then if k is degenerate with respect to C, k is geodesic and shearfree.*

PROOF OUTLINE. If a complex null tetrad frame is established, with k as one of the basis vectors, equation (6) may be examined in a particularly simple canonical form. The degeneracy in C then forces the connection coefficients corresponding to geodesy and shear to be zero. (See [9].)

4. A static generalization of the Reissner-Nordström metric. If one persists in looking at Einstein-Maxwell systems involving symmetry he is tempted to impose the condition that the Killing vector field be time-like and normal; i.e.,

$$\xi \cdot \xi < 0 \quad \text{and} \quad d\xi \wedge \xi = 0.$$

Geometrically this means that along the flow of ξ a one-parameter family of orthogonal hypersurfaces evolve, each of which is isometric with respect to the mapping defined by the flow of ξ. For example, if P and Q lie on the hypersurface $\tau = \tau_0$ and their images P' and Q' lie on $\tau = \tau_1$ (τ is a parameter of the ξ congruence) then $g(P) = g(P')$ and $g(Q) = g(Q')$. Indeed, this condition does give rise to a class of generalizations of the Reissner-Nordström metric (but not the "charged-Kerr" since the Killing field there is not normal).

The following theorem proved useful in the above context.

THEOREM 6. *Let F be a nondegenerate Maxwell field satisfying $\delta F = 0 = dF$ and C the conformal curvature tensor for the Einstein-Maxwell space-time. Let F possess a geodesic null eigenvector, k. If ξ is a Killing vector field admitted by the space, satisfying*

$$\text{(i)} \quad (d\xi : d\xi) < 0 \qquad \textit{and} \qquad \text{(ii)} \quad d\xi \wedge \xi = 0,$$

then

(a) *$d\xi$ and F have identical null eigenvectors k and l.*
(b) *l is a geodesic.*
(c) *k is shearfree $\Leftrightarrow$ l is shearfree.*
(d) *$dk \wedge k = 0 = dl \wedge l$.*
(e) *The Petrov type must be* [1,1,1,1], [2,2], *or* [−].

PROOF OUTLINE. Using a complex null tetrad basis equations (5), (6), and the Maxwell equations prove (a), (b), (d), (e). The first and second structure equations prove the coupling in (c).

Letting k and l be geodesic and shearfree under the conditions of Theorem 6 we obtain the case which is Petrov type [2, 2]. After solving the Einstein-Maxwell equations and choosing coordinates $\zeta, \bar{\zeta}, r, \sigma$ ($\zeta, \bar{\zeta}$ are complex conjugates; r, σ are real) we may write the general metric

$$g = 2r^2 \exp(2p) d\zeta\, d\bar{\zeta} + 2dr\, d\sigma - 2K\, d\sigma^2. \tag{7}$$

Here $K \equiv K_0 - c/r + \alpha\bar{\alpha}/r^2$, where K_0, c, α are constants (K_0 and c are real). Also, the function $p = p(\zeta, \bar{\zeta})$ is arbitrary up to the condition

$$\exp(-2p)\partial^2 p/\partial\zeta\, \partial\bar{\zeta} = -K_0. \tag{8}$$

The electromagnetic field F is given by

$$F = 2(\alpha + \bar{\alpha})/r^2\, d\sigma \wedge dr \tag{9}$$

and is independent of the choice of the function p. The metric on a two-manifold $g_{(2)} = \exp(2p)\, d\zeta\, d\bar{\zeta}$ with curvature $-K_0$ is actually the choice given to us by equation (8). Hence, our solution (7) involves the essentially different choices $K_0 > 0$ $K_0 < 0$, and $K_0 = 0$ with arbitrariness in the coordinate choices $\zeta, \bar{\zeta}$.

Such solutions have appeared at other times in the literature (see, for example, Robinson and Trautman [**10**], and Witten [**11**]). In fact, this case could be identified as one of the degenerate "charged Weyl/Levi-Civita" metrics.

Solutions of the Einstein-Maxwell equations falling into the [2,2] category have been fairly important in the recent past since many of these have lent themselves to useful physical interpretations. Even though it is claimed that all type [2,2] metrics which are Ricci-flat are known (see [**12**]), no such progress has been made for the electromagnetic case. In fact, the problem of finding all Einstein-Maxwell metrics of type [2,2] with the two principal null eigenvectors for both C and F being coincident is still outstanding.

5. A discussion of the structure equations for an Einstein-Maxwell space-time. In order actually to get differential equations to solve one relies heavily upon choosing an appropriate group of frames. There is no guarantee that one *can* solve the equations he derives for his particular case, but a proper choice of frames does assure the elimination of what might be called "spurious nonlinearities" along the way. Reduction of certain tensor quantities to some kind of canonical form has its obvious advantages; one major step is to choose a set of frames $\{\varepsilon^a\}$ so that $g = g_{ab}\, \varepsilon^a \otimes \varepsilon^b$ has constant coefficients g_{ab}.[3]

If $\{e_a\}$ and $\{\varepsilon^a\}$ are bases for the tangent and cotangent planes respectively, and if $\Gamma^a{}_b \equiv \Gamma^a{}_{bc}\, \varepsilon^c$ are the torsion-free connection forms compatible with the metric, the Cartán structure equations may be written

$$d\varepsilon^a = \varepsilon^b \wedge \Gamma^a{}_b, \tag{10}$$

$$d\Gamma^a{}_b + \Gamma^a{}_m \wedge \Gamma^m{}_b = \tfrac{1}{2}\mathscr{R}^a{}_b; \tag{11}$$

$\mathscr{R}^a{}_b \equiv R^a{}_{bcd}\, \varepsilon^c \wedge \varepsilon^d$ is the curvature two-form. The first integrability condition for (10) is the first Bianchi identity

$$\mathscr{R}^a{}_b \wedge \varepsilon^b = 0. \tag{12}$$

The first integrability condition for (11) is the second Bianchi identity

$$d\mathscr{R}^a{}_b = \mathscr{R}^a{}_m \wedge \Gamma^m{}_b - \Gamma^a{}_m \wedge \mathscr{R}^m{}_b. \tag{13}$$

In general equations (12) and (13) provide useful simplifications to put back into equations (10) and (11). If we define $\Gamma_{abc} \equiv g_{am}\Gamma^m{}_{bc}$, the "compatibility" of the connection gives $\Gamma_{abc} = -\Gamma_{bac}$. Since the dimension is 4 this leaves 24 components of the connection, precisely the number of coefficients $\Gamma^a{}_{[bc]}$ in equation (10) $(\Gamma^a{}_{[bc]} \equiv \frac{1}{2}\Gamma^a{}_{bc} - \frac{1}{2}\Gamma^a{}_{cb})$. Hence, for a given frame, the connection forms are completely determined. In a *complex null* frame ("complex tetrad" of Appendix A)

[3] Latin indices $a,b,\ldots$ here and throughout refer to components with respect to a frame.

complex conjugation reduces equations (10) to three two-form equations and equation (11) also to three two-form equations.

For the Einstein-Maxwell system $\mathrm{tr}\,(T) = 0$ so that the Ricci scalar is zero. The conformal tensor C must then have the general form $C(g) = \mathrm{Riem}(g) +$ [terms in $g \times \mathrm{Ric}(g)$]. One can prove that in this case the following "anti-self-dual" quantities are equal:

$$(C + iC^*) = (\mathrm{Riem} + i\,\mathrm{Riem}^*).$$

The Petrov classification *in general* only classifies $(C + iC^*)$. Hence one is able to identify components of the two-form $\mathscr{R}^a{}_b$ in equation (11) with desired canonical components of $(C + iC^*)$. Of course equation (11) contains more on its right-hand side than just the five complex Riemannian scalars of the Petrov classification since this equation is not anti-self-dual. In fact, the remaining terms of $\mathscr{R}^a{}_b$ are just coefficients of the Ricci tensor so that equation (11) contains the Einstein field equations as well. In summary, the structure equations (11) contain only the five Riemannian scalars of the Petrov classification and the six (as a result of complexification) components of the Ricci tensor.

If one takes advantage of the theorem which says that for a nondegenerate field F the null eigenvectors for F are also null eigenvectors for T, he may align his frame accordingly and reduce all components of the Ricci tensor to zero except for two (even those two are negatives of one another). But therein lies the difference from Ricci-flat!

Appendix A

For a Lorentz 4-manifold M with signature $(+++-)$ let $\{e_a | a = 1, \dots, 4\}$ and $\{\varepsilon^a | a = 1, \dots, 4\}$ be a local orthonormal tangent basis and its dual, respectively. This demands that the metric $g = g_{ab}\, \varepsilon^a \otimes \varepsilon^b$ be of the form diag $(1,1,1,-1)$ over a neighborhood. The group preserving this form of g_{ab} is the Lorentz group on the frame $\{e_a\}$. For what must be regarded as convenience at later stages the frames are "complexified" by the process $(i = \sqrt{-1})$

$$\text{(A.1)} \qquad \begin{array}{ll} \hat{e}_1 \equiv (\sqrt{2})^{-1}(e_1 + ie_2), & \hat{e}_2 \equiv (\sqrt{2})^{-1}(e_1 - ie_2), \\ \hat{e}_3 \equiv (\sqrt{2})^{-1}(e_3 + e_4), & \hat{e}_4 \equiv (\sqrt{2})^{-1}(e_3 - e_4); \end{array}$$

if bar $(\overline{\ \ })$ denotes formal complex conjugation and "real" quantities are those which are identical with their complex conjugates we see that $(\hat{e}_2)^- = \hat{e}_1$ and $\hat{e}_3$, $\hat{e}_4$ are real. If $\hat{g}(\hat{e}_a, \hat{e}_b) \equiv \hat{g}_{ab}$ is defined through formal expansion of $\hat{g}_{ab} = \hat{e}_a \cdot \hat{e}_b$ in terms of the earlier g_{ab} we find that the metric $\hat{g}$ has the form

$$\text{(A.2)} \qquad (\hat{g}_{ab}) = \begin{bmatrix} 0 & 1 & 0 & 0 \\ 1 & 0 & 0 & 0 \\ 0 & 0 & 0 & 1 \\ 0 & 0 & 1 & 0 \end{bmatrix}$$

The duals $\{\hat{\varepsilon}^a\}$ are defined as usual so that the original (real) metric is written simply as $g = 2\varepsilon^1 \otimes \varepsilon^2 + 2\varepsilon^3 \otimes \varepsilon^4$. (We remove the " ^ ", work with the complex Lorentz group, and consider all geometric objects in terms of this complexifie dset of frames.) Notice that since $e_i \cdot e_i = 0$ (no sum on i) each vector basis field is formally null. The name commonly used for such a scheme is the "complex null tetrad." Briefly, the algorithms for raising and lowering indices are $1 \leftrightarrow 2$, $3 \leftrightarrow 4$ and $(\bar{1}) = 2$, $(\bar{3}) = 3$, etc. for complex conjugation on components of objects and their indices; e.g., $(\overline{T^{13}}) = T_{23}$ for a "real" tensor $T = (\bar{T})$, and $T^2{}_3 = T_1{}^4$.

For the local fields of two-forms $\wedge^2(M)$ the following choice is made for a basis (in terms of the complex null tetrad):

$$\text{(A.3)} \qquad \begin{aligned} \varepsilon^{\mathrm{I}} &\equiv 2\varepsilon^4 \wedge \varepsilon^2 & \varepsilon^{\mathrm{IV}} &\equiv \overline{\varepsilon^{\mathrm{I}}} \\ \varepsilon^{\mathrm{II}} &\equiv 2(\varepsilon^1 \wedge \varepsilon^2 + \varepsilon^3 \wedge \varepsilon^4) & \varepsilon^{\mathrm{V}} &\equiv \overline{\varepsilon^{\mathrm{II}}} \\ \varepsilon^{\mathrm{III}} &\equiv 2\varepsilon^3 \wedge \varepsilon^1 & \varepsilon^{\mathrm{VI}} &\equiv \overline{\varepsilon^{\mathrm{III}}}; \end{aligned}$$

hence, any two-form $B = B_{ab}\, \varepsilon^a \wedge \varepsilon^b = B_A \varepsilon^A$ ($A = \mathrm{I}, \ldots, \mathrm{VI}$). Let C be the Weyl conformal curvature tensor; its components satisfy

$$C_{abcd} = R_{abcd} + g_{a[c}R_{d]b} - g_{b[c}R_{d]a} - \tfrac{1}{6}(g_{ac}\, g_{bd} - g_{ad}\, g_{bc})R$$

where R_{abcd}, R_{ab}, R are components of the Riemann tensor, Ricci tensor, and Ricci scalar, respectively. (Note that Ricci-flat $\Rightarrow C =$ Riem.)

The Weyl tensor (and the Riemann tensor) may be expressed by

$$C = C_{AB}\, \varepsilon^A \otimes \varepsilon^B = C_{BA}\, \varepsilon^A \otimes \varepsilon^B.$$

By expansion of the Weyl tensor and making use of its symmetries one can prove that (see **[13]**)

$$\text{(A.4)} \qquad (C_{AB}) = \left[\begin{array}{c|c} Q & 0 \\ \hline 0 & \bar{Q} \end{array}\right], \qquad (C_{AB} + i\overset{*}{C}_{AB}) = \left[\begin{array}{c|c} 2Q & 0 \\ \hline 0 & 0 \end{array}\right],$$

where Q is a 3×3 symmetric matrix and $\bar{Q}$ is its complex conjugate. If one makes the further notational changes $C^{(5)} \equiv 2C_{\mathrm{I\,I}}$, $C^{(4)} \equiv 2C_{\mathrm{I\,II}}$, $C^{(3)} \equiv 2C_{\mathrm{I\,III}}$, $C^{(2)} \equiv 2C_{\mathrm{II\,III}}$, and $C^{(1)} \equiv 2C_{\mathrm{III\,III}}$, it can be shown that $C_{\mathrm{II\,II}} = C_{\mathrm{I\,III}}$ so that

$$2Q = \begin{bmatrix} C^{(5)} & C^{(4)} & C^{(3)} \\ C^{(4)} & C^{(3)} & C^{(2)} \\ C^{(3)} & C^{(2)} & C^{(1)} \end{bmatrix}.$$

The "conformal scalars" $C^{(i)}$ in Q determine the Petrov type as the following examples show:

Conformal Scalars	Null Eigenvector(s)	Petrov Type
$C^{(5)} = C^{(1)} = 0; C^{(4)}\, C^{(2)} \neq 0$	e_3, e_4	[1, 1, 1, 1]
$C^{(5)} = C^{(4)} = 0; C^{(3)}\, C^{(2)} \neq 0$	e_4 (degenerate)	[1, 1, 2]
$C^{(5)} = C^{(4)} = 0 = C^{(1)} = C^{(2)}; C^{(3)} \neq 0$	e_3, e_4 (both degenerate)	[2, 2]
$C^{(5)} = C^{(4)} = C^{(3)} = 0; C^{(2)} \neq 0$	e_4 (degenerate)	[1, 3]
$C^{(5)} = C^{(4)} = C^{(3)} = C^{(2)} = 0; C^{(1)} \neq 0$	e_4 (degenerate)	[4]
All $C^{(i)} = 0$ ($\Leftrightarrow C = 0$)	indeterminate	[–]

Without regard to any field equations the second structure equations take the form ($\Gamma_A \equiv \Gamma_{Ac}\,\varepsilon^c$ are the connection one-forms)

$$\text{(A.5)}\quad \begin{aligned} d\Gamma_{\mathrm{I}} + 2\Gamma_{\mathrm{I}} \wedge \Gamma_{\mathrm{II}} &= \tfrac{1}{4}[C^{(5)}\varepsilon^{\mathrm{I}} + C^{(4)}\varepsilon^{\mathrm{II}} + (C^{(3)} - \tfrac{1}{6}R)\varepsilon^{\mathrm{III}} \\ &\qquad - R_{44}\varepsilon^{\mathrm{IV}} - R_{42}\varepsilon^{\mathrm{V}} - R_{22}\varepsilon^{\mathrm{VI}}], \\ d\Gamma_{\mathrm{II}} + \Gamma_{\mathrm{I}} \wedge \Gamma_{\mathrm{III}} &= \tfrac{1}{4}[C^{(4)}\varepsilon^{\mathrm{I}} + (C^{(3)} - \tfrac{1}{6}R)\varepsilon^{\mathrm{II}} + C^{(2)}\varepsilon^{\mathrm{III}} \\ &\qquad - R_{41}\varepsilon^{\mathrm{IV}} - \tfrac{1}{2}(R_{12} - R_{34})\varepsilon^{\mathrm{V}} + R_{32}\varepsilon^{\mathrm{VI}}], \\ d\Gamma_{\mathrm{III}} + 2\Gamma_{\mathrm{II}} \wedge \Gamma_{\mathrm{III}} &= \tfrac{1}{4}[(C^{(3)} - \tfrac{1}{6}R)\varepsilon^{\mathrm{I}} + C^{(2)}\varepsilon^{\mathrm{II}} + C^{(1)}\varepsilon^{\mathrm{III}} \\ &\qquad - R_{11}\varepsilon^{\mathrm{IV}} + R_{31}\varepsilon^{\mathrm{V}} - R_{33}\varepsilon^{\mathrm{VI}}]. \end{aligned}$$

Suppose $F = F_A\varepsilon^A$ is an electromagnetic field which is nondegenerate; i.e., $F_A\,F^A \neq 0$ or $F^*_A\,F^A \neq 0$. Then a theorem based on canonical forms of F (see **[14]**) implies there exists an alignment of e_3 and e_4 (as null eigenvectors) so that $F = F_{\mathrm{II}}\,\varepsilon^{\mathrm{II}} + F_{\mathrm{V}}\,\varepsilon^{\mathrm{V}}$. In this frame define $F^{(-)} = (F + iF^*)$; then $F^{(-)} = 2F_{\mathrm{II}}\,\varepsilon^{\mathrm{II}}$. The Maxwell source-free equations become $dF^{(-)} = 0$. Hence, $dF_{\mathrm{II}} \wedge \varepsilon^{\mathrm{II}} + F_{\mathrm{II}}\,d\varepsilon^{\mathrm{II}} = 0$ so that the components are given by[4]

$$\text{(A.6)}\quad \begin{aligned} e_1(F_{\mathrm{II}}) &= -\,2F_{\mathrm{II}}\,\Gamma_{\mathrm{III}\;4}, \\ e_2(F_{\mathrm{II}}) &= -\,2F_{\mathrm{II}}\,\Gamma_{\mathrm{I}\;3}, \\ e_3(F_{\mathrm{II}}) &= 2F_{\mathrm{II}}\,\Gamma_{\mathrm{III}\;2}, \\ e_4(F_{\mathrm{II}}) &= 2F_{\mathrm{II}}\Gamma_{\mathrm{I}\;1}. \end{aligned} \qquad [F_{\mathrm{II}} = \tfrac{1}{2}(F_{12} + F_{34})]$$

Under this same alignment of basis the only nonzero components of the Maxwell stress-energy tensor T are $T_{12} = -\,T_{34} = (4\pi)^{-1}(|F_{12}|^2 + |F_{34}|^2)$.

As an example, consider an Einstein-Maxwell space-time under the following restrictions: (i) F is nondegenerate, (ii) F and C have precisely the same null eigenvectors, (iii) $C^{(3)} \neq 0$, all other $C^{(i)} = 0$ and $F^{(-)} = 2F_{\mathrm{II}}\,\varepsilon^{\mathrm{II}}$. (This implies the Petrov type is [2, 2].) Since F_{12} is pure imaginary and F_{34} is real, introduce the notation $F_{12} = ib$ and $F_{34} = a$. Maxwell's equations (A.6) become

$$\text{(A.7)}\quad \begin{aligned} e_1(a + ib) &= -\,2(a + ib)\Gamma_{\mathrm{III}\;4}, \\ e_2(a + ib) &= -\,2(a + ib)\Gamma_{\mathrm{I}\;3}, \\ e_3(a + ib) &= 2(a + ib)\Gamma_{\mathrm{III}\;2}, \\ e_4(a + ib) &= 2(a + ib)\Gamma_{\mathrm{I}\;1}. \end{aligned}$$

With the relationships in equation (2) added in, (A.5) reduces to

$$\text{(A.8)}\quad \begin{aligned} 2d\Gamma_{\mathrm{I}} + 4\Gamma_{\mathrm{I}} \wedge \Gamma_{\mathrm{II}} &= C^{(3)}\,\varepsilon^3 \wedge \varepsilon^1, \\ 2d\Gamma_{\mathrm{II}} + 2\Gamma_{\mathrm{I}} \wedge \Gamma_{\mathrm{III}} &= [C^{(3)} - (a^2 + b^2)]\,\varepsilon^1 \wedge \varepsilon^2 + [C^{(3)} + (a^2 + b^2)]\varepsilon^3 \wedge \varepsilon^4, \\ 2d\Gamma_{\mathrm{III}} + 4\Gamma_{\mathrm{II}} \wedge \Gamma_{\mathrm{III}} &= C^{(3)}\,\varepsilon^4 \wedge \varepsilon^2. \end{aligned}$$

The equations (A.7) and (A.8) represent the *entire* Einstein-Maxwell system in this case, i.e., equations (2) – (4).

---

[4] $e_a(F_{\mathrm{II}})$ is the ordinary action of a vector field on a scalar, i.e., the directional derivative in the e_a-direction.

The special [2, 2] case above has never been solved in general. In fact, as it stands, the null eigenvectors are not necessarily geodesic and shearfree (although this latter condition can be shown to be equivalent to $\nabla_{[\mu}T_{\nu]\sigma}=0$ from the Bianchi identities). The charged-Kerr (and therefore the Reissner-Nordström) metric is an example where this condition *does* hold.

References

1. A. S. Eddington, *The mathematical theory of relativity*, Cambridge Univ. Press, Cambridge, 1922.
2. E. T. Newman, et al., *Metric of a rotating, charged mass*, J. Mathematical Phys. **6** (1965), 918–919. MR **31** #3201.
3. G. Debney, R. P. Kerr and A. Schild, J. Mathematical Phys. **10** (1969), 1842.
4. R. Sachs, *Gravitational waves in general relativity*. VI. *The outgoing radiation condition*, Proc. Roy. Soc. Ser. A **264** (1961), 309–338. MR **27** #6598.
5. I. Robinson and A. Schild, *Generalization of a theorem by Goldberg and Sachs*, J. Mathematical Phys. **4** (1963), 484–489. MR **26** #7421.
6. J. N. Goldberg and R. K. Sachs, *A theorem on Petrov types*, Acta Phys. Polon. **22** (1962), 13–23. MR **27** #6599.
7. H. S. Ruse, Proc. London Math. Soc. **41** (1936), 302; J. L. Synge, *Relativity: The special theory*, 2nd ed., North-Holland, Amsterdam, 1965.
8. I. Robinson, J. Mathematical Phys. **2** (1961), 290–291.
9. G. Debney, J. Mathematical Phys. **13** (1972), 1469.
10. I. Robinson and A. Trautman, *Some spherical gravitational waves in general relativity*, Proc. Roy. Soc. Ser. A **265** (1961/62), 463–473. MR **24** #B1970.
11. L. Witten, *A geometric theory of the electromagnetic and gravitational fields*, Gravitation: An Introduction to Current Research, Wiley, New York, 1962, pp. 382–411. MR **26** #1184.
12. W. Kinnersley, *Type D vacuum metrics*, J. Mathematical Phys. **10** (1969), 1195–1203. MR **40** #1122.
13. G. Debney, J. Mathematical Phys. **12** (1971), 1088.
14. G. Debney and J. D. Zund, *A note on the classification of electromagnetic fields*, Tensor **22** (1971), 333–340. MR **45** #1549.

Virginia Polytechnic Institute and State University

Proceedings of Symposia in Pure Mathematics
Volume 27, 1975

LORENTZIAN MANIFOLDS OF NONPOSITIVE CURVATURE

F. J. FLAHERTY

Introduction. In a Riemannian manifold with nonpositive sectional curvature, the geodesics emanating from a point spread apart. In particular, the manifold cannot have any conjugate points. These observations form the basis for the classical theorem of Hadamard and Cartan: In a complete Riemannian manifold of nonpositive sectional curvature, the exponential map is a covering map. Thus, for simply connected manifolds, the exponential map is a diffeomorphism.

Our purpose here is to find a generalization of this important theorem to Lorentzian manifolds. A Lorentzian manifold M is a paracompact smooth (infinitely differentiable) manifold with a smooth nondegenerate symmetric tensor field of type (0, 2), whose signature is of the form $(+, -, \cdots, -)$. The value of the tensor field at two tangent vectors u, v is denoted by (u, v). From this pairing a tangent vector u in M_p is called timelike, spacelike, or null depending on whether or not the Lorentzian norm (u, u) is positive, negative, or zero. The null cone at p is the set of all null vectors in M_p and it clearly separates the timelike vectors into two components. The Lorentzian manifold M is called time-orientable iff it is possible to choose a continuous field of components. Tangent vectors in the chosen component are called future-timelike and the set of all such future-timelike vectors at p is called the future cone at p. Similarly a smooth curve is called a future-timelike curve iff its tangent vector lies in the future cone. There is a corresponding notion of homotopy and we say M is future 1-connected iff any two future-timelike curves from p to q are homotopic through future-timelike curves with fixed endpoints p and q.

AMS (MOS) subject classifications (1970). Primary 53B30; Secondary 83F05.

Finally, two comments about geodesic completeness and sectional curvature. A time-oriented Lorentzian manifold M is called future complete iff the exponential map at each point is defined on the closed future cone, that is, the future cone plus its boundary. A plane σ in M_p is said to be a space-time plane when and only when σ is spanned by a space vector and a time vector.

Let us state the main theorem:

Let M be a time-oriented, future-complete Lorentzian manifold which is also future 1-*connected. Further suppose that the sectional curvature is nonpositive for all space-time planes. Then the exponential map regularly embeds the future cone at each point into M.*

1. Lorentzian geometry. Here we construct the simply connected, geodesically complete Lorentzian manifolds of constant curvature. These examples are well known and are included only to strengthen our intuition for Lorentzian manifolds of variable sectional curvature.

Of course R^n with the Lorentzian pairing

$$(x, y) = x^1y^1 - x^2y^2 - \cdots - x^ny^n$$

is the flat model.

Provide R^{n+1} with the pairing

$$(x, y) = x^1y^1 - x^2y^2 - \cdots - x^ny^n + \varepsilon x^{n+1}y^{n+1}$$

where $\varepsilon = \pm 1$ and consider the quadric Q defined by the relation

$$(x, x) = \varepsilon/k, \qquad k > 0.$$

The induced pairing on Q is Lorentzian and the curvature tensor is given by

$$R(u, v, w, x) = (R(u, v)\, w, x) = \varepsilon k((u, x)\,(v, w) - (v, x)\,(u, w))$$

where $R(u, v)w$ is the curvature transformation. Note that care must be exercised in defining sectional curvature, that is, the Lorentz structure cannot degenerate on the plane.

Following Penrose the quadric determined by putting $\varepsilon = -1$ will be called the de Sitter space, denoted by $Q(-k)$. Thus, $Q(-k)$ is diffeomorphic to $R^1 \times S^{n-1}$ and for nondegenerate planes, has constant sectional curvature $-k$.

Anti-de Sitter space is defined to be the universal cover of the quadric obtained by putting $\varepsilon = 1$. This space has curvature k.

The geodesics of the model spaces are gotten by intersecting planes through 0 in R^{n+1} with the quadric. Hence in $Q(-k)$ the timelike geodesics are branches of hyperbolas, the spacelike geodesics are ellipses and the null geodesics are straight lines.

This is all the information we need about the model spaces, but first a small warning. Often a Lorentzian manifold is defined by a pairing of signature $(-, +, \cdots, +)$. The names of the models remain the same, but the curvature changes sign.

For example the work of Calabi and Markus, cited in the References to this paper, uses this convention.

2. Conjugate points and the exponential map. Conjugate points, Jacobi fields and the exponential map are all tied up together. We briefly summarize the relevant facts. Recall that a vector field X along a geodesic c is said to be a Jacobi field iff

$$X'' + R(X, \dot{c})\dot{c} = 0.$$

Also we say that $c(0)$ and $c(t_0)$ are conjugate along c iff there is a nonzero Jacobi field along c such that $X(0) = X(t_0) = 0$, where t_0 is in the parameter interval of c. Further it is well known that the dimension of the kernel of the derivative of $\exp_p$ at $t_0\dot{c}(0)$ equals the dimension of the space of Jacobi fields along c vanishing at 0 and t_0.

PROPOSITION 2.1. *Let M be a time-oriented Lorentzian manifold such that the sectional curvature is nonpositive for all space-time planes. Then there are no conjugate points along geodesics initially tangent to the closed future cone.*

PROOF. Suppose that X is a Jacobi field along c such that $(X, \dot{c}) = 0$. Consider first the case where c is a future timelike geodesic. Clearly X and its derivative X' have to be spacelike vector fields. Hence

$$d(X, X')/dt = (X', X') - (R(X, \dot{c})\dot{c}, X) \leqq 0.$$

Thus (X, X') is decreasing, but if $X(0) = X(t_0) = 0$ for some $t_0 > 0$ then (X, X') vanishes on $[0, t_0]$. So $X'(0) = 0$ because otherwise (X, X') would be strictly decreasing.

Now consider the case where c is a null geodesic and $\dot{c}(0) = v$ is in the boundary of the future cone at p. We can always choose an orthonormal basis $e_1, \cdots, e_n$ of M_p so that e_n points into the future cone and e_{n-1}, e_n span the plane σ generated by e_n and v. The plane σ is also spanned by v and w, a null vector, such that $(w, v) = 1$. Without loss of generality suppose that the component of X in the direction $\dot{c}$ vanishes. The vector field $X(t)$ can be expressed as

$$X(t) = b(t)W(t) + \sum x^i(t)e_i(t), \qquad 1 \leqq i \leqq n-2,$$

where the fields W and e_i are obtained by parallel translation. Since $(X, \dot{c}) = 0$, the function $b(t)$ vanishes and X and X' are spacelike. The argument proceeds as before after we observe that any space-null plane (here spanned by $X(t)$ and $\dot{c}(t)$) is the limit of space-time planes; so, by continuity, $(R(X(t), \dot{c}(t))\dot{c}(t), X(t))$ remains nonpositive.

PROPOSITION 2.2. *Let M be a time-oriented future-complete Lorentzian manifold with nonpositive sectional curvature for all space-time planes. Then $\exp_p$ decreases the Lorentzian norm $u \to (u, u)$ at any vector of the closed future cone.*

PROOF. Here again we consider two cases. First u a future timelike vector. If w is tangent to M_p at u then decompose $w = w_1 + w_2$ orthogonally into future timelike and spacelike components respectively. By the Gauss lemma

$$\|\exp_{p^*u}(w)\|^2 = \|\exp_{p^*u}(w_1)\|^2 + \|\exp_{p^*u}(w_2)\|^2.$$

The fact that $\exp_p$ is a radial isometry and Proposition 2.1 yield

$$\|\exp_{p^*u}(w)\|^2 \leqq \|w_1\|^2 + \|w_2\|^2 = \|w\|^2.$$

For the other case, let u be a null vector and choose v so that $(v,v) = 0$ and $(u, v) = 1$, then $w = au + bv + w_2$ where w_2 is spacelike and orthogonal to the plane spanned by u and v. Again by the Gauss lemma

$$\|\exp_{p^*u}(w)\|^2 = ab + \|\exp_{p^*u}(w_2)\|^2.$$

Similarly

$$\|\exp_{p^*u}(w)\|^2 \leqq ab + \|w_2\|^2 = \|w\|^2.$$

REMARK. In the terminology of E. C. Zeeman (*Causality implies the Lorentz group*, J. Math. Phys. **5** (1964), 490) there is a new causal structure on M_p so that $\exp_p$ is a causal automorphism on the future cone. Indeed, the ordering on the tangent space is defined by $u < v$ iff $v - u$ is timelike and oriented to the future. The new ordering is defined by saying that $u <' v$ iff $v - u$ is timelike and oriented to the future with respect to the pullback Lorentzian structure, under the exponential map.

3. Proof of the main theorem. For this theorem we use the notation $I^+(v)$ for the future cone at v, $J^+(v)$ for the closed future cone at any point v in the tangent space. Similarly the past cones are denoted by I^-, J^-. Suppose that $c_0(s) = \exp_p(sv)$ and $c_1(s) = \exp_p(sw)$ with $c_0(0) = c_1(0) = p$ and $c_1(1) = c_0(1) = q$. Because M is future 1-connected there is a (p, q)-homotopy H_t of future-timelike curves connecting c_0 to c_1, where t is in the parameter interval [0, 1]. Now because $\exp_p$ has maximal rank along the ray tv, the exponential map has an inverse along this segment for t in [0, 1]. With the aid of this inverse we can start lifting the curves H_t to curves ϕ_t in the tangent space. Further the lifts ϕ_t are timelike curves as follows from Proposition 2.2. In fact, $\exp_{p*}(\dot{\phi}_t) = \dot{H}_t$ where the $\cdot$ signifies differentiation with respect to the horizontal parameter, and since $\|\dot{H}_t\|^2 > 0$, Proposition 2.2 yields $\|\dot{\phi}_t\|^2 > 0$. As a result the image of ϕ_t lies in $I^+(0) \cap I^-(v)$. Next consider the subinterval S of [0, 1] for which H_t can be lifted. Certainly S is open. If $t_0 = \sup S$ and $t_k \to t_0$, we obtain $\lim k\, \phi_{t_k}(s)$ is in $J^+(0) \cap J^-(v)$, a compact set in the tangent space. The worst that could happen here is that the limit be a null vector at which, by Proposition 2.1, $\exp_p$ has maximal rank and so H_{t_0} can be lifted to ϕ_{t_0}. It remains to show that the image of ϕ_{t_0} is contained in $I^+(0) \cap I^-(v)$, but this follows from Proposition 2.2 since H_{t_0} is future-timelike. Finally all of the H_t can be lifted by the inverse of the exponential map and so c_1 as well, but then $v = w$.

COROLLARY. *With the same hypothesis as the theorem M cannot be compact.*

COROLLARY. *Same hypothesis. There are no closed future-timelike curves in M.*

PROOF. The theorem yields the fact that M is future distinguishing.

4. Conclusion. Several comments are in order here. First some remarks about the physical aspect of our theorem. A Lorentzian manifold satisfying the hypotheses of our theorem cannot have closed trips in the sense of Penrose (see the References). Secondly, the curvature assumption implies that Ricci$(u, u) \geqq 0$ for all timelike u, and this condition has no physical meaning (again consult Penrose 7.21 (1972)).

Finally, the connectedness assumption (future 1-connected) is implied by the vanishing of the Lorentzian fundamental group.

REFERENCES

Two general references on differential geometry in the same spirit as this work

D. Gromoll, W. Klingenberg and W. Meyer, *Riemannsche Geometrie im Grossen*, Lecture Notes in Math., no. 55, Springer-Verlag, Berlin and New York, 1968. MR **37** #4751.

S. Kobayashi and K. Nomizu, *Foundations of differential geometry*. Vol. II, Interscience Tracts in Pure and Appl. Math., no. 15, Interscience, New York, 1969. MR **38** #6501.

Papers dealing specifically with Lorentzian geometry and pseudo-Riemannian geometry

R. Abraham, *Piecewise differentiable manifolds and the space-time of general relativity*, J. Math. Mech. **11** (1962), 553–592. MR **25** #2895.

S. Helgason, *Some remarks on the exponential mapping for an affine connection*, Math. Scand. **9** (1961), 129–146. MR **24** #A1688.

Detailed information on space forms for Lorentzian and pseudo-Riemannian geometry can be found in

E. Calabi and L. Markus, *Relativistic space forms*, Ann. of Math. (2) **75** (1962), 63–76. MR **24** #A3614.

J. A.Wolf, *Spaces of constant curvature*, McGraw-Hill, New York, 1967. MR **36** #829.

The notion of the Lorentzian fundamental group is introduced and discussed in

J. W. Smith, *Fundamental groups on a Lorentz manifold*, Amer. J. Math. **82** (1960), 873–890. MR **22** #11350.

There are two articles by Penrose that discuss relativity without a great deal of knowledge in physics.

R. Penrose, *Techniques of differential topology in relativity*, Regional Conference Series in Applied Math., SIAM, Philadelphia, 1972.

———, *Structure of space time*, Batelle Rencontres, Benjamin, New York, 1968.

OREGON STATE UNIVERSITY

Proceedings of Symposia in Pure Mathematics
Volume 27, 1975

GENERAL RELATIVITY*

ROBERT GEROCH

1. Introduction. The general theory of relativity is a physical theory of the gravitational field which, for reasons perhaps not altogether understood, happens also to touch on the structure of space and time. General relativity is, at least in my view, much more a branch of physics than of mathermatics. For example, the issue of what is a fruitful question in general relativity is decided, more often than not, by an external criterion: observations on, or at least expectations regarding, nature. The mathematical structure of the theory is expressed most concisely in the language of differential geometry. For this reason, interaction between the fields of general relativity and differential geometry can provide a frutiful stimulation to both. The natural point of interaction is where the mathematical structure of the theory meets the physical underpinnings.

The purpose of these lectures is to describe this interaction region.

In § 2, we give an elementary, qualitative survey of the general theory of relativity. Our purpose is to state the mathematical structure of the theory, and to indicate why this structure is felt to be appropriate for the description of space, time, and gravitation. In particular, we give some examples in § 2 of how physical considerations suggest which conditions can safely be absorbed as hypotheses in a theorem without destroying the usefulness of the theorem to the theory.

In each of §§ 3 and 4, we describe an area of recent interest in general relativity: the sign of gravitational energy, and singularities, respectively. These particular topics have been selected because in them one arrives fairly quickly at relatively clean mathematical questions. Our purpose is to indicate how these

---

AMS (MOS) subject classifications (1970). Primary 83C99.

*General lecture given at the Institute.

mathematical questions arise from the theory, and to indicate what results are available.

2. General relativity. A fundamental notion in general relativity is that of an event. An event is an idealized occurrence (in the physical world) having extension in neither space nor time. For example, the snapping of one's fingers, or the exploding of a firecracker, would describe an event. Denote by M the collection of all possible events in the universe—those which have happened in the past, which are happening now, or which will happen in the future.

The idea is to introduce certain additional structure on M, structure which is suggested by idealizations of various experiments one could imagine actually carrying out in the physical world.

Fix some subset of M (e.g. all events which occur in a given room during a given interval of time). There are various physical constructions by which one could obtain a one-to-one mapping from this subset to $\boldsymbol{R}^4$. For example, one could fill the room, from floor to ceiling, with people during the given time interval. Each person is to be assigned three numbers (e.g., distance of that person from the front wall, from a side wall, and from the floor) which gives his position in the room. Furthermore, each person is to wear a watch. Then, if we select any event in our subset (e.g., if a firecracker explodes in the room during the given time interval), the person at the occurrence of this event records the time as read by his watch, and the three numbers describing which person he is. Alternatively, one could have four pilots fly about the room during the time interval, with each pilot wearing a watch. An event (exploding firecracker) could then be described by the following four numbers: the time each pilot sees on his watch at the moment he hears the explosion.

We now suppose that constructions of this type yield (C^∞)[1] compatible charts on M. These remarks suggest that it is perhaps reasonable to endow M with the structure of a 4-dimensional manifold. We emphasize that, in questions such as this one (the introduction of differentiable structure on M), the issue is not one of proving anything. Rather, one is trying to construct a mathematical model, and to introduce structure into that model which appears, in some sense, to reflect possibilities in the physical world.

We regard an event as "a basic occurrence in the physical world," and adopt the point of view that more complicated occurrences are to be described, whenever possible, in terms of the basic ones. Some examples of this viewpoint follow.

We wish to describe a particle in terms of M. Consider the collection of all events which occur where the particle is. Our intuitive picture of what a particle is suggests that this set should be a "one-parameter family of points of M," since such an event could occur "where the particle is, but at any time." It seems reasonable, therefore, to describe a particle (a physical thing) in terms of the mathematical

[1]We shall assume, hereafter, that all manifolds, submanifolds, curves, and tensor fields are C^∞ unless otherwise stated.

model by a curve in the manifold M. Similarly, the physical thing "a piece of string" would be idealized and carried into the model as a 2-dimensional submanifold of M, and "a piece of paper" as a 3-dimensional submanifold.

As a second example, let us describe within our model (i.e., in terms of the manifold M) the more conventional picture of the world, in which "space" and "time" are more or less distinct entities. Consider again the room filled with people wearing watches. Assign to each event in this region a real number, the time read by the watch of the person experiencing the event. We are thus mapping events to numbers: the idealization in the model would be a real function t on M. For each real number t_0, the subset of M on which $t = t_0$ (say, a 3-dimensional submanifold of M) would represent "all space at the time t_0." Two events would be described as "happening at the same time" if their t-values were the same. It turns out, unfortunately, that there is no physical construction which yields a preferred t: the t which results from a construction such as that above depends on the details of how people locate their positions, how they attempt to synchronize their watches, etc. For this reason, one is apparently forced into a model involving a 4-dimensional manifold M rather than a family of 3-dimensional manifolds labeled by t.

We now have the basic object in our model of the physical world: a certain 4-dimensional manifold M. We next argue that it is reasonable to endow M with a certain (indefinite) metric g.

We suppose that it is feasible to build clocks which are, for all practical purposes, identical, and whose internal workings are affected neither by external influences nor by the past history of the clock. Each of these clocks has a dial which, at any moment, reads a real number, the time according to that clock.[2] We wish to introduce structure on M which represents these clocks. One of the clocks (idealized to be very small) would be represented within M as a curve. The "time", as read by such a clock, assigns a real number to each event which occurs in the immediate presence of the clock. Thus, within the model, we replace "the time reading of the clock" by a parameter t along the curve describing the clock. In short, the idealization of such a clock is a parametrized curve on M.

Fix two events, p and q, and consider a clock γ which passes through p. It is observed in nature that, if p and q are sufficiently close, there are exactly two light rays each of which has the property that it meets the clock and also the event q (Figure 1). Physically, one light ray is sent from the clock at just the right moment (event u_1) and in just the right direction so that it arrives at the event q. The second light ray is sent from q in the right direction so that it returns to the clock (event u_2). Set

$$D(p, q, \gamma) = [t(u_2) - t(p)]\,[t(p) - t(u_1)], \tag{1}$$

where, for r an event which occurs in the immediate presence of the clock, $t(r)$ is the

[2]This "time" is not to be cyclic, as with a watch, but rather monotonic, as with a watch and calendar.

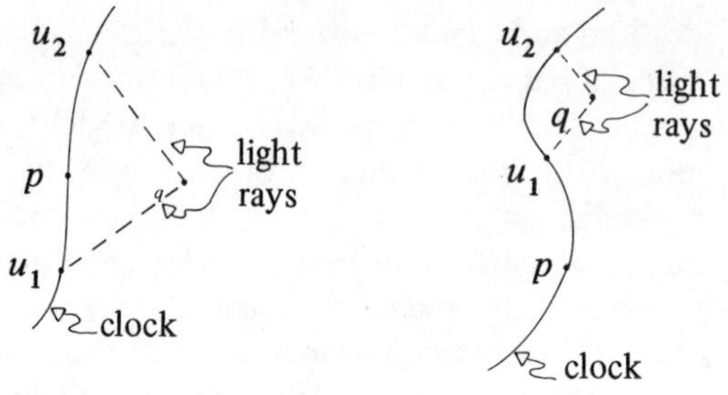

FIGURE 1

clock reading at r. It is further observed in nature that "as q approaches p (γ and p fixed) $D(p, q, \gamma)$ becomes independent of γ, and quadratic in the displacement of q from p." The observations claimed above are actually predicated in the special theory of relativity, which is well established, so our suppositions above require that "locally the world is like that of special relativity."

What we now wish to do is translate the physical observations described above into additional requirements on our mathematical model. First of all, we suppose that, in addition to a certain family of parametrized curves on M (clock-curves), there is another family of curves (light-curves). We require that these families have the following property: given any point p of M, there is a neighborhood U of p such that, for any $q \in U$ and any clock-curve $\gamma(t)$ which passes through p and not q, there are precisely two light-curves which pass through q and intersect γ. Let u_1 and u_2 be these points of intersection, and define $D(p, q, \gamma)$ by (1). Then the limiting process can be formulated within the model as the following requirement: there is a (metric) tensor field g_{ab} on M such that, for any curve $\Gamma(\lambda)$ on M, and any clock-curve $\gamma(t)$ in M through $p = \Gamma(0)$, $\frac{1}{2}(d^2/d\lambda^2)D(P, \Gamma(\lambda), \gamma)|_0 = g_{ab}\xi^a\xi^b$, where ξ^a is the tangent vector to $\Gamma(\lambda)$ at p. This g_{ab} on M, the metric of general relativity, is clearly unique.

It is also clear from the construction that any tangent vector to a light-curve has norm zero (with respect to g_{ab}),[3] and that the tangent vector to any clock-curve $\gamma(t)$ has norm -1, so the parameter t along a clock-curve is just length with respect to g_{ab}. (In fact, these properties could be used as a more concise, but less physical, definition of g_{ab} on M.) This model does not look even roughly like the physical world unless g_{ab} has signature $(-, +, +, +)$.

Thus, our mathematical model now consists of a 4-dimensional manifold M with metric g_{ab} of signature $(-, +, +, +)$.

We must introduce one more important relationship between the experiences of observers in the physical world and the mathematical model. We imagine building an instrument called an accelerometer, which consists of a small rigid frame inside of which a mass is suspended by springs (Figure 2). At any given instant of time,

[3]When we speak of "norm positive", "norm zero", or "norm negative" for a vector ξ^a, we refer to the sign of $g_{ab}\,\xi^a\xi^b$.

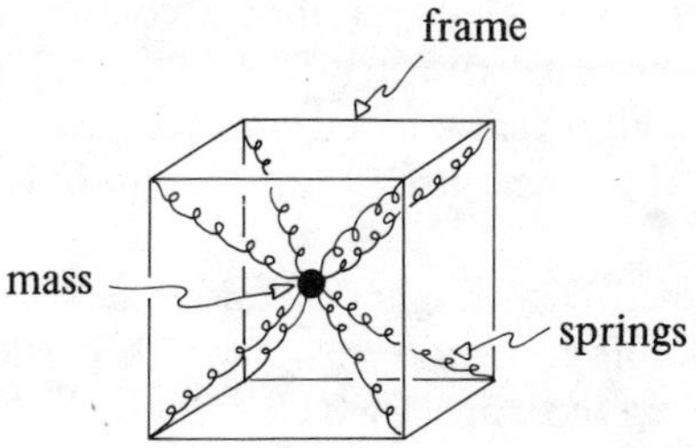

FIGURE 2

the mass may occupy a position within the frame displaced from its normal position at the center. For example, if we set the accelerometer on the earth, the mass will be displaced downward, while if we accelerate it, the object will be displaced to the rear. We wish, as usual, to introduce an idealized version of this instrument within the mathematical model. The center of the frame can be represented by a curve γ in M, while the mass would be represented by a second curve γ' (so the mass's occupying the center of the frame would correspond to a coincidence of γ' and γ). We want to take the limit of a "small" accelerometer, i.e., the limit as γ' approaches γ. In this limit, we can replace γ' by a vector A^a along γ, where A^a is minus (conventionally) the perpendicular displacement of γ' from γ.

Thus, the apparatus of Figure 2 is idealized as follows: a curve γ in M at each point of which there is specified a vector A^a orthogonal to the tangent vector to γ. This A^a is called the acceleration of the curve γ. The acceleration of a curve describes part of what is experienced by an object characterized by that curve. Note that we have, as yet, no way of determining the acceleration of a given curve on a given M with given g_{ab}. One could well imagine introducing additional structure on M which would determine the accelerations of curves. In general relativity, however, one introduces, instead of additional structure, an additional assumption: the acceleration A^a of a curve γ is just the curvature of γ (with respect to g_{ab}).

This assumption could, at least in principle, be tested experimentally (although, as far as I am aware, it never has been to any precision). One could measure the metric in some region of M, compute the curvature of a curve on M, and compare with the observed acceleration.

One adopts the point of view that a nonzero A^a is to be interpreted in terms of the action of some "external force." Thus, one would expect a nonzero A^a for a charged accelerometer in the presence of an electric field, and for an accelerometer sitting on a table. In the first case, the external force is an electric force, in the second, the upward force of the table on the accelerometer. In particular, a free particle (a particle not acted upon by external forces) must be described by a curve with zero acceleration, i.e., by a curve of zero curvature, i.e., by a geodesic.

One does not include a "gravitational force" in the external forces in the discussion above. The reason is that one does not know of a way to measure a thing which could be called a gravitational force. If we place an accelerometer on a table on the surface of the earth, the acceleration is accounted for by the upward force of the table; if we drop an accelerometer above the surface of the Earth, it reads zero acceleration. The point is that gravitation affects the mass inside the frame in the same way it affects the frame itself, and so this instrument cannot measure a "gravitational force."

We have now completed what might be called the kinematical part of general relativity. We have a manifold M in terms of which particles, light rays, etc., are represented, and a metric g in terms of which certain physical experiences (elapsed time and acceleration) can be described. However, we do not yet have what could be called a physical theory. The problem is that, although the present set-up already makes predictions (e.g., that if each of two clocks passes from event p to q, along different paths, the elapsed time between p and q, as measured by the two clocks, will in general be different) there still remains too much freedom in the metric g. In order to make the detailed quantitative predictions characteristic of a physical theory, one must somehow tie down g to the distribution of matter in the universe. The potential φ in Newtonian gravitational theory is analogous to the metric g in general relativity, e.g., because each determines the motion of free particles. What is needed is something analogous to the equation $\nabla^2\varphi = \rho$ in Newtonian theory, the equation which relates the potential φ to the density ρ of matter.

The analogous equation in general relativity, called Einstein's equation, is[4]

$$R_{ab} - \tfrac{1}{2}Rg_{ab} = T_{ab} \tag{2}$$

where R_{ab} is the Ricci tensor and R the scalar curvature of g_{ab}, and T_{ab} is a certain symmetric tensor field on M (analogous to ρ in Newtonian theory) called the stress-energy tensor.

What the stress-energy tensor is depends on the details of what matter is present. Each type of matter (e.g., fluids, solids, electromagnetic fields) is described within the model by certain tensor fields on M, subject to certain differential equations. With each type of matter, there is also associated an expression for the stress-energy T_{ab} of that type of matter in terms of the fields which characterize the matter. This process—taking the observed properties of a type of matter, idealizing the description in terms of M, and writing down the expression for T_{ab} for that type of matter—is much the same as the process we have already discussed of idealizing physical experiments into M.

We give one example of a type of matter. Consider (the idealized limit of) a dense swarm of free particles. Each particle could be represented by a curve in M whose tangent vector has negative norm. Hence, the motion of the entire swarm can be represented by the unit ($\xi^a\xi^b g_{ab} = -1$) tangent vector field ξ^a to this family of curves. The second field is a certain scalar field ρ on M, the mass density of the swarm. These two fields are subject to

$$\nabla_a(\rho\xi^a) = 0, \tag{3}$$

$$\xi^b\nabla_b\xi^a = 0, \tag{4}$$

[4]CONVENTIONS. $\nabla_{[a}\nabla_{b]}k_c = \frac{1}{2}R_{abc}{}^m k_m$, $R_{ab} = R_{amb}{}^m$.

where ∇_a is the covariant derivative on M. Equation (3) is interpreted physically as the conservation of mass: it asserts that particles are neither gained nor lost from the swarm during its motion. Equation (4) (the geodesic equation) is just the statement that our swarm consists of free particles.

The stress-energy of this type of matter is[5]

$$T_{ab} = \rho \xi_a \xi_b. \tag{5}$$

Note that, taking the divergence of both sides of (2), the left side vanishes by Bianchi's identity, while, in this case, the right side also vanishes by (5), (3), and (4). This is in fact a general feature of the description of matter: the stress-energy is so chosen as a function of the matter fields that its divergence vanishes by virtue of the equations on those fields.

To summarize, the set-up in general relativity is this. The arena is a 4-dimensional manifold M. On this manifold there is a metric g of signature $(-, +, +, +)$ in terms of which physical effects on particles, etc. are described, and a tensor field T_{ab} describing the distribution of matter. The two are connected by (2). Gravitational effects result because matter requires curvature by Einstein's equation, with this metric then influencing other objects in the universe.

It was remarked in the introduction that one motivation for considering the physical foundations of general relativity is that it permits one to decide whether conditions can be included in the hypotheses of theorems without destroying their applicability to the theory. We give three examples.

Is it reasonable to assume that M is connected? Suppose that M consisted of two connected components, M' and M'', and that our immediate environment was in M'. Then, since physical phenomena are described by curves, fields, differential equations, etc., no communication with M'' would be possible. But it has not turned out to be fruitful in physics to discuss things which cannot, even in principle, be observed. Hence, one can feel free, where it is convenient, to assume that M is connected.

Is it reasonable to assume that M is compact?

LEMMA [1]. *Let M be a compact 4-dimensional manifold with metric g of signature $(-, +, +, +)$. Then there exists in M a curve γ whose tangent vector has negative norm, and which intersects itself.*

The curvature of γ is the acceleration one must undergo (e.g., with a rocket) in order to be described by γ. But a person following this curve would (since the curve intersects itself) acquire the ability to influence his own past. (E.g., in Figure 3, one might, at the event q, decide that when he reaches p, he will inform himself which stocks will go up between p and q.) Since this is a possibility which, at least, does not seem entirely natural physically, it is normally regarded as somewhat unphysical to assume M compact.

Is it reasonable to assume that (M, g) has no conjugate points? A small mass,

---

[5]Tensor indices will be raised and lowered with the metric: $v_a = g_{ab}v^b$, $v^b = g^{ab}v_a$, where $g^{ac}g_{bc} = \delta^a_b$.

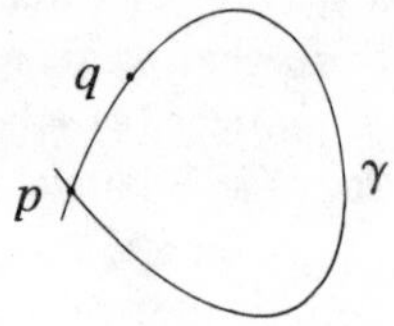

FIGURE 3

thrown from a satellite in a circular orbit around the Earth, will again strike the satellite after half an orbit. This situation would be represented within the model by nearby geodesics which intersect twice. Hence, it is not reasonable to assume the absence of conjugate points.

A pair (M, g), where M is a 4-dimensional manifold without boundary, and g is a metric on M of signature $(-, +, +, +)$, will be called a *space-time*.

3. The sign of gravitational energy. Let (M, g) be a space-time, and let S be a 3-dimensional submanifold of M whose normal has negative norm, i.e., whose induced metric is positive-definite. (Physically, S represents a choice of "space at one instant of time".) Denote by h_{ab} the induced metric on S, by Π^{ab} the second fundamental form of S, and by Π, $\Pi^{ab} h_{ab}$. Then, eliminating the Ricci tensor of M from the Gauss-Codazzi equations via (2), we have

$$\mathscr{R} - \Pi^{ab}\Pi^{cd}h_{ac}h_{bd} + \Pi^2 = 2\mu, \tag{6}$$

$$D_b(\Pi^{ab} - \Pi h^{ab}) = J^a, \tag{7}$$

where $\mathscr{R}$ is the scalar curvature of h_{ab} on S, D_b is the covariant derivative on S defined by h_{ab}, and where $\mu = +T_{ab}n^an^b$, $J^a = T_{mn}n^n(g^{ma} + n^mn^a)$, n^a the unit normal to S.

Physically, μ represents the mass density, and J^a the momentum density, of matter (as seen by an observer as his curve in M passes through S with tangent vector n^a). These physical interpretations suggest that we impose on μ and J^a the following two inequalities

$$\mu \geqq 0, \tag{8}$$

$$\mu^2 \geqq J^aJ^bh_{ab}. \tag{9}$$

Equation (8) requires that the local mass density of matter be nonnegative. Equation (9) requires that the square of the momentum density not exceed the square of the mass density. In order to violate (9), one would have to find matter having large momentum density for given mass density, i.e., to find matter moving at a large speed. Equation (9) requires, essentially, that matter not move faster than the speed of light. The best justification for (8), (9), however, is the fact that matter has yet to be observed violating these inequalities. (Note, for example, that (8) and (9) are satisfied for (5) provided $\rho \geqq 0$.)

More generally, an (S, h, Π, μ, J), where S is a 3-dimensional manifold, h_{ab} a

Riemannian metric on S, Π^{ab} a symmetric tensor field, μ a scalar field, and J^a a vector field on S, subject to (6)—(9), will be called an *initial-data set*. Thus, every 3-submanifold, the norm of whose normal is negative, of a physically reasonable (in the sense of (8), (9)) space-time satisfying Einstein's equation defines an initial-data set. The notion of an initial-data set is of interest because there is a sort of converse of the above statement. We state it, for simplicity, in the case when matter is absent:

THEOREM 1 [2]. *Let (S, h, Π, μ, J) be an initial-data set with $\mu = 0$ and $J^a = 0$. Then there exists a space-time (M, g) satisfying Einstein's equation* (2) *with $T_{ab} = 0$, along with an isometric embedding of S in M as a submanifold such that Π^{ab} on S is mapped by the embedding to the second fundamental form of the submanifold. Furthermore, given two such space-times their embedded S's have isometric neighborhoods.*

Physically, Theorem 1 is interpreted to mean that "an initial-data set is just the information which must be specified about a space-time at an initial instant in order to predict what will happen in the future." It asserts that the equation of general relativity is predicative, just as are other equations in physics.

There exists, in general relativity, the notion of an asymptotically flat initial-data set [3], [4]. The definition requires, roughly speaking, that "h approaches a flat metric, and Π, μ, J approach zero, at appropriate rates, asymptotically." Furthermore, associated with any asymptotically flat initial-data set, (S, h, Π, μ, J), there is a number E which depends explicitly only on h (although implicitly on Π, μ, J via (6)—(9)) [3], [4]. This quantity E, it turns out, has the physical interpretation of "the total energy, including contributions (which cannot be defined individually) from the matter itself and from the gravitational field."

We have:

CONJECTURE 2 [3]. *Let (S, h, Π, μ, J) be an asymptotically flat initial-data set. Then $E \geqq 0$ and, furthermore, $E = 0$ when and only when $\mu = 0$, $J^a = 0$, and (S, h, Π) results, via Theorem* 1, *in a space-time with vanishing Riemann tensor.*

In physical terms, the conjecture states that "the total energy is nonnegative, and vanishes precisely when there is neither matter nor gravitational field." One reason for interest in this question is that the existence of a lower bound for the energy of a physical system often serves as the starting point for a proof that the system is stable. Since other physical theories lead to nonnegative energies, one would hope that the same will be true in general relativity.

Unfortunately, there are some technical complications in the definition of asymptotic flatness and of E which tend to obscure the underlying simplicity of Conjecture 2. However, the essential features of the conjecture can be seen in a number of special cases, designed to avoid the detailed definitions of asymptotic flatness and of E. That is to say, it is widely felt that proofs of several of these special cases could be generalized to a proof of the full conjecture. These special cases are the following:

CONJECTURE 3. *Let (S, h) be a Riemannian 3-manifold, which is Euclidean outside a compact set (i.e., which, outside of some compact, is isometric to Euclidean space minus some compact set). Let $\mathscr{R} \geqq 0$, where $\mathscr{R}$ is the scalar curvature of h, and let the Riemann tensor of h have compact support. Then (S, h) is flat.*

CONJECTURE 4. *Let (S, h) be a Riemannian 3-manifold which is Euclidean outside a compact set C. Let ϕ be a positive scalar field on S which approaches 1 at infinity, and which satisfies $D^2\phi - \frac{1}{8}\mathscr{R}\phi = 0$ on S, where D^2 is the Laplacian on (S, h). Set $E = +\int_K (D_a\phi)\, dS^a$, where $K = S^2$ is a 2-submanifold of S surrounding C (so E is independent of K). Then $E \geqq 0$, and $E = 0$ when and only when $\phi^4 h$ is a flat metric on S.*

CONJECTURE 5. *Let (S, h) be Euclidean 3-space, and let Π^{ab} be a symmetric tensor field on S, with compact support. Let the μ and J^a defined by* (6) *and* (7) *(with $\mathscr{R} = 0$ in* (6)) *satisfy* (8) *and* (9). *Then $\Pi^{ab} = 0$.*

Conjecture 3 is the special case of 2 in which $\Pi^{ab} = 0$ and $J^a = 0$. Then (7) and (9) are satisfied identically, and (6) and (8) together are equivalent to $\mathscr{R} \geqq 0$. The assumption "Euclidean outside a compact set" ensures asymptotic flatness, and that $E = 0$. But, if $E = 0$, the Riemannian manifold (S, h) should, by Conjecture 2, be flat. Conjecture 3 has very much the flavor of a 3-dimensional Gauss-Bonnet theorem. Conjecture 4 is the special case $\Pi^{ab} = 0$, $J^a = 0$, and $\mu = 0$, but differs from 3 in that a less stringent asymptotic condition is imposed. The equation $(D^2 - \frac{1}{8}\mathscr{R})\phi = 0$ ensures that $(S, \phi^4 h, 0, 0)$ is an initial-data set. Asymptotic flatness is guaranteed by the asymptotic conditions on h and φ. The quantity E in this case is just the integral in Conjecture 4. Conjecture 5 is the special case in which h is flat. Then $E = 0$, and asymptotic flatness is guaranteed by the assumption that Π^{ab} has compact support.

Although Conjectures 3, 4 and 5 remain open, certain special cases of these have in fact been proven. It is known **[5]**, for example, that Conjecture 3 is true with the additional assumption that (S, h) possesses a nonzero Killing vector. It is also known **[6]** that Conjecture 5 is true with the additional assumption that $J^a = 0$. We sketch this proof. Set $K_{ab} = \Pi_{ab} - D_a D_b \phi$, where D_a is the covariant derivative on (S, h), and where ϕ is the function on S which satisfies $D^2\phi = \Pi$ and which approaches zero at infinity. Then equation (7), with $J^a = 0$, implies $D_a K^{ab} = 0$. Substituting $K_{ab} + D_a D_b \phi$ for Π_{ab} in (6) and integrating over S, we obtain

$$(10) \qquad -\int_S [K^{ab}K_{ab} + 2K^{ab}D_aD_b\phi + (D^aD^b\phi)(D_aD_b\phi) - (D^2\phi)^2] = 2\int_S \mu.$$

Integrating by parts and using $D_b K^{ab} = 0$, one sees that the second term in the integrand on the left does not contribute. Similarly, the third term cancels the fourth. Thus, the left side of (10) is nonpositive, while, by (8), the right side is nonnegative. So, $K^{ab} = 0$ and $\mu = 0$. Therefore, $\Pi_{ab} = D_a D_b \phi$, where ϕ is a function of compact support on S satisfying (by (6)) $(D^aD^b\phi)(D_aD_b\phi) = (D^2\phi)^2$. It is not difficult to show that the only such ϕ is $\phi = 0$. [NOTE ADDED IN PROOF. Conjecture 5 has been proven by P. S. Jang.]

4. Singularities. Let (M, g) be a space-time.

DEFINITION. A curve in M will be called *timelike* if its tangent vector has negative norm.

DEFINITION. A *Cauchy surface* [7] is a 3-dimensional submanifold S of M, whose normal vector has negative norm, and which has the following property: every timelike curve in M which cannot be assigned an endpoint intersects S once and only once.

THEOREM 6 **[8]**, **[9]**. *Let (M, g) be a space-time, and let S be a Cauchy surface in (M, g). Suppose that* (i) *there is a positive number c such that $\Pi \geqq c$, where Π is the trace of the second fundamental form of S* (*with respect to some system of normals to* S), (ii) *for every vector ξ^a on M with $\xi^a \xi^b g_{ab} < 0$, $\xi^a \xi^b R_{ab} \geqq 0$. Then every timelike curve in M, starting at S in a direction opposite that of the system of normals to S, has finite length.*

We sketch a proof of Theorem 6. Let γ be a timelike curve starting at S in a direction opposite the system of normals to S, and let p be a point of γ, and let L be the length of γ from p to S. Denote by $\mathscr{C}$ the collection of all C^0 curves γ' from p to S having the property that any two points of γ' can be joined by a C^∞ curve whose tangent vector is nonzero and has nonpositive norm. Impose on $\mathscr{C}$ the compact-open topology, and take the quotient by reparametrizations of the curves in $\mathscr{C}$. The result is a topological space $\hat{\mathscr{C}}$. The assumption that S is a Cauchy surface allows one to find an accumulation point for any net in $\hat{\mathscr{C}}$. Hence, $\hat{\mathscr{C}}$ is compact. The length function, $\mathscr{L}$, on $\hat{\mathscr{C}}$, defined by the usual limiting procedure, is upper semicontinuous, and hence achieves its maximum. This maximum must be the length of a timelike geodesic Γ from p to S which meets S orthogonally, and which, from the construction of p, has length from p to S, L or greater.

Now let $L > 3/c$. Then the conditions (i) and (ii) of the theorem imply that the normal geodesics to S near Γ focus before p is reached, whence Γ cannot be the longest timelike curve from p to S. This contradiction implies $L \leqq 3/c$.

Theorem 6 is the most straightforward of a collection of theorems in general relativity called singularity theorems. The interest in these results stems not only from the fact that they are true, but also from the fact that their hypotheses and conclusions are subject to physical interpretation. In this section, we describe the interpretation of Theorem 6, as an example, and then discuss some problems which have arisen in general relativity as a consequence of such theorems.

We must first decide that the various conditions of Theorem 6 are to be expected to hold for a class of reasonable models of our own universe.

That S be a Cauchy surface is interpreted physically to mean that "initial-data on S completely determines the space-time." That is to say, the existence of a Cauchy surface is a precise way of stating the physical notion that the space-time is "predictive." One can see intuitively how this notion leads to the definition of a Cauchy surface, for "signals are propagated along timelike curves" so the requirement that every such curve meets S ensures that "all signals were recorded on S." (There is also a generalization of Theorem 1 which reflects this feature **[10]**: Let

$(S, h, \Pi, 0, 0)$ be an initial-data set, and consider the space-times (M, g) of Theorem 1 for which the embedded S is a Cauchy surface. Then one of these (and only one) is maximal, in the sense that every other one is isometric to an open subset of it.)

Condition (i) of Theorem 6 has the following physical interpretation. The trace of the second fundamental form of S, Π, is the divergence of the normals of S. Our universe is filled with galaxies, each of which can be idealized and introduced into the model as a timelike curve. Conside a 3-dimensional submanifold S of M which, to a good approximation, cuts this family of curves orthogonally. Then the Π of this S can be interpreted as "the rate at which the galaxies are diverging from each other," i.e., as the "rate of expansion of the universe." It is in fact observed that such an expansion is present, and there is as yet no indication that its rate becomes small in distant regions of the universe. Thus, condition (i) requires, physically, that "the universe is expanding at a rate bounded below."

To make sense out of (ii), one restates it in terms of the stress-energy using Einstein's equation (2): if $\xi^a\xi^b g_{ab} < 0$, $(T_{ab} - \frac{1}{2} T^m_{m} g_{ab}) \xi^a\xi^b \geqq 0$. The question is: is it reasonable to require that the matter in the universe satisfy this condition? The answer appears to be yes. The condition has somewhat the flavor of (8) and (9), and is, as are those inequalities, interpreted to require "positive local energy density of matter." It is this interpretation, together with the fact that no matter has been observed which violates (ii) that suggests that this condition be accepted as physically reasonable. It can easily be checked directly that, for example, the T_{ab} of (5) satisfies (ii) when and only when $\rho \geqq 0$.

We have decided that a space-time (M, g) which reflects certain observations and prejudices about our own universe—i.e., which is "predictive," "expanding at a rate bounded below," and which contains "resaonable matter"—satisfies the hypotheses of Theorem 6. Hence, it also satisfies the conclusion. We must now translate the conclusion into physical terms—and thus obtain a prediction from the theory. At a point of S, consider the unit normal which is future-directed, i.e., which is in the direction in which the time-function t alorg clock-curves is increasing. Then, from the discussion above, it is the Π on S with respect to this set of normals that one expects to be positive. But the focusing property used in the proof of Theorem 6 yields incomplete geodesics which are emitted from S in the opposite direction from that of this normal. A particle described by an incomplete geodesic such as this (or, indeed, by a timelike curve of finite length) has, physically, the property that it "has only had available to it a finite amount of time in the past." Thus, the conclusion of the theorem states physically, that "there is an upper bound to the ages of particles in the universe," i.e., that "the universe was created at a finite time in the past."

One might imagine that the incompleteness required by Theorem 6 simply results from some region's having been excised, i.e., that completeness could be restored by replacing (M, g) by a suitable extension.[6] However, Theorem 6 itself asserts that no extension of (M, g) can be complete unless it either "violates predictivity"

[6] (M', g') is said to be an *extension* of (M, g) if (M, g) is isometric to an open submanifold of (M', g').

or "contains unphysical matter." Thus, no physically acceptable extension of (M, g) could be complete. This state of affairs is interpreted to indicate the presence of a "singularity" in the past (i.e., "the singularity from which the big bang originated").

In fact, there exist a number of other theorems along the general lines of Theorem 6, but which establish incompleteness of certain (not in general all) future-directed timelike geodesics, i.e., which establish the presence of "singularities in the future." The existence of such theorems suggests that a clock (or, for that matter, you or I) in free fall could have only a finite amount of time available. "After one's allotted time," there are simply no more events in M for one to occupy.

Whereas singularities in the past are usually regarded as a phenomenon worthy of further study, those in the future are considered as perhaps a bit unpleasant from a physical viewpoint. But, in both cases, one would like to establish theorems which give information about the structure of singularities. For example, one would like to make judgments concerning the extent to which the physics near a singularity is like the physics in our own local region. A natural first step in such a program would be to assign somehow to (M, g) an appropriate boundary of "singular points." One might think that this step could easily be accomplished by a slight change in our mathematical model. Instead of pairs (M, g), where M is a smooth manifold, one could consider triples $(\hat{M}, M, g)$, where $(\hat{M}, M)$ is some sort of "manifold with singularities at $\hat{M} - M$," and with g a metric on M. However, one does not proceed in this way, for the following reason: whereas one has a more or less clear physical prescription for obtaining the points of M ($\subset \hat{M}$) (in terms of events), it is very difficult to give a physical interpretation for the points of $\hat{M} - M$. For example, two triples, $(\hat{M}, M, g)$ and $(\hat{M}', M', g')$ with (M, g) isometric with (M', g'), would, apparently, be identical as far as the physics of the model is concerned.

One therefore adopts, in light of the comments above, the viewpoint that the "singular points" are somehow to be constructed from the space-time (M, g). That is to say, one would like to give a prescription, given (M, g) for constructing a (say, topological) space $\hat{M}$ with M embedded in $\hat{M}$. One might require that certain reasonable properties hold, e.g., that M is dense in $\hat{M}$, or that every incomplete timelike geodesic in (M, g) have an endpoint in $\hat{M}$.

In fact, one such construction has been found recently **[11]**, and is now under active study. Let (M, g) be a space-time, and let $B\ M \xrightarrow{\pi}$ be the principal bundle over M whose fibre over $p \in M$ is the collection of ordered 4-tuples of linearly independent tangent vectors at p, and whose group is $GL(4, \boldsymbol{R})$. We introduce a Riemannian metric on B. Let ξ be a tangent vector to B at $\alpha \in B$. Then $\pi_*\xi$ is a tangent vector in M at $\pi(\alpha)$, while $\xi - l\pi_*\xi$ is a vertical vector in the fibre over $\pi(\alpha)$, where l is the lifting from tangent vectors in M to B induced by parallel transport. Let $G(\xi, \xi)$ be the sum of the squares of the components of $\pi^*\xi$ with respect to the basis at p defined by α, plus the norm of $(\xi - l\pi^*\xi)$ with respect to any left-invariant Riemannian metric on $GL(4, \boldsymbol{R})$. This $G(\xi, \xi)$ defines the desired Riemannian metric on B. Denote by $\hat{B}$ the completion of B as a metric space.

(This $\hat{B}$ is independent of the choice of metric on $GL(4, \boldsymbol{R})$.) Since B is a principal bundle, $GL(4, \boldsymbol{R})$ acts on B. Although this action on B is not metric-preserving, it does take Cauchy sequences to Cauchy sequences. Hence, there is a natural extension of this action of $GL(4, \boldsymbol{R})$ from B to $\hat{B}$. Denote by $\hat{M}$ the topological space obtained by taking the quotient of $\hat{B}$ by this action. This $\hat{M}$, in which M is densely and homeomorphically embedded, is the candidate for "the space-time with singularities attached." We call $(\hat{M}, M, g)$ the *Schmidt completion* of (M, g).[7]

It is known that the Schmidt completion has certain attractive features. For example, if g is a Riemannian metric on a manifold M, then the $\hat{M}$ obtained by the construction above is just the completion of (M, g) as a metric space. Furthermore, every incomplete geodesic in a space-time has an endpoint in the Schmidt completion. On the other hand, the construction also has some unattractive features. For example, the Schmidt completion of a space-time need not be Hausdorff. It is also known that, if K is any separable metric space (e.g., a Riemannian manifold of high dimension, or a separable Banach space), then there exists a space-time (M, g) such that K can be homeomorphically embedded in $\hat{M} - M$.

What is needed at this stage is an understanding of whether or not the Schmidt construction yields singular points which will be reasonable and useful physically. It would be of great help if we had a few nontrivial examples, but, unfortunately, almost no explicit examples are available. The problem is in the passage from B to $\hat{B}$; it is difficult in practice to construct the completion of a given Riemannian manifold.

REFERENCES

1. R. W. Bass and L. Witten, *Remark on cosmological models*, Rev. Mod. Phys. **29** (1957), 452–453. MR **21** #1214.

2. Y. Fourès-Bruhat (Choquet-Bruhat), *Théorème d'existence pour certains systèmes d'équations aux dérivées partielles non linéaires*, Acta Math. **88** (1952), 141–225. MR **14**, 756.

3. R. Arnowitt, S. Deser and C. W. Misner, *Wave zone in general relativity*, Phys. Rev. (2) **121** (1961), 1556–1566. MR **22** #10728.

4. ———, *Energy and the criteria for radiation in general relativity*, Phys. Rev. (2) **118** (1960), 1100–1104. MR **23** #B990.

———, *Coordinate invariance and energy expressions in general relativity*, Phys. Rev. (2) **122** (1961), 997–1006. MR **23** #B991.

5. D. R. Brill, *Time-symmetric gravitational waves*, Les théories relativistes de la gravitation, Centre National de la Recherche Scientifique, Paris, 1962, pp. 147–153. MR **29** #2005.

6. D. Brill and S. Deser, Ann. Phys. **50** (1968), 548.

7. R. Geroch, *Domain of dependence*, J. Mathematical Phys. **11** (1970), 437–449. MR **42** #5585.

8. S. W. Hawking, *The occurrence of singularities in cosmology*. I, Proc. Roy. Soc. Ser. A **294** (1966), 511–521. MR **34** #8786.

9. R. Geroch, Article in *Relativity*, L. Witten, Ed., Plenum Press, New York, 1970.

10. Y. Choquet-Bruhat and R. Geroch, *Global aspects of the Cauchy problem in general relativity*, Comm. Math. Phys. **14** (1969), 329–335. MR **40** #3872.

11. B. Schmidt, J. Relativity and Gravity **1** (1970), 269–280.

ENRICO FERMI INSTITUTE, UNIVERSITY OF CHICAGO

[7]Note that the construction works for an n-manifold with a connection.

Proceedings of Symposia in Pure Mathematics
Volume 27, 1975

GRAVITATIONAL WAVES AND AVERAGED LAGRANGIANS*

A. H. TAUB

1. Introduction. The term "gravitational waves" is used to describe particular types of exact and approximate solutions of the Einstein field equations. The approximate solutions considered are assumed to be of the form

$$g_{\mu\nu} = \mathring{g}_{\mu\nu} + eg'_{\mu\nu} + \dots \tag{1.1}$$

where e is a parameter whose square is neglected and the $\mathring{g}_{\mu\nu}$ satisfy the Einstein field equations. The equations satisfied by the $g'_{\mu\nu}$ are the perturbed Einstein field equations. They involve the perturbed Ricci tensor. This quantity is defined as the coefficient of e in the expansion of the Ricci tensor computed from $g_{\mu\nu}$. Thus $R_{\mu\nu} = \mathring{R}_{\mu\nu} + eR'_{\mu\nu} + \dots$ where $\mathring{R}_{\mu\nu}$ is the Ricci tensor computed from $\mathring{g}_{\mu\nu}$. It may be verified that

$$R'_{\mu\nu} = (\Gamma'^{\tau}_{\mu\tau}\delta^{\sigma}_{\nu} - \Gamma'^{\sigma}_{\mu\nu})_{;\sigma}$$

where the semicolon denotes the covariant derivative with respect to $\mathring{g}_{\mu\nu}$ and $\Gamma'^{\sigma}_{\mu\nu} = \frac{1}{2}g^{\sigma\tau}(g'_{\mu\tau;\nu} + g'_{\nu\tau;\mu} - g'_{\mu\nu;\tau})$. Thus

$$R'_{\mu\nu} = \tfrac{1}{2}(g^{\sigma\tau}g'_{\mu\nu;\tau} - (g'^{\sigma}_{\mu} - \tfrac{1}{2}\delta^{\sigma}_{\mu}g')_{;\nu} - (g'^{\sigma}_{\mu} - \tfrac{1}{2}\delta^{\sigma}_{\mu})_{;\nu})_{;\sigma} \tag{1.2}$$

where $g' = \mathring{g}^{\mu\nu}g'_{\mu\nu}$.

It should be noticed that when

$$k^{\sigma}_{\mu;\sigma} = (g'^{\sigma}_{\mu} - \delta^{\sigma}_{\mu}g')_{;\sigma} = 0 \tag{1.3}$$

AMS (MOS) subject classifications (1970). Primary 83CXX.

*Supported in part by NSF grant GP 31358.

we may write $2R'_{\mu\nu} = (\Delta g')_{\mu\nu}$ where $(\Delta g')_{\mu\nu}$ is the Lichnerowicz Laplacian [1] applied to the symmetric tensor $g'_{\mu\nu}$. Further, it is no restriction to assume that (1.2) holds for under the coordinate transformation

$$x^{*\mu} = x^\mu + e\xi^\mu,$$
$$g_{\mu\nu} \to g_{\mu\nu} + e(g'_{\mu\nu} - \xi_{\mu;\nu} - \xi_{\nu;\mu}).$$

That is, $g'_{\mu\nu}$ and $g'_{\mu\nu} - \xi_{\mu;\nu} - \xi_{\nu;\mu}$ represent the same gravitational field in two coordinate systems, the x^μ one and the $x^{*\mu}$ one respectively. We may always choose the ξ_μ so that equation (1.2) is satisfied. This does not determine the coordinate system completely for if ξ^μ is a solution of the equations $\mathring{g}^{\mu\nu}\xi^\rho_{\mu;\nu} - \xi^\lambda \mathring{R}^\rho_\lambda = 0$, the value of the left-hand side of equation (1.2) is unaltered when $g'_{\mu\nu}$ is replaced by $g'_{\mu\nu} - \xi_{\mu;\nu} - \xi_{\nu;\mu}$.

2. Weak gravitational waves. An insight to the nature of gravitational waves may be obtained by considering the case where the $g_{\mu\nu}$ represent a weak gravitational field in vacuo. Then we may take

$$\mathring{g}_{\mu\nu} = \eta_{\mu\nu} = \delta_{\mu\nu} - 2\delta^0_\mu\delta^0_\nu$$

where the δ's are Kronecker symbols and we have chosen a coordinate system which is "almost galilean" (cf. Landau and Lifshitz [2]). In this case the Einstein field equations reduce to $\eta^{\sigma\tau} g'_{\mu\nu,\sigma\tau} = 0$, where the comma denotes the ordinary partial derivative and we have imposed the normalisation

$$k^\sigma_{\mu,\sigma} = (g'^\sigma_\mu - \tfrac{1}{2}\delta^\sigma_\mu g')_{,\sigma} = 0.$$

Thus the $g'_{\mu\nu}$ satisfy the wave equation. Solutions of the form

$$g'_{\mu\nu} = g'_{\mu\nu}(x^0 - x^1)$$

represent plane waves traveling with the velocity of light in the x^1 direction.

It may be shown that the freedom in the choice of the vector ξ_μ discussed above may be used to reduce $g'_{\mu\nu}$ to the form

$$g'_{\mu\nu} = \left\| \begin{matrix} 0 & 0 & 0 & 0 \\ 0 & 0 & 0 & 0 \\ 0 & 0 & g'_{22} & g'_{23} \\ 0 & 0 & g'_{23} & g'_{33} \end{matrix} \right\|$$

with $g'_{22} + g'_{33} = 0$.

Thus a weak plane gravitational wave is determined by two functions g'_{23} and $g'_{22} = -g'_{33}$. That is, such a wave is a transverse wave whose polarization is determined by a traceless symmetric second order tensor in the plane perpendicular to the direction of motion.

The Riemann curvature tensor computed from $g_{\mu\nu}$ given by equation (1.1) when $\mathring{g}_{\mu\nu}$ is the metric in Minkowski space and $g'_{\mu\nu}$ is given by the plane wave solution discussed above does not vanish. Thus gravitational waves are associated with curvature in space-time. However the curvature tensor has very special algebraic properties. These may be described in terms of the algebraic classification of curvature tensors when these are considered as mappings of antisymmetric two index tensors into such tensors. Pirani [3] pointed out the relevance of this algebraic classification to gravitational wave theory. In case the Ricci tensor vanishes, the curvature tensor reduces to the conformal tensor defined by the equation

$$C_{\mu\nu\sigma\tau} = R_{\mu\nu\sigma\tau} - \tfrac{1}{2}(g_{\mu\sigma}R_{\nu\tau} - g_{\mu\tau}R_{\nu\sigma} + g_{\nu\sigma}R_{\mu\tau} - g_{\nu\tau}R_{\mu\sigma})$$
$$- \tfrac{1}{6} R(g_{\mu\tau}g_{\nu\sigma} - g_{\mu\sigma}g_{\nu\tau}),$$

where R is the scalar curvature tensor and $R_{\mu\nu}$ is the Ricci tensor.

In the case of a weak gravitational wave it may be verified that

$$C_{\mu\nu\sigma\tau}l^{\tau} = R_{\mu\nu\sigma\tau}l^{\tau} = 0$$

where $l_{\sigma} = \delta_{\sigma} - \delta^{1}_{\sigma}$ is the null vector which describes the normal to the surfaces of constant phase of the wave.

Exact solutions of the vacuum Einstein field equations are known which represent a generalisation of the weak gravitational wave propagating in one direction. Landau and Lifshitz [2] discuss one such solution where

$$g_{\mu\nu}dx^{\mu}dx^{\nu} = 2d\theta dx^{1} + g_{ab}(\theta)dx^{a}dx^{b}, \qquad a,b = 2,3,$$

and

$$g_{ab} = x^{2}\gamma_{ab}, \qquad \det\|\gamma_{ab}\| = 1,$$
$$\ddot{x} + \tfrac{1}{8}\,(\dot{\gamma}_{ac}\gamma^{bc})\,(\dot{\gamma}_{bd}\gamma^{ad})x = 0;$$

the dot refers to differentiation with respect to θ and $\gamma_{ab}\gamma^{bc} = \delta^{c}_{a}$. The curvature tensor computed from this metric also satisfies the condition

$$C_{\mu\nu\sigma\tau}l^{\tau} = R_{\mu\nu\sigma\tau}l^{\tau} = 0$$

where $l_{\sigma} = \eta_{,\sigma}$ is the null vector normal to the surfaces of constant phase.

We therefore characterize gravitational waves in vacuum by metric tensors which satisfy the Einstein vacuum field equations and are such that the conformal curvature tensor they determine is algebraically special.

3. The averaged Lagrangian. We shall next discuss high frequency gravitational waves. We shall assume that the metric tensor of space-time undergoes changes that depend on two scales as does a wave train which shows rapid oscillations and a slow change in amplitude, frequency and wave number. We shall be treating the "geometric optics" approximation to gravitational waves in contrast to the "physical optics" discussion.

The method used by Y. Choquet-Bruhat [4] is the two-timing method. It consists in assuming that the dependent variables of a problem say ϕ^{A} may be written as

(3.1) $$\psi^A = \psi^A(X^\mu, \theta)$$

where $X^\mu = \varepsilon x^\mu$ and $\theta = \varepsilon^{-1}\Theta(X^\mu)$ and it is assumed that derivatives of ψ^A with respect to X^μ and θ are of equal magnitude (which we may take as order unity). The small parameter ε then measures the ratio of the fast scale to the slow one. It should be noted that rapid variations only occur in the direction of the vector

(3.2) $$l_\mu = \partial\theta/\partial x^\mu = \partial\Theta/\partial X^\mu = \Theta_{,\mu}.$$

We introduce the notation that if $f = f(X^\mu, \theta)$, then

(3.3) $$f_{,\mu} = \partial f/\partial X^\mu, \qquad \dot{f} = \partial f/\partial\theta$$

so that

(3.4) $$\partial\psi^A/\partial x^\sigma = \varepsilon\psi^A_{,\sigma} + l_\sigma\dot{\psi}^A.$$

When equations (3.1) to (3.4) are substituted into the equations satisfied by the ψ^A, differential equations in the five variables X^μ and θ are obtained. These variables are treated as independent variables. After a solution is obtained the substitution $\theta = \varepsilon^{-1}\Theta(X^\mu)$ is made and a solution to the original problem is obtained. The usual procedure in the two-timing method is to assume that

(3.5) $$\psi^A(X^\mu, \theta) = \sum_0^\infty \varepsilon^n \psi^A_{(n)}(X^\mu, \theta),$$

substitute this into the governing equations, and equate to zero the coefficients of the successive powers of ε. As will be seen below further restrictions on the $\psi^A_{(n)}$ must be imposed in order that the $\psi^A_{(n+1)}$ remain bounded.

When the field equations satisfied by the ψ^A are derived from a variational principle with the Lagrangian density $\mathscr{L}(\psi^A, \partial\psi^A/\partial x^\sigma)$, that is, are of the form

(3.6) $$\partial\Pi^\sigma_A/\partial x^\sigma - \partial\mathscr{L}/\partial\psi^A = 0$$

where

(3.7) $$\Pi^\sigma_A = \frac{\partial\mathscr{L}}{\partial(\partial\psi^A/\partial x^\sigma)},$$

the two-timing method rewrites these equations as

(3.7′) $$\varepsilon\Pi^\sigma_{A,\sigma} + l_\sigma\dot{\Pi}^\sigma_A - \partial\mathscr{L}/\partial\psi^A = 0$$

where

(3.8) $$\mathscr{L} = \mathscr{L}(\psi^A, \phi^A_\sigma), \qquad \Pi^\sigma_A = \Pi^\sigma_A(\psi^A, \phi^A_\sigma)$$

with $\phi^A_\sigma = l_\sigma\dot{\psi}^A + \varepsilon\psi^A_{,\sigma}$.

It is a consequence of equations (3.6) and the assumption that $\mathscr{L}$ does not depend explicitly on the x^σ, that

(3.9) $$\partial t^\sigma_\xi/\partial x^\sigma = 0$$

where

$$t^{\sigma}_{\xi} = \Pi^{\sigma}_{A}(\partial\psi^{A}/\partial x^{\xi}) - \delta^{\sigma}_{\xi}\mathscr{L} \tag{3.10}$$

and is the canonical, not necessarily symmetric, stress-energy tensor. Equations (3.9) and (3.10) may be written as

$$\mathscr{L}_{\sigma}i^{\sigma}_{\tau} + \varepsilon t^{\sigma}_{\tau,\sigma} = 0 \tag{3.11}$$

and

$$t^{\sigma}_{\tau} = l_{\tau}\Pi^{\sigma}_{A}\dot{\psi}^{A} - \delta^{\sigma}_{\tau}\mathscr{L} + \varepsilon\Pi^{\sigma}_{A}\psi^{A}_{,\tau}. \tag{3.12}$$

Equation (3.11) describes conservation theorems in the two-timing formulation.

There is another conservation theorem which follows from equation (3.7), namely

$$(l_{\sigma}\Pi^{\sigma}_{A}\dot{\psi}^{A} - \mathscr{L})^{\cdot} + \varepsilon(\Pi^{\sigma}_{A}\dot{\psi}^{A})_{,\sigma} = 0. \tag{3.13}$$

When ψ^{A} is written as an expansion in terms of ε as in equation (3.5), equations (3.11) and (3.13) imply that

$$l_{\sigma}t^{\sigma}_{(0)\tau} = l_{\sigma}l_{\tau}\Pi^{\sigma}_{(0)A}\dot{\psi}^{A}_{(0)} - l_{\tau}\mathscr{L}_{(0)} = P_{\tau}(X) \tag{3.14}$$

and

$$l_{\sigma}\Pi^{\sigma}_{(0)A}\dot{\psi}^{A}_{(0)} - \mathscr{L}_{(0)} = A(X) \tag{3.15}$$

where the subscript (0) denotes the fact that various quantities are evaluated to the lowest order in ε. It follows from these equations that

$$l_{\sigma}t^{\sigma}_{(0)\tau} = P_{\tau}(X) = A(X)l_{\tau}. \tag{3.16}$$

Equations (3.11) and (3.13) have further consequences: Thus

$$l_{\sigma}i^{\sigma}_{(1)\tau} = -\varepsilon t^{\sigma}_{(0)\tau,\sigma}$$

and

$$H^{\cdot}_{(1)} = (l_{\sigma}\Pi^{\sigma}_{A}\dot{\psi}^{A} - \mathscr{L})^{\cdot}_{(1)} = -\varepsilon[(\Pi^{\sigma}_{A}\dot{\psi}^{A})_{(0)}]_{,\sigma}$$

where the subscript (1) denotes the fact that the expression labeled with it has been evaluated to first order in ε. If the solution we are seeking is to be periodic in θ and if there are to be no secular terms in $t^{\sigma}_{(1)}$ or $H_{(1)}$ we must have

$$\frac{1}{2\pi}\int_{0}^{2\pi} t^{\sigma}_{(0)\tau,\sigma}\, d\theta = 0 \tag{3.17}$$

and

$$\frac{1}{2\pi}\int_{0}^{2\pi} [(\Pi^{\sigma}_{A}\dot{\psi}^{A})_{(0)}]_{,\sigma}\, d\theta = 0. \tag{3.18}$$

The latter equations are conditions that must be imposed on $\psi^{A}_{(0)}$ in addition to the requirement that they satisfy the zero order form of equations (3.7′). They are the conditions referred to above.

Whitham [5] made the important observation that (3.7) are the Euler equations of the five-dimensional variational principle

$$I = \int \mathscr{L}(\phi^A, \phi^A_\sigma)\, d^4x\, d\theta$$

with ϕ^A_σ given as above. The case where the functions $\phi^A(X^\mu, \theta)$ are periodic in θ one may define

$$I = \frac{1}{2\pi}\int_0^{2\pi} \mathscr{L}(\phi^A, \phi^A_\sigma)\, d\theta \tag{3.19}$$

so that, to lowest order in ε,

$$I = \int \bar{L}_{(0)}(l_\sigma, A) d^4x \tag{3.20}$$

where

$$\bar{L}_{(0)}(l_\sigma, A) = \frac{1}{2\pi}\int_0^{2\pi} l_\sigma(\Pi^\sigma_A \dot{\phi}^A)_{(0)}\, d\theta - A(X)$$

or

$$\bar{L}_{(0)}(l_\sigma, A) = \frac{l_\sigma}{2\pi}\oint \Pi^\sigma_{(0)A}\, d\phi^A_{(0)} - A(X). \tag{3.21}$$

In the case where the lowest order solution of the two-timing method is periodic in θ, varying θ and hence l_σ in (3.20) gives rise to equations (3.18). Varying A in (3.20) gives rise to the dispersion relation for the waves described by these periodic solutions. We may interpret equation (3.18) which has been seen to come from the variation of equation (3.20) as the statement that adiabatic invariants exist for the lowest order solutions.

The function $\bar{L}$ defined by equation (3.19) is the averaged Lagrangian. In many problems it may be evaluated to various orders, particularly the zeroth one by using the integral of the motion described by equation (3.15). However, in the problems with which we are concerned, $\mathscr{L}_{(0)}$ vanishes when evaluated for the zeroth order solution and another method must be found for determining the averaged Lagrangian.

4. High-frequency gravitational waves. MacCallum and Taub [6] applied the method described above to the treatment of such waves by making use of the fact that the Einstein vacuum field equations may be derived from the variational principle

$$J = \int (-g)^{1/2} R\, d^4x$$

where R is the scalar curvature computed from $g_{\mu\nu}$ and g is the determinant of this tensor. The two-timing method will then involve the variational principle determined by the integral

$$I = \frac{1}{2\pi}\int\left(\int_0^{2\pi}(-g)^{1/2}R\,d\theta\right)d^4x.$$

If we write

$$g_{\mu\nu}(X,\theta) = \mathring{g}_{\mu\nu}(X,\theta) + eg'_{\mu\nu}(X,\theta) + \tfrac{1}{2}e^2 g''_{\mu\nu}(X,\theta) + \cdots$$

then for general e and ε, I may be expressed as

$$I = I_0 + eI_1 + \tfrac{1}{2}e^2 I_2 + \cdots \tag{4.1}$$

where I_0, I_1 and I_2 are in turn second degree polynomials in ε, as follows from the results of [7] where the following results are given:

$$\begin{aligned} I_0 &= \frac{1}{2\pi}\iint_0^{2\pi}(-\mathring{g})^{1/2}R_0\,d\theta\,d^4x, \\ I_1 &= -\frac{1}{2\pi}\iint_0^{2\pi}(-\mathring{g})^{1/2}\,G_0^{\mu\nu}(g'_{\mu\nu}+\tfrac{1}{2}eg''_{\mu\nu})\,d\theta\,d^4x, \\ I_2 &= \frac{1}{2\pi}\iint_0^{2\pi}(-\mathring{g})^{1/2}(\mathcal{M}(\mathring{g}_{\mu\nu}, g'_{\mu\nu}) + \mathcal{G}(\mathring{g}_{\mu\nu}, g'_{\mu\nu}))d\theta\,d^4x, \end{aligned} \tag{4.2}$$

R_0 and $G_0^{\mu\nu}$ are the scalar curvature and Einstein tensor computed from $\mathring{g}_{\mu\nu}$,

$$\begin{aligned} \mathcal{M} &= (2\mathring{g}^{\sigma\alpha}G_0^{\tau\beta} - G_0^{\sigma\tau}\mathring{g}^{\alpha\beta} + \tfrac{1}{2}R_0(\mathring{g}^{\sigma\alpha}\mathring{g}^{\tau\beta} - \tfrac{1}{2}\mathring{g}^{\sigma\tau}\mathring{g}^{\alpha\beta}))k_{\alpha\beta}k_{\sigma\tau}, \\ \mathcal{G} &= k^{\sigma\rho;\kappa}k_{\rho\kappa;\sigma} - \tfrac{1}{2}k^{\kappa\rho;\sigma}k_{\kappa\rho;\sigma} + \tfrac{1}{4}k^{;\sigma}k_{;\sigma}, \end{aligned} \tag{4.3}$$

the semicolon refers to covariant differentiation with respect to $\mathring{g}_{\mu\nu}$, it and $\mathring{g}^{\mu\nu}$ are used to lower and raise indices, and

$$k_{\mu\nu} = g'_{\mu\nu} + \tfrac{1}{2}k\mathring{g}_{\mu\nu}, \qquad k = -\mathring{g}^{\mu\nu}g'_{\mu\nu}. \tag{4.4}$$

We may take the point of view that equations (4.1) through (4.4) determine a variational principle for the determination of the $\mathring{g}_{\mu\nu}$ and the $g'_{\mu\nu}$. That is, we regard the $k_{\mu\nu}$ (the $g'_{\mu\nu}$) as a tensor field over the space-time given by the $\mathring{g}_{\mu\nu}$ whose field equations are the Euler equations obtained by varying the $k_{\mu\nu}$, and whose stress-energy tensor is the source of the gravitational field determined by the $\mathring{g}_{\mu\nu}$. This stress-energy tensor is obtained by varying $\mathring{g}_{\mu\nu}$ in I_2. When we deal with the five-dimensional integrals we are in a position to use Whitham's technique for carrying out the two-timing procedure.

In applying this technique we must evaluate I_0, I_1, and I_2 by replacing the derivatives of $g_{\mu\nu}$ and $g'_{\mu\nu}$ by the appropriate linear combinations of their derivatives with respect to θ and the X^μ. Thus each of the integrals I_0, I_1, and I_2 becomes a quadratic polynomial in ε, the constant terms being associated with the derivatives with respect to θ, the linear ones with the mixed derivatives with respect to θ and X^μ and the quadratic ones with the second derivatives with respect to the X^μ.

If we now assume as was done in reference [6] that

$$e = \varepsilon \tag{4.5}$$

and that

(4.6) $$\mathring{g}_{\mu\nu} = \mathring{g}_{\mu\nu}(X)$$

that is, the $\mathring{g}_{\mu\nu}$ are independent of θ, various terms in the evaluation of equation (4.1) disappear or may be neglected. Thus as a consequence of equation (4.5) only the term independent of ε need be retained in the evaluation of I_2; a similar term and a term linear in ε must be retained in the evaluation of I_1. All these terms must be kept in the evaluation of I_0. However it follows from equation (4.6) that R_0 and $\underline{G}_0^{\mu\nu}$ are of the order of $\varepsilon^2 = e^2$. Hence I_1 may be completely neglected.

The resulting five-dimensional integral is a functional of $\mathring{g}_{\mu\nu}(X)$ and $g'_{\mu\nu}(X, \theta)$. Its variations with respect to $\mathring{g}_{\mu\nu}(X)$ give equations which determine the X^μ dependence of these variables and its variations with respect to $g'_{\mu\nu}$ give as Euler equations the differential equations in θ and X^μ satisfied by the latter variables. When the Euler equations are considered as differential equations in θ for the $g'_{\mu\nu}$ we note that these are linear differential equations with coefficients which depend on the $\mathring{g}_{\mu\nu}(X)$ and hence are independent of θ. Hence their general solution is expressible as a linear combination of solutions of the form

(4.7) $$g'_{\mu\nu}(X, \theta) = \alpha_{\mu\nu}(X)e^{i\theta} + \bar{\alpha}_{\mu\nu}(X)e^{-i\theta}$$

where the bar denotes the complex conjugate.

When equation (4.7) is substituted into equation (4.1), the latter equation becomes

(4.8) $$I = \varepsilon^2 \int [R_0 + A^\sigma_\alpha \bar{A}^{\alpha\tau} l_\sigma l_\tau - \tfrac{1}{2} l^\sigma l_\sigma (A^{\alpha\beta}\bar{A}_{\alpha\beta} - \tfrac{1}{2} A\bar{A})](-\mathring{g})^{1/2}\, d^4x$$

plus terms of higher order in ε, where

(4.9) $$A_{\alpha\nu} = \alpha_{\mu\nu} + \tfrac{1}{2} A\mathring{g}_{\mu\nu}, \quad A = A_{\mu\nu}\mathring{g}^{\mu\nu}, \quad l_\sigma = \Theta_{,\sigma}.$$

The variational principle based on the integral given by equation (4.8) is the following one: The $A_{\mu\nu}(X)$ are to be varied, $\Theta(X)$ is to be varied, and the $\mathring{g}_{\mu\nu}(X)$ are to be varied. From our discussion of the two-timing method, it is evident that the Euler equations obtained from the first two variations provide the necessary conditions that ensure that the higher order terms in the expansion of $g_{\mu\nu}(X, \theta)$ be bounded, that is, not contain secular terms. The variation of the $\mathring{g}_{\mu\nu}$ gives Euler equations which determine the $\mathring{g}_{\mu\nu}$ as functions of X^ν. These equations contain the "back-reaction" due to the perturbation described by the $g'_{\mu\nu}(X, \theta)$.

The variational principle described above leads to the following results

(4.10) $$\begin{aligned} l^\sigma l_\sigma &= 0, \\ A_{\tau\nu} l^\nu &= 0, \\ R_0^{\nu\mu} - \tfrac{1}{2}\mathring{g}^{\mu\nu} R &= \tfrac{1}{2} N l^\mu l^\nu, \\ N &= A^{\rho\sigma}\bar{A}_{\rho\sigma} - \tfrac{1}{2} A\bar{A} \geqq 0, \\ (N l^\mu)_{;\mu} &= 0, \end{aligned}$$

where the tensor indices are manipulated by use of the tensor $\mathring{g}_{\mu\nu}$. The detailed derivation of these results, their discussion and comparison with the work of Choquet-Bruhat [4], Isaacson [8], and Madore [9] is given in [6].

When the $\mathring{g}_{\mu\nu}$ are not independent of θ, the two-timing method shows [10] that the θ dependence of these variables is given by the exact solutions described by Landau and Lifshitz [2] and referred to in §2. The constants of integration occurring in those solutions are to be interpreted as functions of X^μ. We may use this information concerning $\mathring{g}_{\mu\nu}$ and the five-dimensional variational principle discussed above in studying the perturbations induced on a gravitational wave propagating in one direction by another gravitational wave.

However, in this case the linear differential equations satisfied by $g'_{\mu\nu}$ have coefficients which are no longer independent of θ. Hence the representation of $g'_{\mu\nu}(X, \theta)$ by expressions of the form of equation (4.7) is not useful and hence the averaged Lagrangian is not easily evaluated. It may be possible in some cases to evaluate the Lagrangian, that is, to carry out the integration of θ and thus write

$$I = \varepsilon^2 \int \bar{L}(l_\mu, A_{\mu\nu}(X), \tilde{g}_{\mu\nu}(X))\, d^4x$$

where the $A_{\mu\nu}(X)$ characterize the perturbation, the $\tilde{g}_{\mu\nu}(X)$ characterize the remaining unknown properties of $\mathring{g}_{\mu\nu}$ and the l_μ give the propagation vector. The variation of I with respect to the $A_{\mu\nu}$, $\tilde{g}_{\mu\nu}$ and $\Theta(X)$ will then lead to Euler equations whose solutions satisfy the equations obtained by two-timing the Einstein field equations.

References

1. A. Lichnerowicz, *Propagateurs et commutateurs en relativité générale*, Inst. Hautes Études Sci. Publ. Math. No. 10 (1961), 56 pp. MR **28** #967.

2. L. D. Landau and E. M. Lifshitz, *The classical theory of fields*, Pergamon Press, New York, 1962. MR **26** #1007.

3. F.A.E. Pirani, *Gravitational radiation*, Gravitation: An Introduction to Current Research, Wiley, New York, 1962, pp. 199–226. MR **26** #1181.

4. Y. Choquet-Bruhat, *Construction de solutions radiatives approachées des équations d'Einstein*, Comm. Math. Phys. **12** (1969), 16–35. MR **39** #2432.

5. G. B. Whitham, *Studies in applied mathematics* (edited by A. H. Taub), Math. Assoc. Amer. Studies in Math., vol. 7, Prentice-Hall, Englewood Cliffs, N. J., 1971.

6. M.A.H. MacCallum and A. H. Taub, Comm. Math. Phys. **30** (1973), 153–169.

7. A. H. Taub, *Fluides et champs gravitationnel in relativité générale*, Proc. 1967 Colloq. Internat. du Centre National de la Recherche Scientifique, Paris, 170, 1969, p. 57.

8. R. A. Isaacson, Phys. Rev. **166** (1968), 1263, 1278.

9. J. Madore, Comm. Math. Phys. **17** (1972), 291.

10. A.H. Taub, *Ondes et radiations gravitationelles*, Proc. 1973 Colloq. Internat. du Centre National de la Recherche Scientifique, No. 220, Editions du Centre National la Recherche Scientifique, Paris, 1974.

University of California, Berkeley

Proceedings of Symposia in Pure Mathematics
Volume 27, 1975

GRAVITATIONAL COLLAPSE

JOHN A. THORPE

My goal here is to present a survey of some recent results in general relativity which I find particularly interesting, in the hope that other Riemannian geometers will become more aware of some of the nice differential geometric ideas floating around among physicists. For those who are sufficiently turned on to seek more details and more results, I recommend the recent book by Hawking and Ellis [**1**]. For those who would like only a mathematically sophisticated presentation of the highlights of singularity theory, I recommend the tract by Penrose [**4**].

Let us start with the Schwarzschild model for the gravitational field outside a spherically symmetric static body of mass m (e.g., a star). We regard $\boldsymbol{R}^4$ as $\boldsymbol{R} \times \boldsymbol{R}^3$, take t as coordinate on the first factor and spherical coordinates (r, θ, φ) as coordinates on the second factor. Then a Lorentz metric is given on the region $r > 2m$ by

$$(1) \qquad ds^2 = -(1 - 2m/r)\, dt^2 + (1 - 2m/r)^{-1} dr^2 + r^2(d\theta^2 + \sin^2\theta\, d\varphi^2).$$

This metric defines the (essentially unique) spherically symmetric solution of the Einstein equations for empty space (Ricc = 0).

The conformal structure of the Schwarzschild model can be described by suppressing the coordinates θ and φ and sketching the field of light cones in the (t, r)-plane (Figure 1). [Note that a tangent vector $a(\partial/\partial t) + b(\partial/\partial r)$ is time-like (has negative norm-squared) if and only if $|b/a| < 1 - 2m/r$.] Each point in this (t, r)-plane represents a Euclidean 2-sphere of radius r.

This model provides a representation for our space-time only in the region $r > r_0$ where $r_0 =$ the radius of the given spherical body. Generally $r_0 \gg 2m$.

AMS (MOS) subject classifications (1970). Primary 83C20; Secondary 53C50.

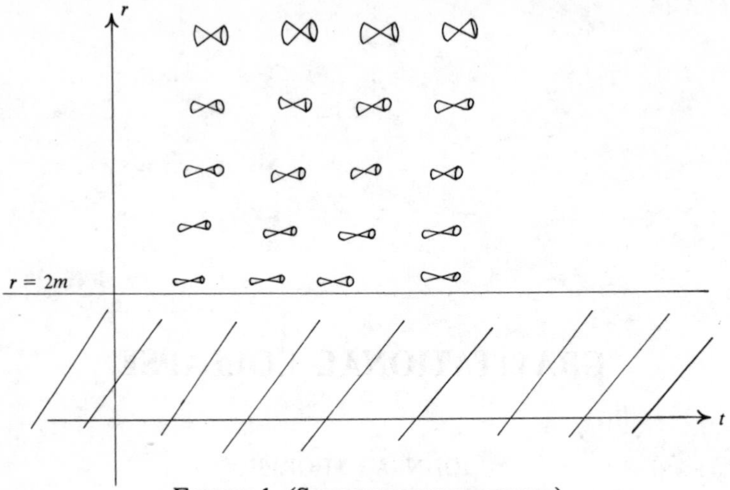

FIGURE 1. (SCHWARZSCHILD MODEL)

However, stars are likely to contract as they burn out and one is then interested in studying the case when $r_0 = 2m$ and even when $r_0 < 2m$. Since the light cones close up near $r = 2m$, it appears that neither time-like curves (paths of people) nor null curves (paths of light signals) could ever pass through the surface $r = 2m$. But this is an illusion. For if one computes the length of a time-like curve asymptotically approaching $r = 2m$ one finds that this length is finite. Physically this says that a person travelling along this curve will reach $r = 2m$ in finite time. So one tries to extend the model so that such curves can continue. Such an extension can be obtained as follows. Setting

$$v = t + r + 2m \log (r - 2m) \tag{2}$$

one can check that (v, r, θ, φ) forms a coordinate system C^∞-related to (t, r, θ, φ) and in the new coordinates the metric takes the Eddington-Finkelstein form

$$ds^2 = -(1 - 2m/r)\, dv^2 + 2\, dv\, dr + r^2(d\theta^2 + \sin^2\theta\, d\varphi^2). \tag{3}$$

But this metric is defined on the region $r > 0$ and so provides an analytic extension of the Schwarzschild solution.

The major effect of this change of coordinates is to "open up" the light cones. (See Figure 2.)

Now it is clear that particles can travel inside the Schwarzschild radius $r = 2m$. (I am implicitly assuming here a time orientation so that $\partial/\partial v$ (and $\partial/\partial t$) are future directed in the region $r > r_0$.) However, this model confronts us with some unpleasant facts of life. First note that the positioning of the light cones in the region $r < 2m$ implies that no person entering this region can ever come back out. Moreover no signal can pass from the region $r < 2m$ into the region $r > 2m$. The region $r < 2m$ is to the observer on the outside a big black hole. You can find out what is inside by following a path which takes you there, but once there you will never

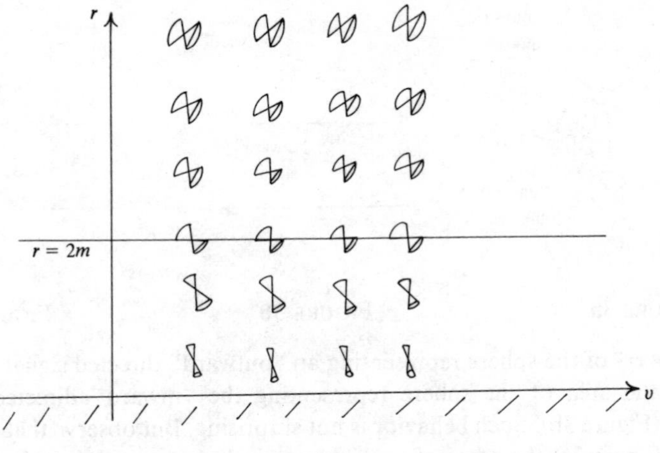

FIGURE 2. (EDDINGTON-FINKELSTEIN MODEL)

be able to come back out to tell your friends what you saw.

An even more distressing feature of this model is that as one approaches $r = 0$ the curvature becomes unbounded so there is no chance of extending the metric beyond $r = 0$. Moreover all time-like curves entering the region $r < 2m$ are forced by the nature of the light cone field to approach the singularity $r = 0$ and in fact must reach it in a finite time. Thus any visitor to the interior of the black hole is hopelessly headed for oblivion.

Recalling that points in Figure 2 represent spheres of radius r, one can regard a time-like geodesic in this diagram as representing the path travelled by the surface of a collapsing star and as describing the ultimate fate of such a star collapsing under the strength of its own gravitational field. This gravitational collapse will be forced upon a star as soon as its density is such that $r_0 < 2m$. Happily, the ratio $r_0/2m$ for our own sun is about 10^6 so there is no immediate danger of seeing it fall into a black hole.

I should mention in passing that a maximal analytic extension of the Schwarzschild model has been obtained by Kruskal [3] which consists essentially of two copies of the Eddington-Finkelstein model appropriately glued together.

One might ask in the Eddington-Finkelstein model what happens as an observer (#1) passes into the region $r < 2m$, say at midnight by his clock. A second observer (#2) in the region $r > 2m$ would observe (#1) slowly fade away and the clock carried by (#1) would appear to asymptotically approach midnight. (#1) would be unaware of this phenomenon and would pass through the surface $r = 2m$ without observing any trauma. There would, however, be a significant change in the properties of space-time about him. For suppose a flash of light is emitted from each point on a sphere v = constant, r = constant (represented by a point in Figure 2). The resulting light signals will radiate as a pair of spheres (Figure 3a) which can be described by a pair of null geodesics in the (v, r)-plane. In the region $r > 2m$,

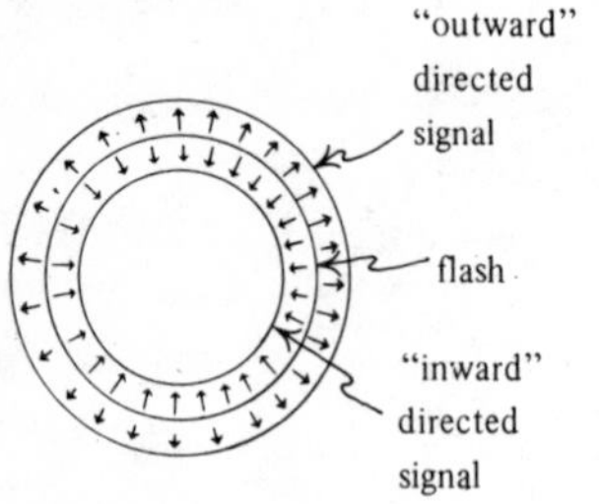

FIGURE 3a

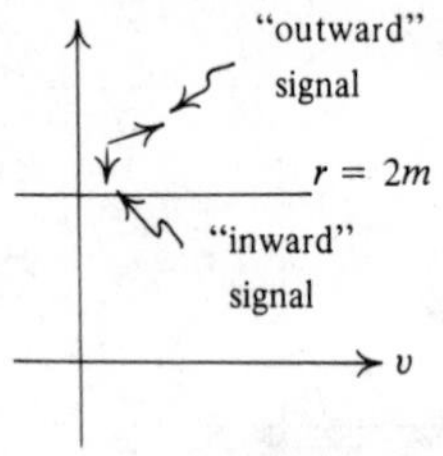

FIGURE 3b

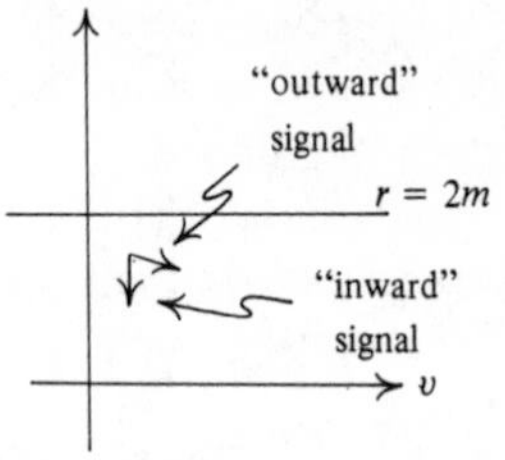

FIGURE 3c

the area $4\pi r^2$ of the sphere representing an "outward" directed signal will increase whereas the area of the sphere representing the "inward" directed signal will decrease (Figure 3b). Such behavior is not surprising. But observe that in the region $r < 2m$ (Figure 3c) the areas of both spheres will decrease. It is this phenomenon which forces the existence of a singularity. For there is the following very general theorem due to Hawking and Penrose [**2**].

THEOREM. *Suppose space-time M satisfies the following four conditions*:
(i) (*energy condition*) $\mathrm{Ricc}(v) \geqq 0$ *for all time-like and null vectors* v;
(ii) (*genericity*) *each causal geodesic* γ *contains a point at which*

$$[R(\langle u,\dot\gamma\rangle v - \langle v,\dot\gamma\rangle u, \dot\gamma)(\dot\gamma)] \wedge \dot\gamma \neq 0 \quad \textit{for some } u, v;$$

(iii) (*causality*) *M has no closed time-like curves; and*
(iv) (*existence of a "trapped" surface*) *there exists a compact space-like 2-dimensional submanifold S of M with the property that the second fundamental forms of S corresponding to the two null directions normal to S both have negative trace everywhere.*

Then there exists a causal geodesic in M which cannot be extended to arbitrary parameter values.

By a causal geodesic is meant a geodesic whose tangent vector is always time-like or null. The energy condition (i) says essentially that energy density should be everywhere nonnegative. The genericity condition (ii) is a rather mild condition which is satisfied for all metrics except extremely special ones. For time-like geodesics, the genericity condition is equivalent with the simpler condition that the tensor $R(\cdot,\dot\gamma,\dot\gamma,\cdot)$ should be nonzero somewhere along γ. Condition (iv) says that the surface S should have the property that null geodesics normal to S tend to converge.

This theorem is a refinement of the Hawking theorem whose proof was sketched by Geroch in his lecture on singularities. The strength of this theorem is that it does not require the existence of a Cauchy surface. The physical content of the theorem is that any space-time which is qualitatively similar to the Schwarzschild model in the sense of properties (i) – (iv) must admit singularities. In particular, the theorem applies to collapsing stars which are neither precisely spherically symmetric nor precisely static.

I conclude by pointing out that there are still open questions in the subject. As Geroch mentioned, the nature of the singularities is still not well understood. For example one would like to know under what conditions will curvature invariants necessarily blow up as one approaches a singularity. There are some known results in this direction (see e.g. [**1**]) but much is yet to be done.

REFERENCES

1. S. W. Hawking and G.F.R. Ellis, *The large-scale structure of space-time*, Cambridge Univ. Press, New York, 1973.

2. S. W. Hawking and R. Penrose, *The singularities of gravitational collapse and cosmology*, Proc. Roy. Soc. London Ser. A **314** (1970), 529–548. MR **41** #9548.

3. M.D. Kruskal, *Maximal extension of Schwarzschild metric*, Phys. Rev. (2) **119** (1960), 1743–1745. MR **22** #6555.

4. R. Penrose, *Techniques of differential topology in relativity*, Regional Conference Series in Applied Math., SIAM, 1972.

STATE UNIVERSITY OF NEW YORK, STONY BROOK

INDEXES

AUTHOR INDEX

Roman numbers refer to pages on which a reference is made to an author or work of an author.

Italic numbers refer to pages on which a complete reference to a work by the author is given.

Boldface numbers indicate the first page of the articles in the book.

SUBJECT INDEX